CLASSICAL DYNAMICS
OF PARTICLES AND SYSTEMS

FOURTH EDITION

CLASSICAL DYNAMICS
OF PARTICLES AND SYSTEMS

FOURTH EDITION

Jerry B. Marion

Late Professor of Physics

University of Maryland

Stephen T. Thornton

Professor of Physics

University of Virginia

SAUNDERS COLLEGE PUBLISHING
Harcourt Brace College Publishers

Fort Worth Philadelphia

San Diego New York Orlando Austin San Antonio

Toronto Montreal London Sydney Tokyo

Text Typeface: Times Roman
Compositor: GTS GRAPHICS, INC.
Vice President/Publisher: John Vondeling
Developmental Editor: Jennifer Bortel
Managing Editor: Carol Field
Project Editor: Nancy Lubars
Copy Editor: Donna Regen
Manager of Art and Design: Carol Bleistine
Art Director: Robin Milicevic
Text Designer: Barbara Gibson
Cover Designer: Nicky Lindeman
Text Artwork: J.A.K. Graphics, Ltd.
Director of EDP: Tim Frelick
Production Manager: Charlene Squibb
Director of Marketing: Marjorie Waldron

Printed in the United States of America

ISBN 0-03-097302-3

Library of Congress Catalog Card Number: 94:061383

4567890123 032 10 987654321

This book was printed on acid-free recycled content paper, containing **MORE THAN 10% POSTCONSUMER WASTE**

To

DR. KATHRYN C. THORNTON
ASTRONAUT AND WIFE

AS SHE SOARS AND WALKS THROUGH SPACE,
MAY HER LIFE BE SAFE AND FULFILLING,
AND LET OUR CHILDREN'S MINDS BE OPEN
FOR ALL THAT LIFE HAS TO OFFER.

PREFACE

In preparing the fourth edition of the late Jerry Marion's text, I have attempted to adhere to his original purpose, as I did in the third edition, of producing a modern and reasonably complete account of the classical mechanics of particles, systems of particles, and rigid bodies for physics students at the advanced undergraduate level. The purpose of this book is threefold:

1. To present a modern treatment of classical mechanical systems in such a way that the transition to the quantum theory of physics can be made with the least possible difficulty.

2. To acquaint the student with new mathematical techniques wherever possible, and to give him/her sufficient practice in solving problems so that the student may become reasonably proficient in their use.

3. To impart to the student, at the crucial period in the student's career between "introductory" and "advanced" physics, some degree of sophistication in handling both the formalism of the theory and the operational technique of problem solving.

After a firm foundation in vector methods is presented in Chapter 1, further mathematical methods are developed in the textbook as the occasion demands. It is advisable for students to continue studying advanced mathematics in a separate course. Mathematical rigor must be learned and appreciated by students of physics, but where the continuity of the physics would be disturbed by insisting on complete generality and mathematical rigor, the *physics* has been given precedence.

Changes to the Fourth Edition

The comments and suggestions of many users of *Classical Dynamics* have been incorporated into this fourth edition. Without the feedback of the many instructors who have used this text, it would not be possible to produce a textbook of significant value to the physics community. Users have requested more numerical calculations. Spread throughout the book, but especially in Chapter 2, numerical examples and end-of-chapter problems have been added. Users have also indicated that students want more examples, and I have responded.

The nonlinear oscillation material formerly in Chapter 3 has been combined with new material on chaos for a new Chapter 4 entitled **Nonlinear Oscillations and Chaos.** Chapters 12 and 13 of the third edition have been combined into a new Chapter 13, and some of the previous material has been omitted. A few more advanced or specialized topics throughout the book have also been omitted to allow new material. Particular effort was made to correct the problem solutions available in the Instructor's and Student Solutions Manuals. I thank the many users who sent comments concerning the various problem solutions. Answers to even-numbered problems have again been included at the end of the book, and the selected and general bibliography has been updated.

Course Suitability

The book is suitable for either a one-semester or two-semester upper level (junior or senior) undergraduate course in classical mechanics taken after an introductory calculus-based physics course. At the University of Virginia we teach a one-semester course based mostly on the first twelve chapters with several omissions of certain sections. Sections that can be omitted without losing continuity are denoted as optional, but the instructor can also choose to skip other sections (or entire chapters) as desired. For example, the new Chapter 4 might be skipped in its entirety for a one-semester course. Some instructors choose not to cover the calculus of variations material in Chapter 6. Other instructors may want to begin with Chapter 2, skip the mathematical introduction of Chapter 1, and introduce the mathematics as needed. This technique of dealing with the mathematics introduction is perfectly acceptable, and the community is divided on this issue with a slight preference for the method used here. The textbook is also suitable for a full academic year course with an emphasis on mathematical and numerical methods as desired by the instructor.

The textbook is appropriate for those who choose to teach in the traditional manner without computer calculations. However, it has been fun adding the numerical calculations, and I must admit that I became much more proficient doing computer calculations as a result of this revision. Practically all students, and most professors, now use computers every day, and they are a useful tool in learning physics. One difficulty is to choose among the many computer techniques available, and I decided to leave this choice to the student and instructor and have not indicated any preferred method.

Special Feature

The author has kept one popular feature of Jerry Marion's original book: the addition of historical footnotes spread throughout. Several users have indicated how valuable these historical comments have been. The history of physics has been almost eliminated from present-day curricula, and as a result, the student is frequently unaware of the background of a particular topic. These footnotes are intended to whet the appetite and to encourage the student to inquire into the history of his/her field.

Teaching Aids

Several teaching aids are available to accompany the textbook. An Instructor's Manual with solutions to all the end-of-chapter problems is available to instructors who adopt this book by contacting the local Saunders (Harcourt Brace) College Publishing sales representative. A separate Student Solutions Manual, with solutions to about 25% of the problems, is available for sale to the students if the instructor decides to order it through the local bookstore. Several instructors have requested that transparencies of the figures be provided. We have appended enlarged versions of many text figures in the Instructor's Manual to allow instructors to make their own transparencies on a copy machine if so desired.

Acknowledgments

I would like to acknowledge the many instructors who provided helpful suggestions through a questionnaire sent out before the revision began. They include

William L. Alford, *Auburn University*
Philip Baldwin, *University of Akron*
Martin Berz, *Michigan State University*
Randy A. Booker, *University of North Carolina–Asheville*
Michael E. Browne, *University of Idaho*
Richard F. Carlson, *University of Redlands*
D. Rae Carpenter, Jr., *Virginia Military Institute*
D. Casavant, *St. Michael's College*
F. Edward Cecil, *Colorado School of Mines*
Albert C. Claus, *Loyola University*
Stan Cloud, *University of Nevada–Las Vegas*
G. T. Condo, *University of Tennessee*
John. E. Crew, *Illinois State University*
Michael De Marco, *Buffalo State College*
George Dixon, *Oklahoma State University*
Warren L. Dumke, *Marshall University*
Cheng-Ming Fou, *University of Delaware*
Norman Fuchs. *Purdue University*
Elsa M. Glover, *Stillman College*
Paul M. Goldbart, *University of Illinois at Urbana-Champaign*
Chris Gould, *North Carolina State University*

Edward Hart, *University of Tennessee*
Richard Heinz, *Indiana University*
Larry D. Johnson, *Northeast Louisiana University*
Thomas Kirkman, *St. John's University*
Carl A. Kocher, *Oregon State University*
Paul L. Lee, *California State University–Northridge*
Charles Leming, *Henderson State University*
Robert R. Marchini, *Memphis State University*
Thomas R. Michalik, *Randolph-Macon Women's College*
R. D. Murphy, *University of Missouri–Kansas City*
Richard P. Olenick, *University of Dallas*
Peter Parker, *Yale University*
J. Pilcher, *University of Chicago*
Jerry Polson, *Southeastern Oklahoma State University*
Dan R. Quisenberry, *Mercer University*
Stephen P. Reynolds, *North Carolina State University*
Albert T. Rosenberger, *University of Alabama–Huntsville*

Robert Sears, Jr., *Austin Peay State University*

Sheridan Simon, *Guilford College*

Jack A. Soules, *Cleveland State University*

S. Sridhar, *Northeastern University*

Paul Stevenson, *Rice University*

N. S. Sullivan, *University of Florida*

Ronald G. Tabak, *Youngstown State University*

Larry Tankersley, *United States Naval Academy*

Philip L. Taylor, *Case Western Reserve University*

Noboru Wada, *Colorado School of Mines*

Bruce Weems, *East Central University*

Hugh D. Young, *Carnegie Mellon University*

Alma C. Zook, *Pomona College*

I would especially like to thank those individuals who either wrote me with suggestions on the text or problems or who reviewed parts of the fourth edition. Their efforts have helped considerably in producing this fourth edition. They include

William L. Alford
Auburn University

Philip Baldwin
University of Akron

Robert P. Bauman
University of Alabama, Birmingham

Michael E. Browne
University of Idaho

Melvin G. Calkin
Dalhousie University

F. Edward Cecil
Colorado School of Mines

Arnold J. Dahm
Case Western Reserve University

Dan de Vries
University of Colorado

George Dixon
Oklahoma State University

John J. Dykla
Loyola University of Chicago

Thomas A. Ferguson
Carnegie Mellon University

Shun-Fu Gao
University of Minnesota, Morris

Reinhard Graetzer
Pennsylvania State University

Thomas M. Helliwell
Harvey Mudd College

Stephen Houk
College of the Sequoias

Joseph Klarmann
Washington University, St. Louis

Kaye D. Lathrop
Stanford University

Robert R. Marchini
Memphis State University

Robert B. Muir
University of North Carolina, Greensboro

Richard P. Olenick
University of Dallas

Tao Pang
University of Nevada, Las Vegas

Peter Parker
Yale University

Peter Rolnick
Northeast Missouri State University

Albert T. Rosenberger
University of Alabama, Huntsville

William E. Slater
University of California, Los Angeles

Herschel Snodgrass
Lewis and Clark College

J. C. Sprott
University of Wisconsin, Madison

Paul Stevenson
Rice University

Joseph S. Tenn
Sonoma State University

Larry Tankersley
United States Naval Academy

In addition, I would like to acknowledge the assistance of Warren Griffith who helped considerably with the problem solutions for the fourth edition and Brian Giambattista who did a similar service for the third edition. Thanks also to Jennie Metz for typing the Instructor's Manual.

The guidance and help of the Saunders College Publishing professional staff is greatly appreciated. These persons include John Vondeling, Vice President and Publisher; Jennifer Bortel, Developmental Editor; Nancy Lubars, Project Editor; Robin Milicevic, Art Director; Marjorie Waldron, Marketing Manager; Randi Misher, Marketing Coordinator.

I would appreciate receiving suggestions or notices of errors in any of these materials. I can be contacted by electronic mail at STT@Virginia.EDU.

Stephen T. Thornton
Charlottesville, Virginia

CONTENTS

7   HAMILTON'S PRINCIPLE–LAGRANGIAN AND HAMILTONIAN DYNAMICS --- 232

8   CENTRAL-FORCE MOTION --- 291

APPENDICES

1

MATRICES, VECTORS, AND VECTOR CALCULUS

1.1 INTRODUCTION

Physical phenomena can be discussed concisely and elegantly through the use of vector methods.* In applying physical "laws" to particular situations, the results must be independent of whether we choose a rectangular or bipolar cylindrical coordinate system. The results must also be independent of the exact choice of origin for the coordinates. The use of vectors gives us this independence. A given physical law will still be correctly represented no matter which coordinate system we decide is most convenient to describe a particular problem. Also, the use of vector notation provides an extremely compact method of expressing even the most complicated results.

In elementary treatments of vectors, the discussion may start with the statement that "a vector is a quantity that can be represented as a directed line segment." To be sure, this type of development will yield correct results, and it is even beneficial to impart a certain feeling for the physical nature of a vector. We assume that the reader is familiar with this type of development, but we forego the approach here because we wish to emphasize the relationship that a vector bears to a coordinate transformation. Therefore, we introduce matrices and matrix notation to describe

---

*Josiah Willard Gibbs (1839–1903) deserves much of the credit for developing vector analysis around 1880–1882. Much of the present-day vector notation was originated by Oliver Heaviside (1850–1925), an English electrical engineer, and dates from about 1893.

not only the transformation but the vector as well. We also introduce a type of notation that is readily adapted to the use of tensors, although we do not encounter these objects until the normal course of events requires their use (see Chapter 11).

We do not attempt a complete exposition of vector methods; instead, we consider only those topics necessary for a study of mechanical systems. Thus in this chapter, we treat the fundamentals of matrix and vector algebra and vector calculus.

1.2 CONCEPT OF A SCALAR

Consider the array of particles shown in Figure 1-1a. Each particle of the array is labeled according to its mass, say, in grams. The coordinate axes are shown so that we can specify a particular particle by a pair of numbers (x, y). The mass M of the particle at (x, y) can be expressed as $M(x, y)$; thus the mass of the particle at $x = 2$, $y = 3$ can be written as $M (x = 2, y = 3) = 4$. Now consider the axes rotated and displaced in the manner shown in Figure 1-1b. The 4 g mass is now located at $x' = 4$, $y' = 3.5$; that is, the mass is specified by $M (x' = 4, y' = 3.5) = 4$. And, in general,

$$M (x, y) = M (x', y') \tag{1.1}$$

because the mass of any particle is not affected by a change in the coordinate axes. Quantities that are *invariant under coordinate transformation*—those that obey an equation of this type—are termed **scalars**.

Although we can describe the mass of a particle (or the temperature, or the speed, etc.) relative to any coordinate system by the same number, some physical properties associated with the particle (such as the direction of motion of the particle or the direction of a force that may act on the particle) cannot be specified in such a simple manner. The description of these more complicated quantities requires the use of **vectors**. Just as a scalar is defined as a quantity that remains invariant under a coordinate transformation, a vector may also be defined in terms of transformation properties. We begin by considering how the coordinates of a point change when the coordinate system rotates around its origin.

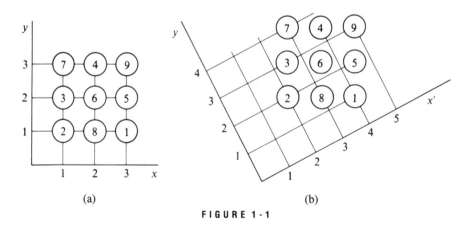

(a)   (b)

FIGURE 1-1

1.3   COORDINATE TRANSFORMATIONS

Consider a point P with coordinates (x_1, x_2, x_3) with respect to a certain coordinate system.* Next consider a different coordinate system, one that can be generated from the original system by a simple rotation; let the coordinates of the point P with respect to the new coordinate system be (x_1', x_2', x_3'). The situation is illustrated for a two-dimensional case in Figure 1-2.

The new coordinate x_1' is the sum of the projection of x_1 onto the x_1'-axis (the line $\overline{Oa}$) plus the projection of x_2 onto the x_1'-axis (the line $\overline{ab} + \overline{bc}$); that is,

$$x_1' = x_1 \cos \theta + x_2 \sin \theta$$

$$= x_1 \cos \theta + x_2 \cos\left(\frac{\pi}{2} - \theta\right) \tag{1.2a}$$

The coordinate x_2' is the sum of similar projections: $x_2' = \overline{Od} - \overline{de}$, but the line $\overline{de}$ is also equal to the line $\overline{Of}$. Therefore

$$x_2' = -x_1 \sin \theta + x_2 \cos \theta$$

$$= x_1 \cos\left(\frac{\pi}{2} + \theta\right) + x_2 \cos \theta \tag{1.2b}$$

Let us introduce the following notation: we write the angle between the x_1'-axis and the x_1-axis as (x_1', x_1), and in general, the angle between the x_i'-axis and the x_j-axis is denoted by (x_i', x_j). Furthermore, we define a set of numbers λ_{ij} by

$$\lambda_{ij} \equiv \cos(x_i', x_j) \tag{1.3}$$

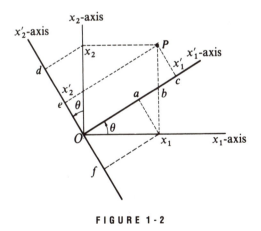

FIGURE 1-2

---

*We label axes as x_1, x_2, x_3 instead of x, y, z to simplify the notation when summations are performed. For the moment, the discussion is limited to Cartesian (or rectangular) coordinate sytems.

Therefore, for Figure 1-2, we have

$$
\left.\begin{aligned}
\lambda_{11} &= \cos(x_1', x_1) = \cos\theta \\
\lambda_{12} &= \cos(x_1', x_2) = \cos\left(\frac{\pi}{2} - \theta\right) = \sin\theta \\
\lambda_{21} &= \cos(x_2', x_1) = \cos\left(\frac{\pi}{2} + \theta\right) = -\sin\theta \\
\lambda_{22} &= \cos(x_2', x_2) = \cos\theta
\end{aligned}\right\} \tag{1.4}
$$

The equations of transformation (Equation 1.2) now become

$$
\begin{aligned}
x_1' &= x_1 \cos(x_1', x_1) + x_2 \cos(x_1', x_2) \\
&= \lambda_{11} x_1 + \lambda_{12} x_2
\end{aligned} \tag{1.5a}
$$

$$
\begin{aligned}
x_2' &= x_1 \cos(x_2', x_1) + x_2 \cos(x_2', x_2) \\
&= \lambda_{21} x_1 + \lambda_{22} x_2
\end{aligned} \tag{1.5b}
$$

Thus, in general, for three dimensions we have

$$
\left.\begin{aligned}
x_1' &= \lambda_{11}x_1 + \lambda_{12}x_2 + \lambda_{13} x_3 \\
x_2' &= \lambda_{21}x_1 + \lambda_{22}x_2 + \lambda_{23} x_3 \\
x_3' &= \lambda_{31}x_1 + \lambda_{32}x_2 + \lambda_{33} x_3
\end{aligned}\right\} \tag{1.6}
$$

or, in summation notation,

$$
\boxed{x_i' = \sum_{j=1}^{3} \lambda_{ij} x_j, \qquad i = 1, 2, 3} \tag{1.7}
$$

The inverse transformation is

$$
\begin{aligned}
x_1 &= x_1' \cos(x_1', x_1) + x_2' \cos(x_2', x_1) + x_3' \cos(x_3', x_1) \\
&= \lambda_{11} x_1' + \lambda_{21} x_2' + \lambda_{31} x_3'
\end{aligned}
$$

or, in general,

$$
\boxed{x_i = \sum_{j=1}^{3} \lambda_{ji} x_j', \qquad i = 1, 2, 3} \tag{1.8}
$$

The quantity λ_{ij} is called the **direction cosine** of the x_i'-axis relative to the x_j-axis. It is convenient to arrange the λ_{ij} into a square array called a **matrix**. The boldface symbol $\boldsymbol{\lambda}$ denotes the totality of the individual elements λ_{ij} when arranged as follows:

$$\boldsymbol{\lambda} = \begin{pmatrix} \lambda_{11} & \lambda_{12} & \lambda_{13} \\ \lambda_{21} & \lambda_{22} & \lambda_{23} \\ \lambda_{31} & \lambda_{32} & \lambda_{33} \end{pmatrix} \tag{1.9}$$

Once we find the direction cosines relating the two sets of coordinate axes, Equations 1.7 and 1.8 give the general rules for specifying the coordinates of a point in either system.

When $\boldsymbol{\lambda}$ is defined this way and when it specifies the transformation properties of the coordinates of a point, it is called a **transformation matrix** or a **rotation matrix**.

EXAMPLE 1.1 ---

A point P is represented in the (x_1, x_2, x_3) system by $P(2, 1, 3)$. Another coordinate system represents the same point as $P(x_1', x_2', x_3')$ but in a system where x_2 has been rotated toward x_3 around the x_1-axis by an angle of 30° (Figure 1-3). Find the rotation matrix and determine $P(x_1', x_2', x_3')$.

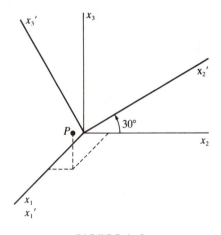

FIGURE 1-3

Solution: The direction cosines λ_{ij} can be determined from Figure 1-3 using the definition of Equation 1.3.

$$\lambda_{11} = \cos(x_1', x_1) = \cos(0°) = 1$$

$$\lambda_{12} = \cos(x_1', x_2) = \cos(90°) = 0$$

$$\lambda_{13} = \cos(x_1', x_3) = \cos(90°) = 0$$

$$\lambda_{21} = \cos(x_2', x_1) = \cos(90°) = 0$$

$$\lambda_{22} = \cos(x_2', x_2) = \cos(30°) = 0.866$$

$$\lambda_{23} = \cos(x_2', x_3) = \cos(90° - 30°) = \cos(60°) = 0.5$$

$$\lambda_{31} = \cos(x_3', x_1) = \cos(90°) = 0$$

$$\lambda_{32} = \cos(x_3', x_2) = \cos(90° + 30°) = -0.5$$

$$\lambda_{33} = \cos(x_3', x_3) = \cos(30°) = 0.866$$

$$\lambda = \begin{pmatrix} 1 & 0 & 0 \\ 0 & 0.866 & 0.5 \\ 0 & -0.5 & 0.866 \end{pmatrix}$$

and using Equation 1.7, $P(x_1', x_2', x_3')$ is

$$x_1' = \lambda_{11} x_1 + \lambda_{12} x_2 + \lambda_{13} x_3 = x_1 = 2$$

$$x_2' = \lambda_{21} x_1 + \lambda_{22} x_2 + \lambda_{23} x_3 = 0.866x_2 + 0.5x_3 = 2.37$$

$$x_3' = \lambda_{31} x_1 + \lambda_{32} x_2 + \lambda_{33} x_3 = -0.5x_2 + 0.866x_3 = 2.10$$

Notice that the rotation operator preserves the length of the position vector.

$$r = \sqrt{x_1^2 + x_2^2 + x_3^2} = \sqrt{x_1'^2 + x_2'^2 + x_3'^2} = 3.74$$

1.4   PROPERTIES OF ROTATION MATRICES *

To begin the discussion of rotation matrices, we must recall two trigonometric results. Consider, as in Figure 1-4a, a line segment extending in a certain direction

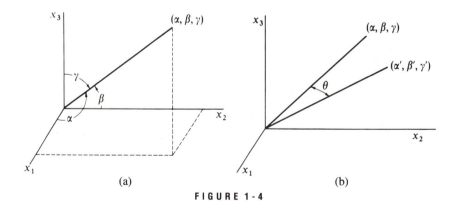

(a)   (b)

FIGURE 1-4

---

in space. We choose an origin for our coordinate system that lies at some point on the line. The line then makes certain definite angles with each of the coordinate axes; we let the angles made with the x_1-, x_2-, x_3-axes be α, β, γ. The quantities of interest are the cosines of these angles; $\cos \alpha$, $\cos \beta$, $\cos \gamma$. These quantities are called the **direction cosines** of the line. The first result we need is the identity (see Problem 1-2)

$$\cos^2\alpha + \cos^2\beta + \cos^2\gamma = 1 \qquad \textbf{(1.10)}$$

Second, if we have two lines with direction cosines $\cos \alpha$, $\cos \beta$, $\cos \gamma$ and $\cos \alpha'$, $\cos \beta'$, $\cos \gamma'$, then the cosine of the angle θ between these lines (see Figure 1-4b) is given (see Problem 1-2) by

$$\cos \theta = \cos \alpha \cos \alpha' + \cos \beta \cos \beta' + \cos \gamma \cos \gamma' \qquad \textbf{(1.11)}$$

With a set of axes x_1, x_2, x_3, let us now perform an arbitrary rotation about some axis through the origin. In the new position, we label the axes x_1', x_2', x_3'. The coordinate rotation may be specified by giving the cosines of all the angles between the various axes, in other words, by the λ_{ij}.

Not all of the nine quantities λ_{ij} are independent; in fact, six relations exist among the λ_{ij}, so only three are independent. We find these six relations by using the trigonometric results stated in Equations 1.10 and 1.11.

First, the x_1'-axis may be considered alone to be a line in the (x_1, x_2, x_3) coordinate system; the direction cosines of this line are $(\lambda_{11}, \lambda_{12}, \lambda_{13})$. Similarly, the direction cosines of the x_2'-axis in the (x_1, x_2, x_3) system are given by $(\lambda_{21}, \lambda_{22}, \lambda_{23})$. Because the angle between the x_1'-axis and the x_2'-axis is $\pi/2$, we have, from Equation 1.11,

$$\lambda_{11} \lambda_{21} + \lambda_{12} \lambda_{22} + \lambda_{13} \lambda_{23} = \cos \theta = \cos(\pi/2) = 0$$

or*

$$\sum_j \lambda_{1j} \lambda_{2j} = 0$$

And, in general,

$$\sum_j \lambda_{ij} \lambda_{kj} = 0, \qquad i \neq k \qquad \textbf{(1.12a)}$$

Equation 1.12a gives three (one for each value of i or k) of the six relations among the λ_{ij}.

Because the sum of the squares of the direction cosines of a line equals unity (Equation 1.10), we have for the x_1'-axis in the (x_1, x_2, x_3) system,

$$\lambda_{11}^2 + \lambda_{12}^2 + \lambda_{13}^2 = 1$$

or

$$\sum_j \lambda_{1j}^2 = \sum_j \lambda_{1j} \lambda_{1j} = 1$$

---

*All summations here are understood to run from 1 to 3.

and, in general,

$$\sum_j \lambda_{ij} \lambda_{kj} = 1, \qquad i = k \tag{1.12b}$$

which are the remaining three relations among the λ_{ij}.

We may combine the results given by Equations 1.12a and 1.12b as

$$\boxed{\sum_j \lambda_{ij} \lambda_{kj} = \delta_{ik}} \tag{1.13}$$

where δ_{ik} is the **Kronecker delta symbol***

$$\delta_{ik} = \begin{cases} 0, & \text{if } i \neq k \\ 1, & \text{if } i = k \end{cases} \tag{1.14}$$

The validity of Equation 1.13 depends on the coordinate axes in each of the systems being mutually perpendicular. Such systems are said to be **orthogonal**, and Equation 1.13 is the **orthogonality condition**. The transformation matrix $\boldsymbol{\lambda}$ specifying the rotation of any orthogonal coordinate system must then obey Equation 1.13.

If we were to consider the x_i-axes as lines in the x_i' coordinate system and perform a calculation analogous to our preceding calculations, we would find the relation

$$\boxed{\sum_i \lambda_{ij} \lambda_{ik} = \delta_{jk}} \tag{1.15}$$

The two orthogonality relations we have derived (Equations 1.13 and 1.15) appear to be different. (*Note*: In Equation 1.13 the summation is over the *second* indices of the λ_{ij}, whereas in Equation 1.15 the summation is over the *first* indices.) Thus, it seems that we have an overdetermined system: twelve equations in nine unknowns.[†] Such is not the case, however, because Equations 1.13 and 1.15 are not actually different. In fact, the validity of either of these equations implies the validity of the other. This is clear on physical grounds (because the transformations between the two coordinate systems in either direction are equivalent), and we omit a formal proof. We regard either Equation 1.13 or 1.15 as providing the orthogonality relations for our systems of coordinates.

In the preceding discussion regarding the transformation of coordinates and the properties of rotation matrices, we considered the point P to be fixed and allowed the coordinate axes to be rotated. This interpretation is not unique; we could equally

---

*Introduced by Leopold Kronecker (1823–1891).

[†]Recall that each of the orthogonality relations represents six equations.

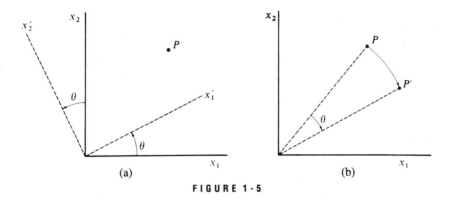

FIGURE 1-5

well have maintained the axes fixed and allowed the point to rotate (always keeping constant the distance to the origin). In either event, the transformation matrix is the same. For example, consider the two cases illustrated in Figure 1-5a and b. In Figure 1-5a, the axes x_1 and x_2 are reference axes, and the x_1'- and x_2'-axes have been obtained by a rotation through an angle θ. Therefore, the coordinates of the point P with respect to the rotated axes may be found (see Equations 1.2a and 1.2b) from

$$\left.\begin{array}{l} x_1' = x_1 \cos \theta + x_2 \sin \theta \\ x_2' = - x_1 \sin \theta + x_2 \cos \theta \end{array}\right\} \tag{1.16}$$

However, if the axes are fixed and the point P is allowed to rotate (as in Figure 1-5b) through an angle θ about the origin (but in the opposite sense from that of the rotated axes), then the coordinates of P' are exactly those given by Equation 1.16. Therefore, we may elect to say either that the transformation acts on the *point* giving a new state of the point expressed with respect to a fixed coordinate system (Figure 1-5b) or that the transformation acts on the *frame of reference* (the coordinate system), as in Figure 1-5a. Mathematically, the interpretations are entirely equivalent.

1.5   MATRIX OPERATIONS*

The matrix $\boldsymbol{\lambda}$ given in Equation 1.9 has equal numbers of rows and columns and is therefore called a **square matrix**. A matrix need not be square. In fact, the coordinates of a point may be written as a **column** matrix

$$\mathbf{x} = \begin{pmatrix} x_1 \\ x_2 \\ x_3 \end{pmatrix} \tag{1.17a}$$

---

*The theory of matrices was first extensively developed by A. Cayley in 1855, but many of these ideas were the work of Sir William Rowan Hamilton (1805–1865), who had discussed "linear vector operators" in 1852. The term *matrix* was first used by J. J. Sylvester in 1850.

or as a **row** matrix

$$\mathbf{x} = (x_1 \quad x_2 \quad x_3) \tag{1.17b}$$

We must now establish rules to multiply two matrices. These rules must be consistent with Equations 1.7 and 1.8 when we choose to express the x_i and the x_i' in matrix form. Let us take a column matrix for the coordinates; then we have the following equivalent expressions:

$$x_i' = \sum_j \lambda_{ij} x_j \tag{1.18a}$$

$$\mathbf{x}' = \boldsymbol{\lambda}\mathbf{x} \tag{1.18b}$$

$$\begin{pmatrix} x_1' \\ x_2' \\ x_3' \end{pmatrix} = \begin{pmatrix} \lambda_{11} & \lambda_{12} & \lambda_{13} \\ \lambda_{21} & \lambda_{22} & \lambda_{23} \\ \lambda_{31} & \lambda_{32} & \lambda_{33} \end{pmatrix} \begin{pmatrix} x_1 \\ x_2 \\ x_3 \end{pmatrix} \tag{1.18c}$$

$$\left. \begin{aligned} x_1' &= \lambda_{11}x_1 + \lambda_{12}x_2 + \lambda_{13}x_3 \\ x_2' &= \lambda_{21}x_1 + \lambda_{22}x_2 + \lambda_{23}x_3 \\ x_3' &= \lambda_{31}x_1 + \lambda_{32}x_2 + \lambda_{33}x_3 \end{aligned} \right\} \tag{1.18d}$$

Equations 1.18a–d completely specify the operation of matrix multiplication for a matrix of three rows and three columns operating on a matrix of three rows and one column. (To be consistent with standard matrix convention we choose $\mathbf{x}$ and $\mathbf{x}'$ to be column matrices; multiplication of the type shown in Equation 1.18c is not defined if $\mathbf{x}$ and $\mathbf{x}'$ are row matrices.)* We must now extend our definition of multiplication to include matrices with arbitrary numbers of rows and columns.

The multiplication of a matrix $\mathbf{A}$ and a matrix $\mathbf{B}$ is defined only if the number of *columns* of $\mathbf{A}$ is equal to the number of *rows* of $\mathbf{B}$. (The number of rows of $\mathbf{A}$ and the number of columns of $\mathbf{B}$ are each arbitrary.) Therefore, in analogy with Equation 1.18a, the product $\mathbf{AB}$ is given by

$$\boxed{ \begin{aligned} \mathbf{C} &= \mathbf{AB} \\ C_{ij} &= [\mathbf{AB}]_{ij} = \sum_k A_{ik} B_{kj} \end{aligned} } \tag{1.19}$$

As an example, let the two matrices $\mathbf{A}$ and $\mathbf{B}$ be

$$\mathbf{A} = \begin{pmatrix} 3 & -2 & 2 \\ 4 & -3 & 5 \end{pmatrix}$$

$$\mathbf{B} = \begin{pmatrix} a & b & c \\ d & e & f \\ g & h & j \end{pmatrix}$$

---

*Although whenever we operate on $\mathbf{x}$ with the $\boldsymbol{\lambda}$ matrix the coordinate matrix $\mathbf{x}$ must be expressed as a column matrix, we may also write $\mathbf{x}$ as a row matrix (x_1, x_2, x_3), for other applications.

We multiply the two matrices by

$$\mathbf{AB} = \begin{pmatrix} 3 & -2 & 2 \\ 4 & -3 & 5 \end{pmatrix} \begin{pmatrix} a & b & c \\ d & e & f \\ g & h & j \end{pmatrix} \tag{1.20}$$

The product of the two matrices, **C**, is

$$\mathbf{C} = \mathbf{AB} = \begin{pmatrix} 3a-2d+2g & 3b-2e+2h & 3c-2f+2j \\ 4a-3d+5g & 4b-3e+5h & 4c-3f+5j \end{pmatrix} \tag{1.21}$$

To obtain the C_{ij} element in the ith row and jth column, we first set the two matrices adjacent as we did in Equation 1.20 in the order **A** and then **B**. We then multiply the individual elements in the ith row of **A**, one by one from left to right, times the corresponding elements in the jth column of **B**, one by one from top to bottom. We add all these products, and the sum is the C_{ij} element. Now it is easier to see why a matrix **A** with m rows and n columns must be multiplied times another matrix **B** with n rows and any number of columns, say p. The result is a matrix **C** of m rows and p columns. Look carefully at the result for Equation 1.21 to see how the result was obtained from the previous equation.

E X A M P L E   1.2   ---

Find the product AB of the two matrices listed below:

$$\mathbf{A} = \begin{pmatrix} 2 & 1 & 3 \\ -2 & 2 & 4 \\ -1 & -3 & -4 \end{pmatrix}$$

$$\mathbf{B} = \begin{pmatrix} -1 & -2 \\ 1 & 2 \\ 3 & 4 \end{pmatrix}$$

Solution: We follow the example of Equations 1.20 and 1.21 to multiply the two matrices together.

$$\mathbf{AB} = \begin{pmatrix} 2 & 1 & 3 \\ -2 & 2 & 4 \\ -1 & -3 & -4 \end{pmatrix} \begin{pmatrix} -1 & -2 \\ 1 & 2 \\ 3 & 4 \end{pmatrix}$$

$$\mathbf{AB} = \begin{pmatrix} -2+1+9 & -4+2+12 \\ 2+2+12 & 4+4+16 \\ 1-3-12 & 2-6-16 \end{pmatrix} = \begin{pmatrix} 8 & 10 \\ 16 & 24 \\ -14 & -20 \end{pmatrix}$$

The result of multiplying a 3 × 3 matrix times a 3 × 2 matrix is a 3 × 2 matrix.

-- ●

It should be evident from Equation 1.19 that matrix multiplication is not commutative. Thus, if **A** and **B** are both square matrices, then the sums

$$\sum_k A_{ik}B_{kj} \quad \text{and} \quad \sum_k B_{ik}A_{kj}$$

are both defined, but, in general, they will not be equal.

EXAMPLE **1.3** -

Show that the multiplication of the matrices A and B in this example is noncommutative.

Solution: If **A** and **B** are the matrices

$$\mathbf{A} = \begin{pmatrix} 2 & 1 \\ -1 & 3 \end{pmatrix}, \quad \mathbf{B} = \begin{pmatrix} -1 & 2 \\ 4 & -2 \end{pmatrix}$$

then

$$\mathbf{AB} = \begin{pmatrix} 2 & 2 \\ 13 & -8 \end{pmatrix}$$

but

$$\mathbf{BA} = \begin{pmatrix} -4 & 5 \\ 10 & -2 \end{pmatrix}$$

thus

$$\mathbf{AB} \neq \mathbf{BA}$$

- ●

1.6   FURTHER DEFINITIONS

A **transposed matrix** is a matrix derived from an original matrix by interchange of rows and columns. We denote the **transpose** of a matrix **A** by $\mathbf{A}^t$. According to the definition, we have

$$\boxed{\lambda_{ij}^t = \lambda_{ji}} \tag{1.22}$$

Evidently,

$$(\boldsymbol{\lambda}^t)^t = \boldsymbol{\lambda} \tag{1.23}$$

Equation 1.8 may therefore be written as any of the following equivalent expressions:

$$x_i = \sum_j \lambda_{ji} x_j' \tag{1.24a}$$

$$x_i = \sum_j \lambda_{ij}^t x_j' \tag{1.24b}$$

$$\mathbf{x} = \boldsymbol{\lambda}^t \mathbf{x}' \tag{1.24c}$$

$$\begin{pmatrix} x_1 \\ x_2 \\ x_3 \end{pmatrix} = \begin{pmatrix} \lambda_{11} & \lambda_{21} & \lambda_{31} \\ \lambda_{12} & \lambda_{22} & \lambda_{32} \\ \lambda_{13} & \lambda_{23} & \lambda_{33} \end{pmatrix} \begin{pmatrix} x_1' \\ x_2' \\ x_3' \end{pmatrix} \tag{1.24d}$$

The **identity matrix** is that matrix which, when multiplied by another matrix, leaves the latter unaffected. Thus

$$\mathbf{1A = A}, \qquad \mathbf{B1 = B} \tag{1.25}$$

that is,

$$\mathbf{1A} = \begin{pmatrix} 1 & 0 \\ 0 & 1 \end{pmatrix} \begin{pmatrix} A_1 \\ A_2 \end{pmatrix} = \begin{pmatrix} A_1 \\ A_2 \end{pmatrix} = \mathbf{A}$$

Let us consider the orthogonal rotation matrix $\boldsymbol{\lambda}$ for the case of two dimensions:

$$\boldsymbol{\lambda} = \begin{pmatrix} \lambda_{11} & \lambda_{12} \\ \lambda_{21} & \lambda_{22} \end{pmatrix}$$

Then

$$\boldsymbol{\lambda\lambda}^t = \begin{pmatrix} \lambda_{11} & \lambda_{12} \\ \lambda_{21} & \lambda_{22} \end{pmatrix} \begin{pmatrix} \lambda_{11} & \lambda_{21} \\ \lambda_{12} & \lambda_{22} \end{pmatrix}$$

$$= \begin{pmatrix} \lambda_{11}^2 + \lambda_{12}^2 & \lambda_{11}^2\lambda_{21}^2 + \lambda_{12}^2\lambda_{22}^2 \\ \lambda_{21}\lambda_{11} + \lambda_{22}\lambda_{12} & \lambda_{21}^2 + \lambda_{22}^2 \end{pmatrix}$$

Using the orthogonality relation (Equation 1.13), we find

$$\lambda_{11}^2 + \lambda_{12}^2 = \lambda_{21}^2 + \lambda_{22}^2 = 1$$

$$\lambda_{21}\,\lambda_{11} + \lambda_{22}\,\lambda_{12} = \lambda_{11}\,\lambda_{21} + \lambda_{12}\,\lambda_{22} = 0$$

so that for the special case of the orthogonal rotation matrix $\boldsymbol{\lambda}$ we have*

$$\boldsymbol{\lambda\lambda}^t = \begin{pmatrix} 1 & 0 \\ 0 & 1 \end{pmatrix} = \mathbf{1} \tag{1.26}$$

The **inverse** of a matrix is defined as that matrix which, when multiplied by the original matrix, produces the identity matrix. The inverse of the matrix $\boldsymbol{\lambda}$ is denoted by $\boldsymbol{\lambda}^{-1}$:

$$\boldsymbol{\lambda\lambda}^{-1} = \mathbf{1} \tag{1.27}$$

By comparing Equations 1.26 and 1.27, we find

$$\boxed{\boldsymbol{\lambda}^t = \boldsymbol{\lambda}^{-1}} \qquad \text{for orthogonal matrices} \tag{1.28}$$

Therefore, the transpose and the inverse of the rotation matrix $\boldsymbol{\lambda}$ are identical. In fact, the transpose of *any* orthogonal matrix is equal to its inverse.

---

*This result is not valid for matrices in general. It is true only for *orthogonal* matrices.

To summarize some of the rules of matrix algebra:

1. Matrix multiplication is not commutative in general:

$$\mathbf{AB} \neq \mathbf{BA} \tag{1.29a}$$

The special case of the multiplication of a matrix and its inverse is commutative:

$$\mathbf{AA}^{-1} = \mathbf{A}^{-1}\mathbf{A} = 1 \tag{1.29b}$$

The identity matrix always commutes:

$$\mathbf{1A} = \mathbf{A1} = \mathbf{A} \tag{1.29c}$$

2. Matrix multiplication is associative:

$$[\mathbf{AB}]\mathbf{C} = \mathbf{A}[\mathbf{BC}] \tag{1.30}$$

3. Matrix addition is performed by adding corresponding elements of the two matrices. The components of $\mathbf{C}$ from the addition $\mathbf{C} = \mathbf{A} + \mathbf{B}$ are

$$C_{ij} = A_{ij} + B_{ij} \tag{1.31}$$

Addition is defined only if $\mathbf{A}$ and $\mathbf{B}$ have the same dimensions.

1.7 GEOMETRICAL SIGNIFICANCE OF TRANSFORMATION MATRICES

Consider coordinate axes rotated counterclockwise* through an angle of 90° about the x_3-axis, as in Figure 1-6. In such a rotation, $x_1' = x_2$, $x_2' = -x_1$, $x_3' = x_3$.

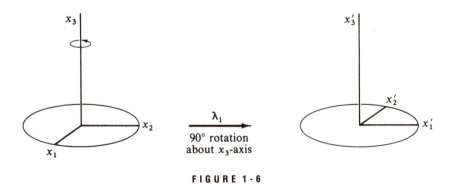

$$\lambda_1$$
90° rotation about x_3-axis

FIGURE 1-6

---

*We determine the sense of the rotation by looking along the positive portion of the axis of rotation at the plane being rotated. This definition is then consistent with the "right-hand rule," in which the positive direction is the direction of advance of a right-hand screw when turned in the same sense.

The only nonvanishing cosines are

$$\cos(x_1', x_2) = \quad 1 = \lambda_{12}$$

$$\cos(x_2', x_1) = -1 = \lambda_{21}$$

$$\cos(x_3', x_3) = \quad 1 = \lambda_{33}$$

so the $\boldsymbol{\lambda}$ matrix for this case is

$$\boldsymbol{\lambda}_1 = \begin{pmatrix} 0 & 1 & 0 \\ -1 & 0 & 0 \\ 0 & 0 & 1 \end{pmatrix}$$

Next consider the counterclockwise rotation through 90° about the x_1-axis, as in Figure 1-7. We have $x_1' = x_1, x_2' = x_3, x_3' = -x_2$, and the transformation matrix is

$$\boldsymbol{\lambda}_2 = \begin{pmatrix} 1 & 0 & 0 \\ 0 & 0 & 1 \\ 0 & -1 & 0 \end{pmatrix}$$

To find the transformation matrix for the combined transformation for rotation about the x_3-axis, followed by rotation about the new x_1'-axis (see Figure 1-8), we have

$$\mathbf{x}' = \boldsymbol{\lambda}_1\mathbf{x} \tag{1.32a}$$

and

$$\mathbf{x}'' = \boldsymbol{\lambda}_2\mathbf{x}' \tag{1.32b}$$

or

$$\mathbf{x}'' = \boldsymbol{\lambda}_2\boldsymbol{\lambda}_1\mathbf{x} \tag{1.33a}$$

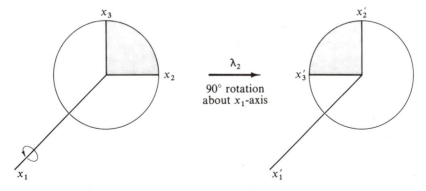

FIGURE 1-7

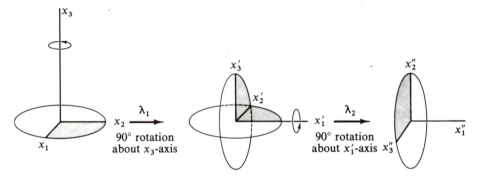

FIGURE 1-8

$$\begin{pmatrix} x_1'' \\ x_2'' \\ x_3'' \end{pmatrix} = \begin{pmatrix} 1 & 0 & 0 \\ 0 & 0 & 1 \\ 0 & -1 & 0 \end{pmatrix} \begin{pmatrix} 0 & 1 & 0 \\ -1 & 0 & 0 \\ 0 & 0 & 1 \end{pmatrix} \begin{pmatrix} x_1 \\ x_2 \\ x_3 \end{pmatrix} = \begin{pmatrix} 0 & 1 & 0 \\ 0 & 0 & 1 \\ 1 & 0 & 0 \end{pmatrix} \begin{pmatrix} x_1 \\ x_2 \\ x_3 \end{pmatrix} = \begin{pmatrix} x_2 \\ x_3 \\ x_1 \end{pmatrix}$$

(1.33b)

Therefore, the two rotations already described may be represented by a single transformation matrix:

$$\lambda_3 = \lambda_2 \lambda_1 = \begin{pmatrix} 0 & 1 & 0 \\ 0 & 0 & 1 \\ 1 & 0 & 0 \end{pmatrix}$$

(1.34)

and the final orientation is specified by $x_1'' = x_2$, $x_2'' = x_3$, $x_3'' = x_1$. Note that the order in which the transformation matrices operate on **x** is important because the multiplication is not commutative. In the other order,

$$\lambda_4 = \lambda_1 \lambda_2$$

$$= \begin{pmatrix} 0 & 1 & 0 \\ -1 & 0 & 0 \\ 0 & 0 & 1 \end{pmatrix} \begin{pmatrix} 1 & 0 & 0 \\ 0 & 0 & 1 \\ 0 & -1 & 0 \end{pmatrix}$$

$$= \begin{pmatrix} 0 & 0 & 1 \\ -1 & 0 & 0 \\ 0 & -1 & 0 \end{pmatrix} \neq \lambda_3$$

(1.35)

and an entirely different orientation results. Figure 1-9 illustrates the different final orientations of a parallelepiped that undergoes rotations corresponding to two rotation matrices λ_A, λ_B when successive rotations are made in different order. The upper portion of the figure represents the matrix product $\lambda_B \lambda_A$, and the lower portion represents the product $\lambda_A \lambda_B$.

Next, consider the coordinate rotation pictured in Figure 1-10 (which is the same as that in Figure 1-2). The elements of the transformation matrix in two

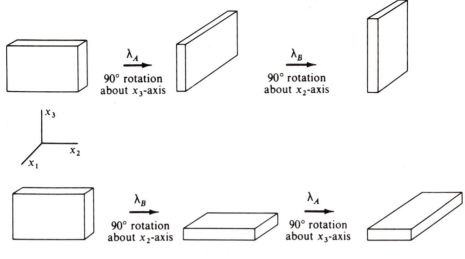

FIGURE 1-9

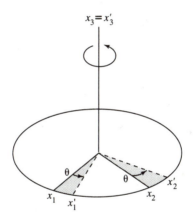

FIGURE 1-10

dimensions are given by the following cosines:

$$\cos(x_1', x_1) = \cos\theta = \lambda_{11}$$

$$\cos(x_1', x_2) = \cos\left(\frac{\pi}{2} - \theta\right) = \sin\theta = \lambda_{12}$$

$$\cos(x_2', x_1) = \cos\left(\frac{\pi}{2} + \theta\right) = -\sin\theta = \lambda_{21}$$

$$\cos(x_2', x_2) = \cos\theta = \lambda_{22}$$

Therefore, the matrix is

$$\lambda_5 = \begin{pmatrix} \cos\theta & \sin\theta \\ -\sin\theta & \cos\theta \end{pmatrix} \tag{1.36a}$$

If this rotation were a three-dimensional rotation with $x_3' = x_3$, we would have the following additional cosines:

$$\cos(x_1', x_3) = 0 = \lambda_{13}$$

$$\cos(x_2', x_3) = 0 = \lambda_{23}$$

$$\cos(x_3', x_3) = 1 = \lambda_{33}$$

$$\cos(x_3', x_1) = 0 = \lambda_{31}$$

$$\cos(x_3', x_2) = 0 = \lambda_{32}$$

and the three-dimensional transformation matrix is

$$\lambda_5 = \begin{pmatrix} \cos\theta & \sin\theta & 0 \\ -\sin\theta & \cos\theta & 0 \\ 0 & 0 & 1 \end{pmatrix} \tag{1.36b}$$

As a final example, consider the transformation that results in the reflection through the origin of all the axes, as in Figure 1-11. Such a transformation is called an **inversion**. In such a case, $x_1' = -x_1, x_2' = -x_2, x_3' = -x_3$, and

$$\lambda_6 = \begin{pmatrix} -1 & 0 & 0 \\ 0 & -1 & 0 \\ 0 & 0 & -1 \end{pmatrix} \tag{1.37}$$

In the preceding examples, we defined the transformation matrix λ_3 to be the result of two successive rotations, each of which was an orthogonal transformation: $\lambda_3 = \lambda_2\lambda_1$. We can prove that the successive application of orthogonal transfor-

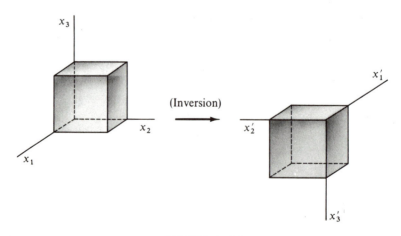

FIGURE 1-11

mations always results in an orthogonal transformation. We write

$$x_i' = \sum_j \lambda_{ij} x_j, \qquad x_k'' = \sum_i \mu_{ki} x_i'$$

Combining these expressions, we obtain

$$x_k'' = \sum_j \left(\sum_i \mu_{ki} \lambda_{ij} \right) x_j$$

$$= \sum_j [\boldsymbol{\mu}\boldsymbol{\lambda}]_{kj} x_j$$

Thus, we accomplish the transformation from x_i to x_i'' by operating on x_i with the $(\boldsymbol{\mu}\boldsymbol{\lambda})$ matrix. The combined transformation will then be shown to be orthogonal if $(\boldsymbol{\mu}\boldsymbol{\lambda})' = (\boldsymbol{\mu}\boldsymbol{\lambda})^{-1}$. The transpose of a product matrix is the product of the transposed matrices taken in reverse order (see Problem 1-4); that is, $(\mathbf{AB})' = \mathbf{B}'\mathbf{A}'$. Therefore

$$(\boldsymbol{\mu}\boldsymbol{\lambda})' = \boldsymbol{\lambda}'\boldsymbol{\mu}' \tag{1.38}$$

But, because $\boldsymbol{\lambda}$ and $\boldsymbol{\mu}$ are orthogonal, $\boldsymbol{\lambda}' = \boldsymbol{\lambda}^{-1}$ and $\boldsymbol{\mu}' = \boldsymbol{\mu}^{-1}$. Multiplying the above equation by $\boldsymbol{\mu}\boldsymbol{\lambda}$ from the right, we obtain

$$(\boldsymbol{\mu}\boldsymbol{\lambda})'\boldsymbol{\mu}\boldsymbol{\lambda} = \boldsymbol{\lambda}'\boldsymbol{\mu}'\boldsymbol{\mu}\boldsymbol{\lambda}$$

$$= \boldsymbol{\lambda}'\mathbf{1}\boldsymbol{\lambda}$$

$$= \boldsymbol{\lambda}'\boldsymbol{\lambda}$$

$$= 1$$

$$= (\boldsymbol{\mu}\boldsymbol{\lambda})^{-1}\boldsymbol{\mu}\boldsymbol{\lambda}$$

Hence

$$(\boldsymbol{\mu}\boldsymbol{\lambda})' = (\boldsymbol{\mu}\boldsymbol{\lambda})^{-1} \tag{1.39}$$

and the $\boldsymbol{\mu}\boldsymbol{\lambda}$ matrix is orthogonal.

The determinants of all the rotation matrices in the preceding examples can be calculated according to the standard rule for the evaluation of determinants of second or third order:

$$|\boldsymbol{\lambda}| = \begin{vmatrix} \lambda_{11} & \lambda_{12} \\ \lambda_{21} & \lambda_{22} \end{vmatrix} = \lambda_{11}\lambda_{22} - \lambda_{12}\lambda_{21} \tag{1.40}$$

$$|\boldsymbol{\lambda}| = \begin{vmatrix} \lambda_{11} & \lambda_{12} & \lambda_{13} \\ \lambda_{21} & \lambda_{22} & \lambda_{23} \\ \lambda_{31} & \lambda_{32} & \lambda_{33} \end{vmatrix}$$

$$= \lambda_{11}\begin{vmatrix} \lambda_{22} & \lambda_{23} \\ \lambda_{32} & \lambda_{33} \end{vmatrix} - \lambda_{12}\begin{vmatrix} \lambda_{21} & \lambda_{23} \\ \lambda_{31} & \lambda_{33} \end{vmatrix} + \lambda_{13}\begin{vmatrix} \lambda_{21} & \lambda_{22} \\ \lambda_{31} & \lambda_{32} \end{vmatrix} \tag{1.41}$$

where the third-order determinant has been expanded in minors of the first row.

Therefore, we find, for the rotation matrices used in this section,

$$|\lambda_1| = |\lambda_2| = \cdots = |\lambda_5| = 1$$

but

$$|\lambda_6| = -1$$

Thus, all those transformations resulting from *rotations starting from the original set of axes* have determinants equal to $+1$. But an *inversion* cannot be generated by any series of rotations, and the determinant of an inversion matrix is equal to -1. Orthogonal transformations, the determinant of whose matrices is $+1$, are called **proper rotations**; those with determinant equal to -1 are called **improper rotations**. *All* orthogonal matrices must have a determinant equal to either $+1$ or -1. Here, we confine our attention to the effect of proper rotations and do not concern ourselves with the special properties of vectors manifest in improper rotations.

EXAMPLE ⬤ 1.4 --

Show that $|\lambda_2| = 1$ and $|\lambda_6| = -1$.

Solution:

$$|\lambda_2| = \begin{vmatrix} 1 & 0 & 0 \\ 0 & 0 & 1 \\ 0 & -1 & 0 \end{vmatrix} = +1 \begin{vmatrix} 0 & 1 \\ -1 & 0 \end{vmatrix} = 0 - (-1) = 1$$

$$|\lambda_6| = \begin{vmatrix} -1 & 0 & 0 \\ 0 & -1 & 0 \\ 0 & 0 & -1 \end{vmatrix} = -1 \begin{vmatrix} -1 & 0 \\ 0 & -1 \end{vmatrix} = -1(1 - 0) = -1$$

-- ⬤

1.8 DEFINITIONS OF A SCALAR AND A VECTOR IN TERMS OF TRANSFORMATION PROPERTIES

Consider a coordinate transformation of the type

$$x_i' = \sum_j \lambda_{ij} x_j \tag{1.42}$$

with

$$\sum_j \lambda_{ij} \lambda_{kj} = \delta_{ik} \tag{1.43}$$

If, under such a transformation, a quantity ϕ is unaffected, then ϕ is called a **scalar** (or **scalar invariant**).

If a set of quantities (A_1, A_2, A_3) is transformed from the x_i system to the x_i' system by a transformation matrix λ with the result

$$A_i' = \sum_j \lambda_{ij} A_j \qquad \textbf{(1.44)}$$

then the quantities A_i transform as the coordinates of a point (i.e., according to Equation 1.42), and the quantity $\mathbf{A} = (A_1, A_2, A_3)$ is termed a **vector**.

1.9 ELEMENTARY SCALAR AND VECTOR OPERATIONS

In the following, $\mathbf{A}$ and $\mathbf{B}$ are vectors (with components A_i and B_i) and ϕ, ψ, and ξ are scalars.

Addition

$$A_i + B_i = B_i + A_i \qquad \text{Commutative law} \qquad \textbf{(1.45)}$$

$$A_i + (B_i + C_i) = (A_i + B_i) + C_i \qquad \text{Associative law} \qquad \textbf{(1.46)}$$

$$\phi + \psi = \psi + \phi \qquad \text{Commutative law} \qquad \textbf{(1.47)}$$

$$\phi + (\psi + \xi) = (\phi + \psi) + \xi \qquad \text{Associative law} \qquad \textbf{(1.48)}$$

Multiplication by a scalar ξ

$$\xi\,\mathbf{A} = \mathbf{B} \qquad \text{is a vector} \qquad \textbf{(1.49)}$$

$$\xi\phi = \psi \qquad \text{is a scalar} \qquad \textbf{(1.50)}$$

Equation 1.49 can be proved as follows:

$$B_i' = \sum_j \lambda_{ij} B_j = \sum_j \lambda_{ij}\, \xi A_j$$

$$= \xi \sum_j \lambda_{ij} A_j = \xi A_i' \qquad \textbf{(1.51)}$$

and $\xi\,\mathbf{A}$ transforms as a vector. Similarly, $\xi\phi$ transforms as a scalar.

1.10 SCALAR PRODUCT OF TWO VECTORS

The multiplication of two vectors $\mathbf{A}$ and $\mathbf{B}$ to form the **scalar product** is defined to be

$$\mathbf{A} \cdot \mathbf{B} = \sum_i A_i B_i \qquad \textbf{(1.52)}$$

where the dot between $\mathbf{A}$ and $\mathbf{B}$ denotes scalar multiplication; this operation is sometimes called the **dot product**.

The vector $\mathbf{A}$ has components A_1, A_2, A_3, and the magnitude (or length) of $\mathbf{A}$ is given by

$$|\mathbf{A}| = +\sqrt{A_1^2 + A_2^2 + A_3^2} \equiv A \qquad \textbf{(1.53)}$$

where the magnitude is indicated by $|\mathbf{A}|$ or, if there is no possibility of confusion, simply by A. Dividing both sides of Equation 1.52 by AB, we have

$$\frac{\mathbf{A} \cdot \mathbf{B}}{AB} = \sum_i \frac{A_i}{A} \frac{B_i}{B} \tag{1.54}$$

A_1/A is the cosine of the angle α between the vector $\mathbf{A}$ and the x_1-axis (see Figure 1-12). In general, A_i/A and B_i/B are the direction cosines Λ_i^A and Λ_i^B of the vectors $\mathbf{A}$ and $\mathbf{B}$:

$$\frac{\mathbf{A} \cdot \mathbf{B}}{AB} = \sum_i \Lambda_i^A \Lambda_i^B \tag{1.55}$$

The sum $\sum_i \Lambda_i^A \Lambda_i^B$ is just the cosine of the angle between $\mathbf{A}$ and $\mathbf{B}$ (see Equation 1.11):

$$\cos(\mathbf{A}, \mathbf{B}) = \sum_i \Lambda_i^A \Lambda_i^B$$

or

$$\boxed{\mathbf{A} \cdot \mathbf{B} = AB \cos(\mathbf{A}, \mathbf{B})} \tag{1.56}$$

That the product $\mathbf{A} \cdot \mathbf{B}$ is indeed a scalar may be shown as follows. $\mathbf{A}$ and $\mathbf{B}$ transform as vectors:

$$A_i' = \sum_j \lambda_{ij} A_j, \qquad B_i' = \sum_k \lambda_{ik} B_k \tag{1.57}$$

Therefore the product $\mathbf{A}' \cdot \mathbf{B}'$ becomes

$$\mathbf{A}' \cdot \mathbf{B}' = \sum_i A_i' B_i'$$

$$= \sum_i \left(\sum_j \lambda_{ij} A_j \right) \left(\sum_k \lambda_{ik} B_k \right)$$

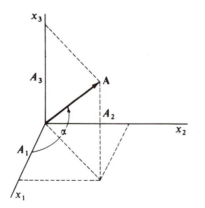

FIGURE 1-12

Rearranging the summations, we can write

$$\mathbf{A}' \cdot \mathbf{B}' = \sum_{j,k} \left(\sum_i \lambda_{ij} \lambda_{ik} \right) A_j B_k$$

But according to the orthogonality condition, the term in parentheses is just δ_{jk}. Thus,

$$\mathbf{A}' \cdot \mathbf{B}' = \sum_j \left(\sum_k \delta_{jk} A_j B_k \right)$$

$$= \sum_j A_j B_j$$

$$= \mathbf{A} \cdot \mathbf{B} \tag{1.58}$$

Because the value of the product is unaltered by the coordinate transformation, the product must be a scalar.

Notice that the distance from the origin to the point (x_1, x_2, x_3) defined by the vector $\mathbf{A}$, called the **position vector**, is given by

$$|\mathbf{A}| = \sqrt{\mathbf{A} \cdot \mathbf{A}} = \sqrt{x_1^2 + x_2^2 + x_3^2} = \sqrt{\sum_i x_i^2}$$

Similarly, the distance from the point (x_1, x_2, x_3) to another point $(\bar{x}_1, \bar{x}_2, \bar{x}_3)$ defined by the vector $\mathbf{B}$ is

$$\sqrt{\sum_i (x_i - \bar{x}_i)^2} = \sqrt{(\mathbf{A} - \mathbf{B}) \cdot (\mathbf{A} - \mathbf{B})} = |\mathbf{A} - \mathbf{B}|$$

That is, we can define the vector connecting any point with any other point as the difference of the position vectors that define the individual points, as in Figure 1-13. The distance between the points is then the magnitude of the difference vector. And because this magnitude is the square root of a scalar product, it is invariant to a coordinate transformation. This is an important fact and can be summarized by the statement that *orthogonal transformations are distance-preserving transformations*. Also, the angle between two vectors is preserved under an orthogonal

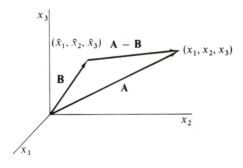

FIGURE 1-13

transformation. These two results are essential if we are successfully to apply transformation theory to physical situations.

The scalar product obeys the commutative and distributive laws:

$$\mathbf{A} \cdot \mathbf{B} = \sum_i A_i B_i = \sum_i B_i A_i = \mathbf{B} \cdot \mathbf{A} \tag{1.59}$$

$$\mathbf{A} \cdot (\mathbf{B} + \mathbf{C}) = \sum_i A_i (B + C)_i = \sum_i A_i (B_i + C_i)$$

$$= \sum_i (A_i B_i + A_i C_i) = (\mathbf{A} \cdot \mathbf{B}) + (\mathbf{A} \cdot \mathbf{C}) \tag{1.60}$$

1.11 UNIT VECTORS

Sometimes we want to describe a vector in terms of the components along the three coordinate axes together with a convenient specification of these axes. For this purpose, we introduce **unit vectors**, which are vectors having a length equal to the unit of length used along the particular coordinate axes. For example, the unit vector along the radial direction described by the vector $\mathbf{R}$ is $\mathbf{e}_R = \mathbf{R}/(|\mathbf{R}|)$. There are several variants of the symbols for unit vectors; examples of the most common sets are $(\mathbf{i}, \mathbf{j}, \mathbf{k})$, $(\mathbf{e}_1, \mathbf{e}_2, \mathbf{e}_3)$, $(\mathbf{e}_r, \mathbf{e}_\theta, \mathbf{e}_\phi)$, and $(\hat{\mathbf{r}}, \hat{\boldsymbol{\theta}}, \hat{\boldsymbol{\phi}})$. The following ways of expressing the vector $\mathbf{A}$ are equivalent:

$$\left. \begin{array}{c} \mathbf{A} = (A_1, A_2, A_3) \quad \text{or} \quad \mathbf{A} = \mathbf{e}_1 A_1 + \mathbf{e}_2 A_2 + \mathbf{e}_3 A_3 = \sum_i \mathbf{e}_i A_i \\ \text{or} \quad \mathbf{A} = A_1 \mathbf{i} + A_2 \mathbf{j} + A_3 \mathbf{k} \end{array} \right\} \tag{1.61}$$

We obtain the components of the vector $\mathbf{A}$ by projection onto the axes:

$$A_i = \mathbf{e}_i \cdot \mathbf{A} \tag{1.62}$$

We have seen (Equation 1.56) that the scalar product of two vectors has a magnitude equal to the product of the individual magnitudes multiplied by the cosine of the angle between the vectors:

$$\mathbf{A} \cdot \mathbf{B} = AB \cos(\mathbf{A}, \mathbf{B}) \tag{1.63}$$

If any two unit vectors are orthogonal, we have

$$\boxed{\mathbf{e}_i \cdot \mathbf{e}_j = \delta_{ij}} \tag{1.64}$$

E X A M P L E   **1.5**   ---

Two position vectors are expressed in Cartesian coordinates as $\mathbf{A} = \mathbf{i} + 2\mathbf{j} - 2\mathbf{k}$ and $\mathbf{B} = 4\mathbf{i} + 2\mathbf{j} - 3\mathbf{k}$. Find the magnitude of the vector from point A to point B, the angle θ between $\mathbf{A}$ and $\mathbf{B}$, and the component of $\mathbf{B}$ in the direction of $\mathbf{A}$.

Solution: The vector from point A to point B is $\mathbf{B} - \mathbf{A}$ (see Figure 1-13).

$$\mathbf{B} - \mathbf{A} = 4\mathbf{i} + 2\mathbf{j} - 3\mathbf{k} - (\mathbf{i} + 2\mathbf{j} - 2\mathbf{k}) = 3\mathbf{i} - \mathbf{k}$$

$$|\mathbf{B} - \mathbf{A}| = \sqrt{9 + 1} = \sqrt{10}$$

From Equation 1.56

$$\cos\theta = \frac{\mathbf{A} \cdot \mathbf{B}}{AB} = \frac{(\mathbf{i} + 2\mathbf{j} - 2\mathbf{k}) \cdot (4\mathbf{i} + 2\mathbf{j} - 3\mathbf{k})}{\sqrt{9}\sqrt{29}}$$

$$\cos\theta = \frac{4 + 4 + 6}{3(\sqrt{29})} = 0.867$$

$$\theta = 30°$$

The component of $\mathbf{B}$ in the direction of $\mathbf{A}$ is $B\cos\theta$ and, from Equation 1.56,

$$B\cos\theta = \frac{\mathbf{A} \cdot \mathbf{B}}{A} = \frac{14}{3} = 4.67$$

1.12   VECTOR PRODUCT OF TWO VECTORS

We next consider another method of combining two vectors—the **vector product** (sometimes called the **cross product**). In most respects, the vector product of two vectors behaves like a vector, and we shall treat it as such.* The vector product of $\mathbf{A}$ and $\mathbf{B}$ is denoted by a bold cross ×,

$$\mathbf{C} = \mathbf{A} \times \mathbf{B} \tag{1.65}$$

where $\mathbf{C}$ is the vector resulting from this operation. The components of $\mathbf{C}$ are defined by the relation

$$\boxed{C_i \equiv \sum_{j,k} \varepsilon_{ijk} A_j B_k} \tag{1.66}$$

where the symbol ε_{ijk} is the **permutation symbol** or (**Levi-Civita density**) and has the following properties:

$$\varepsilon_{ijk} = \left.\begin{array}{ll} 0, & \text{if any index is equal to any other index} \\ +1, & \text{if } i, j, k \text{ form an } even \text{ permutation of } 1, 2, 3 \\ -1, & \text{if } i, j, k \text{ form an } odd \text{ permutation of } 1, 2, 3 \end{array}\right\} \tag{1.67}$$

----

*The product actually produces an *axial* vector, but the term *vector product* is used to be consistent with popular usage.

An even permutation has an even number of exchanges of position of two symbols. Cyclic permutations (for example, $123 \to 231 \to 312$) are always even. Thus

$$\varepsilon_{122} = \varepsilon_{313} = \varepsilon_{211} = 0, \quad \text{etc.}$$

$$\varepsilon_{123} = \varepsilon_{231} = \varepsilon_{312} = +1$$

$$\varepsilon_{132} = \varepsilon_{213} = \varepsilon_{321} = -1$$

Using the preceding notation, the components of $\mathbf{C}$ can be explicitly evaluated. For the first subscript equal to 1, the only nonvanishing ε_{ijk} are ε_{123} and ε_{132}—that is, for $j, k = 2, 3$ in either order. Therefore

$$C_1 = \sum_{j,k} \varepsilon_{1jk} A_j B_k = \varepsilon_{123} A_2 B_3 + \varepsilon_{132} A_3 B_2$$

$$= A_2 B_3 - A_3 B_2 \tag{1.68a}$$

Similarly,

$$C_2 = A_3 B_1 - A_1 B_3 \tag{1.68b}$$

$$C_3 = A_1 B_2 - A_2 B_1 \tag{1.68c}$$

Consider now the expansion of the quantity $[AB \sin(\mathbf{A}, \mathbf{B})]^2 = (AB \sin \theta)^2$:

$$A^2 B^2 \sin^2 \theta = A^2 B^2 - A^2 B^2 \cos^2 \theta$$

$$= \left(\sum_i A_i^2\right)\left(\sum_i B_i^2\right) - \left(\sum_i A_i B_i\right)^2$$

$$= (A_2 B_3 - A_3 B_2)^2 + (A_3 B_1 - A_1 B_3)^2 + (A_1 B_2 - A_2 B_1)^2 \tag{1.69}$$

where the last equality requires some algebra. Identifying the components of $\mathbf{C}$ in the last expression, we can write

$$(AB \sin \theta)^2 = C_1^2 + C_2^2 + C_3^2 = |\mathbf{C}^2| = C^2 \tag{1.70}$$

If we take the positive square root of both sides of this equation,

$$C = AB \sin \theta \tag{1.71}$$

This equation states that if $\mathbf{C} = \mathbf{A} \times \mathbf{B}$, the magnitude of $\mathbf{C}$ is equal to the product of the magnitudes of $\mathbf{A}$ and $\mathbf{B}$ multiplied by the sine of the angle between them. Geometrically, $AB \sin \theta$ is the area of the parallelogram defined by the vectors $\mathbf{A}$ and $\mathbf{B}$ and the angle between them, as in Figure 1-14.

E X A M P L E 1.6

Show by using Equations 1.52 and 1.66 that

$$\mathbf{A} \cdot (\mathbf{B} \times \mathbf{C}) = \mathbf{C} \cdot (\mathbf{A} \times \mathbf{B}) \tag{1.72}$$

Solution: Using Equation 1.66, we have

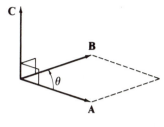

FIGURE 1-14

$$(\mathbf{B} \times \mathbf{C})_i = \sum_{j,k} \varepsilon_{ijk} B_j C_k$$

Using Equation 1.52, we have

$$\mathbf{A} \cdot (\mathbf{B} \times \mathbf{C}) = \sum_{i,j,k} \varepsilon_{ijk} A_i B_j C_k \qquad (1.73)$$

Similarly, for the right-hand side of Equation 1.72, we have

$$\mathbf{C} \cdot (\mathbf{A} \times \mathbf{B}) = \sum_{i,j,k} \varepsilon_{ijk} C_i A_j B_k$$

From the definition (Equation 1.67) of ε_{ijk}, we can interchange two adjacent indices of ε_{ijk}, which changes the sign.

$$\mathbf{C} \cdot (\mathbf{A} \times \mathbf{B}) = \sum_{i,j,k} - \varepsilon_{jik} C_i A_j B_k$$

$$= \sum_{i,j,k} \varepsilon_{jki} A_j B_k C_i \qquad (1.74)$$

Because the indices i, j, k are dummy and can be renamed, the right-hand sides of Equations 1.73 and 1.74 are identical, and Equation 1.72 is proved. Equation 1.72 can also be written as $\mathbf{A} \cdot (\mathbf{B} \times \mathbf{C}) = (\mathbf{A} \times \mathbf{B}) \cdot \mathbf{C}$, indicating that the scalar and vector products can be interchanged as long as the vectors stay in the order $\mathbf{A}$, $\mathbf{B}$, $\mathbf{C}$. Notice that, if we let $\mathbf{B} = \mathbf{A}$, we have

$$\mathbf{A} \cdot (\mathbf{A} \times \mathbf{C}) = \mathbf{C} \cdot (\mathbf{A} \times \mathbf{A}) = 0$$

showing that $\mathbf{A} \times \mathbf{C}$ must be perpendicular to $\mathbf{A}$.

--- ●

$\mathbf{A} \times \mathbf{B}$ (i.e., $\mathbf{C}$) is perpendicular to the plane defined by $\mathbf{A}$ and $\mathbf{B}$ because $\mathbf{A} \cdot (\mathbf{A} \times \mathbf{B}) = 0$ and $\mathbf{B} \cdot (\mathbf{A} \times \mathbf{B}) = 0$. Because a plane area can be represented by a vector normal to the plane and of magnitude equal to the area, $\mathbf{C}$ is evidently such a vector. The positive direction of $\mathbf{C}$ is chosen to be the direction of advance of a right-hand screw when rotated from $\mathbf{A}$ to $\mathbf{B}$.

The definition of the vector product is now complete; components, magnitude, and geometrical interpretation have been given. We may therefore reasonably expect that $\mathbf{C}$ is indeed a vector. The ultimate test, however, is to examine the transformation properties of $\mathbf{C}$, and $\mathbf{C}$ does, in fact, transform as a vector under a proper rotation.

We should note the following properties of the vector product that result from the definitions:

$$(a) \qquad \mathbf{A} \times \mathbf{B} = - \mathbf{B} \times \mathbf{A} \qquad\qquad (1.75)$$

but, in general,

$$(b) \qquad \mathbf{A} \times (\mathbf{B} \times \mathbf{C}) \neq (\mathbf{A} \times \mathbf{B}) \times \mathbf{C} \qquad\qquad (1.76)$$

Another important result (see Problem 1-22) is

$$\mathbf{A} \times (\mathbf{B} \times \mathbf{C}) = (\mathbf{A} \cdot \mathbf{C})\mathbf{B} - (\mathbf{A} \cdot \mathbf{B})\mathbf{C} \qquad\qquad (1.77)$$

E X A M P L E   **1.7**   --

Find the product of $(\mathbf{A} \times \mathbf{B}) \cdot (\mathbf{C} \times \mathbf{D})$.

Solution:

$$(\mathbf{A} \times \mathbf{B})_i = \sum_{j,k} \varepsilon_{ijk} A_j B_k$$

$$(\mathbf{C} \times \mathbf{D})_i = \sum_{l,m} \varepsilon_{ilm} C_l D_m$$

The scalar product is then computed according to Equation 1.52:

$$(\mathbf{A} \times \mathbf{B}) \cdot (\mathbf{C} \times \mathbf{D}) = \sum_i \left(\sum_{j,k} \varepsilon_{ijk} A_j B_k \right) \left(\sum_{l,m} \varepsilon_{ilm} C_l D_m \right)$$

Rearranging the summations, we have

$$(\mathbf{A} \times \mathbf{B}) \cdot (\mathbf{C} \times \mathbf{D}) = \sum_{\substack{l,m \\ j,k}} \left(\sum_i \varepsilon_{jki} \varepsilon_{lmi} \right) A_j B_k C_l D_m$$

where the indices of the ε's have been permuted (twice each so that no sign change occurs) to place in the third position the index over which the sum is carried out. We can now use an important property of the ε_{ijk} (see Problem 1-22):

$$\boxed{\sum_k \varepsilon_{ijk} \varepsilon_{lmk} = \delta_{il} \delta_{jm} - \delta_{im} \delta_{jl}} \qquad\qquad (1.78)$$

We therefore have

$$(\mathbf{A} \times \mathbf{B}) \cdot (\mathbf{C} \times \mathbf{D}) = \sum_{\substack{j,k \\ l,m}} (\delta_{jl} \delta_{km} - \delta_{jm} \delta_{kl}) A_j B_k C_l D_m$$

Carrying out the summations over j and k, the Kronecker deltas reduce the expression to

$$(\mathbf{A} \times \mathbf{B}) \cdot (\mathbf{C} \times \mathbf{D}) = \sum_{l,m} (A_l B_m C_l D_m - A_m B_l C_l D_m)$$

This equation can be rearranged to obtain

$$(\mathbf{A} \times \mathbf{B}) \cdot (\mathbf{C} \times \mathbf{D}) = \left(\sum_l A_l C_l \right) \left(\sum_m B_m D_m \right) - \left(\sum_l B_l C_l \right) \left(\sum_m A_m D_m \right)$$

Because each term in parentheses on the right-hand side is just a scalar product, we have, finally,

$$(\mathbf{A} \times \mathbf{B}) \cdot (\mathbf{C} \times \mathbf{D}) = (\mathbf{A} \cdot \mathbf{C})(\mathbf{B} \cdot \mathbf{D}) - (\mathbf{B} \cdot \mathbf{C})(\mathbf{A} \cdot \mathbf{D})$$

--- ●

The orthogonality of the unit vectors $\mathbf{e}_i$ requires the vector product to be

$$\mathbf{e}_i \times \mathbf{e}_j = \mathbf{e}_k \qquad i, j, k \text{ in cyclic order} \tag{1.79a}$$

We can now use the permutation symbol to express this result as

$$\boxed{\mathbf{e}_i \times \mathbf{e}_j = \sum_k \mathbf{e}_k \, \varepsilon_{ijk}} \tag{1.79b}$$

The vector product $\mathbf{C} = \mathbf{A} \times \mathbf{B}$, for example, can now be expressed as

$$\boxed{\mathbf{C} = \sum_{i,j,k} \varepsilon_{ijk} \, \mathbf{e}_i A_j B_k} \tag{1.80a}$$

By direct expansion and comparison with Equation 1.80a, we can verify a determinantal expression for the vector product:

$$\mathbf{C} = \mathbf{A} \times \mathbf{B} = \begin{vmatrix} \mathbf{e}_1 & \mathbf{e}_2 & \mathbf{e}_3 \\ A_1 & A_2 & A_3 \\ B_1 & B_2 & B_3 \end{vmatrix} \tag{1.80b}$$

We state the following identities without proof:

$$\mathbf{A} \cdot (\mathbf{B} \times \mathbf{C}) = \mathbf{B} \cdot (\mathbf{C} \times \mathbf{A}) = \mathbf{C} \cdot (\mathbf{A} \times \mathbf{B}) \equiv \mathbf{ABC} \tag{1.81}$$

$$\mathbf{A} \times (\mathbf{B} \times \mathbf{C}) = (\mathbf{A} \cdot \mathbf{C})\mathbf{B} - (\mathbf{A} \cdot \mathbf{B})\mathbf{C} \tag{1.82}$$

$$\begin{aligned} (\mathbf{A} \times \mathbf{B}) \cdot (\mathbf{C} \times \mathbf{D}) &= \mathbf{A} \cdot [\mathbf{B} \times (\mathbf{C} \times \mathbf{D})] \\ &= \mathbf{A} \cdot [(\mathbf{B} \cdot \mathbf{D})\mathbf{C} - (\mathbf{B} \cdot \mathbf{C})\mathbf{D}] \\ &= (\mathbf{A} \cdot \mathbf{C})(\mathbf{B} \cdot \mathbf{D}) - (\mathbf{A} \cdot \mathbf{D})(\mathbf{B} \cdot \mathbf{C}) \end{aligned} \right\} \tag{1.83}$$

$$\begin{aligned} (\mathbf{A} \times \mathbf{B}) \times (\mathbf{C} \times \mathbf{D}) &= [(\mathbf{A} \times \mathbf{B}) \cdot \mathbf{D}]\mathbf{C} - [(\mathbf{A} \times \mathbf{B}) \cdot \mathbf{C}]\mathbf{D} \\ &= (\mathbf{ABD})\mathbf{C} - (\mathbf{ABC})\mathbf{D} = (\mathbf{ACD})\mathbf{B} - (\mathbf{BCD})\mathbf{A} \end{aligned} \right\} \tag{1.84}$$

1.13 DIFFERENTIATION OF A VECTOR WITH RESPECT TO A SCALAR

If a scalar function $\phi = \phi(s)$ is differentiated with respect to the scalar variable s, then, because neither part of the derivative can change under a coordinate transformation, the derivative itself cannot change and must therefore be a scalar; that is, in the x_i and x_i' coordinate systems, $\phi = \phi'$ and $s = s'$, so $d\phi = d\phi'$ and $ds = ds'$. Hence

$$d\phi/ds = d\phi'/ds' = (d\phi/ds)'$$

Similarly, we can formally define the differentiation of a vector $\mathbf{A}$ with respect to a scalar s. The components of $\mathbf{A}$ transform according to

$$A_i' = \sum_j \lambda_{ij} A_j \tag{1.85}$$

Therefore, on differentiation, we obtain (because the λ_{ij} are independent of s')

$$\frac{dA_i'}{ds'} = \frac{d}{ds'} \sum_j \lambda_{ij} A_j = \sum_j \lambda_{ij} \frac{dA_j}{ds'}$$

Because s and s' are identical, we have

$$\frac{dA_i'}{ds'} = \left(\frac{dA_i}{ds}\right)' = \sum_j \lambda_{ij} \left(\frac{dA_j}{ds}\right)$$

Thus the quantities dA_j/ds transform as do the components of a vector and hence *are* the components of a vector, which we can write as $d\mathbf{A}/ds$.

We can give a geometrical interpretation to the vector $d\mathbf{A}/ds$ as follows. First, for $d\mathbf{A}/ds$ to exist, $\mathbf{A}$ must be a continuous function of the variable s: $\mathbf{A} = \mathbf{A}(s)$. Suppose this function is represented by the continuous curve Γ in Figure 1-15; at the point P, the variable has the value s, and at Q it has the value $s+\Delta s$. The derivative of $\mathbf{A}$ with respect to s is then given in standard fashion by

$$\frac{d\mathbf{A}}{ds} = \lim_{\Delta s \to 0} \frac{\Delta \mathbf{A}}{\Delta s} = \lim_{\Delta s \to 0} \frac{\mathbf{A}(s + \Delta s) - \mathbf{A}(s)}{\Delta s} \tag{1.86a}$$

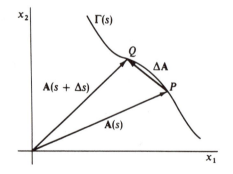

FIGURE 1-15

The derivatives of vector sums and products obey the rules of ordinary vector calculus. For example,

$$\frac{d}{ds}(\mathbf{A} + \mathbf{B}) = \frac{d\mathbf{A}}{ds} + \frac{d\mathbf{B}}{ds} \tag{1.86b}$$

$$\frac{d}{ds}(\mathbf{A} \cdot \mathbf{B}) = \mathbf{A} \cdot \frac{d\mathbf{B}}{ds} + \frac{d\mathbf{A}}{ds} \cdot \mathbf{B} \tag{1.86c}$$

$$\frac{d}{ds}(\mathbf{A} \times \mathbf{B}) = \mathbf{A} \times \frac{d\mathbf{B}}{ds} + \frac{d\mathbf{A}}{ds} \times \mathbf{B} \tag{1.86d}$$

$$\frac{d}{ds}(\phi\mathbf{A}) = \phi\frac{d\mathbf{A}}{ds} + \frac{d\phi}{ds}\mathbf{A} \tag{1.86e}$$

and similarly for total differentials and for partial derivatives.

1.14 EXAMPLES OF DERIVATIVES — VELOCITY AND ACCELERATION

Of particular importance in the development of the dynamics of point particles (and of systems of particles) is the representation of the motion of these particles by vectors. For such an approach, we require vectors to represent the position, velocity, and acceleration of a given particle. It is customary to specify the *position* of a particle with respect to a certain reference frame by a vector $\mathbf{r}$, which is in general a function of time: $\mathbf{r} = \mathbf{r}(t)$. The *velocity* vector $\mathbf{v}$ and the *acceleration* vector $\mathbf{a}$ are defined according to

$$\mathbf{v} \equiv \frac{d\mathbf{r}}{dt} = \dot{\mathbf{r}} \tag{1.87}$$

$$\mathbf{a} \equiv \frac{d\mathbf{v}}{dt} = \frac{d^2\mathbf{r}}{dt^2} = \ddot{\mathbf{r}} \tag{1.88}$$

where a single dot above a symbol denotes the first time derivative, and two dots denote the second time derivative. In rectangular coordinates, the expressions for $\mathbf{r}$, $\mathbf{v}$, and $\mathbf{a}$ are

$$\mathbf{r} = x_1\mathbf{e}_1 + x_2\mathbf{e}_2 + x_3\mathbf{e}_3 = \sum_i x_i\,\mathbf{e}_i \qquad \text{Position}$$

$$\mathbf{v} = \dot{\mathbf{r}} = \sum_i \dot{x}_i\mathbf{e}_i = \sum_i \frac{dx_i}{dt}\mathbf{e}_i \qquad \text{Velocity} \tag{1.89}$$

$$\mathbf{a} = \dot{\mathbf{v}} = \ddot{\mathbf{r}} = \sum_i \ddot{x}_i\mathbf{e}_i = \sum_i \frac{d^2x_i}{dt^2}\mathbf{e}_i \qquad \text{Acceleration}$$

Calculating these quantities in rectangular coordinates is straightforward because the unit vectors $\mathbf{e}_i$ are constant in time. In nonrectangular coordinate systems, however, the unit vectors at the position of the particle as it moves in space are not necessarily constant in time, and the components of the time derivatives of **r** are no longer simple relations, as in Equation 1.89. We do not discuss general curvilinear coordinate systems here, but *plane polar* coordinates, *spherical* coordinates, and *cylindrical* coordinates are of sufficient importance to warrant a discussion of velocity and acceleration in these coordinate systems.*

To express **v** and **a** in plane polar coordinates, consider the situation in Figure 1-16. A point moves along the curve $s(t)$ and in the time interval $t_2 - t_1 = dt$ moves from $P^{(1)}$ to $P^{(2)}$. The unit vectors, $\mathbf{e}_r$ and $\mathbf{e}_\theta$, which are orthogonal, change from $\mathbf{e}_r^{(1)}$ to $\mathbf{e}_r^{(2)}$ and from $\mathbf{e}_\theta^{(1)}$ to $\mathbf{e}_\theta^{(2)}$. The change in $\mathbf{e}_r$ is

$$\mathbf{e}_r^{(2)} - \mathbf{e}_r^{(1)} = d\mathbf{e}_r \tag{1.90}$$

which is a vector normal to $\mathbf{e}_r$ (and, therefore, in the direction of $\mathbf{e}_\theta$). Similarly, the change in $\mathbf{e}_\theta$ is

$$\mathbf{e}_\theta^{(2)} - \mathbf{e}_\theta^{(1)} = d\mathbf{e}_\theta \tag{1.91}$$

which is a vector normal to $\mathbf{e}_\theta$. We can then write

$$d\mathbf{e}_r = d\theta \mathbf{e}_\theta \tag{1.92}$$

and

$$d\mathbf{e}_\theta = -d\theta \mathbf{e}_r \tag{1.93}$$

where the minus sign enters the second relation because $d\mathbf{e}_\theta$ is directed *opposite* to $\mathbf{e}_r$ (see Figure 1-16).

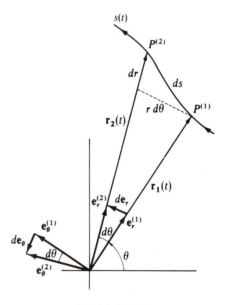

FIGURE 1-16

---

*Refer to the figures in Appendix F for the geometry of these coordinate systems.

Equations 1.92 and 1.93 are perhaps easier to see by referring to Figure 1-16. In this case, $d\mathbf{e}_r$ subtends an angle $d\theta$ with unit sides, so it has a magnitude of $d\theta$. It also points in the direction of $\mathbf{e}_\theta$, so we have $d\mathbf{e}_r = d\theta\mathbf{e}_\theta$. Similarly, $d\mathbf{e}_\theta$ subtends an angle $d\theta$ with unit sides, so it also has a magnitude of $d\theta$, but from Figure 1-16 we see that $d\mathbf{e}_\theta$ points in the direction of $-\mathbf{e}_r$, so we have $d\mathbf{e}_\theta = -d\theta\mathbf{e}_r$.

Dividing each side of Equations 1.92 and 1.93 by dt, we have

$$\dot{\mathbf{e}}_r = \dot\theta\mathbf{e}_\theta \tag{1.94}$$

$$\dot{\mathbf{e}}_\theta = -\dot\theta\mathbf{e}_r \tag{1.95}$$

If we express $\mathbf{v}$ as

$$\mathbf{v} = \frac{d\mathbf{r}}{dt} = \frac{d}{dt}(r\mathbf{e}_r)$$

$$= \dot{r}\mathbf{e}_r + r\dot{\mathbf{e}}_r \tag{1.96}$$

we have immediately, using Equation 1.94,

$$\mathbf{v} = \dot{\mathbf{r}} = \dot{r}\mathbf{e}_r + r\dot\theta\mathbf{e}_\theta \tag{1.97}$$

so that the velocity is resolved into a *radial* component $\dot{r}$ and an *angular* (or *transverse*) component $r\dot\theta$.

A second differentiation yields the acceleration:

$$\mathbf{a} = \frac{d}{dt}(\dot{r}\mathbf{e}_r + r\dot\theta\mathbf{e}_\theta)$$

$$= \ddot{r}\mathbf{e}_r + \dot{r}\dot{\mathbf{e}}_r + \dot{r}\dot\theta\mathbf{e}_\theta + r\ddot\theta\mathbf{e}_\theta + r\dot\theta\dot{\mathbf{e}}_\theta$$

$$= (\ddot{r} - r\dot\theta^2)\mathbf{e}_r + (r\ddot\theta + 2\dot{r}\dot\theta)\mathbf{e}_\theta \tag{1.98}$$

so that the acceleration is resolved into a radial component $(\ddot{r} - r\dot\theta^2)$ and an angular (or transverse) component $(r\ddot\theta + 2\dot{r}\dot\theta)$.

The expressions for ds, ds^2, v^2, and $\mathbf{v}$ in the three most important coordinate systems (see also Appendix F) are

Rectangular coordinates (x, y, z)

$$\left.\begin{aligned} d\mathbf{s} &= dx_1\,\mathbf{e}_1 + dx_2\mathbf{e}_2 + dx_3\mathbf{e}_3 \\ ds^2 &= dx_1^2 + dx_2^2 + dx_3^2 \\ v^2 &= \dot{x}_1^2 + \dot{x}_2^2 + \dot{x}_3^2 \\ \mathbf{v} &= \dot{x}_1\mathbf{e}_1 + \dot{x}_2\mathbf{e}_2 + \dot{x}_3\mathbf{e}_3 \end{aligned}\right\} \tag{1.99}$$

Spherical coordinates (r, θ, ϕ)

$$
\left.
\begin{aligned}
d\mathbf{s} &= dr\,\mathbf{e}_r + rd\theta\,\mathbf{e}_\theta + r\sin\theta\,d\phi\,\mathbf{e}_\phi \\
ds^2 &= dr^2 + r^2 d\theta^2 + r^2\sin^2\theta\,d\phi^2 \\
v^2 &= \dot{r}^2 + r^2\dot{\theta}^2 + r^2\sin^2\theta\dot{\phi}^2 \\
\mathbf{v} &= \dot{r}\,\mathbf{e}_r + r\dot{\theta}\,\mathbf{e}_\theta + r\sin\theta\dot{\phi}\,\mathbf{e}_\phi
\end{aligned}
\right\}
$$

(1.100)

(The expressions for plane polar coordinates result from Equations 1.100 by setting $d\phi = 0$.)

Cylindrical coordinates (r, ϕ, z)

$$
\left.
\begin{aligned}
d\mathbf{s} &= dr\,\mathbf{e}_r + rd\phi\,\mathbf{e}_\phi + dz\,\mathbf{e}_z \\
ds^2 &= dr^2 + r^2 d\phi^2 + dz^2 \\
v^2 &= \dot{r}^2 + r^2\dot{\phi}^2 + \dot{z}^2 \\
\mathbf{v} &= \dot{r}\,\mathbf{e}_r + r\dot{\phi}\,\mathbf{e}_\phi + \dot{z}\,\mathbf{e}_z
\end{aligned}
\right\}
$$

(1.101)

EXAMPLE 1.8

Find the components of the acceleration vector a in cylindrical coordinates.

Solution: The velocity components in cylindrical coordinates were given in Equation 1.101. The acceleration is determined by taking the time derivative of **v**.

$$
\mathbf{a} = \frac{d}{dt}\mathbf{v} = \frac{d}{dt}(\dot{r}\,\mathbf{e}_r + r\dot{\phi}\,\mathbf{e}_\phi + \dot{z}\,\mathbf{e}_z)
$$

$$
= \ddot{r}\,\mathbf{e}_r + \dot{r}\,\dot{\mathbf{e}}_r + \dot{r}\dot{\phi}\,\mathbf{e}_\phi + r\ddot{\phi}\,\mathbf{e}_\phi + r\dot{\phi}\,\dot{\mathbf{e}}_\phi + \ddot{z}\,\mathbf{e}_z + \dot{z}\,\dot{\mathbf{e}}_z
$$

We need to find the time derivative of the unit vectors $\mathbf{e}_r$, $\mathbf{e}_\phi$, and $\mathbf{e}_z$. The cylindrical coordinate system is shown in Figure 1-17, and in terms of the (x, y, z) components, the unit vectors $\mathbf{e}_r$, $\mathbf{e}_\phi$, and $\mathbf{e}_z$ are

$$
\mathbf{e}_r = (\cos\phi,\ \sin\phi,\ 0)
$$

$$
\mathbf{e}_\phi = (-\sin\phi,\ \cos\phi,\ 0)
$$

$$
\mathbf{e}_z = (0,\ 0,\ 1)
$$

The time derivatives of the unit vectors are found by taking the derivatives of the components.

$$
\dot{\mathbf{e}}_r = (-\dot{\phi}\sin\phi,\ \dot{\phi}\cos\phi,\ 0) = \dot{\phi}\,\mathbf{e}_\phi
$$

$$
\dot{\mathbf{e}}_\phi = (-\dot{\phi}\cos\phi,\ -\dot{\phi}\sin\phi,\ 0) = -\dot{\phi}\,\mathbf{e}_r
$$

$$
\dot{\mathbf{e}}_z = 0
$$

We substitute the unit vector time derivatives into the above expression for **a**.

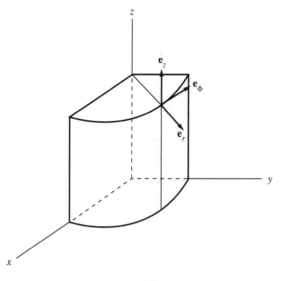

FIGURE 1-17

$$\mathbf{a} = \ddot{r}\mathbf{e}_r + \dot{r}\dot{\phi}\mathbf{e}_\phi + \dot{r}\dot{\phi}\mathbf{e}_\phi + r\ddot{\phi}\mathbf{e}_\phi - r\dot{\phi}^2\mathbf{e}_r + \ddot{z}\mathbf{e}_z$$

$$= (\ddot{r} - r\dot{\phi}^2)\mathbf{e}_r + (r\ddot{\phi} + 2\dot{r}\dot{\phi})\mathbf{e}_\phi + \ddot{z}\mathbf{e}_z$$

1.15   ANGULAR VELOCITY

A point or a particle moving arbitrarily in space may always be considered, *at a given instant*, to be moving in a plane, circular path about a certain axis; that is, the path a particle describes during an infinitesimal time interval δt may be represented as an infinitesimal arc of a circle. The line passing through the center of the circle and perpendicular to the instantaneous direction of motion is called the **instantaneous axis of rotation**. As the particle moves in the circular path, the rate of change of the angular position is called the **angular velocity**:

$$\omega = \frac{d\theta}{dt} = \dot{\theta} \tag{1.102}$$

Consider a particle that moves instantaneously in a circle of radius R about an axis perpendicular to the plane of motion, as in Figure 1-18. Let the position vector $\mathbf{r}$ of the particle be drawn from an origin located at an arbitrary point O on the axis of rotation. The time rate of change of the position vector is the linear velocity vector of the particle, $\dot{\mathbf{r}} = \mathbf{v}$. For motion in a circle of radius R, the instantaneous *magnitude* of the linear velocity is given by

$$v = R\frac{d\theta}{dt} = R\omega \tag{1.103}$$

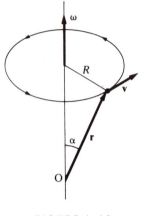

FIGURE 1-18

The *direction* of the linear velocity **v** is perpendicular to **r** and in the plane of the circle.

It would be very convenient if we could devise a vector representation of the angular velocity (say, **ω**) so that all the quantities of interest in the motion of the particle could be described on a common basis. We can define a *direction* for the angular velocity in the following manner. If the particle moves instantaneously in a plane, the normal to that plane defines a precise direction in space—or, rather—*two* directions. We may choose as *positive* that direction corresponding to the direction of advance of a right-hand screw when turned in the same sense as the rotation of the particle (see Figure 1-18). We can also write the magnitude of the linear velocity by noting that $R = r \sin \alpha$. Thus

$$v = r\omega \sin \alpha \qquad (1.104)$$

Having defined a direction and a magnitude for the angular velocity, we note that if we write

$$\boxed{\mathbf{v} = \boldsymbol{\omega} \times \mathbf{r}} \qquad (1.105)$$

then both of these definitions are satisfied, and we have the desired vector representation of the angular velocity.

We should note at this point an important distinction between finite and infinitesimal rotations. An **infinitesimal rotation** can be represented by a vector (actually, an *axial* vector), but a **finite rotation** cannot. The impossibility of describing a finite rotation by a vector results from the fact that such rotations do not commute (see the example of Figure 1-9), and therefore, in general, different results will be obtained depending on the order in which the rotations are made. To illustrate this statement, consider the successive application of two finite rotations described by the rotation matrices $\boldsymbol{\lambda}_1$ and $\boldsymbol{\lambda}_2$. Let us associate the vectors **A** and **B** in a one-to-one manner with these rotations. The vector sum is $\mathbf{C} = \mathbf{A} + \mathbf{B}$, which is equivalent to the matrix $\lambda_3 = \lambda_2\lambda_1$. But because vector addition is commutative, we also have $C = B + A$, with $\lambda_4 = \lambda_1\lambda_2$. But we know that matrix operations are not commutative, so that in general $\lambda_3 \neq \lambda_4$. Hence, the vector C is not unique, and therefore we cannot associate a vector with a finite rotation.

Infinitesimal rotations do not suffer from this defect of noncommutation. We are therefore led to expect that an infinitesimal rotation can be represented by a vector. Although this expectation is, in fact, fulfilled, the ultimate test of the vector nature of a quantity is contained in its transformation properties. We give only a qualitative argument here.

Refer to Figure 1-19. If the position vector of a point changes from r to $r + \delta r$, the geometrical situation is correctly represented if we write

$$\delta r = \delta\theta \times r \qquad\qquad (1.106)$$

where $\delta\theta$ is a quantity whose magnitude is equal to the infinitesimal rotation angle and which has a direction along the instantaneous axis of rotation. The mere fact that Equation 1.106 correctly describes the situation illustrated in Figure 1-19 is not sufficient to establish that $\delta\theta$ is a vector. (We reiterate that the true test must be based on the transformation properties of $\delta\theta$.) But if we show that two infinitesimal rotation "vectors"—$\delta\theta_1$ and $\delta\theta_2$—actually *commute*, the sole objection to representing a finite rotation by a vector will have been removed.

Let us consider that a rotation $\delta\theta_1$ takes r into $r + \delta r_1$, where $\delta r_1 = \delta\theta_1 \times r$. If this is followed by a second rotation $\delta\theta_2$ around a different axis, the initial position vector for this rotation is $r + \delta r_1$. Thus

$$\delta r_2 = \delta\theta_2 \times (r + \delta r_1)$$

and the final position vector for $\delta\theta_1$ followed by $\delta\theta_2$ is

$$r + \delta r_{12} = r + [\delta\theta_1 \times r + \delta\theta_2 \times (r + \delta r_1)]$$

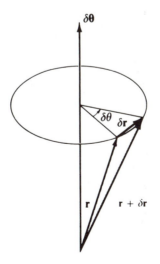

FIGURE 1-19

Neglecting second-order infinitesimals, then,

$$\delta \mathbf{r}_{12} = \delta \boldsymbol{\theta}_1 \times \mathbf{r} + \delta \boldsymbol{\theta}_2 \times \mathbf{r} \tag{1.107}$$

Similarly, if $\delta \boldsymbol{\theta}_2$ is followed by $\delta \boldsymbol{\theta}_1$, we have

$$\mathbf{r} + \delta \mathbf{r}_{21} = \mathbf{r} + [\delta \boldsymbol{\theta}_2 \times \mathbf{r} + \delta \boldsymbol{\theta}_1 \times (\mathbf{r} + \delta \mathbf{r}_2)]$$

or

$$\delta \mathbf{r}_{21} = \delta \boldsymbol{\theta}_2 \times \mathbf{r} + \delta \boldsymbol{\theta}_1 \times \mathbf{r} \tag{1.108}$$

Rotation vectors $\delta \mathbf{r}_{12}$ and $\delta \mathbf{r}_{21}$ are equal, so the rotation "vectors" $\delta \boldsymbol{\theta}_1$ and $\delta \boldsymbol{\theta}_2$ do commute. It therefore seems reasonable that $\delta \boldsymbol{\theta}$ in Equation 1.106 is indeed a vector.

It is the fact that $\delta \boldsymbol{\theta}$ is a vector that allows angular velocity to be represented by a vector, because angular velocity is the ratio of an infinitesimal rotation angle to an infinitesimal time:

$$\boldsymbol{\omega} = \frac{\delta \boldsymbol{\theta}}{\delta t}$$

Therefore, dividing Equation 1.106 by δt, we have

$$\frac{\delta \mathbf{r}}{\delta t} = \frac{\delta \boldsymbol{\theta}}{\delta t} \times \mathbf{r}$$

or, in passing to the limit, $\delta t \rightarrow 0$,

$$\mathbf{v} = \boldsymbol{\omega} \times \mathbf{r}$$

as before.

1.16   GRADIENT OPERATOR

We now turn to the most important member of a class called **vector differential operators**—the **gradient operator**.

Consider a scalar ϕ that is an explicit function of the coordinates x_i and, moreover, is a continuous, single-valued function of these coordinates throughout a certain region of space. Under a coordinate transformation that carries the x_i into the x_i', $\phi'(x_1', x_2', x_3') = \phi(x_1, x_2, x_3)$, and by the chain rule of differentiation, we can write

$$\frac{\partial \phi'}{\partial x_1'} = \sum_j \frac{\partial \phi}{\partial x_j} \frac{\partial x_j}{\partial x_1'} \tag{1.109}$$

The case is similar for $\partial \phi'/\partial x_2'$ and $\partial \phi'/\partial x_3'$, so in general we have

$$\frac{\partial \phi'}{\partial x_i'} = \sum_j \frac{\partial \phi}{\partial x_j} \frac{\partial x_j}{\partial x_i'} \tag{1.110}$$

The inverse coordinate transformation is

$$x_j = \sum_k \lambda_{kj} x_k' \tag{1.111}$$

Differentiating,

$$\frac{\partial x_j}{\partial x_i'} = \frac{\partial}{\partial x_i'} \left(\sum_k \lambda_{kj} x_k' \right) = \sum_k \lambda_{kj} \left(\frac{\partial x_k'}{\partial x_i'} \right) \tag{1.112}$$

But the term in the last parentheses is just δ_{ik}, so

$$\frac{\partial x_j}{\partial x_i'} = \sum_k \lambda_{kj} \, \delta_{ik} = \lambda_{ij} \tag{1.113}$$

Substituting Equation 1.113 into Equation 1.110, we obtain

$$\frac{\partial \phi'}{\partial x_i'} = \sum_j \lambda_{ij} \frac{\partial \phi}{\partial x_j} \tag{1.114}$$

Because it follows the correct transformation equation of a vector (Equation 1.44), the function $\partial \phi / \partial x_j$ is the jth component of a vector termed the **gradient** of the function ϕ. Note that even though ϕ is a scalar, the *gradient* of ϕ is a *vector*. The gradient of ϕ is written either as **grad** ϕ or as $\nabla \phi$("del" ϕ).

Because the function ϕ is an arbitrary scalar function, it is convenient to define the differential operator described in the preceding in terms of the *gradient operator*:

$$(\textbf{grad})_i = \nabla_i = \frac{\partial}{\partial x_i} \tag{1.115}$$

We can express the complete vector gradient operator as

$$\boxed{\textbf{grad} = \nabla = \sum_i \textbf{e}_i \frac{\partial}{\partial x_i}} \qquad \text{Gradient} \tag{1.116}$$

The gradient operator can (a) operate directly on a scalar function, as in $\nabla \phi$; (b) be used in a scalar product with a vector function, as in $\nabla \cdot \textbf{A}$ (the *divergence of* $\textbf{A}$); or (c) be used in a vector product with a vector function, as in $\nabla \times \textbf{A}$ (the *curl* of $\textbf{A}$).

To see a physical interpretation of the gradient of a scalar function, consider the three-dimensional and topographical maps of Figure 1-20. The closed loops of part b represent lines of constant height. Let ϕ denote the height at any point $\phi = \phi(x_1, x_2, x_3)$. Then

$$d\phi = \sum_i \frac{\partial \phi}{\partial x_i} dx_i = \sum_i (\nabla \phi)_i \, dx_i \tag{1.117}$$

The components of the displacement vector $d\textbf{s}$ are the incremental displacements in the direction of the three orthogonal axes:

$$d\textbf{s} = (dx_1, dx_2, dx_3) \tag{1.118}$$

Therefore

$$\boxed{d\phi = (\nabla \phi) \cdot d\textbf{s}} \tag{1.119}$$

(a)

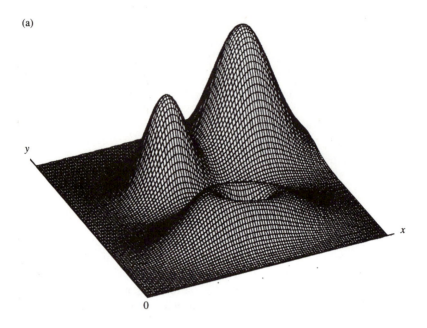

(b)

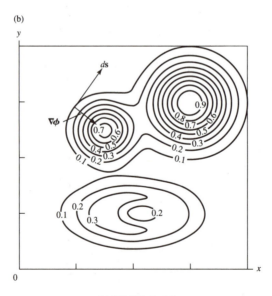

FIGURE 1-20

Let $d\mathbf{s}$ be directed tangentially along one of the isolatitude lines (i.e., along a line for which ϕ = const.), as indicated in Figure 1-20. Because ϕ = const. for this case, $d\phi = 0$. But, because neither $\nabla\phi$ nor $d\mathbf{s}$ is in general zero, they must therefore be perpendicular to each other. Thus $\nabla\phi$ is normal to the line (or in three dimensions, to the surface) for which ϕ = const.

The maximum value of $d\phi$ results when $\nabla\phi$ and $d\mathbf{s}$ are in the same direction; then,

$$(d\phi)_{\max} = |\nabla\phi| \, ds, \qquad \text{for} \qquad \nabla\phi \parallel d\mathbf{s}$$

or

$$|\nabla\phi| = \left(\frac{d\phi}{ds}\right)_{\max} \tag{1.120}$$

Therefore, $\nabla\phi$ is in the direction of the greatest change in ϕ.

We can summarize these results as follows:

1. The vector $\nabla\phi$ is, at any point, normal to the lines or surfaces for which $\phi = \text{const.}$

2. The vector $\nabla\phi$ has the direction of the maximum change in ϕ.

3. Because any direction in space can be specified in terms of the unit vector $\mathbf{n}$ in that direction, the rate of change of ϕ in the direction of $\mathbf{n}$ (the *directional derivative* of ϕ) can be found from $\mathbf{n} \cdot \nabla\phi \equiv \partial\phi/\partial n$.

The successive operation of the gradiant operator produces

$$\nabla \cdot \nabla = \sum_i \frac{\partial}{\partial x_i} \frac{\partial}{\partial x_i} = \sum_i \frac{\partial^2}{\partial x_i^2} \tag{1.121}$$

This important product operator, called the **Laplacian,*** is also written

$$\nabla^2 = \sum_i \frac{\partial^2}{\partial x_i^2} \tag{1.122}$$

When the Laplacian operates on a scalar, we have, for example,

$$\nabla^2\psi = \sum_i \frac{\partial^2\psi}{\partial x_i^2} \tag{1.123}$$

1.17 INTEGRATION OF VECTORS

The vector resulting from the volume integration of a vector function $\mathbf{A} = \mathbf{A}(x_i)$ throughout a volume V is given by[†]

$$\int_V \mathbf{A} \, dv = \left(\int_V A_1 \, dv, \int_V A_2 \, dv, \int_V A_3 \, dv\right) \tag{1.124}$$

Thus, we integrate the vector $\mathbf{A}$ throughout V simply by performing three separate, ordinary integrations.

---

*After Pierre Simon Laplace (1749–1827); the notation ∇^2 is ascribed to Sir William Rowan Hamilton.

[†]The symbol $\int_V$ actually represents a *triple* integral over a certain volume V. Similarly, the symbol $\int_S$ stands for a *double* integral over a certain surface S.

The integral over a surface S of the projection of a vector function $\mathbf{A} = \mathbf{A}(x_i)$ onto the normal to that surface is defined to be

$$\int_S \mathbf{A} \cdot d\mathbf{a}$$

where $d\mathbf{a}$ is an element of area of the surface (Figure 1-21). We write $d\mathbf{a}$ as a vector quantity because we may attribute to it not only a magnitude da but also a direction corresponding to the normal to the surface at the point in question. If the unit normal vector is $\mathbf{n}$, then

$$d\mathbf{a} = \mathbf{n}\,da \tag{1.125}$$

Thus, the components of $d\mathbf{a}$ are the projections of the element of area on the three mutually perpendicular planes defined by the rectangular axes:

$$da_1 = dx_2\,dx_3, \qquad \text{etc.} \tag{1.126}$$

Therefore, we have

$$\int_S \mathbf{A} \cdot d\mathbf{a} = \int_S \mathbf{A} \cdot \mathbf{n}\,da \tag{1.127}$$

or

$$\int_S \mathbf{A} \cdot d\mathbf{a} = \int_S \sum_i A_i\,da_i \tag{1.128}$$

Equation 1.127 states that the integral of $\mathbf{A}$ over the surface S is the integral of the normal component of $\mathbf{A}$ over this surface.

The normal to a surface may be taken to lie in either of two possible directions ("up" or "down"); thus the sign of $\mathbf{n}$ is ambiguous. If the surface is *closed*, we adopt the convention that the *outward* normal is positive.

The **line integral** of a vector function $\mathbf{A} = \mathbf{A}(x_i)$ along a given path extending from the point B to the point C is given by the integral of the component of $\mathbf{A}$ along the path

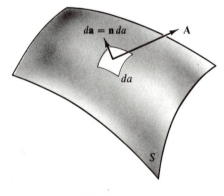

FIGURE 1-21

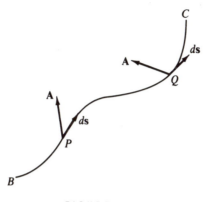

FIGURE 1-22

$$\int_{BC} \mathbf{A} \cdot d\mathbf{s} = \int_{BC} \sum_i A_i \, dx_i \qquad (1.129)$$

The quantity $d\mathbf{s}$ is an element of length along the given path (Figure 1-22). The direction of $d\mathbf{s}$ is taken to be positive along the direction the path is traversed. In Figure 1-22 at point P, the angle between $d\mathbf{s}$ and $\mathbf{A}$ is less than $\pi/2$, so $\mathbf{A} \cdot d\mathbf{s}$ is positive at this point. At point Q, the angle is greater than $\pi/2$, and the contribution to the integral at this point is negative.

It is often useful to relate certain surface integrals to either volume integrals (**Gauss's theorem**) or line integrals (**Stokes's theorem**). Consider Figure 1-23, which shows a closed volume V enclosed by the surface S. Let the vector $\mathbf{A}$ and its first derivatives be continuous throughout the volume. Gauss's theorem states that the surface integral of $\mathbf{A}$ over the closed surface S is equal to the volume integral of the divergence of $\mathbf{A}$ ($\nabla \cdot \mathbf{A}$) throughout the volume V enclosed by the surface S. We write this mathematically as

$$\int_S \mathbf{A} \cdot d\mathbf{a} = \int_V \nabla \cdot \mathbf{A} \, dv \qquad (1.130)$$

Gauss's theorem is sometimes also called the *divergence theorem*. The theorem is particularly useful in dealing with the mechanics of continuous media.

See Figure 1-24 for the physical description needed for Stokes's theorem, which applies to an open surface S and the contour path C that defines the surface.

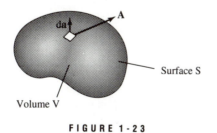

FIGURE 1-23

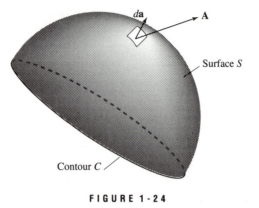

FIGURE 1-24

The curl of the vector $\mathbf{A}$ ($\nabla \times \mathbf{A}$) must exist and be integrable over the entire surface S. Stokes's theorem states that the line integral of the vector $\mathbf{A}$ around the contour path C is equal to the surface integral of the curl of $\mathbf{A}$ over the surface defined by C. We write it mathematically as

$$\int_C \mathbf{A} \cdot d\mathbf{s} = \int_S (\nabla \times \mathbf{A}) \cdot d\mathbf{a} \tag{1.131}$$

where the line integral is around the closed contour path C. Stokes's theorem is particularly useful in reducing certain surface integrals (two dimensional) to a hopefully simpler line integral (one dimensional). Both Gauss's and Stokes's theorems have wide application in vector calculus. In addition to mechanics, they are also useful in electromagnetic applications and in potential theory.

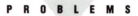

P R O B L E M S

1-1. Find the transformation matrix that rotates the axis x_3 of a rectangular coordinate system $45°$ toward x_1 around the x_2-axis.

1-2. Prove Equations 1.10 and 1.11 from trigonometric considerations.

1-3. Find the transformation matrix that rotates a rectangular coordinate system through an angle of $120°$ about an axis making equal angles with the original three coordinate axes.

1-4. Show
(a) $(\mathbf{AB})^t = \mathbf{B}^t\mathbf{A}^t$ (b) $(\mathbf{AB})^{-1} = \mathbf{B}^{-1}\mathbf{A}^{-1}$

1-5. Show by direct expansion that $|\boldsymbol{\lambda}|^2 = 1$. For simplicity, take $\boldsymbol{\lambda}$ to be a two-dimensional transformation matrix.

1-6. Show that Equation 1.15 can be obtained by using the requirement that the transformation leave unchanged the length of a line segment.

1-7. Consider a unit cube with one corner at the origin and three adjacent sides lying along the three axes of a rectangular coordinate system. Find the vectors describing the diagonals of the cube. What is the angle between any pair of diagonals?

1-8. Let A be a vector from the origin to a point P fixed in space. Let r be a vector from the origin to a variable point $Q(x_1, x_2, x_3)$. Show that

$$\mathbf{A} \cdot \mathbf{r} = A^2$$

is the equation of a plane perpendicular to A and passing through the point P.

1-9. For the two vectors

$$\mathbf{A} = \mathbf{i} + 2\mathbf{j} - \mathbf{k}, \qquad \mathbf{B} = -2\mathbf{i} + 3\mathbf{j} + \mathbf{k}$$

find
(a) $\mathbf{A} - \mathbf{B}$ and $|\mathbf{A} - \mathbf{B}|$ (b) component of B along A (c) angle between A and B
(d) $\mathbf{A} \times \mathbf{B}$ (e) $(\mathbf{A} - \mathbf{B}) \times (\mathbf{A} + \mathbf{B})$

1-10. A particle moves in a plane elliptical orbit described by the position vector

$$\mathbf{r} = 2b \sin \omega t \, \mathbf{i} + b \cos \omega t \, \mathbf{j}$$

(a) Find v, a, and the particle speed.
(b) What is the angle between v and a at time $t = \pi/2\omega$?

1-11. Show that the *triple scalar product* $(\mathbf{A} \times \mathbf{B}) \cdot \mathbf{C}$ can be written as

$$(\mathbf{A} \times \mathbf{B}) \cdot \mathbf{C} = \begin{vmatrix} A_1 & A_2 & A_3 \\ B_1 & B_2 & B_3 \\ C_1 & C_2 & C_3 \end{vmatrix}$$

Show also that the product is unaffected by an interchange of the scalar and vector product operations or by a change in the order of A, B, C, as long as they are in cyclic order; that is,

$$(\mathbf{A} \times \mathbf{B}) \cdot \mathbf{C} = \mathbf{A} \cdot (\mathbf{B} \times \mathbf{C}) = \mathbf{B} \cdot (\mathbf{C} \times \mathbf{A}) = (\mathbf{C} \times \mathbf{A}) \cdot \mathbf{B}, \qquad \text{etc.}$$

We may therefore use the notation **ABC** to denote the triple scalar product. Finally, give a geometric interpretation of **ABC** by computing the volume of the parallelepiped defined by the three vectors A, B, C.

1-12. Let a, b, c be three constant vectors drawn from the origin to the points A, B, C. What is the distance from the origin to the plane defined by the points A, B, C? What is the area of the triangle ABC?

1-13. If X is an unknown vector satisfying the following relations involving the known vectors A and B and the scalar ϕ,

$$\mathbf{A} \times \mathbf{X} = \mathbf{B}, \qquad \mathbf{A} \cdot \mathbf{X} = \phi$$

Express X in terms of A, B, ϕ, and the magnitude of A.

1-14. Consider the following matrices:

$$\mathbf{A} = \begin{pmatrix} 1 & 2 & -1 \\ 0 & 3 & 1 \\ 2 & 0 & 1 \end{pmatrix}, \qquad \mathbf{B} = \begin{pmatrix} 2 & 1 & 0 \\ 0 & -1 & 2 \\ 1 & 1 & 3 \end{pmatrix}, \qquad \mathbf{C} = \begin{pmatrix} 2 & 1 \\ 4 & 3 \\ 1 & 0 \end{pmatrix}$$

Find the following
(a) $|\mathbf{AB}|$ (b) **AC** (c) **ABC** (d) $\mathbf{AB} - \mathbf{B'A'}$

1-15. Find the values of α needed to make the following transformation orthogonal.

$$\begin{pmatrix} 1 & 0 & 0 \\ 0 & \alpha & -\alpha \\ 0 & \alpha & \alpha \end{pmatrix}$$

1-16. What surface is represented by $\mathbf{r} \cdot \mathbf{a} = $ const. that is described if $\mathbf{a}$ is a vector of constant magnitude and direction from the origin and $\mathbf{r}$ is the position vector to the point $P(x_1, x_2, x_3)$ on the surface?

1-17. Obtain the cosine law of plane trigonometry by interpreting the product $(\mathbf{A} - \mathbf{B}) \cdot (\mathbf{A} - \mathbf{B})$ and the expansion of the product.

1-18. Obtain the sine law of plane trigonometry by interpreting the product $\mathbf{A} \times \mathbf{B}$ and the alternate representation $(\mathbf{A} - \mathbf{B}) \times \mathbf{B}$.

1-19. Derive the following expressions by using vector algebra:
(a) $\cos(\alpha - \beta) = \cos \alpha \cos \beta + \sin \alpha \sin \beta$
(b) $\sin(\alpha - \beta) = \sin \alpha \cos \beta - \cos \alpha \sin \beta$

1-20. Show that
(a) $\sum_{i,j} \varepsilon_{ijk} \delta_{ij} = 0$ (b) $\sum_{j,k} \varepsilon_{ijk} \varepsilon_{ljk} = 2\delta_{il}$ (c) $\sum_{i,j,k} \varepsilon_{ijk} \varepsilon_{ijk} = 6$

1-21. Show (see also Problem 1-11) that

$$\mathbf{ABC} = \sum_{i,j,k} \varepsilon_{ijk} A_i B_j C_k$$

1-22. Evaluate the sum $\sum_{k} \varepsilon_{ijk} \varepsilon_{lmk}$ (which contains 81 terms) by considering the result for all possible combinations of i, j, l, m; that is,
(a) $i = j$ (b) $i = l$ (c) $i = m$ (d) $j = l$ (e) $j = m$ (f) $l = m$
(g) $i \neq l$ or m (h) $j \neq l$ or m
Show that

$$\sum_{k} \varepsilon_{ijk} \varepsilon_{lmk} = \delta_{il} \delta_{jm} - \delta_{im} \delta_{jl}$$

and then use this result to prove

$$\mathbf{A} \times (\mathbf{B} \times \mathbf{C}) = (\mathbf{A} \cdot \mathbf{C})\mathbf{B} - (\mathbf{A} \cdot \mathbf{B})\mathbf{C}$$

1-23. Use the ε_{ijk} notation and derive the identity

$$(\mathbf{A} \times \mathbf{B}) \times (\mathbf{C} \times \mathbf{D}) = (\mathbf{ABD})\mathbf{C} - (\mathbf{ABC})\mathbf{D}$$

1-24. Let $\mathbf{A}$ be an arbitrary vector, and let $\mathbf{e}$ be a unit vector in some fixed direction. Show that

$$\mathbf{A} = \mathbf{e}(\mathbf{A} \cdot \mathbf{e}) + \mathbf{e} \times (\mathbf{A} \times \mathbf{e})$$

What is the geometrical significance of each of the two terms of the expansion?

1-25. Find the components of the acceleration vector $\mathbf{a}$ in spherical coordinates.

1-26. A particle moves with $v = $ const. along the curve $r = k(1 + \cos \theta)$ (a *cardioid*). Find $\ddot{\mathbf{r}} \cdot \mathbf{e}_r = \mathbf{a} \cdot \mathbf{e}_r$, $|\mathbf{a}|$, and $\dot{\theta}$.

1-27. If $\mathbf{r}$ and $\dot{\mathbf{r}} = \mathbf{v}$ are both explicit functions of time, show that

$$\frac{d}{dt}[\mathbf{r} \times (\mathbf{v} \times \mathbf{r})] = r^2 \mathbf{a} + (\mathbf{r} \cdot \mathbf{v})\mathbf{v} - (v^2 + \mathbf{r} \cdot \mathbf{a})\mathbf{r}$$

1-28. Show that

$$\nabla(\ln|\mathbf{r}|) = \frac{\mathbf{r}}{r^2}$$

1-29. Find the angle between the surfaces defined by $r^2 = 9$ and $x + y + z^2 = 1$ at the point $(2, -2, 1)$.

1-30. Show that $\nabla(\phi\psi) = \phi\nabla\psi + \psi\nabla\phi$.

1-31. Show that

(a) $\nabla r^n = nr^{(n-2)}\mathbf{r}$ **(b)** $\nabla f(r) = \dfrac{\mathbf{r}}{r}\dfrac{df}{dr}$ **(c)** $\nabla^2(\ln r) = \dfrac{1}{r^2}$

1-32. Show that

$$\int(2a\mathbf{r}\cdot\dot{\mathbf{r}} + 2b\dot{\mathbf{r}}\cdot\ddot{\mathbf{r}})\,dt = ar^2 + b\dot{r}^2 + \text{const.}$$

where $\mathbf{r}$ is the vector from the origin to the point (x_1, x_2, x_3). The quantities r and $\dot{r}$ are the magnitudes of the vectors $\mathbf{r}$ and $\dot{\mathbf{r}}$, respectively, and a and b are constants.

1-33. Show that

$$\int\left(\frac{\dot{\mathbf{r}}}{r} - \frac{\mathbf{r}\dot{r}}{r^2}\right)dt = \frac{\mathbf{r}}{r} + \mathbf{C}$$

where $\mathbf{C}$ is a constant vector.

1-34. Evaluate the integral

$$\int\mathbf{A}\times\ddot{\mathbf{A}}\,dt$$

1-35. Show that the volume common to the intersecting cylinders defined by $x^2 + y^2 = a^2$ and $x^2 + z^2 = a^2$ is $V = 16a^3/3$.

1-36. Find the value of the integral $\int_S \mathbf{A}\cdot d\mathbf{a}$, where $\mathbf{A} = x\mathbf{i} - y\mathbf{j} + z\mathbf{k}$ and S is the closed surface defined by the cylinder $c^2 = x^2 + y^2$. The top and bottom of the cylinder are at $z = d$ and 0, respectively.

1-37. Find the value of the integral $\int_S \mathbf{A}\cdot d\mathbf{a}$, where $\mathbf{A} = (x^2 + y^2 + z^2)(x\mathbf{i} + y\mathbf{j} + z\mathbf{k})$ and the surface S is defined by the sphere $R^2 = x^2 + y^2 + z^2$. Do the integral directly and also by using Gauss's theorem.

1-38. Find the value of the integral $\int_S (\nabla\times\mathbf{A})\cdot d\mathbf{a}$ if the vector $\mathbf{A} = y\mathbf{i} + z\mathbf{j} + x\mathbf{k}$ and S is the surface defined by the paraboloid $z = 1 - x^2 - y^2$, where $z \geq 0$.

2

NEWTONIAN MECHANICS—
SINGLE PARTICLE

2.1 INTRODUCTION

The science of mechanics seeks to provide a precise and consistent description of the dynamics of particles and systems of particles, that is, a set of physical laws mathematically describing the motions of bodies and aggregates of bodies. For this, we need certain fundamental concepts such as distance and time. The combination of the concepts of distance and time allows us to define the **velocity** and **acceleration** of a particle. The third fundamental concept, **mass**, requires some elaboration, which we give when we discuss Newton's laws.

Physical laws must be based on experimental fact. We cannot expect *a priori* that the gravitational attraction between two bodies must vary exactly as the inverse square of the distance between them. But experiment indicates that this is so. Once a set of experimental data has been correlated and a postulate has been formulated regarding the phenomena to which the data refer, then various implications can be worked out. If these implications are all verified by experiment, we may believe that the postulate is generally true. The postulate then assumes the status of a **physical law**. If some experiments disagree with the predictions of the law, the theory must be modified to be consistent with the facts.

Newton provided us with the fundamental laws of mechanics. We state here these laws in modern terms, discuss their meaning, and then derive the implications of the laws in varous situations.* But the logical structure of the science of mechanics is not straightforward. Our line of reasoning in interpreting Newton's laws is

---

* Truesdell (Tr68) points out that Leonhard Euler (1707–1783) clarified and developed the Newtonian concepts. Euler "put most of mechanics into its modern form" and "made mechanics simple and easy" (p. 106).

not the only one possible.* We do not pursue in any detail the philosophy of mechanics but rather give only sufficient elaboration of Newton's laws to allow us to continue with the discussion of classical dynamics. We devote our attention in this chapter to the motion of a single particle, leaving systems of particles to be discussed in Chapters 9 and 11–13.

2.2 NEWTON'S LAWS

We begin by simply stating in conventional form Newton's laws of mechanics[†]:

I. *A body remains at rest or in uniform motion unless acted upon by a force.*

II. *A body acted upon by a force moves in such a manner that the time rate of change of momentum equals the force.*

III. *If two bodies exert forces on each other, these forces are equal in magnitude and opposite in direction.*

These laws are so familiar that we sometimes tend to lose sight of their true significance (or lack of it) as physical laws. The First Law, for example, is meaningless without the concept of "force," a word Newton used in all three laws. In fact, standing alone, the First Law conveys a precise meaning only for *zero force*; that is, a body remaining at rest or in uniform (i.e., unaccelerated, rectilinear) motion is subject to no force whatsoever. A body moving in this manner is termed a **free body** (or **free particle**). The question of the frame of reference with respect to which the "uniform motion" is to be measured is discussed in the following section.

In pointing out the lack of content in Newton's First Law, Sir Arthur Eddington[‡] observed, somewhat facetiously, that all the law actually says is that "every particle continues in its state of rest or uniform motion in a straight line except insofar as it doesn't." This is hardly fair to Newton, who meant something very definite by his statement. But it does emphasize that the First Law by itself provides us with only a qualitative notion regarding "force."

The Second Law provides an explicit statement: Force is related to the time rate of change of *momentum*. Newton appropriately defined **momentum** (although he used the term *quantity of motion*) to be the product of mass and velocity, such

---

* Ernst Mach (1838–1916) expressed his view in his famous book first published in 1883; E. Mach, *Die Mechanic in ihrer Entwicklung historisch-kritisch dargestellt* [The science of mechanics] (Prague, 1883). A translation of a later edition is available (Ma60). Interesting discussions are also given by R. B. Lindsay and H. Margenau (Li36) and N. Feather (Fe59).

[†] Enunciated in 1687 by Sir Isaac Newton (1642–1727) in his *Philosophiae naturalis principia mathematica* [*Mathematical principles of natural philosophy*, normally called *Principia*] (London, 1687). Previously, Galileo (1564–1642) generalized the results of his own mathematical experiments with statements equivalent to Newton's First and Second Laws. But Galileo was unable to complete the description of dynamics because he did not appreciate the significance of what would become Newton's Third Law—and therefore lacked a precise meaning of force.

[‡] Sir Arthur Eddington (Ed30, p. 124).

that

$$\mathbf{p} \equiv m\mathbf{v} \tag{2.1}$$

Therefore, Newton's Second Law can be expressed as

$$\mathbf{F} = \frac{d\mathbf{p}}{dt} = \frac{d}{dt}(m\mathbf{v}) \tag{2.2}$$

The definition of force becomes complete and precise only when "mass" is defined. Thus the First and Second Laws are not really "laws" in the usual sense; rather, they may be considered *definitions*. Because length, time, and mass are concepts normally already understood, we use Newton's First and Second Laws as the operational definition of force. Newton's Third Law, however, is indeed a *law*. It is a statement concerning the real physical world and contains all of the physics in Newton's laws of motion.*

We must hasten to add, however, that the Third Law is not a *general* law of nature. The law does apply when the force exerted by one (point) object on another (point) object is directed along the line connecting the objects. Such forces are called **central forces**; the Third Law applies whether a central force is attractive or repulsive. Gravitational and electrostatic forces are central forces, so Newton's laws can be used in problems involving these types of forces. Sometimes, elastic forces (which are actually macroscopic manifestations of microscopic electrostatic forces) are central. For example, two point objects connected by a straight spring or elastic string are subject to forces that obey the Third Law. Any force that depends on the velocities of the interacting bodies is noncentral, and the Third Law may not apply. Velocity-dependent forces are characteristic of interactions that propagate with finite velocity. Thus the force between *moving* electric charges does not obey the Third Law, because the force propagates with the velocity of light. Even the gravitational force between *moving* bodies is velocity dependent, but the effect is small and difficult to detect. The only observable effect is the precession of the perihelia of the inner planets (see Section 8.9). We will return to a discussion of Newton's Third Law in Chapter 9.

To demonstrate the significance of Newton's Third Law, let us paraphrase it in the following way, which incorporates the appropriate definition of mass:

III′. *If two bodies constitute an ideal, isolated system, then the accelerations of these bodies are always in opposite directions, and the ratio of the magnitudes of the accelerations is constant. This constant ratio is the inverse ratio of the masses of the bodies.*

With this statement, we can give a practical definition of mass and therefore give precise meaning to the equations summarizing Newtonian dynamics. For two iso-

---

* The reasoning presented here, viz., that the First and Second Laws are actually definitions and that the Third Law contains the physics, is not the only possible interpretation. Lindsay and Margenau (Li36), for example, present the first two Laws as physical laws and then derive the Third Law as a consequence.

lated bodies, 1 and 2, the Third Law states that

$$\mathbf{F}_1 = -\mathbf{F}_2 \tag{2.3}$$

Using the definition of force as given by the Second Law, we have

$$\frac{d\mathbf{p}_1}{dt} = -\frac{d\mathbf{p}_2}{dt} \tag{2.4a}$$

or, with constant masses,

$$m_1\left(\frac{d\mathbf{v}_1}{dt}\right) = m_2\left(-\frac{d\mathbf{v}_2}{dt}\right) \tag{2.4b}$$

and, because acceleration is the time derivative of velocity,

$$m_1\,(\mathbf{a}_1) = m_2(-\mathbf{a}_2) \tag{2.4c}$$

Hence,

$$\frac{m_2}{m_1} = -\frac{a_1}{a_2} \tag{2.5}$$

where the negative sign indicates only that the two acceleration vectors are oppositely directed. Mass is taken to be a positive quantity.

We can always select, say, m_1 as the *unit mass*. Then, by comparing the ratio of accelerations when m_1 is allowed to interact with any other body, we can determine the mass of the other body. To measure the accelerations, we must have appropriate clocks and measuring rods; also, we must choose a suitable coordinate system or reference frame. The question of a "suitable reference frame" is discussed in the next section.

One of the more common methods of determining the mass of an object is by *weighing*—for example, by comparing its weight to that of a standard by means of a beam balance. This procedure makes use of the fact that in a gravitational field the weight of a body is just the gravitational force acting on the body; that is, Newton's equation $\mathbf{F} = m\mathbf{a}$ becomes $\mathbf{W} = m\mathbf{g}$, where $\mathbf{g}$ is the acceleration due to gravity. The validity of using this procedure rests on a fundamental assumption: that the mass m appearing in Newton's equation and defined according to Statement III′ is equal to the mass m that appears in the gravitational force equation. These two masses are called the **inertial mass** and **gravitational mass**, respectively. The definitions may be stated as follows: .

Inertial Mass: *That mass determining the acceleration of a body under the action of a given force.*

Gravitational Mass: *That mass determining the gravitational forces between a body and other bodies.*

Galileo was the first to test the equivalence of inertial and gravitational mass in his (perhaps apocryphal) experiment with falling weights at the Tower of Pisa. Newton also considered the problem and measured the periods of pendula of equal lengths but with bobs of different materials. Neither Newton nor Galileo found any difference, but the methods were quite crude.* In 1890 Eötvös[†] devised an ingenious method to test the equivalence of inertial and gravitational masses. Using two objects made of different materials, he compared the effect of the earth's gravitational force (i.e., the weight) with the effect of the inertial force caused by the Earth's rotation. The experiment involved a *null* method using a sensitive torsion balance and was therefore highly accurate. More recent experiments (notably those of Dicke[‡]), using essentially the same method, have improved the accuracy, and we know now that inertial and gravitational mass are identical to within a few parts in 10^{12}. This result is considerably important in the general theory of relativity.[#] The assertion of the *exact* equality of inertial and gravitational mass is termed the **principle of equivalence**.

Newton's Third Law is stated in terms of two bodies that constitute an isolated system. It is impossible to achieve such an ideal condition; every body in the universe interacts with every other body, although the force of interaction may be far too weak to be of any practical importance if great distances are involved. Newton avoided the question of how to disentangle the desired effects from all the extraneous effects. But this practical difficulty only emphasizes the enormity of Newton's assertion made in the Third Law. It is a tribute to the depth of his perception and physical insight that the conclusion, based on limited observations, has successfully borne the test of experiment for 300 years. Only within this century have measurements of sufficient detail revealed certain discrepancies with the predictions of Newtonian theory. The pursuit of these details led to the development of relativity theory and quantum mechanics.[§]

Another interpretation of Newton's Third Law is based on the concept of momentum. Rearranging Equation 2.4a gives

$$\frac{d}{dt}(\mathbf{p}_1 + \mathbf{p}_2) = 0$$

or

$$\mathbf{p}_1 + \mathbf{p}_2 = \text{constant} \qquad (2.6)$$

The statement that momentum is conserved in the isolated interaction of two particles is a special case of the more general **conservation of linear momentum**. Physicists cherish general conservation laws, and the conservation of linear momentum is believed always to be obeyed. Later we shall modify our definition of momentum from Equation 2.1 for high velocities approaching the speed of light.

---

* In Newton's experiment, he could have detected a difference of only one part in 10^3.

[†] Roland von Eötvös (1848–1919), a Hungarian baron; his research in gravitational problems led to the development of a gravimeter, which was used in geological studies.

[‡] P. G. Roll, R. Krotkov, and R. H. Dicke, Ann. Phys. (N.Y.) **26**, 442 (1964). See also Braginsky and Pavov, Sov. Phys.-JETP **34**, 463 (1972).

[#] See, for example, the discussions by P. G. Bergmann (Be46) and J. Weber (We61). Weber's book also provides an analysis of the Eötvös experiment.

[§] See also Section 2.8.

2.3 FRAMES OF REFERENCE

Newton realized that, for the laws of motion to have meaning, the motion of bodies must be measured relative to some reference frame. A reference frame is called an **inertial frame** if Newton's laws are indeed valid in that frame; that is, if a body subject to no external force moves in a straight line with constant velocity (or remains at rest), then the coordinate system establishing this fact is an inertial reference frame. This is a clear-cut operational definition and one that also follows from the general theory of relativity.

If Newton's laws are valid in one reference frame, then they are also valid in any reference frame in uniform motion (i.e., not accelerated) with respect to the first system.* This is a result of the fact that the equation $\mathbf{F} = m\ddot{\mathbf{r}}$ involves the second time derivative of $\mathbf{r}$: A change of coordinates involving a constant velocity does not influence the equation. This result is called **Galilean invariance** or the **principle of Newtonian relativity**.

Relativity theory has shown us that the concepts of *absolute rest* and an *absolute* inertial reference frame are meaningless. Therefore, even though we conventionally adopt a reference frame described with respect to the "fixed" stars—and, indeed, in such a frame the Newtonian equations are valid to a high degree of accuracy—such a frame is, in fact, not an absolute inertial frame. We may, however, consider the "fixed" stars to define a reference frame that approximates an "absolute" inertial frame to an extent quite sufficient for our present purposes.

Although the fixed-star reference frame is a conveniently definable system and one suitable for many purposes, we must emphasize that the fundamental definition of an inertial frame makes no mention of stars, fixed or otherwise. If a body subject to no force moves with constant velocity in a certain coordinate system, that system is, by definition, an inertial frame. Because precisely describing the motion of a real physical object in the real physical world is normally difficult, we usually resort to idealizations and approximations of varying degree; that is, we ordinarily neglect the lesser forces on a body if these forces do not significantly affect the body's motion.

If we wish to describe the motion of, say, a free particle and if we choose for this purpose some coordinate system in an inertial frame, then we require that the (vector) equation of motion of the particle be independent of the *position* of the origin of the coordinate system and independent of its *orientation* in **space**. We further require that **time** be homogeneous; that is, a free particle moving with a certain constant velocity in the coordinate system during a certain time interval must not, during a later time interval, be found to move with a different velocity.

We can illustrate the importance of these properties by the following example. Consider, as in Figure 2-1, a free particle moving along a certain path AC. To describe the particle's motion, let us choose a rectangular coordinate system whose origin moves in a circle, as shown. For simplicity, we let the orientation of the axes

---

* In Chapter 10, we discuss the modification of Newton's equations that must be made if it is desired to describe the motion of a body with respect to a *noninertial* frame of reference, that is, a frame that is accelerated with respect to an inertial frame.

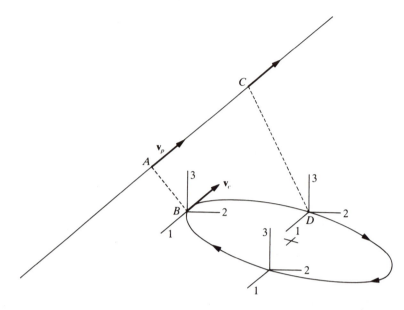

FIGURE 2-1

be fixed in space. The particle moves with a velocity $\mathbf{v}_p$ relative to an inertial reference frame. If the coordinate system moves with a linear velocity $\mathbf{v}_c$ when at the point B, and if $\mathbf{v}_c = \mathbf{v}_p$, then to an observer in the moving coordinate system the particle (at A) will appear to be *at rest*. At some later time, however, when the particle is at C and the coordinate system is at D, the particle will appear to accelerate with respect to the observer. We must, therefore, conclude that the rotating coordinate system does not qualify as an inertial reference frame.

These observations are not sufficient to decide whether time is homogeneous. To reach such a conclusion, repeated measurements must be made in identical situations at various times; identical results would indicate the homogeneity of time.

Newton's equations do not describe the motion of bodies in noninertial systems. We can devise a method to describe the motion of a particle by a rotating coordinate system, but, as we shall see in Chapter 10, the resulting equations contain several terms that do not appear in the simple Newtonian equation $\mathbf{F} = m\mathbf{a}$. For the moment, then, we restrict our attention to inertial reference frames to describe the dynamics of particles.

2.4 THE EQUATION OF MOTION FOR A PARTICLE

Newton's equation $\mathbf{F} = d\mathbf{p}/dt$ can be expressed alternatively as

$$\mathbf{F} = \frac{d}{dt}(m\mathbf{v}) = m\frac{d\mathbf{v}}{dt} = m\ddot{\mathbf{r}} \qquad (2.7)$$

if we assume that the mass m does not vary with time. This is a second-order differential equation that may be integrated to find $\mathbf{r} = \mathbf{r}(t)$ if the function $\mathbf{F}$ is known.

Specifying the initial values of **r** and $\dot{\mathbf{r}} = \mathbf{v}$ then allows us to evaluate the two arbitrary constants of integration. We then determine the motion of a particle by the force function **F** and the initial values of position **r** and velocity **v**.

The force **F** may be a function of any combination of position, velocity, and time and is generally denoted as $\mathbf{F}(\mathbf{r}, \mathbf{v}, t)$. For a given dynamic system, we normally want to know **r** and **v** as a function of time. Solving Equation 2.7 will help us do this by solving for $\ddot{\mathbf{r}}$. Applying Equation 2.7 to physical situations is an important part of mechanics.

In this chapter, we examine several examples in which the force function is known. We begin by looking at simple force functions (either constant or dependent on only one of **r**, **v**, and t) in only one spatial dimension as a refresher of earlier physics courses. It is important to form good habits in problem solving. Here are some useful problem-solving techniques.

1. Make a sketch of the problem, indicating forces, velocities, and so forth.

2. Write down the given quantities.

3. Write down useful equations and what is to be determined.

4. The equations describing the problem normally must be manipulated to find the quantity sought. Algebraic manipulation as well as differentiation or integration is usually required. Sometimes numerical calculations using a computer are the easiest, if not the only, method of solution.

5. Finally, put in the actual values for the assumed variable names to determine the quantity sought.

Let us first consider the problem of a block sliding on an inclined plane. Let the angle of the inclined plane be θ and the mass of the block be 100 g. The sketch of the problem is shown in Figure 2-2a.

E X A M P L E   **2.1**   –

If a block slides without friction down a fixed, inclined plane with $\theta = 30°$, what is the block's acceleration?

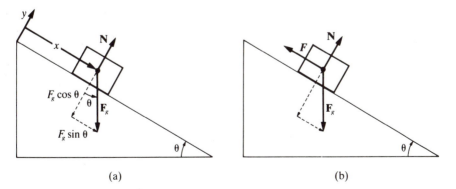

(a)   (b)

FIGURE 2-2

Solution: Two forces act on the block (see Figure 2-2a): the gravitational force $\mathbf{F}_g$ and the plane's normal force $\mathbf{N}$ pushing upward on the block (no friction in this example). The block is constrained to be on the plane, and the only direction the block can move is the x-direction, up and down the plane. We take the $+x$-direction to be down the plane. The total force $\mathbf{F}$ is constant; Equation 2.7 becomes

$$\mathbf{F} = \mathbf{F}_g + \mathbf{N}$$

and because $\mathbf{F}$ is the net resultant force acting on the block,

$$\mathbf{F} = m\ddot{\mathbf{r}}$$

or

$$\mathbf{F}_g + \mathbf{N} = m\ddot{\mathbf{r}} \tag{2.8}$$

This vector must be applied in two directions: x and y (perpendicular to x). The component of force in the y-direction is zero, because no acceleration occurs in this direction. The force $\mathbf{F}_g$ is divided vectorially into its x- and y-components (dashed lines in Figure 2-2a). Equation 2.8 becomes

y-direction

$$- F_g \cos \theta + N = 0 \tag{2.9}$$

x-direction

$$F_g \sin \theta = m\ddot{x} \tag{2.10}$$

with the required result

$$\ddot{x} = \frac{F_g}{m} \sin \theta = \frac{mg \sin \theta}{m} = g \sin \theta$$

$$\ddot{x} = g \sin(30°) = \frac{g}{2} = 4.9 \text{ m/s}^2 \tag{2.11}$$

Therefore the acceleration of the block is a constant.

We can find the velocity of the block after it moves from rest a distance x_0 down the plane by multiplying Equation 2.11 by $2\dot{x}$ and integrating

$$2\dot{x}\ddot{x} = 2\dot{x}g \sin \theta$$

$$\frac{d}{dt}(\dot{x}^2) = 2g \sin \theta \frac{dx}{dt}$$

$$\int_0^{v_0^2} d(\dot{x}^2) = 2g \sin \theta \int_0^{x_0} dx$$

At $t = 0$, both $x = \dot{x} = 0$, and, at $t = t_{\text{final}}$, $x = x_0$, and the velocity $\dot{x} = v_0$.

$$v_0^2 = 2g \sin \theta \, x_0$$

$$v_0 = \sqrt{2g \sin \theta \, x_0}$$

●

EXAMPLE **2.2**

If the coefficient of static friction between the block and plane in the previous example is $\mu_s = 0.4$, at what angle θ will the block start sliding if it is initially at rest?

Solution: We need a new sketch to indicate the additional frictional force f (see Figure 2-2b). The static frictional force has the approximate maximum value

$$f_{max} = \mu_s N \tag{2.12}$$

and Equation 2.7 becomes, in component form,

y-direction

$$- F_g \cos \theta + N = 0 \tag{2.13}$$

x-direction

$$-f_s + F_g \sin \theta = m\ddot{x} \tag{2.14}$$

The static frictional force f_s will be some value $f_s \leq f_{max}$ required to keep $\ddot{x} = 0$—that is, to keep the block at rest. However, as the angle θ of the plane increases, eventually the static frictional force will be unable to keep the block at rest. At that angle θ', f_s becomes

$$f_s(\theta = \theta') = f_{max} = \mu_s N = \mu_s F_g \cos \theta$$

and

$$m\ddot{x} = F_g \sin \theta - f_{max}$$
$$m\ddot{x} = F_g \sin \theta - \mu_s F_g \cos \theta$$
$$\ddot{x} = g(\sin \theta - \mu_s \cos \theta) \tag{2.15}$$

Just before the block starts to slide, the acceleration $\ddot{x} = 0$, so

$$\sin \theta - \mu_s \cos \theta = 0$$
$$\tan \theta = \mu_s = 0.4$$
$$\theta = \tan^{-1}(0.4) = 22°$$

EXAMPLE **2.3**

After the block in the previous example begins to slide, the coefficient of kinetic (sliding) friction becomes $\mu_k = 0.3$. Find the acceleration for the angle $\theta = 30°$.

Solution: Similarly to Example 2.2, the kinetic friction becomes (approximately)

$$f_k = \mu_k N = \mu_k F_g \cos \theta \tag{2.16}$$

and

$$m\ddot{x} = F_g \sin\theta - f_k = mg(\sin\theta - \mu_k \cos\theta) \tag{2.17}$$
$$\ddot{x} = g(\sin\theta - \mu_k \cos\theta) = 0.24\,g \tag{2.18}$$

Generally, the force of static friction ($f_{max} = \mu_s N$) is greater than that of kinetic friction ($f_k = \mu_k N$). This can be observed in a simple experiment. If we lower the angle θ below 16.7°, we find that $\ddot{x} < 0$, and the block eventually stops. If we raise the block back up above $\theta = 16.7°$, we find that the block does not start sliding again until $\theta \geq 22°$ (Example 2.2). The static friction determines when it starts moving again. There is not a discontinuous acceleration as the block starts moving, because of the difference between μ_s and μ_k. For small speeds, the coefficient of friction changes rather quickly from μ_s to μ_k.

The subject of friction is still an interesting and important area of research. There are still surprises. For example, even though we calculate the absolute value of the frictional force as $f = \mu N$, research has shown that the frictional force is directly proportional, not to the load, but to the microscopic area of contact between the two objects (as opposed to the apparent contact area). We use μN as an approximation because, as N increases, so does the actual contact area on a microscopic level. For hundreds of years before the 1940s, it was accepted that the load—and not the area—was directly responsible. We also believe that the static frictional force is larger than that of kinetic friction because the bonding of atoms between the two objects does not have as much time to develop in kinetic motion.

Effects of Retarding Forces

We should emphasize that the force **F** in Equation 2.7 is not necessarily constant, and indeed, it may consist of several distinct parts, as seen in the previous examples. For example, if a particle falls in a constant gravitational field, the gravitational force is $\mathbf{F}_g = m\mathbf{g}$, where **g** is the acceleration of gravity. If, in addition, a retarding force $\mathbf{F}_r$ exists that is some function of the instantaneous speed, then the total force is

$$\mathbf{F} = \mathbf{F}_g + \mathbf{F}_r$$

$$= m\mathbf{g} + \mathbf{F}_r(v) \tag{2.19}$$

It is frequently sufficient to consider that $\mathbf{F}_r(v)$ is simply proportional to some power of the speed. In general, *real* retarding forces are more complicated, but the power-law approximation is useful in many instances in which the speed does not vary greatly. Even more to the point, if $F_r \propto v^n$, then the equation of motion can usually be integrated directly, whereas, if the true velocity dependence were used, numerical integration would probably be necessary. With the power-law approximation, we can then write

$$\mathbf{F} = m\mathbf{g} - mkv^n \frac{\mathbf{v}}{v} \tag{2.20}$$

where k is a positive constant that specifies the strength of the retarding force and where $\mathbf{v}/v$ is a unit vector in the direction of $\mathbf{v}$. Experimentally, we find that, for a relatively small object moving in air, $n \cong 1$ for velocities less than about 24 m/s ($\sim$ 80 ft/s). For higher velocities but below the velocity of sound ($\sim$ 330 m/s or 1,100 ft/s), the retarding force is approximately proportional to the square of the velocity.* For simplicity, the v^2 dependence is usually taken for speeds up to the speed of sound.

The effect of air resistance is important for a ping-pong ball smashed to an opponent, a high flying softball hit deep to the outfield, a golfer's chip shot, and a mortar shell lofted against an enemy. Extensive tabulations have been made for military ballistics of projectiles of various sorts for the velocity as a function of flight time. There are several forces on an actual projectile in flight. The air resistance force is called the drag $\mathbf{W}$ and is opposite to the projectile's velocity as shown in Figure 2-3a. The velocity $\mathbf{v}$ is normally not along the symmetry axis of the shell. The component of force acting perpendicular to the drag is called the lift $\mathbf{L}_a$. There may also be various other forces due to the projectile's spin and oscillation, and a calculation of a projectile's ballistic trajectory is quite complex. The Prandtl expression for the air resistance[†] is

$$W = \frac{1}{2}c_W \rho A v^2 \qquad (2.21)$$

where c_W is the dimensionless drag coefficient, ρ is the air density, v is the velocity, and A is the cross-sectional area of the object (projectile) measured perpendicularly to the velocity. In Figures 2-3b, we plot some typical values for c_W, and in Figures 2-3c and d we display the calculated air resistance W using Equation 2.21 for a projectile diameter of 10 cm and using the values of c_W shown. The air resistance increases dramatically near the speed of sound (Mach number M = speed/speed of sound). Below speeds of about 400 m/s it is evident that an equation of at least second degree is necessary to describe the resistive force. For higher speeds, the retarding force varies approximately linearly with speed.

Several examples of the motion of a particle subjected to various forces are given below. These examples are particularly good to begin computer calculations using any of the available commercial math programs and spreadsheets or for the students to write their own programs. The computer results, especially the plots, can often be compared with the analytical results presented here. Some of the figures shown in this section were produced using a computer, and several end-of-chapter problems are meant to develop the student's computer experience if so desired by the instructor or student.

---

* The motion of a particle in a medium in which there is a resisting force proportional to the speed or to the square of the speed (or to a linear combination of the two) was examined by Newton in his *Principia* (1687). The extension to any power of the speed was made by Johann Bernoulli in 1711. The term *Stokes' law of resistance* is sometimes applied to a resisting force proportional to the speed; Newton's law of resistance is a retarding force proportional to the square of the speed.

[†] See the article by E. Melchior and H. Reuschel in *Handbook on Weaponry* (Rh82, p. 137).

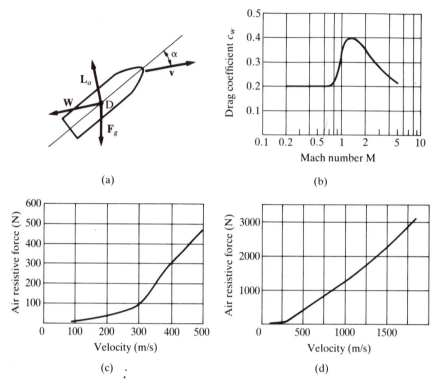

FIGURE 2-3   (a) Aerodynamic forces acting on projectile. **W** is the drag (air resistive force) and is opposite the velocity of the projectile **v**. Notice that **v** may be at an angle α from the symmetry axis of projectile. The component of force acting perpendicular to the drag is called the lift $\mathbf{L_a}$. The point D is the center of pressure. Finally, the gravitational force $\mathbf{F_g}$ acts down. If the center of pressure is not at the projectile's center of mass, there is also a torque about the center of mass. (b) The drag coefficient c_W, from the Rheinmetall resistance law (Rh82), is plotted versus the mach number M. Notice the large change near the speed of sound where M = 1. (c) The air resistive force W (drag) is shown as a function of velocity for a projectile diameter of 10 cm. Notice the inflection near the speed of sound. (d) Same as (c) for higher velocities.

EXAMPLE  **2.4** -

As the simplest example of the resisted motion of a particle, find the displacement and velocity of horizontal motion in a medium in which the retarding force is proportional to the velocity.

Solution: A sketch of the problem is shown in Figure 2-4. The Newtonian equation $F = ma$ provides us with the equation of motion:

x-direction

$$ma = m\frac{dv}{dt} = -kmv \tag{2.22}$$

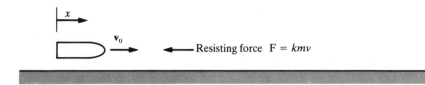

FIGURE 2-4

where kv is the magnitude of the resisting force (k = constant). Then

$$\int \frac{dv}{v} = -k \int dt$$

$$\ln v = -kt + C_1 \qquad (2.23)$$

The integration constant in Equation 2.23 can be evaluated if we prescribe the initial condition $v(t = 0) \equiv v_0$. Then $C_1 = \ln v_0$, and

$$v = v_0 e^{-kt} \qquad (2.24)$$

We can integrate this equation to obtain the displacement x as a function of time:

$$v = \frac{dx}{dt} = v_0 e^{-kt}$$

$$x = v_0 \int e^{-kt} dt = -\frac{v_0}{k} e^{-kt} + C_2 \qquad (2.25a)$$

The initial condition $x(t = 0) \equiv 0$ implies $C_2 = v_0/k$. Therefore

$$x = \frac{v_0}{k}(1 - e^{-kt}) \qquad (2.25b)$$

This result shows that x asymptotically approaches the value v_0/k as $t \to \infty$.
We can also obtain the velocity as a function of displacement by writing

$$\frac{dv}{dx} = \frac{dv}{dt}\frac{dt}{dx} = \frac{dv}{dt} \cdot \frac{1}{v}$$

so that

$$v\frac{dv}{dx} = \frac{dv}{dt} = -kv$$

or

$$\frac{dv}{dx} = -k$$

from which we find, by using the same initial conditions,

$$v = v_0 - kx \qquad (2.26)$$

Therefore, the velocity decreases linearly with displacement.

E X A M P L E **2.5** –

Find the displacement and velocity of a particle undergoing vertical motion in a medium having a retarding force proportional to the velocity.

Solution: Let us consider that the particle is falling downward with an initial velocity v_0 from a height h in a constant gravitational field (Figure 2-5). The equation of motion is

z-direction

$$F = m\frac{dv}{dt} = -mg - kmv \tag{2.27}$$

where $-kmv$ represents a positive *upward* force since we take z and $v = \dot{z}$ to be positive upward, and the motion is downward—that is, $v < 0$, so that $-kmv > 0$. From Equation 2.27, we have

$$\frac{dv}{kv + g} = -dt \tag{2.28}$$

Integrating Equation 2.28 and setting $v(t = 0) \equiv v_0$, we have (noting that $v_0 < 0$)

$$\frac{1}{k}\ln(kv_0 + g) = -t + c$$

$$kv + g = e^{-kt + c}$$

$$v = \frac{dz}{dt} = -\frac{g}{k} + \frac{kv_0 + g}{k}e^{-kt} \tag{2.29}$$

Integrating once more and evaluating the constant by setting $z(t = 0) \equiv h$, we find

$$z = h - \frac{gt}{k} + \frac{kv_0 + g}{k^2}(1 - e^{-kt}) \tag{2.30}$$

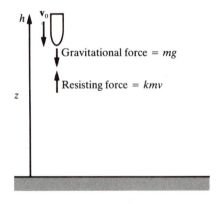

Gravitational force = mg

Resisting force = kmv

FIGURE 2-5

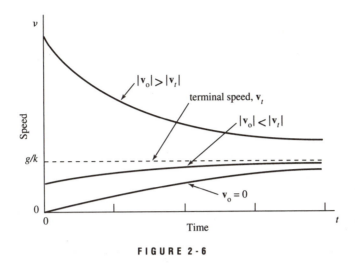

FIGURE 2-6

Equation 2.29 shows that as the time becomes very long, the velocity approaches the limiting value $-g/k$; this is called the **terminal velocity**, v_t. Equation 2.27 yields the same result, because the force will vanish—and hence no further acceleration will occur—when $v = -g/k$. If the initial velocity exceeds the terminal velocity in magnitude, then the body immediately begins to slow down and v approaches the terminal speed from the opposite direction. Figure 2-6 illustrates these results for the downward speeds (positive values).

EXAMPLE 2.6

Next, we treat projectile motion in two dimensions, first without considering air resistance. Let the muzzle velocity of the projectile be v_0 and the angle of elevation be θ (Figure 2-7). Calculate the projectile's displacement, velocity, and range.

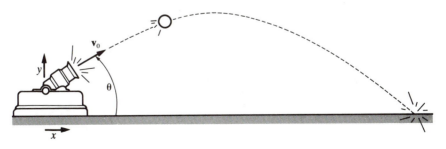

FIGURE 2-7

Solution: Using $\mathbf{F} = m\mathbf{g}$, the force components become

x-direction

$$0 = m\ddot{x} \tag{2.31a}$$

y-direction

$$-mg = m\ddot{y} \tag{2.31b}$$

Neglect the height of the gun, and assume $x = y = 0$ at $t = 0$. Then

$$\ddot{x} = 0$$

$$\dot{x} = v_0 \cos\theta$$

$$x = v_0 t \cos\theta \tag{2.32}$$

and

$$\ddot{y} = -g$$

$$\dot{y} = -gt + v_0 \sin\theta$$

$$y = \frac{-gt^2}{2} + v_0 t \sin\theta \tag{2.33}$$

The speed and total displacement as functions of time are found to be

$$v = \sqrt{\dot{x}^2 + \dot{y}^2} = (v_0^2 + g^2 t^2 - 2v_0 gt \sin\theta)^{1/2} \tag{2.34}$$

and

$$r = \sqrt{x^2 + y^2} = \left(v_0^2 t^2 + \frac{g^2 t^4}{4} - v_0 g t^3 \sin\theta\right)^{1/2} \tag{2.35}$$

We can find the range by determining the value of x when the projectile falls back to ground, that is, when $y = 0$

$$y = t\left(\frac{-gt}{2} + v_0 \sin\theta\right) = 0 \tag{2.36}$$

One value of $y = 0$ occurs for $t = 0$ and the other one for $t = T$.

$$\frac{-gT}{2} + v_0 \sin\theta = 0$$

$$T = \frac{2v_0 \sin\theta}{g} \tag{2.37}$$

The range R is found from

$$x(t = T) = \text{range} = \frac{2v_0^2}{g} \sin\theta \cos\theta \tag{2.38}$$

$$R = \text{range} = \frac{v_0^2}{g} \sin 2\theta \tag{2.39}$$

Notice that the maximum range occurs for $\theta = 45°$.

Let us use some actual numbers in these calculations. The Germans used a long-range gun named Big Bertha in World War I to bombard Paris. Its muzzle velocity was 1,450 m/s. Find its predicted range, maximum projectile height, and projectile time of flight if $\theta = 55°$. We have $v_0 = 1,450$ m/s and $\theta = 55°$, so the range (from Equation 2.39) becomes

$$R = \frac{(1450 \text{ m/s})^2}{9.8 \text{ m/s}^2} [\sin(110°)] = 202 \text{ km}$$

Big Bertha's actual range was 120 km. The difference is a result of the real effect of air resistance.

To find the maximum predicted height, we need to calculate y for the time $T/2$ where T is the projectile time of flight:

$$T = \frac{(2)(1450 \text{ m/s})(\sin 55°)}{9.8 \text{ m/s}^2} = 242 \text{ s}$$

$$y_{max}\left(t = \frac{T}{2}\right) = \frac{-gT^2}{8} + \frac{v_0 T}{2} \sin \theta$$

$$= \frac{-(9.8 \text{ m/s})(242 \text{ s})^2}{8} + \frac{(1450 \text{ m/s})(242 \text{ s})\sin(55°)}{2}$$

$$= 72 \text{ km}$$

E X A M P L E 2.7

Next, we add the effect of air resistance to the motion of the projectile in the previous example. Calculate the decrease in range under the assumption that the force caused by air resistance is directly proportional to the projectile's velocity.

Solution: The initial conditions are the same as in the previous example.

$$\left.\begin{array}{l} x(t = 0) = 0 = y(t = 0) \\ \dot{x}(t = 0) = v_0 \cos \theta \equiv U \\ \dot{y}(t = 0) = v_0 \sin \theta \equiv V \end{array}\right\} \tag{2.40}$$

However, the equations of motion, Equation 2.31, become

$$m\ddot{x} = -km\dot{x} \tag{2.41}$$

$$m\ddot{y} = -km\dot{y} - mg \tag{2.42}$$

Equation 2.41 is exactly that used in Example 2.4. The solution is therefore

$$x = \frac{U}{k}(1 - e^{-kt}) \tag{2.43}$$

Similarly, Equation 2.42 is the same as the equation of the motion in Example 2.5. We can use the solution found in that example by letting $h = 0$. (The fact that we considered the particle to be projected *downward* in Example 2.5 is of no consequence. The sign of the initial velocity automatically takes this into account.) Therefore

$$y = -\frac{gt}{k} + \frac{kV + g}{k^2}(1 - e^{-kt}) \tag{2.44}$$

The trajectory is shown in Figure 2-8 for several values of the retarding force constant k for a given projectile flight.

The range R', which is the range including air resistance, can be found as previously by calculating the time T required for the entire trajectory and then substituting this value into Equation 2.43 for x. The time T is found as previously by finding $t = T$ when $y = 0$. From Equation 2.44, we find

$$T = \frac{kV + g}{gk}(1 - e^{-kt}) \tag{2.45}$$

This is a transcendental equation, and therefore we cannot obtain an analytic expression for T. Nonetheless, we still have powerful methods to use to solve such

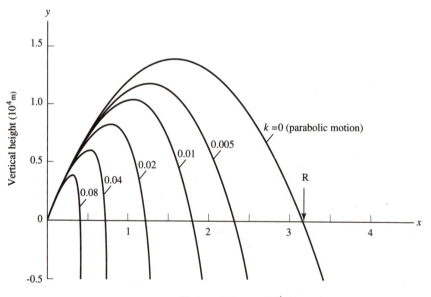

FIGURE 2-8   The calculated trajectories of a particle in air resistance ($F_{res} = -kmv$) for various values of k (in units of s^{-1}). The calculations were performed for values of $\theta = 60°$ and $v_0 = 600$ m/s. The values of y (Equation 2.44) are plotted versus x (Equation 2.43).

problems. We present two of them here: (1) a *perturbation method* to find an approximate solution, and (2) a *numerical method*, which can normally be as accurate as desired. We will compare the results.

Perturbation Method. To use the perturbation method, we find an *expansion parameter* or *coupling constant* that is normally small. In the present case, this parameter is the retarding force constant k, because we have already solved the present problem with $k = 0$, and now we would like to turn on the retarding force, but let k be small. We therefore expand the exponential term of Equation 2.45 (see Equation D.34 of Appendix D) in a power series with the intention of keeping only the lowest terms of k^n, where k is our expansion parameter.

$$T = \frac{kV + g}{gk} \left(kT - \frac{1}{2}k^2T^2 + \frac{1}{6}k^3T^3 - \cdots \right) \tag{2.46}$$

If we keep only terms in the expansion through k^3, this equation can be rearranged to yield

$$T = \frac{2V/g}{1 + kV/g} + \frac{1}{3}kT^2 \tag{2.47}$$

We now have the expansion parameter k in the denominator of the first term on the right-hand side of this equation. We need to expand this term in a power series (Taylor series, see Equation D.8 of Appendix D):

$$\frac{1}{1 + kV/g} = 1 - (kV/g) + (kV/g)^2 - \cdots \tag{2.48}$$

where we have kept only terms through k^2, because we only have terms through k in Equation 2.47. If we insert this expansion of Equation 2.48 into the first term on the right-hand side of Equation 2.47 and keep only the terms in k to first order, we have

$$T = \frac{2V}{g} + \left(\frac{T^2}{3} - \frac{2V^2}{g^2} \right)k + O(k^2) \tag{2.49}$$

where we choose to neglect $O(k^2)$, the terms of order k^2 and higher. In the limit $k \to 0$ (no air resistance), Equation 2.49 gives us the same result as in the previous example:

$$T(k = 0) = T_0 = \frac{2V}{g} = \frac{2v_0 \sin \theta}{g}$$

Therefore, if k is small (but nonvanishing), the flight time will be *approximately* equal to T_0. If we then use this approximate value for $T = T_0$ in the right-hand side of Equation 2.49, we have

$$T \cong \frac{2V}{g} \left(1 - \frac{kV}{3g} \right) \tag{2.50}$$

which is the desired approximate expression for the flight time.

Next, we write the equation for x (Equation 2.43) in expanded form:

$$x = \frac{U}{k}\left(kt - \frac{1}{2}k^2t^2 + \frac{1}{6}k^3t^3 - \cdots\right) \tag{2.51}$$

Because $x(t = T) \equiv R'$, we have approximately for the range

$$R' \cong U\left(T - \frac{1}{2}kT^2\right) \tag{2.52}$$

where again we keep terms only through the first order of k. We can now evaluate this expression by using the value of T from Equation 2.50. If we retain only terms linear in k, we find

$$R' \cong \frac{2UV}{g}\left(1 - \frac{4kV}{3g}\right) \tag{2.53}$$

The quantity $2UV/g$ can now be written (using Equations 2.40) as

$$\frac{2UV}{g} = \frac{2v_0^2}{g}\sin\theta\cos\theta = \frac{v_0^2}{g}\sin 2\theta = R \tag{2.54}$$

which will be recognized as the range R of the projectile when air resistance is neglected. Therefore

$$R' \cong R\left(1 - \frac{4kV}{3g}\right) \tag{2.55}$$

Over what range of values for k would we expect our perturbation method to be correct? If we look at the expansion in Equation 2.48, we see that the expansion will not converge unless $kV/g < 1$ or $k < g/V$, and in fact, we would like $k \ll g/V = g/v_0 \sin\theta$.

Numerical Method. Equation 2.45 can be solved numerically using a computer by a variety of methods. We set up a loop to solve the equation for T for many values of k up to 0.08 s^{-1}: $T_i(k_i)$. These values of T_i and k_i are inserted into Equation 2.43 to find the range R'_i, which is displayed in Figure 2-9. The range drops rapidly for increased air resistance, just as one would expect, but it does not display the linear dependence suggested by the perturbation method solution of Equation 2.55.

For the projectile motion described in Figures 2-8 and 2-9, the linear approximation is inaccurate for k values as low as 0.01 s^{-1} and incorrectly shows the range is zero for all values of k larger than 0.014 s^{-1}. This disagreement with the perturbation method is not surprising because the linear result for the range R' was dependent on $k \ll g/(v_0 \sin\theta) = 0.02$ s^{-1}, which is hardly true for even $k = 0.01$ s^{-1}. The agreement should be adequate for $k = 0.005$ s^{-1}. The results shown in Figure 2-8 indicate that for values of $k > 0.005$ s^{-1}, the drag can hardly be considered a perturbation. In fact, for $k > 0.01$ s^{-1} the drag becomes the dominant factor in the projectile motion.

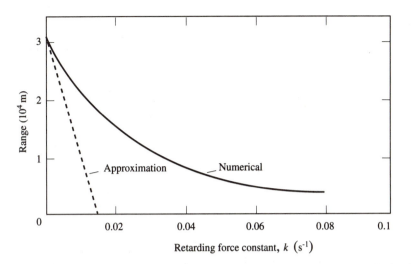

FIGURE 2-9   The range values calculated approximately and numerically for the projectile data given in Figure 2-8 are plotted as a function of the retarding force constant k.

The previous example indicates how complicated the real world can be. In that example, we still had to make assumptions that were nonphysical—in assuming, for example, that the retarding force is always linearly proportional to the velocity. Even our numerical calculation is not accurate, because Figure 2-3 shows us that a better assumption would be to include a v^2 retarding term as well. Adding such a term would not be difficult with the numerical calculation, and we shall do a similar calculation in the next example. We have included the author's Mathcad file that produced Figures 2-8 and 2-9 in Appendix H for those students who might want to reproduce the calculation. We emphasize that there are many ways to perform numerical calculations with computers, and the student will probably want to become proficient with several.

EXAMPLE 2.8 ---

Use the data shown in Figure 2-3 to calculate the trajectory for an actual projectile. Assume a muzzle velocity of 600 m/s, gun elevation of 45°, and a projectile mass of 30 kg. Plot the height y versus the horizontal distance x and plot y, $\dot{x}$, and $\dot{y}$ versus time both with and without air resistance. Include only the air resistance and gravity, and ignore other possible forces such as the lift.

Solution: First, we make a table of retarding force versus velocity by reading Figure 2-3. Read the force every 50 m/s for Figure 2-3c and every 100 m/s for Figure 2-3d. We can then use a straight line interpolation between the tabular values. We use the coordinate system shown in Figure 2-7. The equations of motion become

$$\ddot{x} = -\frac{F_x}{m} \tag{2.56}$$

$$\ddot{y} = -\frac{F_y}{m} - g \tag{2.57}$$

where F_x and F_y are the retarding forces. Assume g is constant. F_x will always be a positive number, but $F_y > 0$ for the projectile going up, and $F_y < 0$ for the projectile coming back down. Let θ be the projectile's elevation angle from the horizontal at any instant.

$$v = \sqrt{\dot{x}^2 + \dot{y}^2} \tag{2.58}$$

$$\tan \theta = \frac{\dot{y}}{\dot{x}} \tag{2.59}$$

$$F_x = F \cos \theta \tag{2.60}$$

$$F_y = F \sin \theta \tag{2.61}$$

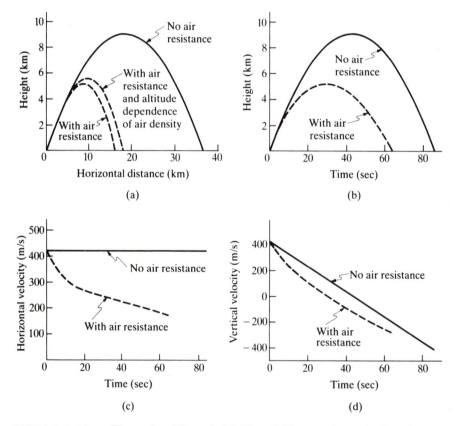

(a)

(b)

(c)

(d)

FIGURE 2-10 The results of Example 2.8. The solid lines are the results if no air resistance is included, whereas the dashed lines include the results of adding the air resistive force. In (a) we also include the effect of the air density dependence getting smaller when the projectile is higher.

We can calculate F_x and F_y at any instant by knowing $\dot{x}$ and $\dot{y}$. Over a small time interval, the next $\dot{x}$ and $\dot{y}$ can be calculated.

$$\dot{x} = \int_0^t \ddot{x}\, dt + v_0 \cos \theta \tag{2.62}$$

$$\dot{y} = \int_0^t \ddot{y}\, dt + v_0 \sin \theta \tag{2.63}$$

$$x = \int_0^t \dot{x}\, dt \tag{2.64}$$

$$y = \int_0^t \dot{y}\, dt \tag{2.65}$$

We wrote a short computer program to contain our table for the retarding forces and to perform the calculations for $\dot{x}$, $\dot{y}$, x, and y as a function of time. We must perform the integrals by summations over small time intervals, because the forces are time dependent. Figure 2-10 shows the results.

Notice the large difference that the air resistance makes. In Figure 2-10a, the horizontal distance (range) that the projectile travels is about 16 km compared to almost 37 km with no air resistance. Our calculation ignored the fact that the air density depends on the altitude. If we take account of the decrease in the air density with altitude, we obtain the third curve with a range of 18 km shown in Figure 2-10a. If we also included the lift, the range would be still greater. Notice that the change in velocities in Figures 2-10c and 2-10d mirror the air resistive force of Figure 2-3. The speeds decrease rapidly until the speed reaches the speed of sound, and then the rate of change of the speeds levels off somewhat.

---- ●

This concludes our subsection on the effects of retarding forces. Much more could be done to include realistic effects, but the method is clear. Normally, one effect is added at a time, and the results are analyzed before another effect is added.

Other Examples of Dynamics

We conclude this section with two additional standard examples of dynamical particle-like behavior.

E X A M P L E **2.9** --------------------------------------

Atwood's machine consists of a smooth pulley with two masses suspended from a light string at each end (Figure 2-11). Find the acceleration of the masses and the tension of the string (a) when the pulley is at rest and (b) when the pulley is descending in an elevator with constant acceleration α.

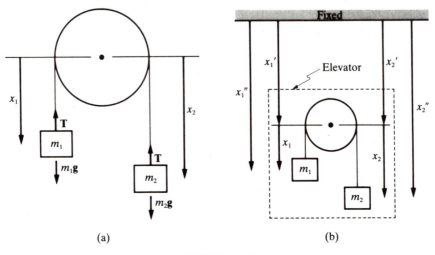

FIGURE 2-11

Solution: We neglect the mass of the string and assume that the pulley is smooth—that is, no friction on the string. The tension T must be the same throughout the string. The equations of motion become, for each mass, for case (a),

$$m_1\ddot{x}_1 = m_1 g - T \tag{2.66}$$
$$m_2\ddot{x}_2 = m_2 g - T \tag{2.67}$$

Notice again the advantage of the force concept: We need only identify the forces acting on each mass. The tension T is the same in both equations. If the string is inextensible, then $\ddot{x}_2 = -\ddot{x}_1$, and Equations 2.66 and 2.67 may be combined

$$m_1\ddot{x}_1 = m_1 g - (m_2 g - m_2\ddot{x}_2)$$
$$= m_1 g - (m_2 g + m_2\ddot{x}_1)$$

Rearranging,

$$\ddot{x}_1 = \frac{g(m_1 - m_2)}{m_1 + m_2} = -\ddot{x}_2 \tag{2.68}$$

If $m_1 > m_2$, then $\ddot{x}_1 > 0$, and $\ddot{x}_2 < 0$. The tension can be obtained from Equations 2.68 and 2.66:

$$T = m_1 g - m_1\ddot{x}_1$$

$$T = m_1 g - m_1 g \frac{(m_1 - m_2)}{m_1 + m_2}$$

$$T = \frac{2m_1 m_2 g}{m_1 + m_2} \tag{2.69}$$

For case (b), in which the pulley is in an elevator, the coordinate system with origins at the pulley center is no longer an inertial system. We need an inertial system with the origin at the top of the elevator shaft (Figure 2-11b). The equations of motion in the inertial system ($x_1'' = x_1' + x_1, x_2'' = x_2' + x_2$) are

$$m_1\ddot{x}_1'' = m_1(\ddot{x}_1' + \ddot{x}_1) = m_1 g - T$$

$$m_2\ddot{x}_2'' = m_2(\ddot{x}_2' + \ddot{x}_2) = m_2 g - T$$

so

$$\left.\begin{array}{l} m_1\ddot{x}_1 = m_1 g - T - m_1\ddot{x}_1' = m_1(g - \alpha) - T \\ m_2\ddot{x}_2 = m_2 g - T - m_2\ddot{x}_2' = m_2(g - \alpha) - T \end{array}\right\} \qquad (2.70)$$

where $\ddot{x}_1' = \ddot{x}_2' = \alpha$. We have $\ddot{x}_2 = -\ddot{x}_1$, so we solve for $\ddot{x}_1$ as before by eliminating T:

$$\ddot{x}_1 = -\ddot{x}_2 = (g - \alpha)\frac{(m_1 - m_2)}{m_1 + m_2} \qquad (2.71)$$

and

$$T = \frac{2m_1 m_2(g - \alpha)}{m_1 + m_2} \qquad (2.72)$$

Notice that the results for the acceleration and tension are just as if the acceleration of gravity were reduced by the amount of the elevator acceleration α. The change for an ascending elevator should be obvious.

--●

E X A M P L E   **2.10**   -

In our last example in this lengthy review of the equations of motion for a particle, let us examine particle motion in an electromagnetic field. Consider a charged particle entering a region of uniform magnetic field B—for example, the Earth's field—as shown in Figure 2-12. Determine its subsequent motion.

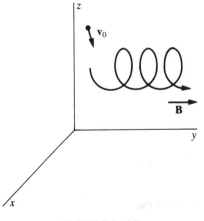

FIGURE 2-12

Solution: Choose a Cartesian coordinate system with its y-axis parallel to the magnetic field. If q is the charge on the particle, $\mathbf{v}$ its velocity, $\mathbf{a}$ its acceleration, and $\mathbf{B}$ the Earth's magnetic field, then

$$\mathbf{v} = \dot{x}\mathbf{i} + \dot{y}\mathbf{j} + \dot{z}\mathbf{k}$$

$$\mathbf{a} = \ddot{x}\mathbf{i} + \ddot{y}\mathbf{j} + \ddot{z}\mathbf{k}$$

$$\mathbf{B} = B_0\mathbf{j}$$

The magnetic force $\mathbf{F} = q\mathbf{v} \times \mathbf{B} = m\mathbf{a}$, so

$$m(\ddot{x}\mathbf{i} + \ddot{y}\mathbf{j} + \ddot{z}\mathbf{k}) = q(\dot{x}\mathbf{i} + \dot{y}\mathbf{j} + \dot{z}\mathbf{k}) \times B_0\mathbf{j} = qB_0(\dot{x}\mathbf{k} - \dot{z}\mathbf{i})$$

Equating like vector components gives

$$\left. \begin{array}{l} m\ddot{x} = -qB_0\dot{z} \\[2mm] m\ddot{y} = 0 \\[2mm] m\ddot{z} = qB_0\dot{x} \end{array} \right\} \tag{2.73}$$

Integrating the second of these equations, $m\ddot{y} = 0$, yields

$$\dot{y} = \dot{y}_0$$

where $\dot{y}_0$ is a constant and is the initial value of $\dot{y}$. Integrating a second time gives

$$y = \dot{y}_0 t + y_0$$

where y_0 is also a constant.

To integrate the first and last equations of Equation 2.73, let $\alpha = qB_0/m$, so that

$$\left. \begin{array}{l} \ddot{x} = -\alpha\dot{z} \\[2mm] \ddot{z} = \alpha\dot{x} \end{array} \right\} \tag{2.74}$$

These coupled, simultaneous differential equations can be easily uncoupled by differentiating one and substituting it into the other, giving

$$\dddot{z} = \alpha\ddot{x} = -\alpha^2\dot{z}$$

$$\dddot{x} = -\alpha\ddot{z} = -\alpha^2\dot{x}$$

so that

$$\left. \begin{array}{l} \dddot{z} = -\alpha^2\dot{z} \\[2mm] \dddot{x} = -\alpha^2\dot{x} \end{array} \right\} \tag{2.75}$$

Both of these differential equations have the same form of solution. Using the technique of Example C.2 of Appendix C, we have

$$x = A \cos \alpha t + B \sin \alpha t + x_0$$

$$z = A' \cos \alpha t + B' \sin \alpha t + z_0$$

where A, A', B, B', x_0, and z_0 are constants of integration that are determined by the particle's initial position and velocity and by the equations of motion, Equation 2.74. These solutions can be rewritten

$$\left.\begin{array}{l} (x - x_0) = A \cos \alpha t + B \sin \alpha t \\ (y - y_0) = \dot{y}_0 t \\ (z - z_0) = A' \cos \alpha t + B' \sin \alpha t \end{array}\right\} \tag{2.76}$$

The x- and z-coordinates are connected by Equation 2.74, so substituting Equations 2.76 into the first equation of Equation 2.74 gives

$$-\alpha^2 A \cos \alpha t - \alpha^2 B \sin \alpha t = -\alpha(-\alpha A' \sin \alpha t + \alpha B' \cos \alpha t) \tag{2.77}$$

Because Equation 2.77 is valid for all t, in particular $t = 0$ and $t = \pi/2\alpha$, Equation 2.77 yields

$$-\alpha^2 A = -\alpha^2 B'$$

so that

$$A = B'$$

and

$$-\alpha^2 B = \alpha^2 A'$$

gives

$$B = -A'$$

We now have

$$\left.\begin{array}{l} (x - x_0) = A \cos \alpha t + B \sin \alpha t \\ (y - y_0) = \dot{y}_0 t \\ (z - z_0) = -B \cos \alpha t + A \sin \alpha t \end{array}\right\} \tag{2.78}$$

If at $t = 0$, $\dot{z} = \dot{z}_0$ and $\dot{x} = 0$, then from Equation 2.78, differentiating and setting $t = 0$ gives

$$\alpha B = 0$$

and

$$\alpha A = \dot{z}_0$$

so

$$(x - x_0) = \frac{\dot{z}_0}{\alpha} \cos \alpha t$$

$$(y - y_0) = \dot{y}_0 t$$

$$(z - z_0) = \frac{\dot{z}_0}{\alpha} \sin \alpha t$$

Finally,

$$
\left.
\begin{aligned}
x - x_0 &= \left(\frac{\dot{z}_0 m}{qB_0}\right) \cos\left(\frac{qB_0 t}{m}\right) \\
(y - y_0) &= \dot{y}_0 t \\
z - z_0 &= \left(\frac{\dot{z}_0 m}{qB_0}\right) \sin\left(\frac{qB_0 t}{m}\right)
\end{aligned}
\right\}
\tag{2.79}
$$

These are the parametric equations of a circular helix of radius $\dot{z}_0 m / qB_0$. Thus, the faster the particle enters the field or the greater its mass, the larger the radius of the helix. And the greater the charge on the particle or the stronger the magnetic field, the tighter the helix. Notice also how the charged particle is captured by the magnetic field—just drifting along the field direction. In this example, the particle had no initial component of its velocity along the x-axis, but even if it had it would not drift along this axis (see Problem 2-31). Finally, notice that the magnetic force on the particle always acts perpendicular to its velocity and hence cannot speed it up. Equation 2.79 verifies this fact.

The Earth's magnetic field is not as simple as the uniform field of this example. Nevertheless, this example gives some insight into one of the mechanisms by which the Earth's magnetic field traps low-energy cosmic rays and the solar wind to create the Van Allen belts.

2.5 CONSERVATION THEOREMS

We now turn to a detailed discussion of the Newtonian mechanics of a single particle and derive the important theorems regarding conserved quantities. We must emphasize that we are not *proving* the conservation of the various quantities. We are merely deriving the consequences of Newton's laws of dynamics. These implications must be put to the test of experiment, and their verification then supplies a measure of confirmation of the original dynamical laws. The fact that these conservation theorems have indeed been found to be valid in many instances furnishes an important part of the proof for the correctness of Newton's laws, at least in classical physics.

The first of the conservation theorems concerns the **linear momentum** of a particle. If the particle is *free*, that is, if the particle encounters no force, then Equation 2.2 becomes simply $\dot{\mathbf{p}} = 0$. Therefore, $\mathbf{p}$ is a vector constant in time, and the first conservation theorem becomes

I. *The total linear momentum $\mathbf{p}$ of a particle is conserved when the total force on it is zero.*

Note that this result is derived from a vector equation, $\dot{\mathbf{p}} = 0$, and therefore applies for each component of the linear momentum. To state the result in other terms, we let **s** be some constant vector such that $\mathbf{F} \cdot \mathbf{s} = 0$, independent of time. Then

$$\dot{\mathbf{p}} \cdot \mathbf{s} = \mathbf{F} \cdot \mathbf{s} = 0$$

or, integrating with respect to time,

$$\mathbf{p} \cdot \mathbf{s} = \text{constant} \tag{2.80}$$

which states that the *component of linear momentum in a direction in which the force vanishes is constant in time*.

The **angular momentum L** of a particle with respect to an origin from which the position vector **r** is measured is defined to be

$$\boxed{\mathbf{L} \equiv \mathbf{r} \times \mathbf{p}} \tag{2.81}$$

The **torque** or **moment of force N** with respect to the same origin is defined to be

$$\boxed{\mathbf{N} \equiv \mathbf{r} \times \mathbf{F}} \tag{2.82}$$

where **r** is the position vector from the origin to the point where the force **F** is applied. Because $\mathbf{F} = m\dot{\mathbf{v}}$ for the particle, the torque becomes

$$\mathbf{N} = \mathbf{r} \times m\dot{\mathbf{v}} = \mathbf{r} \times \dot{\mathbf{p}}$$

Now

$$\dot{\mathbf{L}} = \frac{d}{dt}(\mathbf{r} \times \mathbf{p}) = (\dot{\mathbf{r}} \times \mathbf{p}) + (\mathbf{r} \times \dot{\mathbf{p}})$$

but

$$\dot{\mathbf{r}} \times \mathbf{p} = \dot{\mathbf{r}} \times m\mathbf{v} = m(\dot{\mathbf{r}} \times \dot{\mathbf{r}}) \equiv 0$$

so

$$\boxed{\dot{\mathbf{L}} = \mathbf{r} \times \dot{\mathbf{p}} = \mathbf{N}} \tag{2.83}$$

If no torques act on a particle (i.e., if $\mathbf{N} = 0$), then $\dot{\mathbf{L}} = 0$ and **L** is a vector constant in time. The second important conservation theorem is

II. *The angular momentum of a particle subject to no torque is conserved.*

We remind the student that a judicious choice of the origin of a coordinate system will often allow a problem to be solved much more easily than a poor choice. For example, the torque will be zero in coordinate systems centered along the resultant line of force. The angular momentum will be conserved in this case.

If work is done on a particle by a force **F** in transforming the particle from Condition 1 to Condition 2, then this **work** is defined to be

$$W_{12} \equiv \int_1^2 \mathbf{F} \cdot d\mathbf{r}$$

(2.84)

If **F** is the net resultant force acting on the particle,

$$\mathbf{F} \cdot d\mathbf{r} = m \frac{d\mathbf{v}}{dt} \cdot \frac{d\mathbf{r}}{dt} dt = m \frac{d\mathbf{v}}{dt} \cdot \mathbf{v} \, dt$$

$$= \frac{m}{2} \frac{d}{dt} (\mathbf{v} \cdot \mathbf{v}) \, dt = \frac{m}{2} \frac{d}{dt} (v^2) \, dt = d(\tfrac{1}{2} m v^2)$$

(2.85)

The integrand in Equation 2.84 is thus an exact differential, and the work done by the total force **F** acting on a particle is equal to its change in kinetic energy:

$$W_{12} = (\tfrac{1}{2} m v^2) \Big|_1^2 = \tfrac{1}{2} m(v_2^2 - v_1^2) = T_2 - T_1$$

(2.86)

where $T \equiv \tfrac{1}{2} m v^2$ is the **kinetic energy** of the particle. If $T_1 > T_2$ then $W_{12} < 0$, and the particle has done work with a resulting decrease in kinetic energy. It is important to realize that the force **F** leading to Equation 2.85 is the **total** (i.e., net resultant) **force** on the particle.

Let us now examine the integral appearing in Equation 2.84 from a different standpoint. In many physical problems, the force **F** has the property that the work required to move a particle from one position to another without any change in kinetic energy depends only on the original and final positions and not on the exact path taken by the particle. For example, assume the work done to move the particle from point 1 in Figure 2-13 to point 2 is independent of the actual paths a, b, or c taken. This property is exhibited, for example, by a constant gravitational force field. Thus, if a particle of mass m is raised through a height h (by *any* path), then an amount of work mgh has been done on the particle, and the particle can do an

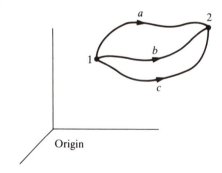

FIGURE 2-13

equal amount of work in returning to its original position. This capacity to do work is called the **potential energy** of the particle.

We may define the potential energy of a particle in terms of the work (done by the force **F**) required to transport the particle from a point 1 to a point 2 (with no net change in kinetic energy):

$$\int_1^2 \mathbf{F} \cdot d\mathbf{r} = U_1 - U_2 \tag{2.87}$$

The work done in moving the particle is thus simply the difference in the potential energy U at the two points. For example, if we lift a suitcase from position 1 on the ground to position 2 in a car trunk, we as the external agent are doing work against the force of gravity. Let the force **F** in Equation 2.87 be the gravitational force, and in raising the suitcase, $\mathbf{F} \cdot d\mathbf{r}$ becomes negative. The result of the integration in Equation 2.87 is that $U_1 - U_2$ is negative, so that the potential energy at position 2 in the car's trunk is greater than that at position 1 on the ground. The change in potential energy $U_2 - U_1$ is the negative of the work done by the gravitational force, as can be seen by multiplying both sides of Equation 2.87 by -1. As the external agent, we do positive work (against gravity) to raise the potential energy of the suitcase.

Equation 2.87 can be reproduced* if we write **F** as the gradient of the scalar function U:

$$\boxed{\mathbf{F} = -\mathbf{grad}\, U = -\nabla U} \tag{2.88}$$

Then

$$\int_1^2 \mathbf{F} \cdot d\mathbf{r} = -\int_1^2 (\nabla U) \cdot d\mathbf{r} = -\int_1^2 dU = U_1 - U_2 \tag{2.89}$$

In most systems of interest, the potential energy is a function of position and, possibly, time: $U = U(\mathbf{r})$ or $U = U(\mathbf{r}, t)$. We do not consider cases in which the potential energy is a function of the velocity.[†]

It is important to realize that the potential energy is defined only to within an additive constant; that is, the force defined by $-\nabla U$ is no different from that defined by $-\nabla(U + \text{constant})$. Potential energy therefore has no absolute meaning; only *differences* of potential energy are physically meaningful (as in Equation 2.87).

If we choose a certain inertial frame of reference to describe a mechanical process, the laws of motion are the same as in any other reference frame in uniform motion relative to the original frame. The velocity of a particle is in general different depending on which inertial reference frame we chose as the basis for

---

* The necessary and sufficient condition that permits a vector function to be represented by the gradient of a scalar function is that the *curl* of the vector function vanish identically.

[†] Velocity-dependent potentials are necessary in certain situations, e.g., in electromagnetism (the so-called Liénard–Wiechert potentials).

describing the motion. We therefore find that it is impossible to ascribe an *absolute kinetic energy* to a particle in much the same way that it is impossible to assign any absolute meaning to potential energy. Both of these limitations are the result of the fact that selecting an *origin* of the coordinate system used to describe physical processes is always arbitrary. The nineteenth-century Scottish physicist James Clerk Maxwell (1831–1879) summarized the situation as follows*:

> We must, therefore, regard the energy of a material system as a quantity of which we may ascertain the increase or diminution as the system passes from one definite condition to another. The absolute value of the energy in the standard condition is unknown to us, and it would be of no value to us if we did know it, as all phenomena depend on the variations of energy and not on its absolute value.

Next, we define the **total energy** of a particle to be the sum of the kinetic and potential energies:

$$\boxed{E \equiv T + U} \tag{2.90}$$

The total time derivative of E is

$$\frac{dE}{dt} = \frac{dT}{dt} + \frac{dU}{dt} \tag{2.91}$$

To evaluate the time derivatives appearing on the right-hand side of this equation, we first note that Equation 2.85 can be written as

$$\mathbf{F} \cdot d\mathbf{r} = d(\tfrac{1}{2} m v^2) = dT \tag{2.92}$$

Dividing through by dt,

$$\frac{dT}{dt} = \mathbf{F} \cdot \frac{d\mathbf{r}}{dt} = \mathbf{F} \cdot \dot{\mathbf{r}} \tag{2.93}$$

We have also

$$\frac{dU}{dt} = \sum_i \frac{\partial U}{\partial x_i} \frac{dx_i}{dt} + \frac{\partial U}{\partial t}$$

$$= \sum_i \frac{\partial U}{\partial x_i} \dot{x}_i + \frac{\partial U}{\partial t}$$

$$= (\nabla U) \cdot \dot{\mathbf{r}} + \frac{\partial U}{\partial t} \tag{2.94}$$

Substituting Equations 2.93 and 2.94 into 2.91, we find

---

* J. C. Maxwell, *Matter and Motion* (Cambridge, 1877), p. 91.

$$\frac{dE}{dt} = \mathbf{F} \cdot \dot{\mathbf{r}} + (\boldsymbol{\nabla} U) \cdot \dot{\mathbf{r}} + \frac{\partial U}{\partial t}$$

$$= (\mathbf{F} + \boldsymbol{\nabla} U) \cdot \dot{\mathbf{r}} + \frac{\partial U}{\partial t}$$

$$= \frac{\partial U}{\partial t} \tag{2.95}$$

because the term $\mathbf{F} + \boldsymbol{\nabla} U$ vanishes in view of the definition of the potential energy (Equation 2.88) if the total force is the conservative force $\mathbf{F} = -\boldsymbol{\nabla} U$.

If U is not an explicit function of the time (i.e., if $\partial U / \partial t = 0$; recall that we do not consider velocity-dependent potentials), the force field represented by $\mathbf{F}$ is **conservative**. Under these conditions, we have the third important conservation theorem:

III. *The total energy E of a particle in a conservative force field is a constant in time.*

It must be reiterated that we have not *proved* the conservation laws of linear momentum, angular momentum, and energy. We have only derived various consequences of Newton's laws; that is, *if* these laws are valid in a certain situation, then momentum and energy will be conserved. But we have become so enamored with these conservation theorems that we have elevated them to the status of laws and we have come to *insist* that they be valid in any physical theory, even those that apply to situations in which Newtonian mechanics is not valid, as, for example, in the interaction of moving charges or in quantum-mechanical systems. We do not actually have conservation laws in such situations, but rather conservation *postulates* that we force on the theory. For example, if we have two isolated moving electric charges, the electromagnetic forces between them are not conservative. We therefore endow the electromagnetic field with a certain amount of energy so that energy conservation will be valid. This procedure is satisfactory only if the consequences do not contradict any experimental fact, and this is indeed the case for moving charges. We therefore extend the usual concept of energy to include "electromagnetic energy" to satisfy our preconceived notion that energy must be conserved. This may seem an arbitrary and drastic step to take, but nothing, it is said, succeeds as does success, and these conservation "laws" have been the most successful set of principles in physics. The refusal to relinquish energy and momentum conservation led Wolfgang Pauli (1900–1958) to postulate in 1930 the existence of the neutrino to account for the "missing" energy and momentum in radioactive β decay. This postulate allowed Enrico Fermi (1901–1954) to construct a successful theory of β decay in 1934, but direct observation of the neutrino was not made until 1953 when Reines and Cowan performed their famous experiment.* By adhering to the conviction that energy and momentum must be conserved, a new elementary particle was discovered, one that is of great importance in modern theories

---

* C. L. Cowan, F. Reines, F. B. Harrison, H. W. Kruse, and A. D. McGuire, Science, **124**, 103 (1956).

D. VENNE

of nuclear physics. This discovery is only one of the many advances in the understanding of the properties of matter that have resulted directly from the application of the conservation laws.

We shall apply these conservation theorems to several physical situations in the remainder of this book, among them Rutherford scattering and planetary motion. A simple example here indicates the usefulness of the conservation theorems.

E X A M P L E ⬤ **2.11** ---

A mouse of mass m jumps on the outside edge of a freely turning ceiling fan of moment of inertia I and radius R. By what ratio does the angular velocity change?

Solution: Angular momentum must be conserved during the process. We are using the concept of moment of inertia learned in elementary physics to relate angular momentum L to angular velocity ω: $L = I\omega$. The initial angular momentum $L_0 = I\omega_0$ must be equal to the angular momentum L (fan plus mouse) after the mouse jumps on. The velocity of the outside edge is $v = \omega R$.

$$L = I\omega + mvR = \frac{v}{R}(I + mR^2)$$

$$L = L_0 = I\omega_0$$

$$\frac{v}{R}(I + mR^2) = I\frac{v_0}{R}$$

$$\frac{v}{v_0} = \frac{I}{I + mR^2}$$

and

$$\frac{\omega}{\omega_0} = \frac{I}{I + mR^2}$$

-- ⬤

2.6 ENERGY

The concept of energy was not nearly as popular in Newton's time as it is today. Later we shall study two new formulations of dynamics, different from Newton's, based on energy—the Lagrangian and Hamiltonian methods.

Early in the nineteenth century, it became clear that heat was another form of energy and not a form of fluid (called "caloric") that flowed between hot and cold bodies. Count Rumford* is generally given credit for realizing that the great amount

---

* Benjamin Thompson (1753–1814) was born in Massachusetts and emigrated to Europe in 1776 as a loyalist refugee. Among the activities of his distinguished military and, later, scientific career, he supervised the boring of cannons as head of the Bavarian war department.

of heat generated during the boring of a cannon was caused by friction and not the caloric. If frictional energy is just heat energy, interchangeable with mechanical energy, then a total conservation of energy can occur.

Throughout the nineteenth century, scientists performed experiments on the conservation of energy, resulting in the prominence given energy today. Hermann von Helmholtz (1821–1894) formulated the general law of conservation of energy in 1847. He based his conclusion largely on the calorimetric experiments of James Prescott Joule (1818–1889) begun in 1840.

Consider a point particle under the influence of a conservative force with potential U. The conservation of energy (actually, mechanical energy, to be precise in this case) is reflected in Equation 2.90.

$$E = T + U = \frac{1}{2} mv^2 + U(x) \tag{2.96}$$

where we consider only the one-dimensional case. We can rewrite Equation 2.96 as

$$v(t) = \frac{dx}{dt} = \pm \sqrt{\frac{2}{m} [E - U(x)]} \tag{2.97}$$

and by integrating

$$t - t_0 = \int_{x_0}^{x} \frac{\pm \, dx}{\sqrt{\frac{2}{m} [E - U(x)]}} \tag{2.98}$$

where $x = x_0$ at $t = t_0$. We have formally solved the one-dimensional case in Equation 2.98; that is, we have found $x(t)$. All that remains is to insert the potential $U(x)$ into Equation 2.98 and integrate, using computer techniques if necessary. We shall study later in some detail the potentials $U = \frac{1}{2} kx^2$ for harmonic oscillations and $U = -k/x$ for the gravitational force.

We can learn a good deal about the motion of a particle simply by examining a plot of an example of $U(x)$ as shown in Figure 2-14. First, notice that, because $\frac{1}{2} mv^2 = T \geq 0$, $E \geq U(x)$ for any real physical motion. We see in Figure 2-14 that the motion is *bounded* for energies E_1 and E_2. For E_1, the motion is *periodic* between the *turning points* x_a and x_b. Similarly, for E_2 the motion is periodic, but there are two possible regions: $x_c \leq x \leq x_d$ and $x_e \leq x \leq x_f$. The particle cannot "jump" from one "pocket" to the other; once in a pocket, it must remain there forever if its energy remains at E_2. The motion for a particle with energy E_0 has only one value, $x = x_0$. The particle is at rest with $T = 0$ [$E_0 = U(x_0)$].

The motion for a particle with energy E_3 is simple: The particle comes in from infinity, stops and turns at $x = x_g$, and returns to infinity—much like a tennis ball bouncing against a practice wall. For the energy E_4, the motion is unbounded and the particle may be at any position. Its speed will change because it depends on the difference between E_4 and $U(x)$. If it is moving to the right, it will speed up and slow down but continue to infinity.

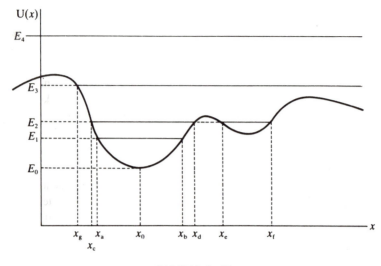

FIGURE 2-14

The motion of a particle of energy E_1 is similar to that of a mass at the end of a spring. The potential in the region $x_a < x < x_b$ can be approximated by $U(x) = \frac{1}{2} k(x - x_0)^2$. A particle with energy barely above E_0 will oscillate about the point $x = x_0$. We refer to such a point as an **equilibrium point**, because if the particle is placed at $x = x_0$ it remains there. Equilibrium may be stable, unstable, or neutral. The equilibrium just discussed is *stable* because if the particle were placed on either side of $x = x_0$ it would eventually return there. We can use a hemispherical mixing bowl with a steel ball as an example. With the bowl right side up, the ball can roll around inside the bowl; but it will eventually settle to the bottom—in other words, there is a stable equilibrium. If we turn the bowl upside down and place the ball precisely outside at $x = x_0$, the ball remains there in equilibrium. If we place the ball on either side of $x = x_0$ on the rounded surface, it rolls off; we call this *unstable* equilibrium. *Neutral* equilibrium would apply when the ball rolls on a flat, smooth, horizontal surface.

In general, we can express the potential $U(x)$ in a Taylor series about a certain equilibrium point. For mathematical simplicity, let us assume that the equilibrium point is at $x = 0$ rather than $x = x_0$ (if not, we can always redefine the coordinate system to make it so). Then we have

$$U(x) = U_0 + x \left(\frac{dU}{dx} \right)_0 + \frac{x^2}{2!} \left(\frac{d^2 U}{dx^2} \right)_0 + \frac{x^3}{3!} \left(\frac{d^3 U}{dx^3} \right)_0 + \cdots \qquad \textbf{(2.99)}$$

The zero subscript indicates that the quantity is to be evaluated at $x = 0$. The potential energy U_0 at $x = 0$ is simply a constant that we can define to be zero without any loss of generality. If $x = 0$ is an equilibrium point, then

$$\left(\frac{dU}{dx} \right)_0 = 0 \qquad \text{Equilibrium point} \qquad \textbf{(2.100)}$$

and Equation 2.99 becomes

$$U(x) = \frac{x^2}{2!}\left(\frac{d^2U}{dx^2}\right)_0 + \frac{x^3}{3!}\left(\frac{d^3U}{dx^3}\right)_0 + \cdots \tag{2.101}$$

Near the equilibrium point $x = 0$, the value of x is small, and each term in Equation 2.101 is considerably smaller than the previous one. Therefore, we keep only the first term in Equation 2.101:

$$U(x) = \frac{x^2}{2}\left(\frac{d^2U}{dx^2}\right)_0 \tag{2.102}$$

We can determine whether the equilibrium at $x = 0$ is stable or unstable by examining $(d^2U/dx^2)_0$. If $x = 0$ is a stable equilibrium, $U(x)$ must be greater (more positive) on either side of $x = 0$. Because x^2 is always positive, the conditions for the equilibrium are

$$\left(\frac{d^2U}{dx^2}\right)_0 > 0 \qquad \text{Stable equilibrium}$$
$$\tag{2.103}$$
$$\left(\frac{d^2U}{dx^2}\right)_0 < 0 \qquad \text{Unstable equilibrium}$$

If $(d^2U/dx^2)_0$ is zero, higher-order terms must be examined (see Problems 2-45 and 2-46).

EXAMPLE **2.12** -

Consider the system of pulleys, masses, and string shown in Figure 2-15. A light string of length b is attached at point A, passes over a pulley at point B located a distance $2d$ away, and finally attaches to mass m_1. Another pulley with mass m_2 attached passes over the string, pulling it down between A and B. Calculate the distance x_1 when the system is in equilibrium, and determine whether the equilibrium is stable or unstable. The pulleys are massless.

Solution: We can solve this example by either using forces (i.e., when $\ddot{x}_1 = 0 = \dot{x}_1$) or energy. We choose the energy method, because in equilibrium the kinetic energy is zero and we need to deal only with the potential energy when Equation 2.100 applies.

We let $U = 0$ along the line AB.

$$U = -m_1gx_1 - m_2g(x_2 + c) \tag{2.104}$$

We assume that the pulley holding mass m_2 is small, so we can neglect the pulley radius. The distance c in Figure 2-15 is constant.

$$x_2 = \sqrt{[(b - x_1)^2/4] - d^2}$$

$$U = -m_1gx_1 - m_2g\sqrt{[(b - x_1)^2/4] - d^2} - m_2gc$$

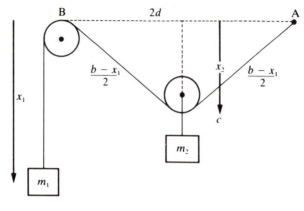

FIGURE 2-15

By setting $dU/dx_1 = 0$, we can determine the equilibrium position $(x_1)_0 \equiv x_0$:

$$\left(\frac{dU}{dx_1}\right)_0 = -m_1g + \frac{m_2g(b - x_0)}{4\sqrt{[(b - x_0)^2/4] - d^2}} = 0$$

$$4m_1\sqrt{[(b - x_0)^2/4] - d^2} = m_2(b - x_0)$$

$$(b - x_0)^2(4m_1^2 - m_2^2) = 16m_1^2d^2$$

$$x_0 = b - \frac{4m_1d}{\sqrt{4m_1^2 - m_2^2}} \tag{2.105}$$

Notice that a real solution exists only when $4m_1^2 > m_2^2$.

Under what circumstances will the mass m_2 pull the mass m_1 up to the pulley B (i.e., $x_1 = 0$)? We can use Equation 2.103 to determine whether the equilibrium is stable or unstable:

$$\frac{d^2U}{dx_1^2} = \frac{-m_2g}{4\{[(b - x_1)^2/4] - d^2\}^{1/2}} + \frac{m_2g(b - x_1)^2}{16\{[(b - x_1)^2/4] - d^2\}^{3/2}}$$

Now insert $x_1 = x_0$.

$$\left(\frac{d^2U}{dx_1^2}\right)_0 = \frac{g(4m_1^2 - m_2^2)^{3/2}}{4m_2^2d}$$

The condition for the equilibrium (real motion) previously was for $4m_1^2 > m_2^2$, so the equilibrium, when it exists, will be stable, because $(d^2U/dx^2)_0 > 0$.

EXAMPLE 2.13

Consider the one-dimensional potential

$$U(x) = \frac{-Wd^2(x^2 + d^2)}{x^4 + 8d^4} \tag{2.106}$$

Sketch the potential and discuss the motion at various values of x. Is the motion bounded or unbounded? Where are the equilibrium values? Are they stable or unstable? Find the turning points for $E = -W/8$. The value of W is a positive constant.

Solution: Rewrite the potential as

$$Z(y) = \frac{U(x)}{W} = \frac{-(y^2 + 1)}{y^4 + 8} \qquad \text{where } y = \frac{x}{d} \qquad (2.107)$$

First, find the equilibrium points, which will help guide us in sketching the potential.

$$\frac{dZ}{dy} = \frac{-2y}{y^4 + 8} + \frac{4y^3(y^2 + 1)}{(y^4 + 8)^2} = 0$$

This is reduced to

$$y(y^4 + 2y^2 - 8) = 0$$

$$y(y^2 + 4)(y^2 - 2) = 0$$

$$y_0^2 = 2, 0$$

so

$$\left.\begin{array}{l} x_{01} = 0 \\[2mm] x_{02} = \sqrt{2}\,d \\[2mm] x_{03} = -\sqrt{2}\,d \end{array}\right\} \qquad (2.108)$$

There are three equilibrium points. We sketch $U(x)/W$ versus x/d in Figure 2-16.

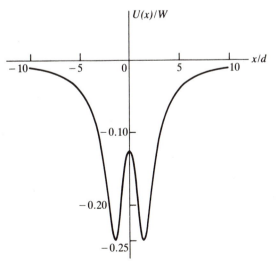

FIGURE 2-16

The equilibrium is stable at x_{02} and x_{03} but unstable at x_{01}. The motion is bounded for all energies $E < 0$. We can determine turning points for any energy E by setting $E = U(x)$.

$$E = -\frac{W}{8} = U(y) = \frac{-W(y^2 + 1)}{y^4 + 8} \tag{2.109}$$
$$y^4 + 8 = 8y^2 + 8$$
$$y^4 = 8y^2$$
$$y = \pm 2\sqrt{2}, 0 \tag{2.110}$$

The turning points for $E = -W/8$ are $x = -2\sqrt{2}d$ and $+2\sqrt{2}d$, as well as $x = 0$ —which is the unstable equilibrium point.

2.7 ROCKET MOTION

Rocket motion is an interesting application of elementary Newtonian dynamics. The two cases we examine are (1) rocket motion in free space and (2) the vertical ascent of rockets under gravity. The first case requires an application of the conservation of linear momentum. The second case requires a more complicated application of Newton's Second Law.

Rocket Motion in Free Space

We assume here that the rocket (space ship) moves under the influence of no external forces. We choose a closed system in which Newton's Second Law can be applied. In outer space, the motion of the space ship must depend entirely on its own energy. It moves by the reaction of ejecting mass at high velocities. To conserve linear momentum, the space ship will have to move in the opposite direction. The diagram of the space ship motion is shown in Figure 2–17. At some time t, the instantaneous total mass of the space ship is m, and the instantaneous speed of the space ship is v with respect to an inertial reference system. We assume that all

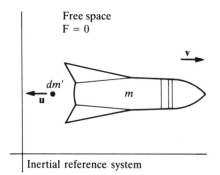

FIGURE 2-17

motion is in the x direction and eliminate the vector notation. During a time interval dt, a positive mass dm' is ejected from the rocket engine with a speed $-u$ with respect to the space ship. Immediately after the mass dm' is ejected, the space ship's mass and speed are $m - dm'$ and $v + dv$, respectively.

$$\text{Initial momentum} = mv \qquad (\text{at time } t) \tag{2.111}$$

$$\text{Final momentum} = (m - dm')(v + dv) + dm'(v - u) \qquad (\text{at time } t + dt)$$
$$\underset{\text{Space ship less } dm'}{} \quad \underset{\text{Rocket exhaust } dm'}{} \tag{2.112}$$

Notice that the speed of the ejected mass dm' with respect to the reference system is $v - u$. The conservation of linear momentum requires that Equations 2.111 and 2.112 be equal. There are no external forces ($F_{\text{ext}} = 0$).

$$p_{\text{initial}} = p_{\text{final}}$$

$$p(t) = p(t + dt)$$

$$mv = (m - dm')(v + dv) + dm'(v - u) \tag{2.113}$$

$$mv = mv + m\,dv - v\,dm' - dm'\,dv + v\,dm' - u\,dm'$$

$$m\,dv = u\,dm'$$

$$dv = u\frac{dm'}{m} \tag{2.114}$$

where we have neglected the product of two differentials $dm'\,dv$. We have considered dm' to be a positive mass ejected from the space ship. The change in mass of the space ship itself is dm, where

$$dm = -dm' \tag{2.115}$$

and

$$dv = -u\frac{dm}{m} \tag{2.116}$$

because dm must be negative. Let m_0 and v_0 be the initial mass and speed of the space ship, respectively, and integrate Equation 2.116 to its final values m and v.

$$\int_{v_0}^{v} dv = -u\int_{m_0}^{m} \frac{dm}{m}$$

$$v - v_0 = u\ln\left(\frac{m_0}{m}\right) \tag{2.117}$$

$$v = v_0 + u\ln\left(\frac{m_0}{m}\right) \tag{2.118}$$

The exhaust velocity u is assumed constant. Thus, to maximize the space ship's speed, we need to maximize the exhaust velocity u and the ratio m_0/m.

Because the terminal speed is limited by the ratio m_0/m, engineers have constructed multistage rockets. The minimum mass (less fuel) of the space ship is limited by structural material. However, if the fuel container itself is jettisoned after its fuel has been burned, the mass of the remaining space ship is less. The space ship can contain two or more fuel containers, each of which can be jettisoned.

For example, let

m_0 = Initial total mass of space ship

$m_1 = m_a + m_b$

m_a = Mass of first-stage payload

m_b = Mass of first-stage fuel containers, etc.

v_1 = Terminal speed of first stage at "burnout" after all

fuel is burned

$$v_1 = v_0 + u \ln\left(\frac{m_0}{m_1}\right) \tag{2.119}$$

At burnout, the terminal speed v_1 of the first stage is reached, and the mass m_b is released into space. Next, the second-stage rocket ignites with the same exhaust velocity, and we have

m_a = Initial total mass of space ship second stage

$m_2 = m_c + m_d$

m_c = Mass of second-stage payload

m_d = Mass of second-stage fuel container, etc.

v_1 = Initial speed of second stage

v_2 = Terminal speed of second stage at burnout

$$v_2 = v_1 + u \ln\left(\frac{m_a}{m_2}\right) \tag{2.120}$$

$$v_2 = v_0 + u \ln\left(\frac{m_0 m_a}{m_1 m_2}\right) \tag{2.121}$$

The product $(m_0 m_a/m_1 m_2)$ can be made much larger than just m_0/m_1. Multistage rockets are more commonly used in ascent under gravity than in free space.

We have seen that the space ship is propelled as a result of the conservation of linear momentum. But engineers and scientists like to refer to the force term as rocket "thrust." If we multiply Equation 2.116 by m and divide by dt, we have

$$m\frac{dv}{dt} = -u\frac{dm}{dt} \tag{2.122}$$

Since the left side of this equation "appears" as ma(force), the right side is called thrust:

$$\text{Thrust} \equiv -u\frac{dm}{dt} \qquad \textbf{(2.123)}$$

Because dm/dt is negative, the thrust is actually positive.

Vertical Ascent Under Gravity

The actual motion of a rocket attempting to leave the Earth's gravitational field is quite complicated. For analytical purposes, we begin by making several assumptions. The rocket will have only vertical motion, with no horizontal component. We neglect air resistance and assume that the acceleration of gravity is constant with height. We also assume that the burn rate of the fuel is constant. All these factors that are neglected can reasonably be included with a numerical analysis by computer.

We can use the results of the previous case of rocket motion in free space, but we no longer have $F_{\text{ext}} = 0$. The geometry is shown in Figure 2-18. We again have dm' as positive, with $dm = -dm'$. The external force F_{ext} is

$$F_{\text{ext}} = \frac{d}{dt}(mv)$$

or

$$F_{\text{ext}}\, dt = d(mv) = dp = p(t + dt) - p(t) \qquad \textbf{(2.124)}$$

over a small differential time.

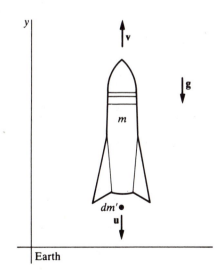

FIGURE 2-18

For the space ship system, we found the initial and final momenta in Equations 2.111–2.116. We now use the results leading up to Equation 2.116 to obtain

$$p(t + dt) - p(t) = mdv + udm \tag{2.125}$$

In free space, $F_{ext} = 0$, but in ascent, $F_{ext} = -mg$. Combining Equations 2.124 and 2.125 gives

$$F_{ext}\, dt = -mg\, dt = m\, dv + u\, dm$$

$$-mg = m\dot{v} + u\dot{m} \tag{2.126}$$

Because the fuel burn rate is constant, let

$$\dot{m} = \frac{dm}{dt} = -\alpha, \qquad \alpha > 0 \tag{2.127}$$

and Equation 2.126 becomes

$$dv = \left(-g + \frac{\alpha}{m} u\right) dt \tag{2.128}$$

This equation, however, has three unknowns (v, m, t), so we use Equation 2.127 to eliminate time, giving

$$dv = \left(\frac{g}{\alpha} - \frac{u}{m}\right) dm \tag{2.129}$$

Assume the initial and final values of the velocity to be 0 and v respectively and of the mass m_0 and m respectively, so that

$$\int_0^v dv = \int_{m_0}^m \left(\frac{g}{\alpha} - \frac{u}{m}\right) dm$$

$$v = -\frac{g}{\alpha}(m_0 - m) + u \ln\left(\frac{m_0}{m}\right) \tag{2.130}$$

We can integrate Equation 2.127 to find the time:

$$\int_{m_0}^m dm = -\alpha \int_0^t dt$$

$$m_0 - m = \alpha t \tag{2.131}$$

Equation 2.130 becomes

$$v = -gt + u \ln\left(\frac{m_0}{m}\right) \tag{2.132}$$

We could continue with Equation 2.130 and integrate once more to determine the height of the rocket (Problem 2-52). Such integrations are tedious, and the problem is more easily handled by computer methods. Even at burnout, the rocket will continue rising because it still has an upward velocity. Eventually, with the preceding assumptions, the gravitational force will stop the rocket (because we assumed a constant g not decreasing with height).

An interesting situation occurs if the exhaust velocity u is not sufficiently great to make v in Equation 2.132 positive. In this case, the rocket would remain on the ground. This situation occurs because of the limits of integration we assumed leading to Equation 2.130. We would need to burn off sufficient fuel before the rocket thrust would lift it off the ground (see Problem 2-54). Of course, rockets are not designed this way; they are made to lift off as the rockets reach a full burn rate.

EXAMPLE 2.14 --

Consider the first stage of a Saturn V rocket used for the Apollo moon program. The initial mass is 2.8×10^6 kg, and the mass of the first-stage fuel is 2.1×10^6 kg. Assume a mean thrust of 37×10^6 N. The exhaust velocity is 2,600 m/s. Calculate the final speed of the first stage at burnout. Using the result of Problem 2-52 (Equation 2.133) also calculate the vertical height at burnout.

Solution: From the thrust (Equation 2.123), we can determine the fuel burn rate:

$$\frac{dm}{dt} = \frac{\text{thrust}}{-u} = \frac{37 \times 10^6 \text{N}}{-2600 \text{ m/s}} = -1.42 \times 10^4 \text{ kg/s}$$

The final rocket mass is $(2.8 \times 10^6 \text{ kg} - 2.1 \times 10^6 \text{ kg})$ or 0.7×10^6 kg. We can determine the rocket speed at burnout (v_b) using Equation 2.130:

$$v_b = -\frac{9.8 \text{ m/s}^2 (2.1 \times 10^6 \text{ kg})}{1.42 \times 10^4 \text{ kg/s}} + 2600 \text{ m/s} \ln\left[\frac{2.8 \times 10^6 \text{ kg}}{0.7 \times 10^6 \text{ kg}}\right]$$

$$v_b = 2.16 \times 10^3 \text{ m/s}$$

The time to burnout t_b, from Equation 2.131, is

$$t_b = \frac{m_0 - m}{\alpha} = \frac{2.1 \times 10^6 \text{ kg}}{1.42 \times 10^4 \text{ kg/s}} = 148 \text{ s}$$

or about $2\frac{1}{2}$ min.

We use the result of Problem 2-52 to obtain the height at burnout y_b:

$$y_b = ut_b - \frac{1}{2} g t_b^2 - \frac{mu}{\alpha} \ln\left(\frac{m_0}{m}\right) \tag{2.133}$$

$$y_b = (2600 \text{ m/s})(148 \text{ s}) - \frac{1}{2}(9.8 \text{ m/s}^2) \cdot (148 \text{ s})^2$$

$$- \frac{(0.7 \times 10^6 \text{ kg}) \cdot (2600 \text{ m/s})}{1.42 \times 10^4 \text{ kg/s}} \ln\left(\frac{2.8 \times 10^6 \text{ kg}}{0.7 \times 10^6 \text{ kg}}\right)$$

$$y_b = 9.98 \times 10^4 \text{m} \approx 100 \text{ km}$$

The actual height is only about two thirds of this value.

E X A M P L E   **2.15**   --

The space shuttle has been described as an amazing flying machine. Although the shuttle is too complex for us to describe fully, we can examine its launch ascent by making various assumptions. Two primary systems lift the orbiter from the ground into orbit. The space shuttle main engines (SSME) consist of three engines within the orbiter, burning liquid hydrogen (LH_2) and liquid oxygen (LO_2) stored in the large external tank attached to the orbiter at lift-off. The main thrust comes from two solid-fuel rocket boosters (SRB) located on the sides of the external tank. Both systems burn during the first 2 min of launch, after which the fuel in the SRB is consumed, and the SRB casings are jettisoned (Figure 2-19) to be retrieved later from the ocean. For the next $6\frac{1}{2}$

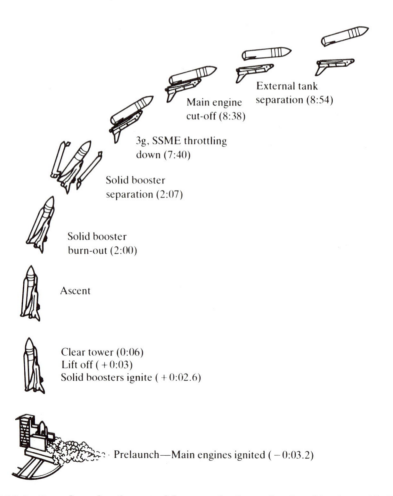

External tank
separation (8:54)

Main engine
cut-off (8:38)

3g, SSME throttling
down (7:40)

Solid booster
separation (2:07)

Solid booster
burn-out (2:00)

Ascent

Clear tower (0:06)
Lift off (+0:03)
Solid boosters ignite (+0:02.6)

Prelaunch—Main engines ignited (−0:03.2)

F I G U R E 2 - 19   Launch and ascent of the space shuttle, putting the orbiter into orbit. The time is listed in parentheses beside the event (minutes:seconds).

min, the SSMEs keep the space shuttle rising and provide most of the horizontal velocity necessary to enter orbit.

Although not strictly correct, we assume that the rocket is moving vertically during the SRB burning (2 min) and calculate the velocity and height of the shuttle. We also assume that during the next $6\frac{1}{2}$ min of flight the shuttle is moving 45° from horizontal, and we calculate its orbital velocity.

The specifications of the three main parts of the space shuttle for a typical flight are*

> Orbiter (actual spacecraft/aircraft):
>> Mass with payload: 105,000 kg
>
> External tank (fuel for SSME):
>> Empty tank: 36,000 kg
>>
>> LH_2 and LO_2 fuel: 7.2×10^5 kg
>>
>> Total loaded mass: 7.56×10^5 kg
>
> SRBs (two):
>> Total empty mass: 1.68×10^5 kg
>>
>> Solid fuel: 1.01×10^6 kg
>>
>> Total loaded mass: 1.18×10^6 kg
>
> Total mass of fueled space shuttle at launch: 2.04×10^6 kg
>
> Thrust of all SSMEs: 5.0×10^6 N
>
> Thrust of all SRBs: 23.6×10^6 N
>
> Total thrust at launch: 28.6×10^6 N

Although the SSMEs burn at various levels of power to keep the acceleration below 3g (because of structural limitations, not because of the astronauts), we assume that the SSMEs burn at a constant rate throughout the entire $8\frac{1}{2}$ min.

Solution: During the first 2 min, when both engine systems are firing, we are not too incorrect if we add the thrust from both systems, find the total dm/dt, and use Equation 2.123 to find an average exhaust velocity u.
First 120 sec:

$$\Delta m(SRB) = 1.01 \times 10^6 \text{ kg}$$

$$\Delta m(SSME) = \left(\frac{2}{8.5}\right) 7.2 \times 10^5 \text{ kg} = 1.7 \times 10^5 \text{ kg}$$

$$\frac{dm}{dt} = -\frac{1.18 \times 10^6 \text{kg}}{120 \text{ s}} = -9800 \text{ kg/s}$$

---

* Specifications are from K. M. Joels, G. P. Kennedy, and D. Larkin, *The Space Shuttle Operator's Manual* (New York: Ballantine, 1982).

From Equation 2.127, $\alpha = 9800$ kg/s.

$$u = \frac{\text{Thrust}}{-dm/dt} = \frac{28.6 \times 10^6 \text{ N}}{9800 \text{ kg/s}} = 2900 \text{ m/s}$$

$$m_0 = 2.04 \times 10^6 \text{ kg}$$

$$\text{Fuel burned} = 1.18 \times 10^6 \text{ kg}$$

$$m = 0.86 \times 10^6 \text{ kg}$$

We calculate the burnout velocity using Equation 2.130:

$$v = -\frac{9.8 \text{ m/s}^2 (1.18 \times 10^6 \text{ kg})}{9800 \text{ kg/s}} + (2900 \text{ m/s}) \ln\left(\frac{2.04}{0.86}\right)$$

$$v = 1325 \text{ m/s} \approx 2900 \text{ MPH}$$

The height can be determined from Equation 2.133:

$$y = 2900 \text{ m/s } (120 \text{ s}) - \left(\frac{1}{2}\right) 9.8 \text{ m/s}^2 (120 \text{ s})^2$$

$$- \frac{(0.86 \times 10^6 \text{ kg}) \cdot (2900 \text{ m/s})}{9800 \text{ kg/s}} \ln\left(\frac{2.04}{0.86}\right) = 58 \text{ km} \approx 36 \text{ miles}$$

The actual figures at SRB burnout are about $v = 1300$ m/s and $y = 45$ km.

By the time the SRB casings are jettisoned, the shuttle is starting to turn around to prepare to enter orbit. In the horizontal direction, the flight occurs as if it were in free space (neglecting air resistance, which is small at an altitude of 50 km). The thrust of the SSMEs is not enough to accelerate the shuttle vertically, but it keeps the shuttle from slowing down rapidly. Let us assume that the ascent angle now averages 45° and use Equation 2.118 to calculate the horizontal velocity.

$$m_0 \text{ (after SRB casings are expelled): } 6.9 \times 10^5 \text{ kg}$$

$$\text{Fuel remaining in external tank: } 5.5 \times 10^5 \text{ kg}$$

$$m \text{ (after SSME shutdown): } 1.4 \times 10^5 \text{ kg}$$

We calculate the SSME exhaust velocity using Equation 2.123:

$$u = -\frac{\text{thrust}}{dm/dt}$$

$$\frac{dm}{dt}(\text{SSME}) = -\frac{5.5 \times 10^5 \text{ kg}}{6.5 \text{ min } \dfrac{60 \text{ s}}{\text{min}}} = -1400 \text{ kg/s}$$

$$u = \frac{-5 \times 10^6 \text{ N}}{-1400 \text{ kg/s}} = 3550 \text{ m/s}$$

$$u_{\text{horz}} = 3550 \text{ m/s } (\cos 45°) = 2500 \text{ m/s}$$

$$u_{horz} = (2500 \text{ m/s}) \ln\left(\frac{6.9}{1.4}\right)$$

$$= 4000 \text{ m/s} \approx 8900 \text{ MPH}$$

After the fuel in the external tank is exhausted, the SSMEs shut down and the external tank is jettisoned. The orbital maneuvering system, consisting of two 27,000-N thrust engines, eventually places the orbiter in the correct orbit at a speed of about 8,000 m/s and altitude of 400 km.

2.8 LIMITATIONS OF NEWTONIAN MECHANICS

In this chapter, we have introduced such concepts as position, time, momentum, and energy. We have implied that these are all measurable quantities and that they can be specified with any desired accuracy, depending only on the degree of sophistication of our measuring instruments. Indeed, this implication appears to be verified by our experience with all macroscopic objects. At any given instant of time, for example, we can measure with great precision the position of, say, a planet in its orbit about the sun. A series of such measurements allows us to determine (also with great precision) the planet's velocity at any given position.

When we attempt to make precise measurements on microscopic objects, however, we find a fundamental limitation in the accuracy of the results. For example, we can conceivably measure the position of an electron by scattering a light photon from the electron. The wave character of the photon precludes an *exact* measurement, and we can determine the position of the electron only within some uncertainty Δx related to the *extent* (i.e., the wavelength) of the photon. By the very act of measurement, however, we have induced a change in the state of the electron, because the scattering of the photon imparts momentum to the electron. This momentum is uncertain by an amount Δp. The product $\Delta x \, \Delta p$ is a measure of the precision with which we can simultaneously determine the electron's position and momentum; $\Delta x \to 0$, $\Delta p \to 0$ implies a measurement with all imaginable precision. It was shown by the German physicist Werner Heisenberg (1901–1976) in 1927 that this product must always be larger than a certain minimum value.* We cannot, then, simultaneously specify both the position *and* momentum of the electron with infinite precision, for if $\Delta x \to 0$, then we must have $\Delta p \to \infty$ for **Heisenberg's uncertainty principle** to be satisfied.

The minimum value of $\Delta x \, \Delta p$ is of the order of 10^{-34} J · s. This is extremely small by macroscopic standards, so for laboratory-scale objects there is no practical difficulty in performing simultaneous measurements of position and momentum. Newton's laws can therefore be applied as if position and momentum were precisely definable. But because of the uncertainty principle, Newtonian mechanics cannot

---

* This result also applies to the measurement of energy at a particular time, in which case the product of the uncertainties is $\Delta E \, \Delta t$ (which has the same dimensions as $\Delta x \, \Delta p$).

be applied to microscopic systems. To overcome these fundamental difficulties in the Newtonian system, a new method of dealing with microscopic phenomena was developed, beginning in 1926. The work of Erwin Schrödinger (1887–1961), Heisenberg, Max Born (1872–1970), Paul Dirac (1902–1984), and others subsequently placed this new discipline on a firm foundation. Newtonian mechanics, then, is perfectly adequate for describing large-scale phenomena. But we need the new mechanics (quantum mechanics) to analyze processes in the atomic domain. As the size of the system increases, quantum mechanics goes over into the limiting form of Newtonian mechanics.

In addition to the fundamental limitations of Newtonian mechanics as applied to microscopic objects, there is another inherent difficulty in the Newtonian scheme—one that rests on the concept of time. In the Newtonian view, time is *absolute*, that is, it is supposed that it is always possible to determine unambiguously whether two events have occurred simultaneously or whether one has preceded the other. To decide on the time sequence of events, the two observers of the events must be in instantaneous communication, either through some system of signals or by establishing two exactly synchronous clocks at the points of observation. But the setting of two clocks into exact synchronism requires the knowledge of the time of transit of a signal *in one direction* from one observer to the other. (We could accomplish this if we already had two synchronous clocks, but this is a circular argument.) When we actually measure signal velocities, however, we always obtain an *average* velocity for propagation in opposite directions. And to devise an experiment to measure the velocity in only *one* direction inevitably leads to the introduction of some new assumption that we cannot verify before the experiment.

We know that instantaneous communication by signaling is impossible: Interactions between material bodies propagate with finite velocity, and an interaction of some sort must occur for a signal to be transmitted. The *maximum* velocity with which any signal can be propagated is that of light in free space: $c \cong 3 \times 10^8$ m/s.*

The difficulties in establishing a time scale between separate points lead us to believe that time is, after all, not absolute and that space and time are somehow intimately related. The solution to the dilemma was found during the period 1904–1905 by Hendrik Lorenz (1853–1928), Henri Poincaré (1854–1912), and Albert Einstein (1879–1955) and is embodied in the **special theory of relativity** (see Chapter 14).

Newtonian mechanics is therefore subject to fundamental limitations when small distances or high velocities are encountered. Difficulties with Newtonian mechanics may also occur when massive objects or enormous distances are involved. A practical limitation also occurs when the number of bodies constituting the system is large. In Chapter 8, we see that we cannot obtain a general solution

---

* The speed of light has now been defined to be 299,792,458.0 m/s to make comparisons of other measurements more standard. The meter is now defined as the distance traveled by light in a vacuum during a time interval of 1/299,792,458 of a second.

in closed form for the motion of a system of more than two interacting bodies even for the relatively simple case of gravitational interaction. To calculate the motion in a three-body system, we must resort to a numerical approximation procedure. Although such a method is in principle capable of any desired accuracy, the labor involved is considerable. The motion in even more complex systems (for example, the system composed of all the major objects in the solar system) can likewise be computed, but the procedure rapidly becomes too unwieldy to be of much use for any larger system. To calculate the motion of the individual molecules in, say, a cubic centimeter of gas containing $\approx 10^{19}$ molecules is clearly out of the question. A successful method of calculating the *average* properties of such systems was developed in the latter part of the nineteenth century by Boltzmann, Maxwell, Gibbs, Liouville, and others. These procedures allowed the dynamics of systems to be calculated from probability theory, and a *statistical mechanics* was evolved. Some comments regarding the formulation of statistical concepts in mechanics are found in Section 7.13.

P R O B L E M S

ADD 3 MORE

2-1. Suppose that the force acting on a particle is factorable into one of the following forms:

(a) $F(x_i, t) = f(x_i)g(t)$ **(b)** $F(\dot{x}_i, t) = f(\dot{x}_i)g(t)$ **(c)** $F(x_i, \dot{x}_i) = f(x_i)g(\dot{x}_i)$

For which cases are the equations of motion integrable?

2-2. A particle of mass m is constrained to move on the surface of a sphere of radius R by an applied force $\mathbf{F}(\theta, \phi)$. Write the equation of motion.

2-3. If a projectile is fired from the origin of the coordinate system with an initial velocity v_0 and in a direction making an angle α with the horizontal, calculate the time required for the projectile to cross a line passing through the origin and making an angle $\beta < \alpha$ with the horizontal.

2-4. A clown is juggling four balls simultaneously. Students use a video tape to determine that it takes the clown 0.9 s to cycle each ball through his hands (including catching, transferring, and throwing) and to be ready to catch the next ball. What is the minimum vertical speed the clown must throw up each ball?

2-5. A jet fighter pilot knows he is able to withstand an acceleration of $9g$ before blacking out. The pilot points his plane vertically down while traveling at Mach 3 speed and intends to pull up in a circular maneuver before crashing into the ground. **(a)** Where does the maximum acceleration occur in the maneuver? **(b)** What is the minimum circle the pilot can take?

2-6. In the blizzard of '88, a rancher was forced to drop hay bales from an airplane to feed her cattle. The plane flew horizontally at 160 km/hr and dropped the bales from a height of 80 m above the flat range. **(a)** She wanted the bales of hay to land 30 m behind the cattle so as to not hit them. Where should she push the bales out of the airplane? **(b)** To not hit the cattle, what is the largest time error she could make while pushing the bales out of the airplane? Ignore air resistance.

2-7. Include air resistance for the bales of hay in the previous problem. A bale of hay has a mass of about 30 kg and an average area of about 0.2 m². Let the resistance be proportional to the square of the speed and let $c_W = 0.8$. Plot the trajectories with a computer if the hay bales land 30 m behind the cattle for both including air resistance and not. If the bales of hay were released at the same time in the two cases, what is the distance between landing positions of the bales?

2-8. A projectile is fired with a velocity v_0 such that it passes through two points both a distance h above the horizontal. Show that if the gun is adjusted for maximum range, the separation of the points is

$$d = \frac{v_0}{g}\sqrt{v_0^2 - 4gh}$$

2-9. Consider a projectile fired vertically in a constant gravitational field. For the same initial velocities, compare the times required for the projectile to reach its maximum height (**a**) for zero resisting force, (**b**) for a resisting force proportional to the instantaneous velocity of the projectile.

2-10. Repeat Example 2.4 by performing a calculation using a computer to solve Equation 2.22. Use the following values: $m = 1$ kg, $v_0 = 10$ m/s, $x_0 = 0$, and $k = 0.1$ s⁻¹. Make plots of v versus t, x versus t, and v versus x. Compare with the results of Example 2.4 to see if your results are reasonable.

2-11. Consider a particle of mass m whose motion starts from rest in a constant gravitational field. If a resisting force proportional to the square of the velocity (i.e., kmv^2) is encountered, show that the distance s the particle falls in accelerating from v_0 to v_1 is given by

$$s(v_0 \to v_1) = \frac{1}{2k} \ln\left[\frac{g - kv_0^2}{g - kv_1^2}\right]$$

2-12. A particle is projected vertically upward in a constant gravitational field with an initial speed v_0. Show that if there is a retarding force proportional to the square of the instantaneous speed, the speed of the particle when it returns to the initial position is

$$\frac{v_0 v_t}{\sqrt{v_0^2 + v_t^2}}$$

where v_t is the terminal speed.

2-13. A particle moves in a medium under the influence of a retarding force equal to $mk(v^3 + a^2 v)$, where k and a are constants. Show that for any value of the initial speed the particle will never move a distance greater than $\pi/2ka$ and that the particle comes to rest only for $t \to \infty$.

2-14. A projectile is fired with initial speed v_0 at an elevation angle of α up a hill of slope β ($\alpha > \beta$).
(**a**) How far up the hill will the projectile land?
(**b**) At what angle α will the range be a maximum?
(**c**) What is the maximum range?

2-15. A particle of mass m slides down an inclined plane under the influence of gravity. If the motion is resisted by a force $f = kmv^2$, show that the time required to move a distance *d* after starting from rest is

$$t = \frac{\cosh^{-1}(e^{kd})}{\sqrt{kg}\,\sin\theta}$$

where θ is the angle of inclination of the plane.

2-16. A particle is projected with an initial velocity v_0 up a slope that makes an angle α with the horizontal. Assume frictionless motion and find the time required for the particle to return to its starting position.

2-17. A strong softball player smacks the ball at a height of 0.7 m above home plate. The ball leaves the player's bat at an elevation angle of 35^0 and travels toward a fence 2 m high and 60 m away in center field. What must the initial speed of the softball be to clear the center field fence? Ignore air resistance.

2-18. Include air resistance proportional to the square of the ball's speed in the previous problem. Let the drag coefficient be $c_W = 0.5$, the softball radius be 5 cm and the mass be 200 g. **(a)** Find the initial speed of the softball needed now to clear the fence. **(b)** For this speed, find the initial elevation angle that allows the ball to most easily clear the fence. By how much does the ball now vertically clear the fence?

2-19. If a projectile moves such that its distance from the point of projection is always increasing, find the maximum angle above the horizontal with which the particle could have been projected. (Assume no air resistance.)

2-20. A gun fires a projectile of mass 10 kg of the type to which the curves of Figure 2-3 apply. The muzzle velocity is 140 m/s. Through what angle must the barrel be elevated to hit a target on the same horizontal plane as the gun and 1000 m away? Compare the results with those for the case of no retardation.

2-21. Show directly that the time rate of change of the angular momentum about the origin for a projectile fired from the origin (constant *g*) is equal to the moment of force (or torque) about the origin.

2-22. The motion of a charged particle in an electromagnetic field can be obtained from the **Lorentz equation*** for the force on a particle in such a field. If the electric field vector is **E** and the magnetic field vector is **B**, the force on a particle of mass *m* that carries a charge *q* and has a velocity **v** is given by

$$F = qE + qv \times B$$

where we assume that $v \ll c$(speed of light).
(a) If there is no electric field and if the particle enters the magnetic field in a direction perpendicular to the lines of magnetic flux, show that the trajectory is a circle with radius

$$r = \frac{mv}{qB} = \frac{v}{\omega_c}$$

where $\omega_c \equiv qB/m$ is the *cyclotron frequency.*
(b) Choose the *z*-axis to lie in the direction of **B** and let the plane containing **E** and **B** be the *yz*−plane. Thus

$$B = Bk, \quad E = E_y e_j + E_z k$$

---

* See, for example, Heald and Marion, *Classical Electromagnetic Radiation* (95, Section 1.7).

Show that the z component of the motion is given by

$$z(t) = z_0 + \dot{z}_0 t + \frac{qE_z}{2m} t^2$$

where

$$z(0) \equiv z_0 \quad \text{and} \quad \dot{z}(0) \equiv \dot{z}_0$$

(c) Continue the calculation and obtain expressions for $\dot{x}(t)$ and $\dot{y}(t)$. Show that the time averages of these velocity components are

$$\langle \dot{x} \rangle = \frac{E_y}{B}, \quad \langle \dot{y} \rangle = 0$$

(Show that the motion is periodic and then average over one complete period.)

(d) Integrate the velocity equations found in (c) and show (with the initial conditions $x(0) = -A/\omega_c$, $\dot{x}(0) = E_y/B$, $y(0) = 0$, $\dot{y}(0) = A$) that

$$x(t) = \frac{-A}{\omega_c} \cos \omega_c t + \frac{E_y}{B} t, \quad y(t) = \frac{A}{\omega_c} \sin \omega_c t$$

These are the parametric equations of a trochoid. Sketch the projection of the trajectory on the xy − plane for the cases (i) $A > |E_y/B|$, (ii) $A < |E_y/B|$, and (iii) $A = |E_y/B|$. (The last case yields a cycloid.)

2-23. A particle of mass $m = 1$ kg is subjected to a one-dimensional force $F(t) = kte^{-\alpha t}$, where $k = 1$N/s and $\alpha = 0.5$ s^{-1}. If the particle is initially at rest, calculate and plot with the aid of a computer the position, speed, and acceleration of the particle as a function of time.

2-24. A skier weighing 90 kg starts from rest down a hill inclined at 17°. He skis 100 m down the hill and then coasts for 70 m along level snow until he stops. Find the coefficient of kinetic friction between the skis and the snow. What velocity does the skier have at the bottom of the hill?

2-25. A block of mass m slides down a frictionless incline (Figure 2-A). The block is released a height h above the bottom of the loop.

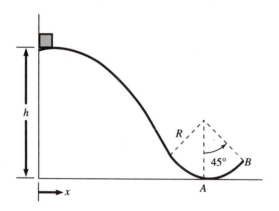

FIGURE 2-A

(a) What is the force of the inclined track on the block at the bottom (point *A*)?
(b) What is the force of the track on the block at point *B*?
(c) At what speed does the block leave the track?
(d) How far away from point *A* does the block land on level ground?
(e) Sketch the potential energy *U(x)* of the block. Indicate the total energy on the sketch.

2-26. A child slides a block of mass 2 kg along a slick kitchen floor. If the initial speed is 4 m/s and the block hits a spring with spring constant 6 N/m, what is the maximum compression of the spring? What is the result if the block slides across 2 m of a rough floor that has $\mu_k = 0.2$?

2-27. A rope having a total mass of 0.4 kg and total length 4 m has 0.6 m of the rope hanging vertically down off a work bench. How much work must be done to place all the rope on the bench?

2-28. A superball of mass *M* and a marble of mass *m* are dropped from a height *h* with the marble just on top of the superball. A superball has a coefficient of restitution of nearly 1 (i.e., its collision is essentially elastic). Ignore the sizes of the superball and marble. The superball collides with the floor, rebounds, and smacks the marble, which moves back up. How high does the marble go if all the motion is vertical? How high does the superball go?

2-29. An automobile driver traveling down an 8% grade slams on his brakes and skids 30 m before hitting a parked car. A lawyer hires an expert who measures the coefficient of kinetic friction between the tires and road to be $\mu_k = 0.45$. Is the lawyer correct to accuse the driver of exceeding the 25-MPH speed limit? Explain.

2-30. A student drops a water-filled balloon from the roof of the tallest building in town trying to hit her roommate on the ground (who is too quick). The first student ducks back but hears the water splash 4.021 s after dropping the balloon. If the speed of sound is 331 m/s, find the height of the building, neglecting air resistance.

2-31. In Example 2.10, the initial velocity of the incoming charged particle had no component along the *x*-axis. Show that, even if it had an *x* component, the subsequent motion of the particle would be the same—that only the radius of the helix would be altered.

2-32. Two blocks of unequal mass are connected by a string over a smooth pulley (Figure 2-B). If the coefficient of kinetic friction is μ_k, what angle θ of the incline allows the masses to move at a constant speed?

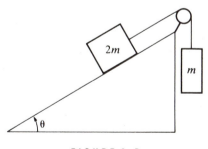

FIGURE 2-B

2-33. Perform a computer calculation for an object moving vertically in air under gravity and experiencing a retarding force proportional to the square of the object's speed (see Equation 2.21). Use variables m for mass and r for the object's radius. All the objects are dropped from rest from the top of a 100-m-tall building. Use a value of $c_W = 0.5$ and make computer plots of height y, speed v, and acceleration a versus t for the following conditions and answer the questions:

(a) A baseball of $m = 0.145$ kg and $r = 0.0366$ m.

(b) A ping-pong ball of $m = 0.0024$ kg and $r = 0.019$ m.

(c) A raindrop of $r = 0.003$ m.

(d) Do all the objects reach their terminal speeds? Discuss the values of the terminal velocities and explain their differences.

(e) Why can a baseball be thrown farther than a ping-pong ball even though the baseball is so much more massive?

(f) Discuss the terminal speeds of big and small raindrops. What are the terminal speeds of raindrops having radii 0.002 m and 0.004 m?

2-34. A particle is released from rest ($y = 0$) and falls under the influence of gravity and air resistance. Find the relationship between v and the distance of falling y when the air resistance is equal to (a) αv and (b) βv^2.

2-35. Perform the numerical calculations of Example 2.7 for the values given in Figure 2-8. Plot both Figures 2-8 and 2-9. Do not duplicate the solution in Appendix H; compose your own solution.

2-36. A gun is located on a bluff of height h overlooking a river valley. If the muzzle velocity is v_0, find the expression for the range as a function of the elevation angle of the gun. Solve numerically for the maximum range out into the valley for a given h and v_0.

2-37. A particle of mass m has speed $v = \alpha/x$, where x is its displacement. Find the force $F(x)$ responsible.

▷▷ **2-38.** The speed of a particle of mass m varies with the distance x as $v(x) = \alpha x^{-n}$. Assume $v(x = 0) = 0$ at $t = 0$. (a) Find the force $F(x)$ responsible. (b) Determine $x(t)$ and (c) $F(t)$.

2-39. A boat with initial speed v_0 is launched on a lake. The boat is slowed by the water by a force $F = -\alpha e^{\beta v}$. (a) Find an expression for the speed $v(t)$. (b) Find the time and (c) distance for the boat to stop.

▷▷ **2-40.** A particle moves in a two-dimensional orbit defined by

$$x(t) = A(2\alpha t - \sin \alpha t)$$
$$y(t) = A(1 - \cos \alpha t)$$

(a) Find the tangential acceleration a_t and normal acceleration a_n as a function of time where the tangential and normal components are taken with respect to the velocity.

(b) Determine at what times in the orbit a_n has a maximum.

2-41. A train moves along the tracks at a constant speed u. A woman on the train throws a ball of mass m straight ahead with a speed v with respect to herself. (a) What is the kinetic energy gain of the ball as measured by a person on the train? (b) by a person standing by the railroad track? (c) How much work is done by the woman throwing the ball and (d) by the train?

2-42. A solid cube of uniform density and sides of b is in equilibrium on top of a cylinder of radius R (Figure 2-C). The planes of four sides of the cube are parallel to the axis of the cylinder. The contact between cube and sphere is perfectly rough. Under what conditions is the equilibrium stable or not stable?

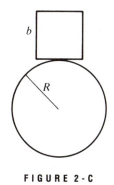

FIGURE 2-C

2-43. A particle is under the influence of a force $F = -kx + kx^3/\alpha^2$, where k and α are constants and k is positive. Determine $U(x)$ and discuss the motion. What happens when $E = (1/4)k\alpha^2$?

2-44. Solve Example 2.12 by using forces rather than energy. How can you determine whether the system equilibrium is stable or unstable?

2-45. Describe how to determine whether an equilibrium is stable or unstable when $(d^2U/dx^2)_0 = 0$.

2-46. Write the criteria for determining whether an equilibrium is stable or unstable when all derivatives up through order n, $(d^nU/dx^n)_0 = 0$.

2-47. Consider a particle moving in the region $x > 0$ under the influence of the potential

$$U(x) = U_0 \left(\frac{a}{x} + \frac{x}{a} \right)$$

where $U_0 = 1$ J and $a = 2$ m. Plot the potential, find the equilibrium points, and determine whether they are maxima or minima.

2-48. A rocket starts from rest in free space by emitting mass. At what fraction of the initial mass is the momentum a maximum?

▷▷ **2-49.** An extremely well-constructed rocket has a mass ratio (m_0/m) of 10. A new fuel is developed that has an exhaust velocity as high as 4,500 m/s. The fuel burns at a constant rate for 300 s. Calculate the maximum velocity of this single-stage rocket, assuming constant acceleration of gravity. If the escape velocity of a particle from the earth is 11.3 km/s, can a similar single-stage rocket with the same mass ratio and exhaust velocity be constructed that can reach the moon?

2-50. A water droplet falling in the atmosphere is spherical. Assume that as the droplet passes through a cloud, it acquires mass at a rate proportional to kA where k is a constant(>0) and A its cross-sectional area. Consider a droplet of initial radius r_0 that enters a cloud with a velocity v_0. Assume no resistive force and show **(a)** that the radius increases linearly with the time, and **(b)** that if r_0 is negligibly small then the speed increases linearly with the time within the cloud.

2-51. A rocket in outer space in a negligible gravitational field starts from rest and accelerates uniformly at a until its final speed is v. The initial mass of the rocket is m_0. How much work does the rocket's engine do?

2-52. Consider a single-stage rocket taking off from the Earth. Show that the height of the rocket at burnout is given by Equation 2.133. How much farther in height will the rocket go after burnout?

2-53. A rocket has an initial mass of m and a fuel burn rate of α (Equation 2.127). What is the minimum exhaust velocity that will allow the rocket to lift off immediately after firing?

2-54. A rocket has an initial mass of 7×10^4 kg and on firing burns its fuel at a rate of 250 kg/s. The exhaust velocity is 2,500 m/s. If the rocket has a vertical ascent from resting on the Earth, how long after the rocket engines fire will the rocket lift off? What is wrong with the design of this rocket?

2-55. Consider a multistage rocket of n stages, each with exhaust speed u. Each stage of the rocket has the same mass ratio at burnout ($k = m_i/m_f$). Show that the final speed of the nth stage is $nu \ln k$.

2-56. To perform a rescue, a lunar landing craft needs to hover just above the surface of the moon, which has a gravitational acceleration of $g/6$. The exhaust velocity is 2,000 m/s, but fuel amounting to only 20% of the total mass may be used. How long can the landing craft hover?

2-57. A new projectile launcher is developed in the year 2013 that can launch a 10^4 kg spherical probe with an initial speed of 6,000 m/s. For testing purposes, objects are launched vertically.
(a) Neglect air resistance and assume the acceleration of gravity is constant. Determine how high the launched object can reach above the surface of the Earth.
(b) If the object has a radius of 20 cm and the air resistance is proportional to the square of the object's speed with $c_W = 0.2$, determine the maximum height reached. Assume the density of air is constant.
(c) Now also include the fact that the acceleration of gravity decreases as the object soars above the Earth. Find the height reached.
(d) Now add the effects of the decrease in air density with altitude to the calculation. We can very roughly represent the air density by $\log_{10}(\rho) = -0.05h + 0.11$ where ρ is the air density in kg/m^3 and h is the altitude above Earth in km. Determine how high the object now goes.

2-58. A new single-stage rocket is developed in the year 2013, having a gas exhaust velocity of 4,000 m/s. The total mass of the rocket is 10^5 kg, with 90% of its mass being fuel. The fuel burns quickly in 100 s at a constant rate. For testing purposes, the rocket is launched vertically at rest from the Earth's surface. Answer parts **(a)** through **(d)** of the previous problem.

3

OSCILLATIONS

3.1 INTRODUCTION

We begin by considering the oscillatory motion of a particle constrained to move in one dimension. We assume that a position of stable equilibrium exists for the particle, and we designate this point as the origin (see Section 2.6). If the particle is displaced from the origin (in either direction), a certain force tends to restore the particle to its original position. An example is an atom in a long molecular chain. The restoring force is, in general, some complicated function of the displacement and perhaps of the particle's velocity or even of some higher time derivative of the position coordinate. We consider here only cases in which the restoring force F is a function only of the displacement: $F = F(x)$.

We assume that the function $F(x)$ that describes the restoring force possesses continuous derivatives of all orders so that the function can be expanded in a Taylor series:

$$F(x) = F_0 + x\left(\frac{dF}{dx}\right)_0 + \frac{1}{2!}x^2\left(\frac{d^2F}{dx^2}\right)_0 + \frac{1}{3!}x^3\left(\frac{d^3F}{dx^3}\right)_0 + \cdots \qquad (3.1)$$

where F_0 is the value of $F(x)$ at the origin ($x = 0$), and $(d^nF/dx^n)_0$ is the value of the nth derivative at the origin. Because the origin is defined to be the equilibrium point, F_0 must vanish, because otherwise the particle would move away from the equilibrium point and not return. If, then, we confine our attention to displacements of the particle that are sufficiently small, we can normally neglect all terms involving x^2 and higher powers of x. We have, therefore, the approximate relation

$$F(x) = -kx \qquad (3.2)$$

where we have substituted $k \equiv -(dF/dx)_0$. Because the restoring force is always directed toward the equilibrium position (the origin), the derivative $(dF/dx)_0$ is negative, and therefore k is a positive constant. Only the first power of the displacement occurs in $F(x)$, so the restoring force in this approximation is a *linear* force.

Physical systems described in terms of Equation 3.2 obey **Hooke's Law**.* One of the classes of physical processes that can be treated by applying Hooke's Law is that involving elastic deformations. As long as the displacements are small and the elastic limits are not exceeded, a linear restoring force can be used for problems of stretched springs, elastic springs, bending beams, and the like. But we must emphasize that such calculations are only approximate, because essentially every real restoring force in nature is more complicated than the simple Hooke's Law force. Linear forces are only useful approximations, and their validity is limited to cases in which the amplitudes of the oscillations are small (but see Problem 3-8).

Damped oscillations, usually resulting from friction, are almost always the type of oscillations that occur in nature. We learn in this chapter how to design an efficiently damped system. This damping of the oscillations may be counteracted if some mechanism supplies the system with energy from an external source at a rate equal to that absorbed by the damping medium. Motions of this type are called **driven** (or **forced**) **oscillations**. Normally sinusoidal, they have important applications in mechanical vibrations as well as in electrical systems.

The extensive discussion of linear oscillatory systems is warranted by the great importance of oscillatory phenomena in many areas of physics and engineering. It is frequently permissible to use the linear approximation in the analysis of such systems. The usefulness of these analyses is due in large measure to the fact that we can usually use *analytical* methods.

When we look more carefully at physical systems, we find that a large number of them are *nonlinear* in general. We will discuss nonlinear systems in Chapter 4.

3.2 SIMPLE HARMONIC OSCILLATOR

The equation of motion for the simple harmonic oscillator may be obtained by substituting the Hooke's Law force into the Newtonian equation $F = ma$. Thus

$$-kx = m\ddot{x} \qquad (3.3)$$

---

* Robert Hooke (1635–1703). The equivalent of this force law was originally announced by Hooke in 1676 in the form of a Latin cryptogram: CEIIINOSSSTTUV. Hooke later provided a translation: *ut tensio sic vis* [the stretch is proportional to the force].

If we define

$$\omega_0^2 \equiv k/m \tag{3.4}$$

Equation 3.3 becomes

$$\boxed{\ddot{x} + \omega_0^2 x = 0} \tag{3.5}$$

According to the results of Appendix C, the solution of this equation can be expressed in either of the forms

$$x(t) = A \sin(\omega_0 t - \delta) \tag{3.6a}$$

$$x(t) = A \cos(\omega_0 t - \phi) \tag{3.6b}$$

where the phases* δ and ϕ differ by $\pi/2$. (An alteration of the phase angle corresponds to a change of the instant that we designate $t = 0$, the origin of the time scale.) Equations 3.6a and b exhibit the well-known sinusoidal behavior of the displacement of the simple harmonic oscillator.

We can obtain the relationship between the total energy of the oscillator and the amplitude of its motion as follows. Using Equation 3.6a for $x(t)$, we find for the kinetic energy,

$$T = \frac{1}{2} m\dot{x}^2 = \frac{1}{2} m\omega_0^2 A^2 \cos^2(\omega_0 t - \delta)$$

$$= \frac{1}{2} kA^2 \cos^2(\omega_0 t - \delta) \tag{3.7}$$

The potential energy may be obtained by calculating the work required to displace the particle a distance x. The incremental amount of work dW necessary to move the particle by an amount dx against the restoring force F is

$$dW = -F\,dx = kx\,dx \tag{3.8}$$

Integrating from 0 to x and setting the work done on the particle equal to the potential energy, we have

$$U = \frac{1}{2} kx^2 \tag{3.9}$$

Then

$$U = \frac{1}{2} kA^2 \sin^2(\omega_0 t - \delta) \tag{3.10}$$

---

* The symbol δ is often used to represent phase angle, and its value is either assigned or determined within the context of an application. Be careful when using equations within this chapter because δ in one application may not be the same as the δ in another. It might be prudent to assign subscripts, for example, δ_1 and δ_2, when using different equations.

Combining the expressions for T and U to find the total energy E, we have

$$E = T + U = \frac{1}{2}kA^2 [\cos^2(\omega_0 t - \delta) + \sin^2(\omega_0 t - \delta)]$$

$$\boxed{E = T + U = \frac{1}{2}kA^2} \tag{3.11}$$

so that the total energy is proportional to the *square of the amplitude*; this is a general result for linear systems. Notice also that E is independent of the time; that is, energy is conserved. (Energy conservation is guaranteed, because we have been considering a system without frictional losses or other external forces.)

The period τ_0 of the motion is defined to be the time interval between successive repetitions of the particle's position and direction of motion. Such an interval occurs when the argument of the sine in Equation 3.6a increases by 2π:

$$\omega_0 \tau_0 = 2\pi \tag{3.12}$$

or

$$\tau_0 = 2\pi \sqrt{\frac{m}{k}} \tag{3.13}$$

From this expression, as well as from Equation 3.6a, it should be clear that ω_0 represents the **angular frequency** of the motion, which is related to the **frequency** ν_0 by*

$$\boxed{\omega_0 = 2\pi\nu_0 = \sqrt{\frac{k}{m}}} \tag{3.14}$$

$$\boxed{\nu_0 = \frac{1}{\tau_0} = \frac{1}{2\pi}\sqrt{\frac{k}{m}}} \tag{3.15}$$

Note that the period of the simple harmonic oscillator is independent of the amplitude (or total energy); a system exhibiting this property is said to be **isochronous**.

For many problems, of which the simple pendulum is the best example, the equation of motion results in $\ddot{\theta} + \omega_0^2 \sin\theta = 0$, where θ is the displacement angle from equilibrium, and $\omega_0 = \sqrt{g/\ell}$, where ℓ is the length of the pendulum arm. We can make this differential equation describe simple harmonic motion by invoking the **small oscillation** assumption. If the oscillations about the equilibrium are small,

---

* Henceforth we shall denote angular frequencies by ω (units: radians per unit time) and frequencies by ν (units: vibrations per unit time or Hertz, Hz). Sometimes ω will be referred to as a "frequency" for brevity, although "angular frequency" is to be understood.

we expand $\sin \theta$ and $\cos \theta$ in power series (see Appendix A) and keep only the lowest terms of importance. This often means $\sin \theta \simeq \theta$ and $\cos \theta \simeq 1 - \theta^2/2$, where θ is measured in radians. If we use the small oscillation approximation for the simple pendulum, the equation of motion above becomes $\ddot{\theta} + \omega_0^2\theta = 0$, an equation that does represent simple harmonic motion. We shall often invoke this assumption throughout this text and in its problems.

3.3 HARMONIC OSCILLATIONS IN TWO DIMENSIONS

We next consider the motion of a particle that is allowed two degrees of freedom. We take the restoring force to be proportional to the distance of the particle from a force center located at the origin and to be directed toward the origin:

$$\mathbf{F} = -k\mathbf{r} \tag{3.16}$$

which can be resolved in polar coordinates into the components

$$\left.\begin{array}{l} F_x = -kr \cos \theta = -kx \\ F_y = -kr \sin \theta = -ky \end{array}\right\} \tag{3.17}$$

The equations of motion are

$$\left.\begin{array}{l} \ddot{x} + \omega_0^2 x = 0 \\ \ddot{y} + \omega_0^2 y = 0 \end{array}\right\} \tag{3.18}$$

where, as before, $\omega_0^2 = k/m$. The solutions are

$$\left.\begin{array}{l} x(t) = A \cos(\omega_0 t - \alpha) \\ y(t) = B \cos(\omega_0 t - \beta) \end{array}\right\} \tag{3.19}$$

Thus, the motion is one of simple harmonic oscillation in each of the two directions, both oscillations having the same frequency but possibly differing in amplitude and in phase. We can obtain the equation for the path of the particle by eliminating the time t between the two equations (Equation 3.19). First we write

$$y(t) = B \cos[\omega_0 t - \alpha + (\alpha - \beta)]$$

$$= B \cos(\omega_0 t - \alpha)\cos(\alpha - \beta) - B \sin(\omega_0 t - \alpha)\sin(\alpha - \beta) \tag{3.20}$$

Defining $\delta \equiv \alpha - \beta$ and noting that $\cos(\omega_0 t - \alpha) = x/A$, we have

$$y = \frac{B}{A} x \cos \delta - B \sqrt{1 - \left(\frac{x^2}{A^2}\right)} \sin \delta$$

or

$$Ay - Bx \cos \delta = - B \sqrt{A^2 - x^2} \sin \delta \tag{3.21}$$

On squaring, this becomes

$$A^2 y^2 - 2ABxy \cos \delta + B^2 x^2 \cos^2 \delta = A^2 B^2 \sin^2 \delta - B^2 x^2 \sin^2 \delta$$

so that

$$B^2x^2 - 2ABxy \cos \delta + A^2y^2 = A^2B^2 \sin^2 \delta \qquad (3.22)$$

If δ is set equal to $\pm\pi/2$, this equation reduces to the easily recognized equation for an ellipse:

$$\frac{x^2}{A^2} + \frac{y^2}{B^2} = 1, \qquad \delta = \pm\pi/2 \qquad (3.23)$$

If the amplitudes are equal, $A = B$, and if $\delta = \pm\pi/2$, we have the special case of circular motion:

$$x^2 + y^2 = A^2, \qquad \text{for } A = B \text{ and } \delta = \pm\pi/2 \qquad (3.24)$$

Another special case results if the phase δ vanishes; then we have

$$B^2x^2 - 2ABxy + A^2y^2 = 0, \qquad \delta = 0$$

Factoring,

$$(Bx - Ay)^2 = 0$$

which is the equation of a straight line:

$$y = \frac{B}{A}x, \qquad \delta = 0 \qquad (3.25)$$

Similarly, the phase $\delta = \pm\pi$ yields the straight line of opposite slope:

$$y = -\frac{B}{A}x, \qquad \delta = \pm\pi \qquad (3.26)$$

The curves of Figure 3-1 illustrate Equation 3.22 for the case $A = B$; $\delta = 90°$ or 270° yields a circle, and $\delta = 180°$ or 360° (0°) yields a straight line. All other values of δ yield ellipses.

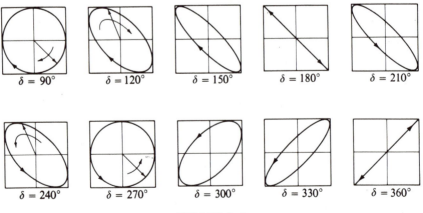

| $\delta = 90°$ | $\delta = 120°$ | $\delta = 150°$ | $\delta = 180°$ | $\delta = 210°$ |

| $\delta = 240°$ | $\delta = 270°$ | $\delta = 300°$ | $\delta = 330°$ | $\delta = 360°$ |

FIGURE 3-1

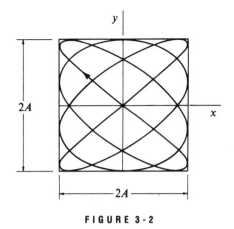

FIGURE 3-2

In the *general* case of two-dimensional oscillations, the angular frequencies for the motions in the x- and y-directions need not be equal, so that Equation 3.19 becomes

$$\left. \begin{array}{l} x(t) = A \cos(\omega_x t - \alpha) \\ y(t) = B \cos(\omega_y t - \beta) \end{array} \right\} \tag{3.27}$$

The path of the motion is no longer an ellipse but a **Lissajous curve**.* Such a curve will be *closed* if the motion repeats itself at regular intervals of time. This will be possible only if the angular frequencies ω_x and ω_y are *commensurable*, that is, if ω_x/ω_y is a rational fraction. Such a case is shown in Figure 3-2, in which $\omega_y = \frac{3}{4}\omega_x$ (also $A = B$ and $\alpha = \beta$). If the ratio of the angular frequencies is not a rational fraction, the curve will be *open*; that is, the moving particle will never pass twice through the same point with the same velocity. In such a case, after a sufficiently long time has elapsed, the curve will pass arbitrarily close to any given point lying within the rectangle $2A \times 2B$ and will therefore "fill" the rectangle.†

The two-dimensional oscillator is an example of a system in which an infinitesimal change can result in a qualitatively different type of motion. The motion will be along a closed path if the two angular frequencies are commensurable. But if the angular frequency ratio deviates from a rational fraction by even an infinitesimal amount, then the path will no longer be closed and it will "fill" the rectangle. For the path to be closed, the angular frequency ratio must be known to be a rational fraction with infinite precision.

If the angular frequencies for the motions in the x- and y-directions are different, the shape of the resulting Lissajous curve strongly depends on the phase difference $\delta \equiv \alpha - \beta$. Figure 3-3 shows the results for the case $\omega_y = 2\omega_x$ for phase differences of 0, $\pi/3$, and $\pi/2$.

---

* First demonstrated in 1857 by the French physicist Jules Lissajous (1822–1880).

† A proof is given, for example, by Haag (Ha62, p. 36).

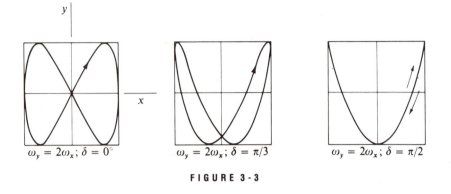

FIGURE 3-3

3.4 PHASE DIAGRAMS

The state of motion of a one-dimensional oscillator, such as that discussed in Section 3.2, will be completely specified as a function of time if *two* quantities are given at one instant of time, that is, the initial conditions $x(t_0)$ and $\dot{x}(t_0)$. (Two quantities are needed because the differential equation for the motion is of second order.) We may consider the quantities $x(t)$ and $\dot{x}(t)$ to be the coordinates of a point in a two-dimensional space, called **phase space**. (In two dimensions, the phase space is a phase *plane*. But for a general oscillator with n degrees of freedom, the phase space is a $2n$-dimensional space.) As the time varies, the point $P(x, \dot{x})$ describing the state of the oscillating particle will move along a certain phase path in the phase plane. For different initial conditions of the oscillator, the motion will be described by different phase paths. Any given path represents the complete time history of the oscillator for a certain set of initial conditions. The totality of all possible phase paths constitutes the **phase portrait** or the **phase diagram** of the oscillator.*

According to the results of the preceding section, we have, for the simple harmonic oscillator,

$$x(t) = A \sin(\omega_0 t - \delta) \tag{3.28a}$$

$$\dot{x}(t) = A\omega_0 \cos(\omega_0 t - \delta) \tag{3.28b}$$

If we eliminate t from these equations, we find for the equation of the path

$$\frac{x^2}{A^2} + \frac{\dot{x}^2}{A^2\omega_0^2} = 1 \tag{3.29}$$

This equation represents a family of ellipses,[†] several of which are shown in Figure 3-4. We know that the total energy E of the oscillator is $\frac{1}{2}kA^2$ (Equation 3.11), and

---

* These considerations are not restricted to oscillating particles or oscillating systems. The concept of phase space is applied extensively in various fields of physics, particularly in statistical mechanics.

[†] The ordinate of the phase plane is sometimes chosen to be $\dot{x}/\omega_0$ instead of $\dot{x}$; the phase paths are then circles.

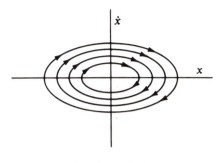

FIGURE 3-4

because $\omega_0^2 = k/m$, Equation 3.29 can be written as

$$\frac{x^2}{2E/k} + \frac{\dot{x}^2}{2E/m} = 1 \tag{3.30}$$

Each phase path, then, corresponds to a definite total energy of the oscillator. This result is expected because the system is conservative (i.e., E = const.).

No two phase paths of the oscillator can cross. If they could cross, this would imply that for a given set of initial conditions $x(t_0)$, $\dot{x}(t_0)$ (i.e., the coordinates of the crossing point), the motion could proceed along different phase paths. But this is impossible because the solution of the differential equation is unique.

If the coordinate axes of the phase plane are chosen as in Figure 3-4, the motion of the **representative point** $P(x, \dot{x})$ will always be in a clockwise direction, because for $x > 0$ the velocity $\dot{x}$ is always decreasing and for $x < 0$ the velocity is always increasing.

To obtain Equations 3.28 for $x(t)$ and $\dot{x}(t)$, we must integrate Equation 3.5, a second-order differential equation:

$$\frac{d^2x}{dt^2} + \omega_0^2 x = 0 \tag{3.31}$$

We can obtain the equation for the phase path, however, by a simpler procedure, because Equation 3.31 can be replaced by the pair of equations

$$\frac{dx}{dt} = \dot{x}, \qquad \frac{d\dot{x}}{dt} = -\omega_0^2 x \tag{3.32}$$

If we divide the second of these equations by the first, we obtain

$$\frac{d\dot{x}}{dx} = -\omega_0^2 \frac{x}{\dot{x}} \tag{3.33}$$

This is a first-order differential equation for $\dot{x} = \dot{x}(x)$, the solution to which is just Equation 3.29. For the simple harmonic oscillator, there is no difficulty in obtaining the general solution for the motion by solving the second-order equation. But in more complicated situations, it is sometimes considerably easier to find directly the equation of the phase path $\dot{x} = \dot{x}(x)$ without proceeding through the calculation of $x(t)$.

3.5 DAMPED OSCILLATIONS

The motion represented by the simple harmonic oscillator is termed a **free oscillation**; once set into oscillation, the motion would never cease. This oversimplifies the actual physical case, in which dissipative or frictional forces would eventually damp the motion to the point that the oscillations would no longer occur. We can analyze the motion in such a case by incorporating into the differential equation a term representing the damping force. It does not seem reasonable that the damping force should, in general, depend on the displacement, but it could be a function of the velocity or perhaps of some higher time derivative of the displacement. It is frequently assumed that the damping force is a linear function of the velocity,* $\mathbf{F}_d = \alpha\mathbf{v}$. We consider here only one-dimensional damped oscillations so that we can represent the damping term by $-b\dot{x}$. The parameter b must be *positive* in order that the force indeed be *resisting*. (A force $-b\dot{x}$ with $b < 0$ would act to *increase* the speed instead of decreasing it as any resisting force must.) Thus, if a particle of mass m moves under the combined influence of a linear restoring force $-kx$ and a resisting force $-b\dot{x}$, the differential equation describing the motion is

$$m\ddot{x} + b\dot{x} + kx = 0 \qquad\qquad (3.34)$$

which we can write as

$$\boxed{\ddot{x} + 2\beta\dot{x} + \omega_0^2 x = 0} \qquad\qquad (3.35)$$

Here $\beta \equiv b/2m$ is the **damping parameter** and $\omega_0 = \sqrt{k/m}$ is the characteristic angular frequency in the absence of damping. The roots of the auxiliary equation are (cf. Equation C.8, Appendix C)

$$\left.\begin{array}{l} r_1 = -\beta + \sqrt{\beta^2 - \omega_0^2} \\ r_2 = -\beta - \sqrt{\beta^2 - \omega_0^2} \end{array}\right\} \qquad\qquad (3.36)$$

The general solution of Equation 3.35 is therefore

$$\boxed{x(t) = e^{-\beta t}\left[A_1 \exp\left(\sqrt{\beta^2 - \omega_0^2}\,t\right) + A_2 \exp\left(-\sqrt{\beta^2 - \omega_0^2}\,t\right)\right]} \qquad (3.37)$$

There are three general cases of interest:

Underdamping: $\omega_0^2 > \beta^2$

Critical damping: $\omega_0^2 = \beta^2$

Overdamping: $\omega_0^2 < \beta^2$

The motion of the three cases is shown schematically in Figure 3-5 for specific initial conditions. We shall see that only the case of underdamping results in oscillatory motion. These three cases are discussed separately.

---

* See Section 2.4 for a discussion of the dependence of resisting forces on velocity.

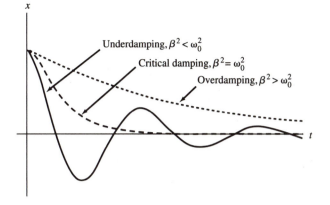

FIGURE 3-5

Underdamped Motion

For the case of underdamped motion, it is convenient to define

$$\omega_1^2 \equiv \omega_0^2 - \beta^2 \tag{3.38}$$

where $\omega_1^2 > 0$; then the exponents in the brackets of Equation 3.37 are imaginary, and the solution becomes

$$x(t) = e^{-\beta t}[A_1 e^{i\omega_1 t} + A_2 e^{-i\omega_1 t}] \tag{3.39}$$

Equation 3.39 can be rewritten as*

$$x(t) = Ae^{-\beta t} \cos(\omega_1 t - \delta) \tag{3.40}$$

We call the quantity ω_1 the *angular frequency* of the damped oscillator. Strictly speaking, we cannot define a frequency when damping is present, because the motion is not periodic—that is, the oscillator never passes twice through a given point with the same velocity. However, because $\omega_1 = 2\pi/(2T_1)$, where T_1 is the time between adjacent zero x-axis crossings, the angular frequency ω_1 has meaning for a given time period. Note that $2T_1$ would be the "period" in this case, not T_1. For simplicity, we refer to ω_1 as the "angular frequency" of the damped oscillator, and we note that this quantity is less than the frequency of the oscillator in the absence of damping (i.e., $\omega_1 < \omega_0$). If the damping is small, then

$$\omega_1 = \sqrt{\omega_0^2 - \beta^2} \cong \omega_0$$

so the term angular *frequency* may be used. But the meaning is not precise unless $\beta = 0$.

The maximum amplitude of the motion of the damped oscillator decreases with time because of the factor $\exp(-\beta t)$, where $\beta > 0$, and the envelope of the displacement *versus* time curve is given by

$$x_{en} = \pm Ae^{-\beta t} \tag{3.41}$$

---

* See Exercise D-6, Appendix D.

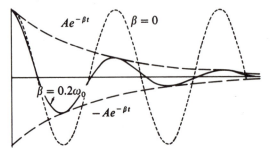

FIGURE 3-6

This envelope and the displacement curve are shown in Figure 3-6 for the case $\delta = 0$. The sinusoidal curve for undamped motion ($\beta = 0$) is also shown in this figure. A close comparison of the two curves indicates that the frequency for the damped case is *less* (i.e., that the period is *longer*) than that for the undamped case.

The ratio of the amplitudes of the oscillation at two successive maxima is

$$\frac{Ae^{-\beta T}}{Ae^{-\beta(T + \tau_1)}} = e^{\beta\tau_1} \tag{3.42}$$

where the first of any pair of maxima occurs at $t = T$ and where $\tau_1 = 2\pi/\omega_1$. The quantity $\exp(\beta\tau_1)$ is called the **decrement** of the motion; the logarithm of $\exp(\beta\tau_1)$—that is, $\beta\tau_1$—is known as the **logarithmic decrement** of the motion.

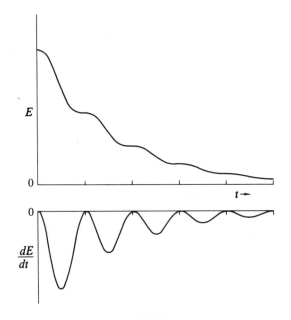

FIGURE 3-7

Unlike the simple harmonic oscillator discussed previously, the energy of the damped oscillator is not constant in time; rather, energy is continually given up to the damping medium and dissipated as heat (or, perhaps, as radiation in the form of fluid waves). The rate of energy loss is proportional to the square of the velocity (see Problem 3-11), so the decrease of energy does not take place uniformly. The loss rate will be a maximum when the particle attains its maximum velocity near (but not exactly at) the equilibrium position, and it will instantaneously vanish when the particle is at maximum amplitude and has zero velocity. Figure 3-7 shows the total energy and the rate of energy loss for the damped oscillator.

EXAMPLE 3.1 $-----------------------------$

Construct a general phase diagram analytically for the damped oscillator. Then, using a computer, make a plot for x and $\dot{x}$ versus t and a phase diagram for the following values: $A = 1$ cm, $\omega_0 = 1$ rad/s, $\beta = 0.2$ s^{-1}, and $\delta = \pi/2$ rad.

Solution: First, we write the expressions for the displacement and the velocity:

$$x(t) = Ae^{-\beta t}\cos(\omega_1 t - \delta)$$

$$\dot{x}(t) = -Ae^{-\beta t}[\beta\cos(\omega_1 t - \delta) + \omega_1\sin(\omega_1 t - \delta)]$$

These equations can be coverted into a more easily recognized form by introducing a change of variables according to the following linear transformations:

$$u = \omega_1 x, \qquad w = \beta x + \dot{x}$$

Then

$$u = \omega_1 Ae^{-\beta t}\cos(\omega_1 t - \delta)$$

$$\omega = -\omega_1 Ae^{-\beta t}\sin(\omega_1 t - \delta)$$

If we represent u and ω in polar coordinates (Figure 3-8), then

$$\rho = \sqrt{u^2 + \omega^2}, \qquad \phi = \omega_1 t$$

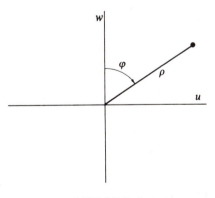

FIGURE 3-8

Thus

$$\rho = \omega_1 A e^{-(\beta/\omega_1)\phi}$$

which is the equation of a logarithmic spiral. Because the transformation from x, $\dot{x}$ to u, w is linear, the phase path has basically the same shape in the u-w plane (Figure 3-9a) and $\dot{x}$-x plane (Figure 3-9b). They both show a spiral phase path of the underdamped oscillator. The continually decreasing magnitude of the radius vector for a representative point in the phase plane always indicates damped motion of the oscillator.

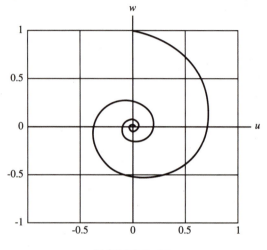

FIGURE 3-9a

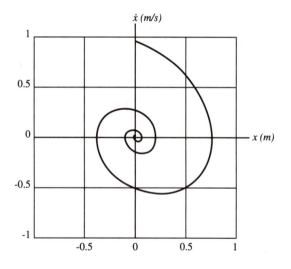

FIGURE 3-9b

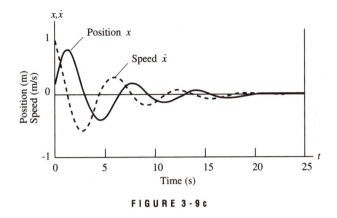

FIGURE 3-9c

The actual calculation using numbers can be done by various means with a computer. We chose to use one of the commercially available numerical programs that has good graphics output. We chose the values $A = 1$, $\beta = 0.2$, $k = 1$, $m = 1$, and $\delta = \pi/2$ in the appropriate units to produce Figure 3-9. For the particular value of δ chosen, the amplitude has $x = 0$ at $t = 0$, but $\dot{x}$ has a large positive value, which causes x to rise to a maximum value of about 0.7 m at 2 s (Figure 3-9c). The weak damping parameter β allows the system to oscillate about zero several times (Figure 3-9c) before the system finally spirals down to zero. The system crosses the $x = 0$ line eleven times before x decreases finally to less than 10^{-3} of its maximum amplitude. The phase diagram of Figure 3-9b displays the actual path.

-- ●

Critically Damped Motion

If the damping force is sufficiently large (i.e., if $\beta^2 > \omega_0^2$), the system is prevented from undergoing oscillatory motion. If zero initial velocity occurs, the displacement decreases monotonically from its initial value to the equilibrium position ($x = 0$). The case of **critical damping** occurs when β^2 is just equal to ω_0^2. The roots of the auxiliary equation are then equal, and the function x must be written as (cf., Equation C.11, Appendix C)

$$x(t) = (A + Bt)e^{-\beta t} \tag{3.43}$$

This displacement curve for critical damping is shown in Figure 3-5 for the case in which the initial velocity is zero. For a given set of initial conditions, a critically damped oscillator will approach equilibrium at a rate more rapid than that for either an overdamped or an underdamped oscillator. This is important in designing certain practical oscillatory systems (e.g., galvanometers) when the system must return to equilibrium as rapidly as possible. A pneumatic-tube screen-door closure system is a good example of a device that should be critically damped. If the closure were underdamped, the door would slam shut as other doors with springs always seem to do. If it were overdamped, it might take an unreasonably long time to close.

Overdamped Motion

If the damping parameter β is even larger than ω_0, then overdamping results. Because $\beta^2 > \omega_0^2$, the exponents in the brackets of Equation 3.37 become real quantities:

$$x(t) = e^{-\beta t}[A_1 e^{\omega_2 t} + A_2 e^{-\omega_2 t}] \tag{3.44}$$

where

$$\omega_2 = \sqrt{\beta^2 - \omega_0^2} \tag{3.45}$$

Note that ω_2 does not represent an angular frequency, because the motion is not periodic. The displacement asymptotically approaches the equilibrium position (Figure 3-5).

Overdamping results in a decrease of the amplitude to zero that may have some strange behavior as shown in the phase space diagram of Figure 3-10. Notice that for all the phase paths of the initial positions shown, the asymptotic paths at longer times are along the dashed curve $\dot{x} = -(\beta - \omega_2)x$. Only a special case (see Problem 3-22) has a phase path along the other dashed curve. Depending on the initial values of the position and the velocity, a change in sign of both x and $\dot{x}$ may occur; for example, see the phase path labeled III in Figure 3-10. Figure 3-11 displays x and $\dot{x}$ as a function of time for the three phase paths labeled I, II, and III in Figure 3-10. All three cases have initial positive displacements, $x(0) \equiv x_0 > 0$. Each of the three phase paths has interesting behavior depending on the initial value, $\dot{x}(0) \equiv \dot{x}_0$, of the velocity:

I. $\dot{x}_0 > 0$, so that $x(t)$ reaches a maximum at some $t > 0$ before approaching zero. The velocity $\dot{x}$ decreases, becomes negative, and then approaches zero.

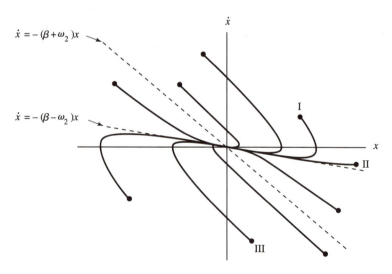

FIGURE 3-10

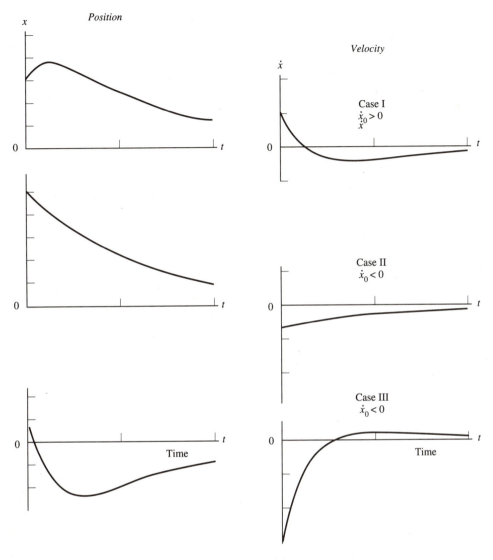

FIGURE 3-11

II. $\dot{x}_0 < 0$, with $x(t)$ and $\dot{x}(t)$ monotonically approaching zero.

III. $\dot{x}_0 < 0$, but below the curve $\dot{x} = -(\beta + \omega_2)x$, that $x(t)$ goes negative before approaching zero, and $\dot{x}(t)$ goes positive before approaching zero. The motion in this case could be considered oscillatory.

The initial points lying between the two dashed curves in Figure 3-10 seem to have phase paths decreasing monotonically to zero, whereas those lying outside those two lines do not. Critical damping has phase paths similar to the overdamping curves shown in Figure 3-10 (see Problem 3-21), rather than the spiral paths of Figure 3-9b.

EXAMPLE **3.2** --

Consider a pendulum of length ℓ and a bob of mass m at its end (Figure 3-12) moving through oil with θ decreasing. The massive bob undergoes small oscillations, but the oil retards the bob's motion with a resistive force proportional to the speed with $F_{res} = 2m\sqrt{g/\ell}\,(\ell\dot\theta)$. The bob is initially pulled back at $t = 0$ with $\theta = \alpha$ and $\dot\theta = 0$. Find the angular displacement θ and velocity $\dot\theta$ as a function of time. Sketch the phase diagram if $\sqrt{g/\ell} = 10$ s^{-1} and $\alpha = 10^{-2}$ rad.

Solution: Gravity produces the restoring force, and the component pulling the bob back to equilibrium is $mg \sin \theta$. Newton's Second Law becomes

$$\text{Force} = m(\ell\ddot\theta) = \text{Restoring force} + \text{Resistive force}$$

$$m\ell\ddot\theta = -mg \sin \theta - 2m\sqrt{g/\ell}\,(\ell\dot\theta) \tag{3.46}$$

Check that the force direction is correct, depending on the signs of θ and $\dot\theta$. For small oscillations $\sin \theta \approx \theta$, and Equation 3.46 becomes

$$\ddot\theta + 2\sqrt{g/\ell}\,\dot\theta + \frac{g}{\ell}\theta = 0 \tag{3.47}$$

Comparing this equation with Equation 3.35 reveals that $\omega_0^2 = g/\ell$, and $\beta^2 = g/\ell$. Therefore, $\omega_0^2 = \beta^2$ and the pendulum is critically damped. After being initially pulled back and released, the pendulum accelerates and then decelerates as θ goes to zero. The pendulum moves only in one direction as it returns to its equilibrium position.

The solution of Equation 3.47 is Equation 3.43. We can determine the values of A and B by substituting Equation 3.43 into Equation 3.47 using the initial conditions.

$$\theta(t) = (A + Bt)e^{-\beta t}$$

$$\theta(t = 0) = \alpha = A \tag{3.43}$$

$$\dot\theta(t) = Be^{-\beta t} - \beta(A + Bt)e^{-\beta t}$$

$$\dot\theta(t = 0) = 0 = B - \beta A$$

$$B = \beta A = \beta\alpha \tag{3.48}$$

$$\theta(t) = \alpha\left(1 + \sqrt{g/\ell}\,t\right)e^{-\sqrt{g/\ell}\,t} \tag{3.49}$$

$$\dot\theta(t) = \frac{-\alpha g}{\ell}te^{-\sqrt{g/\ell}\,t} \tag{3.50}$$

If we calculate $\theta(t)$ and $\dot\theta(t)$ for several values of time up to about 0.5 s, we can sketch the phase diagram of Figure 3-13. Notice that Figure 3-13 is consistent with the typical paths shown in Figure 3-10. The angular velocity is always neg-

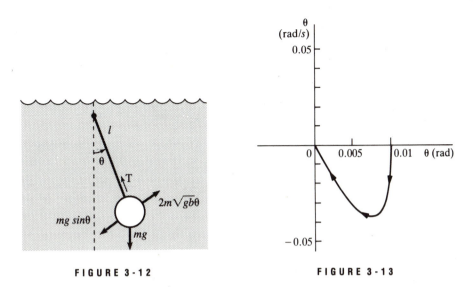

FIGURE 3-12   FIGURE 3-13

ative after the bob starts until it returns to equilibrium. The bob speeds up quickly and then slows down.

3.6 SINUSOIDAL DRIVING FORCES

The simplest case of driven oscillation is that in which an external driving force varying harmonically with time is applied to the oscillator. The total force on the particle is then

$$F = -kx - b\dot{x} + F_0 \cos \omega t \qquad (3.51)$$

where we consider a restoring linear force and a viscous damping force in addition to the driving force. The equation of motion becomes

$$m\ddot{x} + b\dot{x} + kx = F_0 \cos \omega t \qquad (3.52)$$

or, using our previous notation,

$$\boxed{\ddot{x} + 2\beta\dot{x} + \omega_0^2 x = A \cos \omega t} \qquad (3.53)$$

where $A = F_0/m$ and where ω is the angular frequency of the driving force. The solution of Equation 3.53 consists of two parts, a **complementary function** $x_c(t)$, which is the solution of Equation 3.53 with the right-hand side set equal to zero, and a **particular solution** $x_p(t)$, which reproduces the right-hand side. The complementary solution is the same as that given in Equation 3.37 (see Appendix C):

$$x_c(t) = e^{-\beta t}\left[A_1 \exp\left(\sqrt{\beta^2 - \omega_0^2}\, t\right) + A_2 \exp\left(-\sqrt{\beta^2 - \omega_0^2}\, t\right)\right] \qquad (3.54)$$

For the particular solution, we try

$$x_p(t) = D \cos(\omega t - \delta) \tag{3.55}$$

Substituting $x_p(t)$ in Equation 3.53 and expanding $\cos(\omega t - \delta)$ and $\sin(\omega t - \delta)$, we obtain

$$\{A - D[(\omega_0^2 - \omega^2)\cos \delta + 2\omega\beta \sin \delta]\} \cos \omega t$$

$$- \{D[(\omega_0^2 - \omega^2)\sin \delta - 2\omega\beta \cos \delta]\} \sin \omega t = 0 \tag{3.56}$$

Because $\sin \omega t$ and $\cos \omega t$ are linearly independent functions, this equation can be satisfied in general only if the coefficient of each term vanishes identically. From the $\sin \omega t$ term, we have

$$\tan \delta = \frac{2\omega\beta}{\omega_0^2 - \omega^2} \tag{3.57}$$

so we can write

$$\left. \begin{aligned} \sin \delta &= \frac{2\omega\beta}{\sqrt{(\omega_0^2 - \omega^2)^2 + 4\omega^2\beta^2}} \\ \cos \delta &= \frac{\omega_0^2 - \omega^2}{\sqrt{(\omega_0^2 - \omega^2)^2 + 4\omega^2\beta^2}} \end{aligned} \right\} \tag{3.58}$$

And from the coefficient of the $\cos \omega t$ term, we have

$$D = \frac{A}{(\omega_0^2 - \omega^2)\cos \delta + 2\omega\beta \sin \delta}$$

$$= \frac{A}{\sqrt{(\omega_0^2 - \omega^2)^2 + 4\omega^2\beta^2}} \tag{3.59}$$

Thus, the particular integral is

$$\boxed{x_p(t) = \frac{A}{\sqrt{(\omega_0^2 - \omega^2)^2 + 4\omega^2\beta^2}} \cos(\omega t - \delta)} \tag{3.60}$$

with

$$\boxed{\delta = \tan^{-1}\left(\frac{2\omega\beta}{\omega_0^2 - \omega^2}\right)} \tag{3.61}$$

The quantity δ represents the phase difference between the driving force and the resultant motion; a real delay occurs between the action of the driving force and the response of the system. For a fixed ω_0, as ω increases from 0, the phase increases from $\delta = 0$ at $\omega = 0$ to $\delta = \pi/2$ at $\omega = \omega_0$ and to π as $\omega \to \infty$. The variation of δ with ω is shown later in Figure 3-15.

The general solution is

$$x(t) = x_c(t) + x_p(t) \tag{3.62}$$

But $x_c(t)$ here represents *transient* effects (i.e., effects that die out), and the terms contained in this solution damp out with time because of the factor $\exp(-\beta t)$. The term $x_p(t)$ represents the steady-state effects and contains all the information for t large compared with $1/\beta$. Thus,

$$x(t \gg 1/\beta) = x_p(t)$$

The steady-state solution is important in many applications and problems (see Section 3.7).

The details of the motion during the period before the transient effects have disappeared (i.e., $t \lesssim 1/\beta$) strongly depend on the oscillator's conditions at the time that the driving force is first applied and also on the relative magnitudes of the driving frequency ω and the damping frequency $\sqrt{\omega_0^2 - \beta^2}$ in the case of underdamped, undriven oscillations. This can be shown by numerically calculating $x_p(t)$, $x_c(t)$, and the sum $x(t)$ (see Equation 3.62) for different values of β and ω as we have done for Figure 3-14. The student may profit from solving Problems 3-24 (underdamped) and 3-25 (critically damped) where such a procedure is suggested. Figure 3-14 illustrates the transient motion of an underdamped oscillator when driving frequencies less than and greater than $\omega_1 = \sqrt{\omega_0^2 - \beta^2}$ are applied. If $\omega < \omega_1$ (Figure 3-14a), the transient response of the oscillator greatly distorts the sinusoidal

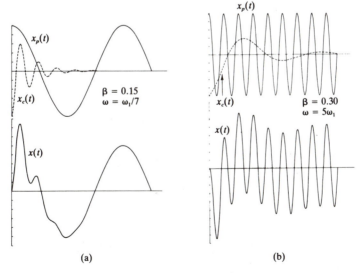

(a) (b)

FIGURE 3-14

shape of the forcing function during the time interval immediately after the application of the driving force, whereas if $\omega > \omega_1$ (Figure 3-14b), the effect is a modulation of the forcing function with little distortion of the high-frequency sinusoidal oscillations.

The steady-state solution (x_p) is widely studied in many applications and problems (see Section 3.7). The transient effects (x_c), although perhaps not as important overall, must be understood and accounted for in many cases, especially in certain types of electrical circuits.

Resonance Phenomena

To find the angular frequency ω_R at which the amplitude D is a maximum (i.e., the **amplitude resonance frequency**), we set

$$\left.\frac{dD}{d\omega}\right|_{\omega \,=\, \omega_R} = 0$$

Performing the differentiation, we find

$$\omega_R = \sqrt{\omega_0^2 - 2\beta^2} \tag{3.63}$$

Thus, the resonance frequency ω_R is lowered as the damping coefficient β is increased. No resonance occurs if $\beta^2 > \omega_0/2$, for then ω_R is imaginary and D decreases monotonically with increasing ω.

We may now compare the oscillation frequencies for the various cases we have considered:

1. Free oscillations, no damping (Equation 3.4):

$$\omega_0^2 = \frac{k}{m}$$

2. Free oscillations, damping (Equation 3.38):

$$\omega_1^2 = \omega_0^2 - \beta^2$$

3. Driven oscillations, damping (Equation 3.63):

$$\omega_R^2 = \omega_0^2 - 2\beta^2$$

and we note that $\omega_0 > \omega_1 > \omega_R$.

We customarily describe the degree of damping in an oscillating system in terms of the "quality factor" Q of the system:

$$Q \equiv \frac{\omega_R}{2\beta} \tag{3.64}$$

If little damping occurs, then Q is very large and the shape of the resonance curve approaches that for an undamped oscillator. But the resonance can be completely destroyed if the damping is large and Q is very small. Figure 3-15 shows the resonance and phase curves for several different values of Q. These curves indicate the lowering of the resonance frequency with a decrease in Q (i.e., with an increase

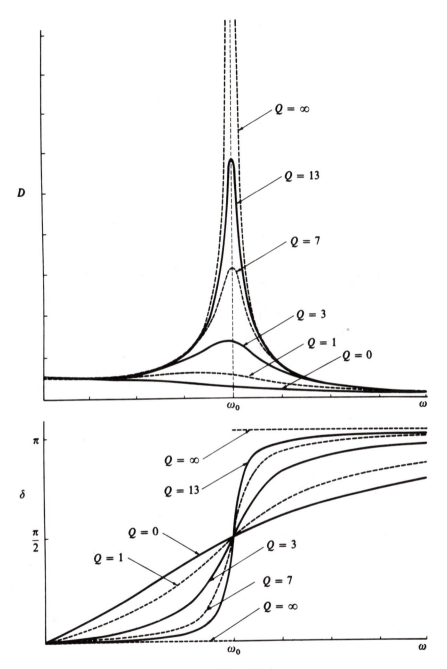

FIGURE 3-15

of the damping coefficient β). The effect is not large, however; the frequency shift is less than 3% even for Q as small as 3 and is about 18% for $Q = 1$.

For a lightly damped oscillator, we can show (see Problem 3-19) that

$$Q \cong \frac{\omega_0}{\Delta\omega} \tag{3.65}$$

where $\Delta\omega$ represents the frequency interval between the points on the amplitude resonance curve that are $1/\sqrt{2} = 0.707$ of the maximum amplitude.

The values of Q found in real physical situations vary greatly. In rather ordinary mechanical systems (e.g., loudspeakers), the values may be in the range from a few to 100 or so. Quartz crystal oscillators or tuning forks may have Qs of 10^4. Highly tuned electrical circuits, including resonant cavities, may have values of 10^4 to 10^5. We may also define Qs for some atomic systems. According to the classical picture, the oscillation of electrons within atoms leads to optical radiation. The sharpness of spectral lines is limited by the damping due to the loss of energy by radiation (**radiation damping**). The minimum width of a line can be calculated classically and is* $\Delta\omega \cong 2 \times 10^{-8}\omega$. The Q of such an oscillator is therefore approximately 5×10^7. Resonances with the largest known Qs occur in the radiation from gas lasers (Light Amplification by Stimulated Emission of Radiation). Measurements with such devices have yielded Qs of approximately 10^{14}.

Equation 3.63 gives the frequency for *amplitude* resonance. We now calculate the frequency for *kinetic energy* resonance—that is, the value of ω for which T is a maximum. The kinetic energy is given by $T = \frac{1}{2}m\dot{x}^2$, and computing $\dot{x}$ from Equation 3.60, we have

$$\dot{x} = \frac{-A\omega}{\sqrt{(\omega_0^2 - \omega^2)^2 + 4\omega^2\beta^2}} \sin(\omega t - \delta) \tag{3.66}$$

so that the kinetic energy becomes

$$T = \frac{mA^2}{2} \cdot \frac{\omega^2}{(\omega_0^2 - \omega^2)^2 + 4\omega^2\beta^2} \sin^2(\omega t - \delta) \tag{3.67}$$

To obtain a value of T independent of the time, we compute the average of T over one complete period of oscillation:

$$\langle T \rangle = \frac{mA^2}{2} \cdot \frac{\omega^2}{(\omega_0^2 - \omega^2)^2 + 4\omega^2\beta^2} \langle \sin^2(\omega t - \delta) \rangle \tag{3.68}$$

The average value of the square of the sine function taken over one period is[†]

$$\langle \sin^2(\omega t - \delta) \rangle = \frac{\omega}{2\pi} \int_0^{2\pi/\omega} \sin^2(\omega t - \delta)dt = \frac{1}{2} \tag{3.69}$$

---

* See Marion and Heald (Ma80).

[†] The reader should prove the important result that the average over a complete period of $\sin^2 \omega t$ or $\cos^2 \omega t$ is equal to $\frac{1}{2}$: $\langle \sin^2 \omega t \rangle = \langle \cos^2 \omega t \rangle = \frac{1}{2}$.

Therefore,

$$\langle T \rangle = \frac{mA^2}{4} \cdot \frac{\omega^2}{(\omega_0^2 - \omega^2)^2 + 4\omega^2\beta^2} \tag{3.70}$$

The value of ω for $\langle T \rangle$ a maximum is labeled ω_E and is obtained from

$$\frac{d\langle T \rangle}{d\omega}\bigg|_{\omega = \omega_E} = 0 \tag{3.71}$$

Differentiating Equation 3.70 and equating the result to zero, we find

$$\omega_E = \omega_0 \tag{3.72}$$

so the kinetic energy resonance occurs at the natural frequency of the system for undamped oscillations.

We see therefore that the amplitude resonance occurs at a frequency $\sqrt{\omega_0^2 - 2\beta^2}$, whereas the kinetic energy resonance occurs at ω_0. Because the potential energy is proportional to the square of the amplitude, the potential energy resonance must also occur at $\sqrt{\omega_0^2 - 2\beta^2}$. That the kinetic and potential energies resonate at different frequencies is a result of the fact that the damped oscillator is not a conservative system. Energy is continually exchanged with the driving mechanism, and energy is being transferred to the damping medium.

3.7 PHYSICAL SYSTEMS

We stated in the introduction to this chapter that linear oscillations apply to more systems than just the small oscillations of the mass–spring and the simple pendulum. The same mathematical formulation applies to a whole host of physical systems. Mechanical systems include the torsion pendulum, vibrating string or membrane, and elastic vibrations of bars or plates. These systems may have overtones, and each overtone can be treated much the same as we did in the previous discussion.

We can apply our mechanical system analog to acoustic systems. In this case, the air molecules vibrate. We can have resonances that depend on the properties and dimensions of the medium. Several factors cause the damping, including friction and sound-wave radiation. The driving force can be a tuning fork or vibrating string, among many sources of sound.

Atomic systems can also be represented classically as linear oscillators. When light (consisting of electromagnetic radiation of high frequency) falls on matter, it causes the atoms and molecules to vibrate. When light having one of the resonant frequencies of the atomic or molecular system falls on the material, electromagnetic energy is absorbed, causing the atoms or molecules to oscillate with large amplitude. Large electromagnetic fields of the same frequency are produced by the oscillating electric charges. Wave mechanics (or quantum mechanics) uses linear oscillator theory to explain many of the phenomena associated with light absorption, dispersion, and radiation.

Even to describe nuclei, linear oscillator theory is used. One of the modes of excitation of nuclei is collective excitation. Neutrons and protons vibrate in various collective motions. Resonances occur and damping exists. The classical mechanical analog is very useful in describing the motion.

Electrical circuits are, however, the most noted examples of nonmechanical oscillations. Indeed, because of its great practical importance, the electrical example has been so thoroughly investigated that the situation is frequently reversed, and mechanical vibrations are analyzed in terms of the "equivalent electrical circuit." Electrical circuits are so important that we devote an entire section to comparing them with their mechanical analogs.

3.8   ELECTRICAL OSCILLATIONS

Consider the simple mechanical oscillator shown in Figure 3-16a, where the mass m slides on a frictionless platform. We know that the equation of motion is

$$m\ddot{x} + kx = 0 \tag{3.73}$$

and that the oscillation frequency is given by

$$\omega_0 = \sqrt{\frac{k}{m}} \tag{3.74}$$

Next consider the electrical circuit shown in Figure 3-16b. At some instant of time t, the charge on the capacitor C is $q(t)$, and the current flowing through the inductor L is $I(t) = \dot{q}$. Applying Kirchhoff's equation to this circuit gives, for the voltage drops around the circuit,

$$L\frac{dI}{dt} + \frac{1}{C}\int I\, dt = 0 \tag{3.75}$$

or, in terms of q,

$$L\ddot{q} + \frac{1}{C}q = 0 \tag{3.76}$$

This equation is of exactly the same form as Equation 3.73; hence, the solution is

$$q(t) = q_0 \cos \omega_0 t \tag{3.77}$$

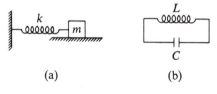

(a)   (b)

FIGURE 3-16

where the frequency is

$$\omega_0 = \frac{1}{\sqrt{LC}} \tag{3.78}$$

and where we have set the phase equal to zero by stipulating that $q(t = 0) = q_0$ and $I(t = 0) = 0$.

By comparing the terms in Equations 3.73 and 3.76, we see that the electric analog of mass (or inertia) is *inductance*, and the *compliance* of the spring, represented by the reciprocal of the spring constant k, is to be identified with the *capacitance* C. Altogether, we have

$$m \rightarrow L, \qquad x \rightarrow q$$
$$\frac{1}{k} \rightarrow C, \qquad \dot{x} \rightarrow I$$

Differentiating the expression for $q(t)$, we find

$$\dot{q} = I(t) = -\omega_0 q_0 \sin \omega_0 t \tag{3.79}$$

Squaring $q(t)$ and $I(t)$, we can write

$$\frac{1}{2} LI^2 + \frac{1}{2} \frac{q^2}{C} = \frac{1}{2} \frac{q_0^2}{C} = \text{constant} \tag{3.80}$$

The term $\frac{1}{2}LI^2$ represents the energy stored in the inductor (and corresponds to mechanical kinetic energy), whereas the term $\frac{1}{2}(q^2/C)$ represents the energy stored in the capacitor (and corresponds to mechanical potential energy). The sum of these two energies is constant, indicating that the system is conservative. We see in the following that an electrical circuit can be conservative only if it does not contain resistance (an ideal situation and one unrealistic from a practical point of view).

The mass–spring combination illustrated in Figure 3-17a differs from that of Figure 3-16a by the addition of a constant force due to the weight of the mass: $F_1 = mg$. Without this gravitational force, the equilibrium position would be at $x = 0$; the addition of the force extends the spring by an amount $h = mg/k$ and displaces the equilibrium position to $x = h$. Therefore, the equation of motion is just Equation 3.73 with x replaced by $x - h$:

$$m\ddot{x} + kx = kh \tag{3.81}$$

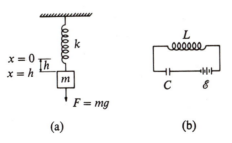

(a) (b)

FIGURE 3-17

with solution

$$x(t) = h + A \cos \omega_0 t \tag{3.82}$$

where we have chosen the initial conditions $x(t = 0) = h + A$ and $\dot{x}(t = 0) = 0$.

In Figure 3-17b, we have added a battery (with emf $\mathscr{E}$) to the circuit of Figure 3-16b. Kirchhoff's equation now becomes

$$L\frac{dI}{dt} + \frac{1}{C}\int I dt = \mathscr{E} = \frac{q_1}{C} \tag{3.83}$$

where q_1 represents the charge that must be applied to C to produce a voltage $\mathscr{E}$. Using $I = \dot{q}$, we have

$$L\ddot{q} + \frac{q}{C} = \frac{q_1}{C} \tag{3.84}$$

If $q = q_0$ and $I = 0$ at $t = 0$, the solution is

$$q(t) = q_1 + (q_0 - q_1) \cos \omega_0 t \tag{3.85}$$

which is the exact electrical analog of Equation 3.82.

The addition of damping to the mechanical oscillator of Figure 3-16a can be represented by a "dash pot" containing some viscous fluid, as in Figure 3-18a. The equation of motion is

$$m\ddot{x} + b\dot{x} + kx = 0 \tag{3.86}$$

Kirchhoff's equation for the analogous electrical circuit of Figure 3-18b is

$$L\ddot{q} + R\dot{q} + \frac{1}{C}q = 0 \tag{3.87}$$

so that the electrical resistance R corresponds to the mechanical damping resistance b. The analogy between mechanical and electrical quantities can be summarized as in Table 3-1. For example, the mass in a mechanical system is analogous to the inductance in an electrical system.

Because of the reciprocal nature of the correspondence between mechanical compliance and electrical capacitance, the addition of springs and capacitors to systems must be made in different ways to produce the same effects. For example, consider the mass in Figure 3-19a, where two springs are attached in tandem. If a

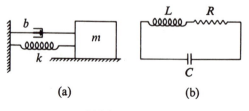

(a)   (b)

FIGURE 3-18

TABLE 3-1

ANALOGOUS MECHANICAL AND ELECTRICAL QUANTITIES

Mechanical		**Electrical**	
x	Displacement	q	Charge
$\dot{x}$	Velocity	$\dot{q}=I$	Current
m	Mass	L	Inductance
b	Damping resistance	R	Resistance
$1/k$	Mechanical compliance	C	Capacitance
F	Amplitude of impressed force	$\mathscr{E}$	Amplitude of impressed emf

force F is applied to the mass, Spring 1 will extend an amount $x_1 = F/k_1$, whereas Spring 2 will extend an amount $x_2 = F/k_2$. The total extension will be

$$x = x_1 + x_2 = F\left(\frac{1}{k_1} + \frac{1}{k_2}\right) \tag{3.88}$$

The electrical analog of this equation (see Figure 3-19b) is

$$q = \mathscr{E}(C_1 + C_2) \tag{3.89}$$

Thus, springs acting in *series* are equivalent to capacitors acting in *parallel*. Similarly, springs in parallel operate in the same way as capacitors in series (see Figures 3-19c and d).

If we replace the battery in Figure 3-17b with an AC generator, the case of driven electrical oscillations occurs. Many of the terms used to describe AC circuits

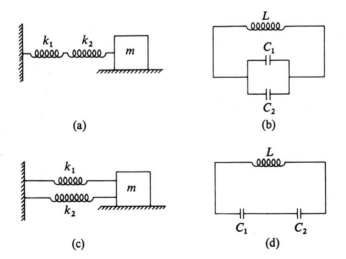

(a)   (b)

(c)   (d)

FIGURE 3-19

(impedance, reactance, inductance, phase angles, power dissipation, bandwidth, etc.) can be applied to other linear oscillator systems. The importance of the analogy with electrical circuits is the ease with which electrical circuits can be used to test mechanical (and other) analogs.

EXAMPLE **3.3**

Consider a series RLC circuit driven by an alternating emf of value $E_0 \sin \omega t$. Find the current, the voltage V_L across the inductor, and the angular frequency ω at which V_L is a maximum.

Solution: We add the alternating emf to Equation 3.87 to determine

$$L\ddot{q} + R\dot{q} + \frac{q}{C} = E_0 \sin \omega t$$

We identify this equation as similar to Equation 3.53, which we have already solved. In addition to the relationships in Table 3-1, we also have $\beta = b/2m \rightarrow R/2L$, $\omega_0 = \sqrt{k/m} \rightarrow 1/\sqrt{LC}$, and $A = F_0/m \rightarrow E_0/L$. The solution for the charge q is given by transcribing Equation 3.60, and the equation for the current I is given by transcribing Equation 3.66, which allows us to write

$$I = \frac{-E_0}{\sqrt{R^2 + \left(\dfrac{1}{\omega C} - \omega L\right)^2}} \sin(\omega t - \delta)$$

where δ can be found by transcribing Equation 3.61.

The voltage across the inductor is found from the time derivative of the current.

$$V_L = L\frac{dI}{dt} = \frac{-\omega L E_0}{\sqrt{R^2 + \left(\dfrac{1}{\omega C} - \omega L\right)^2}} \cos(\omega t - \delta)$$

$$= V(\omega) \cos(\omega t - \delta)$$

To find the driving frequency ω_{max}, which makes V_L a maximum, we must take the derivative of V_L with respect to ω and set the result equal to zero. We only need to consider the amplitude $V(\omega)$ and not the time dependence.

$$\frac{dV(\omega)}{d\omega} = \frac{LE_0\left(R^2 - \dfrac{2L}{C} + \dfrac{2}{\omega^2 C^2}\right)}{\left[R^2 + \left(\dfrac{1}{\omega C} - \omega L\right)^2\right]^{3/2}}$$

We have skipped a few intermediate steps to arrive at this result. We determine the value ω_{max} sought by setting the term in parentheses in the numerator equal to zero.

By doing so and solving for $\omega_{\max}$ gives

$$\omega_{\max} = \frac{1}{\sqrt{LC - \dfrac{R^2 C^2}{2}}}$$

which is the result we need. Note the difference between this frequency and those given by the natural frequency, $\omega_0 = 1/\sqrt{LC}$, and the charge resonance frequency (given by transcribing Equation 3.63), $\omega_R = \sqrt{1/LC - 2R^2/L^2}$.

-- ●

3.9 PRINCIPLE OF SUPERPOSITION—FOURIER SERIES

The oscillations we have been discussing obey a differential equation of the form

$$\left(\frac{d^2}{dt^2} + a\frac{d}{dt} + b \right) x(t) = A \cos \omega t \tag{3.90}$$

The quantity in parentheses on the left-hand side is a **linear operator**, which we may represent by **L**. If we generalize the time-dependent forcing function on the right-hand side, we can write the equation of motion as

$$\mathbf{L}\, x(t) = F(t) \tag{3.91}$$

An important property of linear operators is that they obey the **principle of superposition**. This property results from the fact that linear operators are distributive; that is,

$$\mathbf{L}(x_1 + x_2) = \mathbf{L}\,(x_1) + \mathbf{L}(x_2) \tag{3.92}$$

Therefore, if we have two solutions, $x_1(t)$ and $x_2(t)$, for two different forcing functions, $F_1(t)$ and $F_2(t)$,

$$\mathbf{L}x_1 = F_1(t), \qquad \mathbf{L}x_2 = F_2(t) \tag{3.93}$$

we can add these equations (multiplied by arbitrary constants α_1 and α_2) and obtain

$$\mathbf{L}(\alpha_1 x_1 + \alpha_2 x_2) = \alpha_1 F_1(t) + \alpha_2 F_2(t) \tag{3.94}$$

We can extend this argument to a set of solutions $x_n(t)$, each of which is appropriate for a given $F_n(t)$:

$$\mathbf{L}\left(\sum_{n=1}^{N} \alpha_n x_n(t) \right) = \sum_{n=1}^{N} \alpha_n F_n(t) \tag{3.95}$$

This equation is just Equation 3.91 if we identify the linear combinations as

$$x(t) = \sum_{n=1}^{N} \alpha_n x_n(t)$$
$$F(t) = \sum_{n=1}^{N} \alpha_n F_n(t) \tag{3.96}$$

If each of the individual functions $F_n(t)$ has a simple harmonic dependence on time, such as $\cos \omega_n t$, we know that the corresponding solution $x_n(t)$ is given by Equation 3.60. Thus, if $F(t)$ has the form

$$F(t) = \sum_n \alpha_n \cos(\omega_n t - \phi_n) \tag{3.97}$$

the steady-state solution is

$$x(t) = \frac{1}{m} \sum_n \frac{\alpha_n}{\sqrt{(\omega_0^2 - \omega_n^2)^2 + 4\omega_n^2\beta^2}} \cos(\omega_n t - \phi_n - \delta_n) \tag{3.98}$$

where

$$\delta_n = \tan^{-1}\left(\frac{2\omega_n\beta}{\omega_0^2 - \omega_n^2}\right) \tag{3.99}$$

We can write down similar solutions where $F(t)$ is represented by a series of terms, $\sin(\omega_n t - \phi_n)$. We therefore arrive at the important conclusion that if some arbitrary forcing function $F(t)$ can be expressed as a series (finite or infinite) of harmonic terms, the complete solution can also be written as a similar series of harmonic terms. This is an extremely useful result, because, according to Fourier's theorem, *any* arbitrary periodic function (subject to certain conditions that are not very restrictive) can be represented by a series of harmonic terms. Thus, in the usual physical case in which $F(t)$ is periodic with period $\tau = 2\pi/\omega$,

$$F(t + \tau) = F(t) \tag{3.100}$$

we then have

$$F(t) = \frac{1}{2}a_0 + \sum_{n=1}^{\infty} (a_n \cos n\omega t + b_n \sin n\omega t) \tag{3.101}$$

where

$$\left. \begin{aligned} a_n &= \frac{2}{\tau} \int_0^\tau F(t')\cos n\omega t' \, dt' \\ b_n &= \frac{2}{\tau} \int_0^\tau F(t')\sin n\omega t' \, dt' \end{aligned} \right\} \tag{3.102a}$$

or, because $F(t)$ has a period τ, we can replace the integral limits 0 and τ by the limits $-\frac{1}{2}\tau = -\pi/\omega$ and $+\frac{1}{2}\tau = +\pi/\omega$:

$$\left. \begin{aligned} a_n &= \frac{\omega}{\pi} \int_{-\pi/\omega}^{+\pi/\omega} F(t')\cos n\omega t' \, dt' \\ b_n &= \frac{\omega}{\pi} \int_{-\pi/\omega}^{+\pi/\omega} F(t')\sin n\omega t' \, dt' \end{aligned} \right\} \tag{3.102b}$$

Before we discuss the response of damped systems to arbitrary forcing functions (in the following section), we give an example of the Fourier representation of periodic functions.

EXAMPLE 3.4 --

A sawtooth driving force function is shown in Figure 3-20. Find the coefficients a_n and b_n, and express $F(t)$ as a Fourier series.

Solution: In this case, $F(t)$ is an *odd* function, $F(-t) = -F(t)$, and is expressed by

$$F(t) = A \cdot \frac{t}{\tau} = \frac{\omega A}{2\pi} t, \qquad -\tau/2 < t < \tau/2 \tag{3.103}$$

Because $F(t)$ is odd, the coefficients a_n all vanish identically. The b_n are given by

$$b_n = \frac{\omega^2 A}{2\pi^2} \int_{-\pi/\omega}^{+\pi/\omega} t' \sin n\omega t' \, dt'$$

$$= \frac{\omega^2 A}{2\pi^2} \left[-\frac{t' \cos n\omega t'}{n\omega} + \frac{\sin n\omega t'}{n^2\omega^2} \right] \Bigg|_{-\pi/\omega}^{+\pi/\omega}$$

$$= \frac{\omega^2 A}{2\pi^2} \cdot \frac{2\pi}{n\omega^2} \cdot (-1)^{n+1} = \frac{A}{n\pi} (-1)^{n+1} \tag{3.104}$$

where the term $(-1)^{n+1}$ takes account of the fact that

$$-\cos n\pi = \begin{cases} +1, & n \text{ odd} \\ -1, & n \text{ even} \end{cases} \tag{3.105}$$

Therefore we have

$$F(t) = \frac{A}{\pi} \left[\sin \omega t - \frac{1}{2} \sin 2\omega t + \frac{1}{3} \sin 3\omega t - \cdots \right] \tag{3.106}$$

Figure 3-21 shows the results for two terms, five terms, and eight terms of this expansion. The convergence toward the sawtooth function is none too rapid.

We should note two features of the expansion. At the points of discontinuity ($t = \pm\tau/2$) the series yields the mean value (zero), and in the region immediately adjacent to the points of discontinuity, the expansion "overshoots" the original function. This latter effect, known as the **Gibbs phenomenon**,* occurs in all orders of

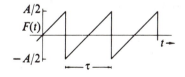

FIGURE 3-20

---

* Josiah Willard Gibbs (1839–1903) discovered this effect empirically in 1898. A detailed discussion is given, for example, by Davis (Da63, pp. 113–118). The amount of overshoot is actually 8.9490···%.

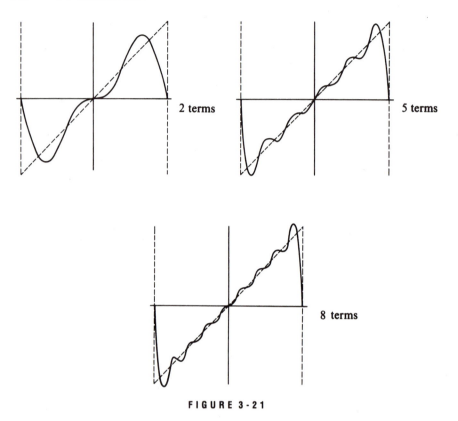

2 terms

5 terms

8 terms

FIGURE 3-21

approximation. The Gibbs overshoot amounts to about 9% on each side of *any* discontinuity, even in the limit of an infinite series.

3.10 THE RESPONSE OF LINEAR OSCILLATORS TO IMPULSIVE FORCING FUNCTIONS (OPTIONAL)

In the previous discussions, we have mainly considered steady-state oscillations. For many types of physical problems (particularly those involving oscillating electrical circuits), the transient effects are quite important. Indeed, the transient solution may be of dominating interest in such cases. In this section, we investigate the transient behavior of a linear oscillator subjected to a driving force that acts discontinuously. Of course, a "discontinuous" force is an idealization, because it always takes a finite time to apply a force. But if the application time is small compared with the natural period of the oscillator, the result of the ideal case is a close approximation to the actual physical situation.

The differential equation describing the motion of a damped oscillator is

$$\ddot{x} + 2\beta\dot{x} + \omega_0^2 x = \frac{F(t)}{m} \tag{3.107}$$

The general solution is composed of the complementary and particular solutions:

$$x(t) = x_c(t) + x_p(t) \tag{3.108}$$

We can write the complementary solution as

$$x_c(t) = e^{-\beta t} (A_1 \cos \omega_1 t + A_2 \sin \omega_1 t) \tag{3.109}$$

where

$$\omega_1 \equiv \sqrt{\omega_0^2 - \beta^2} \tag{3.110}$$

The particular solution $x_p(t)$ depends on the nature of the forcing function $F(t)$.

Two types of idealized discontinuous forcing functions are of considerable interest. These are the **step function** (or **Heaviside function**) and the **impulse function**, shown in Figures 3-22a and b, respectively. The step function H is given by

$$H(t_0) = \begin{cases} 0, & t < t_0 \\ a, & t > t_0 \end{cases} \tag{3.111}$$

where a is a constant with the dimensions of acceleration and where the argument t_0 indicates that the time of application of the force is $t = t_0$.

The impulse function I is a positive step function applied at $t = t_0$, followed by a negative step function applied at some later time t_1. Thus

$$I(t_0, t_1) = H(t_0) - H(t_1)$$

$$I(t_0, t_1) = \begin{cases} 0, & t < t_0 \\ a, & t_0 < t < t_1 \\ 0, & t > t_1 \end{cases} \tag{3.112}$$

Although we write the Heaviside and impulse functions as $H(t_0)$ and $I(t_0, t_1)$ for simplicity, these functions depend on the time t and are more properly written as $H(t; t_0)$ and $I(t; t_0, t_1)$.

Response to a Step Function

For step functions, the differential equation that describes the motion for $t > t_0$ is

$$\ddot{x} + 2\beta\dot{x} + \omega_0^2 x = a, \qquad t > t_0 \tag{3.113}$$

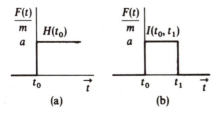

FIGURE 3-22

We consider the initial conditions to be $x(t_0) = 0$ and $\dot{x}(t_0) = 0$. The particular solution is just a constant, and examination of Equation 3.113 shows that it must be a/ω_0^2. Thus, the general solution for $t > t_0$ is

$$x(t) = e^{-\beta(t-t_0)}[A_1 \cos \omega_1(t-t_0) + A_2 \sin \omega_1(t-t_0)] + \frac{a}{\omega_0^2} \qquad (3.114)$$

Applying the initial conditions yields

$$A_1 = -\frac{a}{\omega_0^2}, \qquad A_2 = -\frac{\beta a}{\omega_1 \omega_0^2} \qquad (3.115)$$

Therefore, for $t > t_0$, we have

$$x(t) = \frac{a}{\omega_0^2}\left[1 - e^{-\beta(t-t_0)} \cos \omega_1(t-t_0) - \frac{\beta e^{-\beta(t-t_0)}}{\omega_1} \sin \omega_1(t-t_0)\right] \qquad (3.116a)$$

and $x(t) = 0$ for $t < t_0$.

If, for simplicity, we take $t_0 = 0$, the solution can be expressed as

$$x(t) = \frac{H(0)}{\omega_0^2}\left[1 - e^{-\beta t} \cos \omega_1 t - \frac{\beta e^{-\beta t}}{\omega_1} \sin \omega_1 t\right] \qquad (3.116b)$$

This response function is shown in Figure 3-23 for the case $\beta = 0.2\omega_0$. It should be clear that the ultimate condition of the oscillator (i.e., the steady-state condition) is simply a displacement by an amount a/ω_0^2.

If no damping occurs, $\beta = 0$ and $\omega_1 = \omega_0$. Then, for $t_0 = 0$, we have

$$x(t) = \frac{H(0)}{\omega_0^2}[1 - \cos \omega_0 t], \qquad \beta = 0 \qquad (3.117)$$

The oscillation is thus sinusoidal with amplitude extremes $x = 0$ and $x = 2a/\omega_0^2$ (see Figure 3-23).

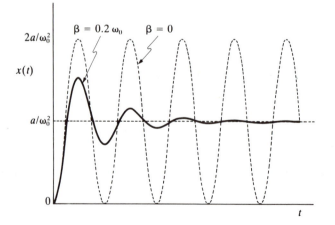

FIGURE 3-23

Response to an Impulse Function

If we consider the impulse function as the difference between two step functions separated by a time $t_1 - t_0 = \tau$, then, because the system is linear, the general solution for $t > t_1$ is given by the superposition of the solutions (Equation 3.116a) for the two step functions taken individually:

$$x(t) = \frac{a}{\omega_0^2}\left[1 - e^{-\beta(t-t_0)}\cos\omega_1(t-t_0) - \frac{\beta e^{-\beta(t-t_0)}}{\omega_1}\sin\omega_1(t-t_0)\right]$$

$$-\frac{a}{\omega_0^2}\left[1 - e^{-\beta(t-t_0-\tau)}\cos\omega_1(t-t_0-\tau) - \frac{\beta e^{-\beta(t-t_0-\tau)}}{\omega_1}\sin\omega_1(t-t_0-\tau)\right]$$

$$= \frac{ae^{-\beta(t-t_0)}}{\omega_0^2}\left[e^{\beta\tau}\cos\omega_1(t-t_0-\tau) - \cos\omega_1(t-t_0)\right.$$

$$\left.+\frac{\beta e^{\beta\tau}}{\omega_1}\sin\omega_1(t-t_0-\tau) - \frac{\beta}{\omega_1}\sin\omega_1(t-t_0)\right], \qquad t > t_1 \qquad \textbf{(3.118)}$$

The *total* response (i.e., Equations 3.116a and 3.118) to an impulse function of duration $\tau = 5 \times 2\pi/\omega_1$ applied at $t = t_0$ is shown in Figure 3-24 for $\beta = 0.2\omega_0$.

If we allow the duration τ of the impulse function to approach zero, the response function will become vanishingly small. But if we allow $a \to \infty$ as $\tau \to 0$ so that the product $a\tau$ is constant, then the response will be finite. This particular limiting case is considerably important, because it approximates the

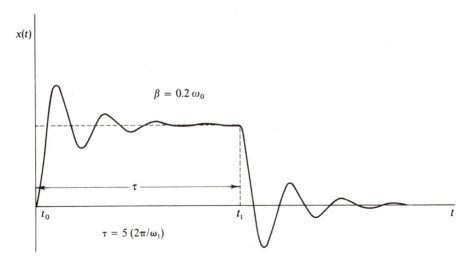

FIGURE 3-24

application of a driving force that is a "spike" at $t = t_0$ (i.e., $\tau \ll 2\pi/\omega_1$).* We want to expand Equation 3.118 by letting $\tau \to 0$, but with $b = a\tau =$ constant. Let $A = t - t_0$ and $B = \tau$, then use Equations D.11 and D.12 (from Appendix D) to obtain

$$x(t) = \frac{ae^{-\beta(t-t_0)}}{\omega_0^2} \left\{ e^{\beta\tau} \left[\cos \omega_1(t - t_0) \cos \omega_1 \tau + \sin \omega_1 (t - t_0)\sin \omega_1 \tau \right] \right.$$

$$- \cos \omega_1(t - t_0) + \frac{\beta e^{\beta\tau}}{\omega_1} \left[\sin \omega_1(t - t_0) \cos \omega_1 \tau - \cos \omega_1(t - t_0)\sin \omega_1 \tau \right]$$

$$\left. - \frac{\beta}{\omega_1} \sin \omega_1(t - t_0) \right\}, \qquad t > t_0$$

Because τ is small, we can expand $e^{\beta\tau}$, $\cos \omega_1\tau$, and $\sin \omega_1\tau$ using Equations D.34, D.29, and D.28, keeping only the first two terms in each. After multiplying out all the terms containing τ, we keep only the lowest-order term of τ.

$$x(t) = \frac{ae^{-\beta(t-t_0)}}{\omega_0^2} \sin \omega_1(t - t_0) \left[\omega_1\tau + \frac{\beta^2\tau}{\omega_1} \right], \qquad t > t_0$$

Using Equation 3.110 for ω_0^2 and $\tau = b/a$ gives us, finally,

$$x(t) = \frac{b}{\omega_1}e^{-\beta(t-t_0)} \sin \omega_1(t-t_0), \qquad t > t_0 \qquad \textbf{(3.119)}$$

This response function is shown in Figure 3-25 for the case $\beta = 0.2\omega_0$. Notice that, as t becomes large, the oscillator returns to its original position of equilibrium.

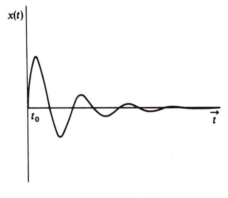

FIGURE 3-25

---

* A "spike" of this type is usually termed a **delta function** and is written $\delta(t - t_0)$. The delta function has the property that $\delta(t) = 0$ for $t \neq 0$ and $\delta(0) = \infty$, but

$$\int_{-\infty}^{+\infty} \delta(t - t_0)dt = 1$$

This is therefore not a proper function in the mathematical sense, but it can be defined as the limit of a well-behaved and highly local function (such as a Gaussian function) as the width parameter approaches zero. See also Marion and Heald (Ma80, Section 1.11).

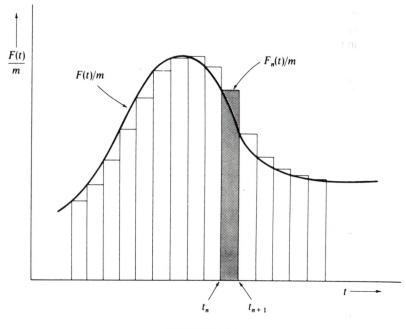

FIGURE 3-26

The fact that the response of a linear oscillator to an impulsive driving force can be represented in the simple manner of Equation 3.119 leads to a powerful technique for dealing with general forcing functions, which was developed by Green.* Green's method is based on representing an arbitrary forcing function as a series of impulses, shown schematically in Figure 3-26. If the driven system is linear, the principle of superposition is valid, and we can express the inhomogeneous part of the differential equation as the sum of individual forcing functions $F_n(t)/m$, which in Green's method are impulse functions:

$$\ddot{x} + 2\beta\dot{x} + \omega_0^2 x = \sum_{n=-\infty}^{\infty} \frac{F_n(t)}{m} = \sum_{n=-\infty}^{\infty} I_n(t) \qquad (3.120)$$

where

$$I_n(t) = I(t_n, t_{n+1})$$

$$= \begin{cases} a_n(t_n), & t_n < t < t_{n+1} \\ 0, & \text{Otherwise} \end{cases} \qquad (3.121)$$

The interval of time over which I_n acts is $t_{n+1} - t_n = \tau$, and $\tau \ll 2\pi/\omega_1$. The solution for the nth impulse is, according to Equation 3.119,

$$x_n(t) = \frac{a_n(t_n)\tau}{\omega_1} e^{-\beta(t-t_n)} \sin \omega_1(t - t_n), \qquad t > t_n + \tau \qquad (3.122)$$

---

* George Green (1793–1841), a self-educated English mathematician.

and the solution for all the impulses up to and including the Nth impulse is

$$x(t) = \sum_{n=-\infty}^{N} \frac{a_n(t_n)\tau}{\omega_1} e^{-\beta(t-t_n)} \sin \omega_1(t-t_n), \qquad t_N < t < t_{N+1} \qquad \textbf{(3.123)}$$

If we allow the interval τ to approach zero and write t_n as t', then the sum becomes an integral:

$$x(t) = \int_{-\infty}^{t} \frac{a(t')}{\omega_1} e^{-\beta(t-t')} \sin \omega_1(t-t')dt' \qquad \textbf{(3.124)}$$

We define

$$G(t, t') \equiv \begin{cases} \dfrac{1}{m\omega_1} e^{-\beta(t-t')} \sin \omega_1(t-t'), & t \geq t' \\ 0, & t < t' \end{cases} \qquad \textbf{(3.125)}$$

Then, because

$$ma(t') = F(t') \qquad \textbf{(3.126)}$$

we have

$$\boxed{x(t) = \int_{-\infty}^{t} F(t')G(t, t')dt'} \qquad \textbf{(3.127)}$$

The function $G(t, t')$ is known as the **Green's function** for the linear oscillator equation (Equation 3.107). The solution expressed by Equation 3.127 is valid only for an oscillator initially at rest in its equilibrium position, because the solution we used for a single impulse (Equation 3.119) was obtained for just such an initial condition. For other initial conditions, the general solution may be obtained in an analogous manner.

Green's method is generally useful for solving linear, inhomogeneous differential equations. The main advantage of the method lies in the fact that the Green's function $G(t, t')$, which is the solution of the equation for an infinitesimal element of the inhomogeneous part, *already contains the initial conditions*—so the general solution, expressed by the integral of $F(t')G(t, t')$, automatically also contains the initial conditions.

E X A M P L E ⬤ 3.5 ---

Find $x(t)$ for an exponentially decaying forcing function beginning at $t = 0$ and having the following form for $t > 0$:

$$F(t) = F_0 e^{-\gamma t}, \qquad t > 0 \qquad \textbf{(3.128)}$$

Solution: The solution for $x(t)$ according to Green's method is

$$x(t) = \frac{F_0}{m\omega_1} \int_{0}^{t} e^{-\gamma t'} e^{-\beta(t-t')} \sin \omega_1(t-t') \, dt' \qquad \textbf{(3.129)}$$

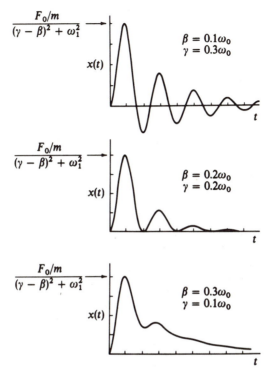

FIGURE 3-27

Making a change of variable to $z = \omega_1(t - t')$, we find

$$x(t) = -\frac{F_0}{m\omega_1^2} \int_{\omega_1 t}^{0} e^{-\gamma t} e^{[(\gamma - \beta)/\omega_1]z} \sin z \, dz$$

$$= \frac{F_0/m}{(\gamma - \beta)^2 + \omega_1^2} \left[e^{-\gamma t} - e^{-\beta t} \left(\cos \omega_1 t - \frac{\gamma - \beta}{\omega_1} \sin \omega_1 t \right) \right] \quad \textbf{(3.130)}$$

This response function is illustrated in Figure 3-27 for three different combinations
of the damping parameters β and γ. When γ is large compared with β, and if both
are small compared with ω_0, then the response approaches that for a "spike"; compare Figure 3-25 with the upper curve in Figure 3-27. When γ is small compared
with β, the response approaches the shape of the forcing function itself—that is,
an initial increase followed by an exponential decay. The lower curve in Figure
3-27 shows a decaying amplitude on which is superimposed a residual oscillation.
When β and γ are equal, Equation 3.130 becomes

$$x(t) = \frac{F_0}{m\omega_1^2} e^{-\beta t}(1 - \cos \omega_1 t), \qquad \beta = \gamma \quad \textbf{(3.131)}$$

Thus, the response is oscillatory with a "period" equal to $2\pi/\omega_1$ but with an exponentially decaying amplitude, as shown in the middle curve of Figure 3-27.

A response of the type given by Equation 3.130 could result, for example, if a quiescent but intrinsically oscillatory electronic circuit were suddenly driven by the decaying voltage on a capacitor.

P R O B L E M S

3-1. A simple harmonic oscillator consists of a 100-g mass attached to a spring whose force constant is 10^4 dyne/cm. The mass is displaced 3 cm and released from rest. Calculate (**a**) the natural frequency v_0 and the period τ_0, (**b**) the total energy, and (**c**) the maximum speed.

3-2. Allow the motion in the preceding problem to take place in a resisting medium. After oscillating for 10 s, the maximum amplitude decreases to half the initial value. Calculate (**a**) the damping parameter β, (**b**) the frequency v_1 (compare with the undamped frequency v_0), and (**c**) the decrement of the motion.

3-3. The oscillator of Problem 3-1 is set into motion by giving it an initial velocity of 1 cm/s at its equilibrium position. Calculate (**a**) the maximum displacement and (**b**) the maximum potential energy.

3-4. Consider a simple harmonic oscillator. Calculate the *time* averages of the kinetic and potential energies over one cycle, and show that these quantities are equal. Why is this a reasonable result? Next calculate the *space* averages of the kinetic and potential energies. Discuss the results.

3-5. Obtain an expression for the fraction of a complete period that a simple harmonic oscillator spends within a small interval Δx at a position x. Sketch curves of this function versus x for several different amplitudes. Discuss the physical significance of the results. Comment on the areas under the various curves.

3-6. Two masses $m_1 = 100$ g and $m_2 = 200$ g slide freely in a horizontal frictionless track and are connected by a spring whose force constant is $k = 0.5$ N/m. Find the frequency of oscillatory motion for this system.

3-7. A body of uniform cross-sectional area $A = 1$ cm^2 and of mass density $\rho = 0.8$ g/cm^3 floats in a liquid of density $\rho_0 = 1$ g/cm^3 and at equilibrium displaces a volume $V = 0.8$ cm^3. Show that the period of small oscillations about the equilibrium position is given by

$$\tau = 2\pi\sqrt{V/gA}$$

where g is the gravitational field strength. Determine the value of τ.

3-8. A pendulum is suspended from the cusp of a cycloid* cut in a rigid support (Figure 3-A). The path described by the the pendulum bob is cycloidal and is given by

$$x = a(\phi - \sin \phi), \qquad y = a(\cos\phi - 1)$$

where the length of the pendulum is $l = 4a$, and where ϕ is the angle of rotation of the

---

* The reader unfamiliar with the properties of cycloids should consult a text on analytic geometry.

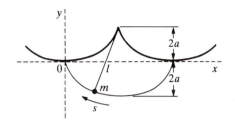

FIGURE 3-A

circle generating the cycloid. Show that the oscillations are exactly isochronous with a frequency $\omega_0 = \sqrt{g/l}$, independent of the amplitude.

3-9. A particle of mass m is at rest at the end of a spring (force constant $= k$) hanging from a fixed support. At $t = 0$, a constant downward force F is applied to the mass and acts for a time t_0. Show that, after the force is removed, the displacement of the mass from its equilibrium position ($x = x_0$, where x is down) is

$$x - x_0 = \frac{F}{k}[\cos \omega_0(t - t_0) - \cos \omega_0 t]$$

where $\omega_0^2 = k/m$.

3-10. If the amplitude of a damped oscillator decreases to $1/e$ of its initial value after n periods, show that the frequency of the oscillator must be approximately $[1 - (8\pi^2 n^2)^{-1}]$ times the frequency of the corresponding undamped oscillator.

3-11. Derive the expressions for the energy and energy-loss curves shown in Figure 3-7 for the damped oscillator. For a lightly damped oscillator, calculate the *average rate* at which the damped oscillator loses energy (i.e., compute a time over one cycle).

3-12. A simple pendulum consists of a mass m suspended from a fixed point by a weightless, extensionless rod of length l. Obtain the equation of motion and, in the approximation that $\sin \theta \cong \theta$, show that the natural frequency is $\omega_0 = \sqrt{g/l}$, where g is the gravitational field strength. Discuss the motion in the event that the motion takes place in a viscous medium with retarding force $2m\sqrt{gl}\,\dot\theta$.

3-13. Show that Equation 3.43 is indeed the solution for critical damping by assuming a solution of the form $x(t) = y(t)\exp(-\beta t)$ and determining the function $y(t)$.

3-14. Express the displacement $x(t)$ and the velocity $\dot x(t)$ for the overdamped oscillator in terms of hyperbolic functions.

3-15. Reproduce Figures 3-9b and c for the same values given in Example 3.1, but instead let $\beta = 0.1$ s^{-1} and $\delta = \pi$ rad. How many times does the system cross the $x = 0$ line before the amplitude finally falls below 10^{-2} of its maximum value? Which plot, b or c, is more useful for determining this number? Explain.

3-16. Discuss the motion of a particle described by Equation 3.34 in the event that $b < 0$ (i.e., the damping resistance is *negative*).

3-17. For a damped, driven oscillator, show that the average kinetic energy is the same at a frequency of a given number of octaves* above the kinetic energy resonance as at a frequency of the same number of octaves below resonance.

---

* An octave is a frequency interval in which the highest frequency is just twice the lowest frequency.

3-18. Show that, if a driven oscillator is only lightly damped and driven near resonance, the Q of the system is approximately

$$Q \cong 2\pi \times \left(\frac{\text{Total energy}}{\text{Energy loss during one period}} \right)$$

3-19. For a lightly damped oscillator, show that $Q \cong \omega_0/\Delta\omega$ ((Equation 3.65).

3-20. Plot a *velocity* resonance curve for a driven, damped oscillator with $Q = 6$, and show that the full width of the curve between the points corresponding to $\dot{x}_{max}/\sqrt{2}$ is approximately equal to $\omega_0/6$.

3-21. Use a computer to produce a phase space diagram similar to Figure 3-10 for the case of critical damping. Show analytically that the equation of the line that the phase paths approach asymptotically is $\dot{x} = -\beta x$. Show the phase paths for at least three initial positions above and below the line.

3-22. Let the initial position and speed of an overdamped, nondriven oscillator be x_0 and v_0, respectively.
(a) Show that the values of the amplitudes A_1 and A_2 in Equation 3.44 have the values

$$A_1 = \frac{\beta_2 x_0 + v_0}{\beta_2 - \beta_1} \text{ and } A_2 = -\frac{\beta_1 x_0 + v_0}{\beta_2 - \beta_1} \text{ where } \beta_1 = \beta - \omega_2 \text{ and } \beta_2 = \beta + \omega_2.$$

(b) Show that when $A_1 = 0$, the phase paths of Figure 3-10 must be along the dashed curve given by $\dot{x} = -\beta_2 x$, otherwise the asymptotic paths are along the other dashed curve given by $\dot{x} = -\beta_1 x$. *Hint:* Note that $\beta_2 > \beta_1$ and find the asymptotic paths when $t \to \infty$.

3-23. To better understand underdamped motion, use a computer to plot $x(t)$ of Equation 3.40 (with $A = 1$ m) and its two components $[e^{-\beta t}$ and $\cos(\omega_1 t - \delta)]$ and comparisons (with $\beta = 0$) on the same plot as in Figure 3-5. Let $\omega_0 = 1$ rad/s and make separate plots for $\beta^2/\omega_0^2 = 0.1$, 0.5, and 0.9 and for δ (in radians) = 0, $\pi/2$, and π. Have only one value of δ and β on each plot (i.e., nine plots). Discuss the results.

3-24. For $\beta = 0.2$ s^{-1}, produce computer plots like those shown in Figure 3-14 for a sinusoidal driven, damped oscillator where $x_p(t)$, $x_c(t)$, and the sum $x(t)$ are shown. Let $k = 1$ kg/s^2 and $m = 1$ kg. Do this for values of ω/ω_1 of 1/9, 1/3, 1.1, 3, and 6. For the $x_c(t)$ solution (Equation 3.40), let the phase angle $\delta = 0$ and the amplitude $A = -1$ m. For the $x_p(t)$ solution (Equation 3.60), let $A = 1$ m/s^2 but calculate δ. What do you observe about the relative amplitudes of the two solutions as ω increases? Why does this occur? For $\omega/\omega_1 = 6$, let $A = 20$ m/s^2 for $x_p(t)$ and produce the plot again.

3-25. For values of $\beta = 1$ s^{-1}, $k = 1$ kg/s^2, and $m = 1$ kg, produce computer plots like those shown in Figure 3-14 for a sinusoidal driven, damped oscillator where $x_p(t)$, $x_c(t)$, and the sum $x(t)$ are shown. Do this for values of ω/ω_0 of 1/9, 1/3, 1.1, 3, and 6. For the critically damped $x_c(t)$ solution of Equation 3.43, let $A = -1$ m and $B = 1$ m/s. For the $x_p(t)$ solution of Equation 3.60, let $A = 1$ m/s^2 and calculate δ. What do you observe about the relative amplitudes of the two solutions as ω increases? Why does this occur? For $\omega/\omega_0 = 6$, let $A = 20$ m/s^2 for $x_p(t)$ and produce the plot again.

3-26. Show that for an R-L-C circuit in which the resistance is small, the logarithmic decrement of the oscillations is approximately $\pi R \sqrt{C/L}$.

3-27. Compute the oscillation frequency for the circuit in Figure 3-18b if $L = 0.1$ H, $C = 10$ μF, $R = 100$ Ω.

3-28. An electrical circuit consists of a resistor R and a capacitor C connected in series to a source of alternating emf. Find the expression for the current as a function of time and show that it decreases to zero as the frequency of the alternating emf approaches zero.

3-29. An $R\text{-}L\text{-}C$ circuit (see Figure 3-18b) contains an inductor of 0.01 H and a resistor of 100 Ω. The oscillation frequency is 1 kHz. If at $t = 0$ the voltage across the capacitor is 10 V and the current is 0, find the current 0.2 ms later.

3-30. Figure 3-B illustrates a mass m_1 driven by a sinusoidal force whose frequency is ω. The mass m_1 is attached to a rigid support by a spring of force constant k and slides on a second mass m_2. The frictional force between m_1 and m_2 is represented by the damping parameter b_1, and the frictional force between m_2 and the support is represented by b_2. Construct the electrical analog of this system and calculate the impedance.

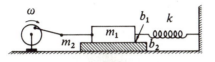

FIGURE 3-B

3-31. Show that the Fourier series of Equation 3.101 can be expressed as

$$F(t) = \frac{1}{2}a_0 + \sum_{n=1}^{\infty} c_n \cos(n\omega t - \phi_n)$$

Relate the coefficients c_n to the a_n and b_n of Equation 3.102a.

3-32. Obtain the Fourier expansion of the function

$$F(t) = \begin{cases} -1, & -\pi/\omega < t < 0 \\ +1, & 0 < t < \pi/\omega \end{cases}$$

in the interval $-\pi/\omega < t < \pi/\omega$, where $\omega = 1$ rad/s. Calculate and plot the sums of the first two terms, the first three terms, and the first four terms to demonstrate the convergence of the series.

3-33. Obtain the Fourier series representing the function

$$F(t) = \begin{cases} 0, & -2\pi/\omega < t < 0 \\ \sin \omega t, & 0 < t < 2\pi/\omega \end{cases}$$

3-34. Obtain the Fourier representation of the output of a full-wave rectifier. Plot the first three terms of the expansion and compare with the exact function.

3-35. A damped linear oscillator, originally at rest in its equilibrium position, is subjected to a forcing function given by

$$\frac{F(t)}{m} = \begin{cases} 0, & t < 0 \\ a \times (t/\tau), & 0 < t < \tau \\ a, & t > \tau \end{cases}$$

Find the response function. Allow $\tau \to 0$ and show that the solution becomes that for a step function.

3-36. Obtain the response of a linear oscillator to a step function and to an impulse function (in the limit $\tau \to 0$) for overdamping. Sketch the response functions.

3-37. Calculate the maximum values of the amplitudes of the response functions shown in Figures 3-23 and 3-25. Obtain numerical values for $\beta = 0.2\omega_0$ when $a = 2$ m/s^2, $\omega_0 = 1$ rad/s, and $t_0 = 0$.

3-38. Consider an undamped linear oscillator with a natural frequency $\omega_0 = 0.5$ rad/s and $a = 1$ m/s^2. Calculate and sketch the response function for an impulse forcing function acting for a time $\tau = 2\pi/\omega_0$. Give a physical interpretation of the results.

3-39. Obtain the response of a linear oscillator to the forcing function

$$\frac{F(t)}{m} = \begin{cases} 0, & t < 0 \\ a \sin \omega t, & 0 < t < \pi/\omega \\ 0, & t > \pi/\omega \end{cases}$$

3-40. Derive an expression for the displacement of a linear oscillator analogous to Equation 3.119 but for the initial conditions $x(t_0) = x_0$ and $\dot{x}(t_0) = \dot{x}_0$.

3-41. Derive the Green's method solution for the response caused by an arbitrary forcing function. Consider the function to consist of a series of step functions—that is, start from Equation 3.116a rather than from Equation 3.119.

3-42. Use Green's method to obtain the response of a damped oscillator to a forcing function of the form

$$F(t) = \begin{cases} 0 & t < 0 \\ F_0 e^{-\gamma t} \sin \omega t & t > 0 \end{cases}$$

3-43. Consider the periodic function

$$F(t) = \begin{cases} \sin \omega t, & 0 < t < \pi/\omega \\ 0, & \pi/\omega < t < 2\pi/\omega \end{cases}$$

which represents the positive portions of a sine function. (Such a function represents, for example, the output of a half-wave rectifying circuit.) Find the Fourier representation and plot the sum of the first four terms.

4

NONLINEAR OSCILLATIONS
AND CHAOS

4.1 INTRODUCTION

The discussion of oscillators in Chapter 3 was limited to linear systems. When pressed to divulge greater detail, however, nature insists on being *nonlinear;* examples are the flapping of a flag in the wind, the dripping of a leaky water faucet, and the oscillations of a double pendulum. The techniques learned thus far for linear systems may not be useful for nonlinear systems, but a large number of techniques have been developed for nonlinear systems, some of which we address in this chapter. We use numerical techniques to solve some of the nonlinear equations in this chapter.

The equation of motion for the damped and driven oscillator of Chapter 3 moving in only one dimension can be written as

$$m\ddot{x} + f(\dot{x}) + g(x) = h(t) \tag{4.1}$$

If $f(\dot{x})$ or $g(x)$ contain powers of $\dot{x}$ or x, respectively, higher than linear, then the physical system is nonlinear. Complete solutions are not always available for Equation 4.1, and sometimes special treatment is needed to solve such equations. For example, we can learn much about a physical system by considering the deviation of the forces from linearity and by examining phase diagrams. Such a system is the simple plane pendulum, a system that is linear only when small oscillations are assumed.

In the beginning of the nineteenth century, the famous French mathematician Pierre Simon de Laplace espoused the view that if we knew the position and velocities of all the particles in the universe, then we would know the future for all time.

This is the *deterministic* view of nature. In recent years, researchers in many disciplines have come to realize that knowing the laws of nature is not enough. Much of nature seems to be **chaotic**. In this case, we refer to **deterministic chaos**, as opposed to *randomness,* to be the motion of a system *whose time evolution has a sensitive dependence on initial conditions.* The deterministic development refers to the way a system develops from one moment to the next, where the present system depends on the one just past in a well-determined way through physical laws. We are not referring to a random process in which the present system has no causal connection to the previous one (e.g., the previous flipping of a coin).

Measurements made on the state of a system at a given time may not allow us to predict the future situation even moderately far ahead, despite the fact that the governing equations are known exactly. Deterministic chaos is always associated with a nonlinear system; nonlinearity is a necessary condition for chaos but not a sufficient one. Chaos occurs when a system depends in a sensitive way on its previous state. Even a tiny effect, such as a butterfly flying nearby, may be enough to vary the conditions such that the future is *entirely* different than what it might have been, *not* just a tiny bit different. The advent of computers has allowed chaos to be studied because we now have the capability of performing calculations of the time evolution of the properties of a system that includes these tiny variations in the initial conditions. Chaotic systems can *only* be solved numerically, and there are no simple, general ways to predict when a system will exhibit chaos.

Chaotic phenomena have been uncovered in practically all areas of science and engineering—in irregular heartbeats, the motion of planets in our solar system, water dripping from a tap, electrical circuits, weather patterns, epidemics, changing populations of insects, birds, and animals, and the motion of electrons in atoms. The list goes on and on. Henri Poincaré* is generally given credit for first recognizing the existence of chaos during his investigation of celestial mechanics at the end of the nineteenth century. He came to the realization that the motion of apparently simple systems, such as the planets in our solar system, can be extremely complicated. Although various investigators also eventually came to understand the existence of chaos, tremendous breakthroughs did not happen until the 1970s when computers were readily available to calculate the long-time histories required to document the behavior.

The study of chaos has become widespread, and we will only be able to look at the rudimentary aspects of the phenomena. Specialized textbooks† on the subject have become abundant for those desiring further study. For example, space does not permit us to discuss the fascinating area of fractals, the complicated patterns that arise from chaotic processes.

---

* Henri Poincaré (1854–1912) was a mathematician who could also be considered a physicist and philosopher. His career spanned the era when classical mechanics was at its height, soon to be overtaken by relativity and quantum mechanics. He searched for precise mathematical formulas that would allow him to understand the dynamic stability of systems.

† Particularly useful books are by Baker and Gollub (Ba90), Moon (Mo92), and Hilborn (Hi94).

4.2 NONLINEAR OSCILLATIONS

Consider a potential energy of the parabolic form

$$U(x) = \frac{1}{2}kx^2 \qquad (4.2)$$

Then the corresponding force is

$$F(x) = -kx \qquad (4.3)$$

This is just the case of simple harmonic motion discussed in Section 3.2. Now, suppose a particle moves in a potential well, which is some arbitrary function of distance (as in Figure 4-1). Then, in the vicinity of the minimum of the well, we usually approximate the potential with a parabola. Therefore, if the energy of the particle is only slightly greater than U_{min}, only small amplitudes are possible and the motion is approximately simple harmonic. If the energy is appreciably greater than U_{min}, so that the amplitude of the motion cannot be considered small, then it may no longer be sufficiently accurate to make the approximation $U(x) \approx \frac{1}{2}kx^2$ and we must deal with a *nonlinear* force.

In many physical situations, the deviation of the force from linearity is *symmetric* about the equilibrium position (which we take to be at $x = 0$). In such cases, the *magnitude* of the force exerted on a particle is the same at $-x$ as at x; the *direction* of the force is opposite in the two cases. Therefore, in a symmetric situation, the first correction to a linear force must be a term proportional to x^3; hence,

$$F(x) \cong -kx + \varepsilon x^3 \qquad (4.4)$$

where ε is usually a small quantity. The potential corresponding to such a force is

$$U(x) = \frac{1}{2}kx^2 - \frac{1}{4}\varepsilon x^4 \qquad (4.5)$$

Depending on the sign of the quantity ε, the force may either be greater or less than the linear approximation. If $\varepsilon > 0$, then the force is less than the linear term alone and the system is said to be *soft*; if $\varepsilon < 0$, then the force is greater and the

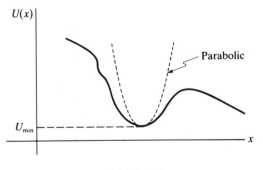

FIGURE 4-1

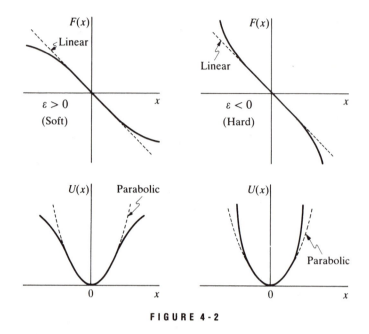

FIGURE 4-2

system is *hard*. Figure 4-2 shows the form of the force and the potential for a soft and a hard system.

E X A M P L E   **4.1**   -

Consider a particle of mass *m* suspended between two identical springs (Figure 4-3). Show that the system is nonlinear. Find the steady-state solution for a driving force F_0 cos ωt.

Solution: If both springs are in their unextended conditions (i.e., there is no tension, and therefore no potential energy, in either spring) when the particle is in its equilibrium position—and if we neglect gravitational forces—then when the particle is displaced from equilibrium (Figure 4-3b), each spring exerts a force $- k(s - l)$ on the particle (*k* is the force constant of each spring). The net (horizontal) force on the particle is

$$F = - 2k(s - l)\sin \theta \tag{4.6}$$

Now,

$$s = \sqrt{l^2 + x^2}$$

so

$$\sin \theta = \frac{x}{s} = \frac{x}{\sqrt{l^2 + x^2}}$$

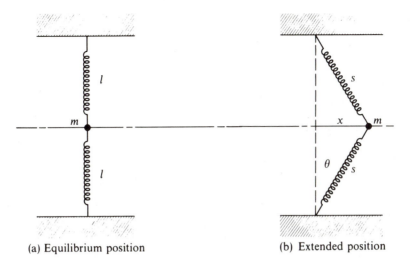

(a) Equilibrium position (b) Extended position

FIGURE 4-3

Hence,

$$F = -\frac{2kx}{\sqrt{l^2 + x^2}} \cdot \left(\sqrt{l^2 + x^2} - l \right) = -2kx \left(1 - \frac{1}{\sqrt{1 + (x/l)^2}} \right) \quad \textbf{(4.7)}$$

If we consider x/l to be a small quantity and expand the radical, we find

$$F = -kl\left(\frac{x}{l}\right)^3 \left[1 - \frac{3}{4}\left(\frac{x}{l}\right)^2 + \cdots \right]$$

If we neglect all terms except the leading term, we have, approximately,

$$F(x) \cong -(k/l^2)x^3 \quad \textbf{(4.8)}$$

Therefore, even if the amplitude of the motion is sufficiently restricted so that x/l is a small quantity, the force is still proportional to x^3. The system is therefore *intrinsically nonlinear*. However, if it had been necessary to stretch each spring a distance d to attach it to the mass when at the equilibrium position, then we would find for the force (see Problem 4-1):

$$F(x) \cong -2(kd/l)x - [k(l - d)/l^3]\, x^3 \quad \textbf{(4.9)}$$

and a linear term is introduced.

From Equation 4.9 we identify

$$\varepsilon' = -k(l - d)/l^3 < 0$$

Thus the system is *hard*, and for oscillations with small amplitude, the motion is approximately simple harmonic.

If we have a driving force $F_0 \cos \omega t$, the equation of motion for the stretched spring (force of Equation 4.9) becomes

$$m\ddot{x} = -\frac{2kd}{l}x - \frac{k(l - d)}{l^3}x^3 + F_0 \cos \omega t \quad \textbf{(4.10)}$$

Let

$$\varepsilon = \frac{\varepsilon'}{m}, \qquad a = \frac{2kd}{ml} \qquad \text{and} \qquad G = \frac{F_0}{m} \tag{4.11}$$

then

$$\ddot{x} = -ax + \varepsilon x^3 + G \cos \omega t \tag{4.12}$$

Equation 4.12 is a difficult differential equation to solve. We can find the important characteristics of the solution by a method of successive approximations (perturbation technique). First, try a solution $x_1 = A \cos \omega t$, and insert x_1 into the right-hand side of Equation 4.12, which becomes

$$\ddot{x}_2 = -aA \cos \omega t + \varepsilon A^3 \cos^3 \omega t + G \cos \omega t \tag{4.13}$$

where the solution of Equation 4.13 is $x = x_2$. This equation can be solved for x_2 using the identity

$$\cos^3 \omega t = \frac{3}{4} \cos \omega t + \frac{1}{4} \cos 3\omega t \tag{4.14}$$

Using Equation 4.14 in Equation 4.13 gives

$$\ddot{x}_2 = -\left(aA - \frac{3}{4}\varepsilon A^3 - G\right)\cos \omega t + \frac{1}{4}\varepsilon A^3 \cos 3\omega t \tag{4.15}$$

Integrating twice (with integration constants set equal to zero) gives

$$x_2 = \frac{1}{\omega^2}\left(aA - \frac{3}{4}\varepsilon A^3 - G\right)\cos \omega t - \frac{\varepsilon A^3}{36\omega^2} \cos 3\omega t \tag{4.16}$$

This is already a complicated solution. Under what conditions for ε, a, and x is x_2 a suitable solution? Numerical techniques with a computer can quickly yield a perturbative solution quite accurately. We have found that the amplitude depends on the driving frequency, but no resonance occurs at the natural frequency of the system.

Further discussion of solution methods for Equation 4.12 would take us too far afield of our present discussion. The result is that for some values of the driving frequency ω, three different amplitudes may occur with "jumps" between the amplitudes. The amplitude may have a different value for a given ω depending on whether ω is increasing or decreasing (hysteresis). We present a simple case of this effect in Section 4.5.

-- ●

In real physical situations, we are often concerned with symmetric forces and potentials. But some cases have asymmetric forms. For example,

$$F(x) = -kx + \lambda x^2 \tag{4.17}$$

The potential for which is

$$U(x) = \frac{1}{2}kx^2 - \frac{1}{3}\lambda x^3 \tag{4.18}$$

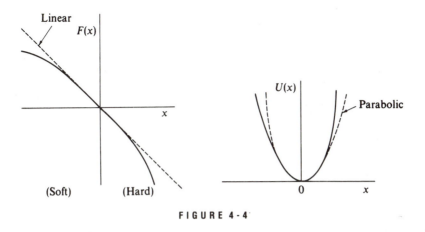

FIGURE 4-4

This case is illustrated in Figure 4-4 for $\lambda < 0$; the system is hard for $x > 0$ and soft for $x < 0$.

4.3 PHASE DIAGRAMS FOR NONLINEAR SYSTEMS

The construction of a phase diagram for a nonlinear system may be accomplished by using Equation 2.97:

$$\dot{x}(x) \propto \sqrt{E - U(x)} \tag{4.19}$$

When $U(x)$ is known, it is relatively easy to make a phase diagram for $\dot{x}(x)$. Computers, with their ever-improving graphics capability, make this a particularly easy task. However, in many cases it is difficult to obtain $U(x)$, and we must resort to approximation procedures to eventually produce the phase diagram. On the other hand, it is relatively easy to obtain a qualitative picture of the phase diagram for the motion of a particle in an arbitrary potential. For example, consider the asymmetric potential shown in the upper portion of Figure 4-5, which represents a system that is soft for $x < 0$ and hard for $x > 0$. If no damping occurs, then because $\dot{x}$ is proportional to $\sqrt{E - U(x)}$, the phase diagram must be of the form shown in the lower portion of the figure. Three of the oval phase paths are drawn, corresponding to the three values of the total energy indicated by the dotted lines in the potential diagram. For a total energy only slightly greater than that of the minimum of the potential, the oval phase paths approach ellipses. If the system is damped, then the oscillating particle will "spiral down the potential well" and eventually come to rest at the equilibrium position, $\dot{x} = 0$. The equilibrium point at $x = 0$ in this case is called an **attractor**. An attractor is a set of points (or one point) in phase space toward which a system is "attracted" when damping is present.

For the case shown in Figure 4-5, if the total energy E of the particle is less than the height to which the potential rises on either side of $x = 0$, then the particle is "trapped" in the potential well (cf., the region $x_a < x < x_b$ in Figure 2-14). The point $x = 0$ is a position of *stable* equilibrium, because $(d^2U(x)/dx^2)_0 > 0$ (see Equation 2.103), and a small disturbance results in locally bounded motion.

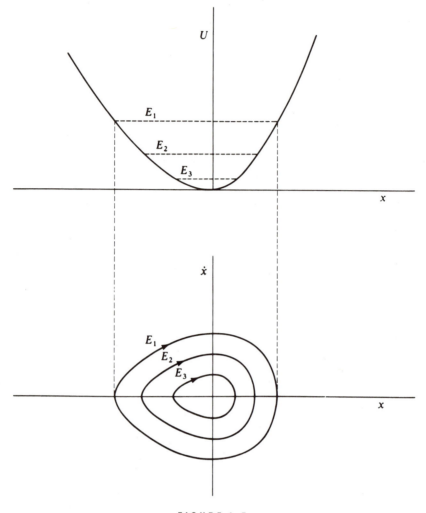

FIGURE 4-5

In the vicinity of the *maximum* of a potential, a qualitatively different type of motion occurs (Figure 4-6). Here the point $x = 0$ is one of *unstable* equilibrium, because if a particle is at rest at this point, then a slight disturbance will result in locally *unbounded* motion.* Similarly, $(d^2U(x)/dx^2)_0 < 0$ gives unstable equilibrium.

If the potential in Figure 4-6 were parabolic—if $U(x) = -\frac{1}{2}kx^2$—then the phase paths corresponding to the energy E_0 would be straight lines and those corresponding to the energies E_1 and E_2 would be hyperbolas. This is therefore the

---

* The definition of instability must be stated in terms of *locally* unbounded motion, for if there are other maxima of the potential greater than the one shown at $x = 0$, the motion will be bounded by these other potential barriers.

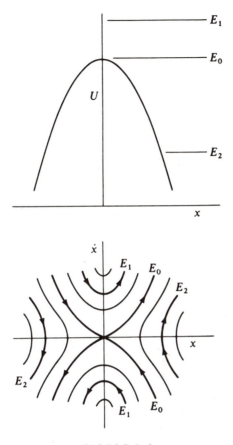

FIGURE 4-6

limit to which the phase paths of Figure 4-6 would approach if the nonlinear term in the expression for the force were made to decrease in magnitude.

By referring to the phase paths for the potentials shown in Figures 4-5 and 4-6, we can rapidly construct a phase diagram for any arbitrary potential (such as that in Figure 2-14).

An important type of nonlinear equation was extensively studied by van der Pol in his investigation of oscillations in vacuum tube circuits.* This equation has the form

$$\ddot{x} - \mu(x_0^2 - x^2)\dot{x} + \omega_0^2 x = 0 \tag{4.20}$$

where μ is a small, positive parameter. A system described by van der Pol's equation has the following interesting property. If the amplitude $|x|$ exceeds the critical

---

* B. van der Pol, *Phil. Mag.* **2**, 978 (1926). Extensive treatments of van der Pol's equation may be found, for example, in Minorsky (Mi47) or in Andronow and Chaikin (An49); brief discussions are given by Lindsay (Li51, pp. 64–66) and by Pipes (Pi46, pp. 606–610).

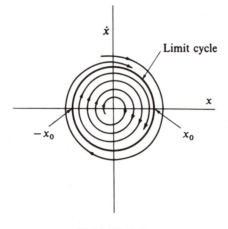

FIGURE 4-7

value $|x_0|$, then the coefficient of $\dot{x}$ is positive and the system is damped. But if $|x|$ $< |x_0|$, then *negative damping* occurs; that is, the amplitude of the motion *increases*. It follows that there must be some amplitude for which the motion neither increases nor decreases with time. Such a curve in the phase plane is called the **limit cycle*** (Figure 4-7) and is the attractor for this system. Phase paths outside the limit cycle spiral *inward*, and those inside the limit cycle spiral *outward*. Inasmuch as the limit cycle defines locally *bounded* motion, we may refer to the situation it represents as *stable*.

A system described by van der Pol's equation is *self-limiting*; that is, once set into motion under conditions that lead to an increasing amplitude, the amplitude is automatically prevented from growing without bound. The system has this property whether the initial amplitude is greater or smaller than the critical (limiting) amplitude x_0.

4.4 PLANE PENDULUM

The solutions of certain types of nonlinear oscillation problems can be expressed in closed form by elliptic integrals.[†] An example of this type is the **plane pendulum**. Consider a particle of mass m constrained by a weightless, extensionless rod to move in a vertical circle of radius l (Figure 4-8). The gravitational force acts downward, but the component of this force influencing the motion is *perpendicular* to the support rod. This force component, shown in Figure 4-9, is simply $F(\theta) = -mg \sin \theta$. The plane pendulum is a nonlinear system with a symmetric restoring force. It is only for small angular deviations that a linear approximation may be used.

---

* The term was introduced by Poincaré and is often called the *Poincaré limit cycle*.

[†] See Appendix B for a list of some elliptic integrals.

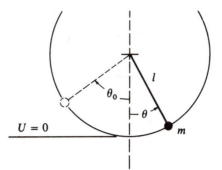

FIGURE 4-8   The angle $\theta > 0$ is in the counterclockwise direction so that $\theta_0 < 0$.

We obtain the equation of motion for the plane pendulum by equating the torque about the support axis to the product of the angular acceleration and the moment of inertia about the same axis:

$$I\ddot{\theta} = lF$$

or, because $I = ml^2$ and $F = -mg \sin \theta$,

$$\boxed{\ddot{\theta} + \omega_0^2 \sin \theta = 0} \tag{4.21}$$

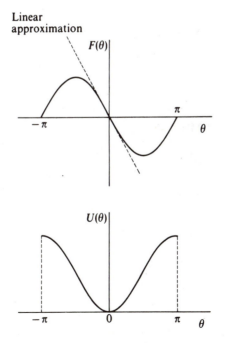

FIGURE 4-9

where

$$\omega_0^2 \equiv \frac{g}{l} \tag{4.22}$$

If the amplitude of the motion is small, we may approximate $\sin\theta \cong \theta$, and the equation of motion becomes identical with that for the simple harmonic oscillator:

$$\ddot{\theta} + \omega^2\theta = 0$$

In this approximation, the period is given by the familiar expression

$$\tau \cong 2\pi\sqrt{\frac{l}{g}}$$

If we wish to obtain the general result for the period in the event that the amplitude is finite, we may begin with Equation 4.21. But because the system is *conservative*, we can use the fact that

$$T + U = E = \text{constant}$$

to obtain a solution by considering the energy of the system rather than by solving the equation of motion.

If we take the zero of potential energy to be the lowest point on the circular path described by the pendulum bob (i.e., $\theta = 0$; see Figure 4-9), the kinetic and potential energies can be expressed as

$$\left.\begin{array}{l} T = \frac{1}{2}I\omega^2 = \frac{1}{2}ml^2\dot{\theta}^2 \\[2mm] U = mgl(1 - \cos\theta) \end{array}\right\} \tag{4.23}$$

If we let $\theta = \theta_0$ at the highest point of the motion, then

$$T(\theta = \theta_0) = 0$$

$$U(\theta = \theta_0) = E = mgl(1 - \cos\theta_0)$$

Using the trigonometric identity

$$\cos\theta = 1 - 2\sin^2(\theta/2)$$

we have

$$E = 2mgl\sin^2(\theta_0/2) \tag{4.24}$$

and

$$U = 2mgl\sin^2(\theta/2) \tag{4.25}$$

Expressing the kinetic energy as the difference between the total energy and the potential energy, we have $T = E - U$,

$$\frac{1}{2}ml^2\dot{\theta}^2 = 2mgl[\sin^2(\theta_0/2) - \sin^2(\theta/2)]$$

or

$$\dot{\theta} = 2\sqrt{\frac{g}{l}} [\sin^2(\theta_0/2) - \sin^2(\theta/2)]^{1/2} \qquad (4.26)$$

from which

$$dt = \frac{1}{2}\sqrt{\frac{l}{g}} [\sin^2(\theta_0/2) - \sin^2(\theta/2)]^{-1/2} d\theta$$

This equation may be integrated to obtain the period τ. Because the motion is symmetric, the integral over θ from $\theta = 0$ to $\theta = \theta_0$ yields $\tau/4$; hence

$$\tau = 2\sqrt{\frac{l}{g}} \int_0^{\theta_0} [\sin^2(\theta_0/2) - \sin^2(\theta/2)]^{-1/2} d\theta \qquad (4.27)$$

That this is actually an *elliptic integral of the first kind** may be seen more clearly by making the substitutions

$$z = \frac{\sin(\theta/2)}{\sin(\theta_0/2)}, \qquad k = \sin(\theta_0/2)$$

Then

$$dz = \frac{\cos(\theta/2)}{2\sin(\theta_0/2)} d\theta = \frac{\sqrt{1-k^2z^2}}{2k} d\theta$$

from which

$$\tau = 4\sqrt{\frac{l}{g}} \int_0^1 [(1-z^2)(1-k^2z^2)]^{-1/2} dz \qquad (4.28)$$

Numerical values for integrals of this type can be found in various tables.

For oscillatory motion to result, $|\theta_0| < \pi$, or, equivalently, $\sin(\theta_0/2) = k$, where $-1 < k < +1$. For this case, we can evaluate the integral in Equation 4.28 by expanding $(1 - k^2z^2)^{-1/2}$ in a power series:

$$(1 - k^2z^2)^{-1/2} = 1 + \frac{k^2z^2}{2} + \frac{3k^4z^4}{8} + \cdots$$

Then, the expression for the period becomes

$$\tau = 4\sqrt{\frac{l}{g}} \int_0^1 \frac{dz}{(1-z^2)^{1/2}} \left[1 + \frac{k^2z^2}{2} + \frac{3k^4z^4}{8} + \cdots \right]$$

$$= 4\sqrt{\frac{l}{g}} \left[\frac{\pi}{2} + \frac{k^2}{2} \cdot \frac{1}{2} \cdot \frac{\pi}{2} + \frac{3k^4}{8} \cdot \frac{3}{8} \cdot \frac{\pi}{2} + \cdots \right]$$

$$= 2\pi\sqrt{\frac{l}{g}} \left[1 + \frac{k^2}{4} + \frac{9k^4}{64} + \cdots \right]$$

---

* Refer to Equations B.2, Appendix B.

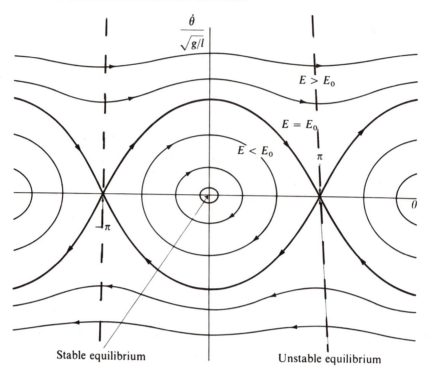

Stable equilibrium — Unstable equilibrium

FIGURE 4-10

If $|k|$ is large (i.e., near 1), then we need many terms to produce a reasonably accurate result. But for small k, the expansion converges rapidly. And because $k = \sin(\theta_0/2)$, then $k \cong (\theta_0/2) - (\theta_0^3/48)$; the result, correct to the fourth order, is

$$\tau \cong 2\pi \sqrt{\frac{l}{g}}\left[1 + \frac{1}{16}\theta_0^2 + \frac{11}{3072}\theta_0^4\right] \qquad (4.29)$$

Therefore, although the plane pendulum is not isochronous, it is very nearly so for small amplitudes of oscillation.*

We may construct the phase diagram for the plane pendulum in Figure 4-10 because Equation 4.26 provides the necessary relationship $\dot{\theta} = \dot{\theta}(\theta)$. The parameter θ_0 specifies the total energy through Equation 4.24. If θ and θ_0 are small angles, then Equation 4.26 can be written as

$$\left(\sqrt{\frac{l}{g}}\dot{\theta}\right)^2 + \theta^2 \cong \theta_0^2 \qquad (4.30)$$

---

* This was discovered by Galileo in the cathedral at Pisa in 1581. The expression for the period of small oscillations was given by Christiaan Huygens (1629–1695) in 1673. Finite oscillations were first treated by Euler in 1736.

If the coordinates of the phase plane are θ and $\dot{\theta}/\sqrt{g/l}$, then the phase paths near $\theta = 0$ are approximately circles. This result is expected, because for small θ_0, the motion is approximately simple harmonic.

For $-\pi < \theta < \pi$ and $E < 2mgl \equiv E_0$, the situation is equivalent to a particle bound in the potential well $U(\theta) = mgl(1 - \cos \theta)$ (see Figure 4-9). The phase paths are therefore closed curves for this region and are given by Equation 4.26. Because the potential is periodic in θ, exactly the same phase paths exist for the regions $\pi < \theta < 3\pi$, $-3\pi < \theta < -\pi$, and so forth. The points $\theta = \cdots$, $-2\pi, 0, 2\pi, \cdots$ along the θ-axis are positions of *stable* equilibrium and are the attractors when the undriven pendulum is damped.

For values of the total energy exceeding E_0, the motion is no longer oscillatory—although it is still periodic. This situation corresponds to the pendulum executing complete revolutions about its support axis. Normally the phase space diagram is plotted for only one complete cycle or a "unit cell," in this case over the interval $-\pi < \theta < \pi$. We denote this region in Figure 4-10 between the dashed lines at angles $-\pi$ and π. One can follow a phase path by noting that motion that exits on the left of the cell re-enters on the right and vice versa.

If the total energy equals E_0, then Equation 4.24 shows that $\theta_0 = \pm \pi$. In this case, Equation 4.26 reduces to

$$\dot{\theta} = \pm 2 \sqrt{\frac{g}{l}} \cos(\theta/2) \qquad (4.31)$$

so the phase paths for $E = E_0$ are just cosine functions (see the heavy curves in Figure 4-10). There are two branches, depending on the direction of motion.

The phase paths for $E = E_0$ do not actually represent possible continuous motions of the pendulum. If the pendulum were at rest at, say, $\theta = \pi$ (which is a point on the $E = E_0$ phase paths), then any small disturbance would cause the motion to follow *closely but not exactly on* one of the phase paths that diverges from $\theta = \pi$, because the total energy would be $E = E_0 + \delta$, where δ is a small but *nonzero* quantity. If the motion were along one of the $E = E_0$ phase paths, the pendulum would reach one of the points $\theta = n\pi$ with exactly zero velocity, but only after an infinite time! (This may be verified by evaluating Equation 4.27 for $\theta_0 = \pi$; the result is $\tau \to \infty$.)

A phase path separating locally bounded motion from locally unbounded motion (such as the path for $E = E_0$ in Figure 4-10) is called a **separatrix**. A separatrix always passes through a point of unstable equilibrium. The motion in the vicinity of such a separatrix is extremely sensitive to initial conditions because points on either side of the separatrix have very different trajectories.

4.5 JUMPS, HYSTERESIS, AND PHASE LAGS

In Example 4.1 we considered a particle of mass m suspended between two springs. We showed that the system was nonlinear and mentioned the phenomena of jumps in amplitude and hysteresis effects. Now, we want to examine such phenomena more carefully. We follow closely the description by Janssen and colleagues* who developed a simple method to investigate such effects.

Consider a harmonic oscillator subjected to an external force $F(t) = F_0 \cos \omega t$ and a resistive viscous force $-r\dot{x}$, where r is a constant. The equation of motion for a particle of mass m connected to a spring with force constant k is

$$m\ddot{x} = -r\dot{x} - kx + F_0 \cos \omega t \tag{4.32}$$

A solution to Equation 4.32 is

$$x(t) = A(\omega)\cos[\omega t - \phi(\omega)] \tag{4.33}$$

where

$$A(\omega) = \frac{F_0}{\left[(k - m\omega^2)^2 + (r\omega)^2\right]^{1/2}} \tag{4.34}$$

and

$$\tan[\phi(\omega)] = \frac{r\omega}{(k - m\omega^2)} \tag{4.35}$$

The reader can verify that Equation 4.33 is a particular solution by substitution into Equation 4.32.

If the spring constant k depends on x as $k(x)$, then we have a nonlinear oscillator. An often used dependence is

$$k(x) = 1 + \beta x^2 \tag{4.36}$$

and the resulting equation of motion in Equation 4.32 is known as the *Duffing equation*. It has been widely studied through perturbation techniques with solutions similar to Equation 4.33 but with complicated results for $A(\omega)$ and $\phi(\omega)$ as shown in Figure 4-11. As ω increases, $A(\omega)$ increases to its peak until it reaches $\omega = \omega_2$, where the amplitude suddenly decreases by a large factor. As ω decreases from large values, the amplitude slowly increases until $\omega = \omega_1$, where the amplitude suddenly approximately doubles. These are the "jumps" referred to earlier. The

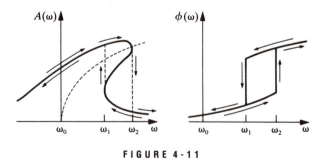

FIGURE 4-11

---

* H. J. Janssen, et al., *Am. J. Phys.*, **51**, 655 (1983).

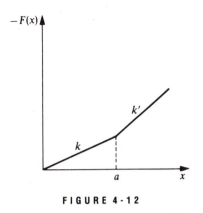

FIGURE 4-12

amplitude between ω_1 and ω_2 depends on whether ω is increasing or decreasing (hysteresis effect). Similarly strange phenomena occur for the phase $\phi(\omega)$ in Figure 4-11. The physical explanation of Figure 4-11 is not very transparent, so we consider a simpler dependence of k as shown in Figure 4-12:

$$F(x) = -kx \qquad x \leq a$$
$$\simeq -k'x \qquad x \geq a \tag{4.37}$$

The Duffing equation represents a situation with many values of a, because $k(x)$ continuously varies in Equation 4.36. Our example of an anharmonic oscillator allows simpler mathematics.

Figure 4-13 shows the harmonic response curves $A(\omega)$ for k and k' (with $k < k'$). For very large values of a ($a \approx \infty$), we have a linear oscillator with force constant k (because $x < a$, see Figure 4-12) and a resonance frequency $\omega_0 = (k/m)^{1/2}$. For very small values of a ($a \approx 0$), the force constant is k' and $\omega_0' = (k'/m)^{1/2}$.

We want to consider intermediate values of a, where both k and k' are effective. We consider the situation in which a is much smaller than the maximum amplitude of $A(\omega)$. If we start at small values of ω, our system has small vibrations that follow

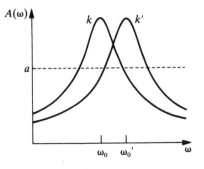

FIGURE 4-13

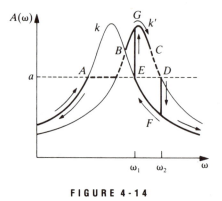

FIGURE 4-14

the amplitude curve for k. The amplitude moves up the tail of the $A(\omega)$ curve for k as shown in Figure 4-14.

However, when the vibration amplitude $A(\omega)$ is larger than the critical amplitude a, the force constant k' is effective. For these larger amplitudes, the system follows $A'(\omega)$ for force constant k'. This is represented by the solid bold line from B to C in Figure 4-14.

Between A and B, as the frequency increases, the system follows the simplified amplitude rise shown by the dashed line in Figure 4-14. Continuing to increase the driving frequency ω at C, we again reach the critical amplitude a at point D. If ω is only slightly increased, the system must follow $A(\omega)$ for k, and the amplitude suddenly jumps down from $A'(\omega)$ at point D to $A(\omega)$ at point F at $\omega = \omega_2$. As ω continues increasing above ω_2, the system follows the $A(\omega)$ curve.

Now let us see what happens if we decrease ω from large values. The system follows $A(\omega)$ until $\omega = \omega_1$, where $A(\omega) = a$. If ω is barely decreased, the amplitude increases above a, and the system must follow $A'(\omega)$. Therefore the amplitude jumps from E to G. As ω continues decreasing, it follows a similar path as before.

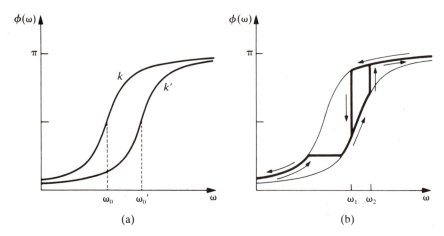

FIGURE 4-15

A hysteresis effect occurs because the system behaves differently depending on whether ω is increasing or decreasing. Two amplitude jumps occur, one for ω increasing and one for ω decreasing. The system's paths are *ABGCDF* (ω increasing) and *FEGBA* (ω decreasing).

Similar phenomena occur for the phase lag $\phi(\omega)$. Figure 4-15a shows the phase curves $\phi(\omega)$ and $\phi'(\omega)$ for the linear harmonic oscillators. Using the same arguments as applied to $A(\omega)$, we depict the system's paths in Figure 4-15b by the bold lines and the arrows. The reader is referred to the article by Janssen et al. for an experiment suitably demonstrating these phenomena.

4.6   CHAOS IN A PENDULUM

We will use the damped and driven pendulum to introduce several chaos concepts. The simple motion of a pendulum is well understood after hundreds of years of study, but its chaotic motion has been extensively studied only in the past few years. Among the motions of pendula that have been found to be chaotic are a pendulum with a forced oscillating support as shown in Figure 4-16a, the double pendulum

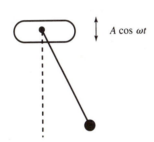

(a) Forced Pivot

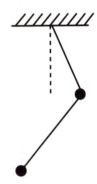

(b) Double pendulum

(c) Coupled pendulums

(d) Magnetic pendulum

FIGURE 4-16   Examples of systems that can have chaotic motion.

(Figure 4-16b), coupled pendula (Figure 4-16c), and a pendulum oscillating between magnets (Figure 4-16d). The damped and driven pendulum that we will consider is driven around its pivot point, and the geometry is displayed in Figure 4-17.

The torque around the pivot point can be written as

$$N = I\frac{d^2\theta}{dt^2} = I\ddot{\theta} = -b\dot{\theta} - mg\ell \sin\theta + N_d \cos\omega_d t \tag{4.38}$$

where I is the moment of inertia, b is the damping coefficient, and N_d is the driving torque of angular frequency ω_d. If we divide by $I = m\ell^2$, we have

$$\ddot{\theta} = -\frac{b}{m\ell^2}\dot{\theta} - \frac{g}{\ell}\sin\theta + \frac{N_d}{m\ell^2}\cos\omega_d t \tag{4.39}$$

We will eventually want to deal with this equation with a computer, and it will be much easier in that case to use dimensionless parameters. Let's divide Equation 4.39 by $\omega_0^2 = g/\ell$ and define the dimensionless time $t' = t/t_0$ with $t_0 = 1/\omega_0$ and the dimensionless driving frequency $\omega = \omega_d/\omega_0$. The new dimensionless variables and parameters are

$$x = \theta \qquad\qquad \text{oscillating variable} \tag{4.40a}$$

$$c = \frac{b}{m\ell^2\omega_0} \qquad\qquad \text{damping coefficient} \tag{4.40b}$$

$$F = \frac{N_d}{m\ell^2\omega_0^2} = \frac{N_d}{mg\ell} \qquad \text{driving force strength} \tag{4.40c}$$

$$t' = \frac{t}{t_0} = \sqrt{\frac{g}{\ell}}\,t \qquad\qquad \text{dimensionless time} \tag{4.40d}$$

$$\omega = \frac{\omega_d}{\omega_0} = \sqrt{\frac{\ell}{g}}\,\omega_d \qquad \text{driving angular frequency} \tag{4.40e}$$

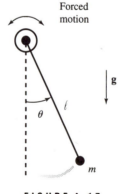

FIGURE 4-17

Note that

$$\dot{x} = \frac{dx}{dt'} = \frac{d\theta}{dt}\frac{dt}{dt'} = \frac{d\theta}{dt}\frac{1}{\omega_0}$$

$$\ddot{x} = \frac{d^2x}{dt'^2} = \frac{d^2\theta}{dt^2}\left(\frac{dt}{dt'}\right)^2 = \frac{d^2\theta}{dt^2}\frac{1}{\omega_0{}^2} = \frac{\ddot{\theta}}{\omega_0{}^2}$$

Using these variables and parameters, Equation 4.39 becomes

$$\ddot{x} = -c\dot{x} - \sin x + F\cos \omega t' \tag{4.41}$$

Equation 4.41 is a nonlinear equation of the form first presented in Equation 4.1. We will use numerical methods to solve this equation for x, given the parameters c, F, and ω. The techniques mentioned in Chapter 3 are used to solve this equation, depending on the accuracy desired and computer speed available, and commercial software programs are available. We use the program *Chaos Demonstrations* by Sprott and Rowlands (Sp92).

Equation 4.41, a second-order differential equation, can be reduced to two first-order equations by making the substitution

$$y = \frac{dx}{dt'} \tag{4.42}$$

Equation 4.41 becomes a first-order differential equation

$$\frac{dy}{dt'} = -cy - \sin x + F\cos z \tag{4.43}$$

where we have also made the substitution $z = \omega t'$. Equations 4.42 and 4.43 are the first-order differential equations.

We present the results of numerical methods solutions in Figure 4-18. We leave the parameters c and ω set at 0.05 and 0.7, respectively, and vary only the driving strength F in steps of 0.1 from 0.4 to 1.0. The results are that the motion is periodic for F values of 0.4, 0.5, 0.8, and 0.9 but is chaotic for 0.6, 0.7, and 1.0. These results indicate the beautiful and surprising results obtained from nonlinear dynamics. The left side of Figure 4-18 displays $y = dx/dt'$ (angular velocity) versus time long after the initial motion (i.e., transient effects have died out). The value of $F = 0.4$ shows simple harmonic motion, but the results for 0.5, 0.8, and 0.9, although periodic, are hardly simple.

We can learn more by examining the phase space plots, shown in the middle column of Figure 4-18 (note that we present only a unit cell of the phase diagram from $-\pi$ to π). As expected, the result for $F = 0.4$ shows the results seen previously in Chapter 3 (Figure 3-1). The phase plot for $F = 0.5$ shows one long cycle that includes two complete revolutions and two oscillations. The entire allowed area in the phase plane is accessed chaotically for $F = 0.6$ and 0.7, but for $F = 0.8$, the motion becomes periodic again with one complete revolution and an oscillation. The result for $F = 0.9$ is interesting, because there appears to be two different

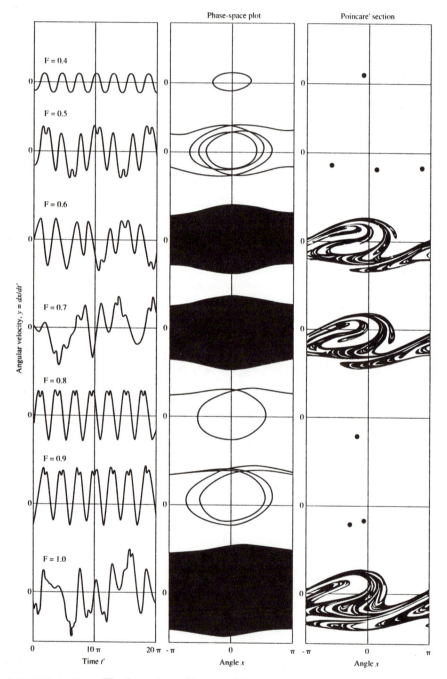

FIGURE 4-18 The damped and driven pendulum for various values of the driving force strength. The angular velocity versus time is shown on the left, and phase diagrams are in the center. Poincaré sections are shown on the right. Note that motion is chaotic for the driving force F values of 0.6, 0.7, and 1.0.

revolutions in one cycle, each similar to the one for $F = 0.8$. This result is called *period doubling* (i.e., the period for $F = 0.9$ is twice the period for $F = 0.8$). After close inspection, this effect can also be observed from the dx/dt' versus time plot, shown on the left column of Figure 4-18.

Poincaré Section

Henry Poincaré invented a technique to simplify the representations of phase space diagrams, which can become quite complicated. It is equivalent to taking a stroboscopic view of the phase space diagram. A three-dimensional phase diagram plots $y (= \dot{x} = \dot{\theta})$ versus $x (= \dot{\theta})$ versus $z (= \omega t')$. The left column of Figure 4-18 is a projection of this plot onto a y-z plane, showing points that correspond to various values of phase angle x. The middle column of Figure 4-18 is a projection onto a y-x plane, showing points belonging to various values of z. In Figure 4-19,

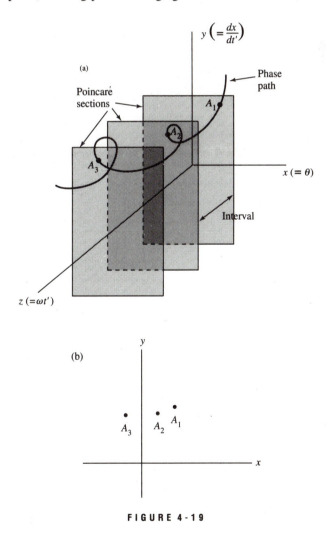

FIGURE 4-19

we show the three-dimensional phase space diagram intersected by a set of y-x planes, perpendicular to the z-axis, at equal z intervals. A *Poincaré section plot* is the sequence of points formed by the intersections of the phase path with these parallel planes in phase space, projected on one of the planes. The phase path pierces the planes as a function of angular speed ($y = \dot{\theta}$), time ($z = \omega t'$), and phase angle ($x = \theta$). The points on the intersections are labeled as A_1, A_2, A_3, etc. This set of points A_i forms a pattern when projected on one of the planes (Figure 4-19b) that sometimes will be a recognizable curve, but sometimes will appear irregular. For simple harmonic motion, such as $F = 0.4$ in Figure 4-18, all the points projected are the same (or in a smooth curve, depending on the z spacing of the y-x planes). Poincaré realized that the simple curves represent motion with possibly analytic solutions, but the many complicated, apparently irregular, curves represent chaos. The Poincaré section curve effectively reduces an N-dimensional diagram to $(N-1)$-dimensions for graphical purposes and often helps visualize the motion in phase space.

For the case of the damped and driven pendulum, the regularity of the dynamical motion is due to the forcing period, and a complete description of the dynamical motion depends on three parameters. We can take those parameters to be x (angle θ), $y = dx/dt'$ (angular frequency), and $z = \omega t'$ (phase of the driving force). A complete description of the motion in phase space would require three-dimensional phase diagrams rather than displaying just two parameters as in Figure 4-18. All the values of z are included in the middle column of Figure 4-18, so we choose to take the stroboscopic sections of the motion for just the values of $z = 2n\pi$ ($n = 0, 1, 2, \ldots$), which is at a frequency equal to that of the driving force.

We show the Poincaré section for the pendulum in the right column of Figure 4-18 for the same systems displayed in the left and middle columns. For the simple motion of $F = 0.4$, the system always comes back to the same position of (x, y) after z goes through 2π. Therefore, we expect the Poincaré section to show only one point, and that is what we find in the top figure of the right column of Figure 4-18. The motion for $F = 0.8$ also shows only one point, but $F = 0.5$ and 0.9 show three and two points, respectively, because of the more complex motion. The number of points n on the Poincaré section here shows that the new period $T = T_0 n/m$, where $T_0 = 2\pi/\omega$ is the period of the driven force and m is an integer ($m = 2$ for the $F = 0.5$ plot and $m = 1$ for the $F = 0.9$ plot). The chaotic motions for $F = 0.6$, 0.7, and 1.0 display the complicated variation of points expected for chaotic motion with a period $T \to \infty$. The Poincaré sections are also rich in structure for chaotic motion.

On three occasions thus far (Figures 4-5, 4-7, and 4-10), we have pointed out *attractors*, a set of points (or a point) on which the motion converges for dissipative systems. The regions traversed in phase space are strictly bounded when there is an attractor. In chaotic motion, nearby trajectories in phase space are continually diverging from one another but must eventually return to the attractor. Because the attractors in these chaotic motions, called *strange* or *chaotic attractors*, are necessarily bounded in phase space, the attractors must fold back into the nearby regions of phase space. Strange attractors create intricate patterns, because the folding and stretching of the trajectories must occur such that no trajectory in phase space intersects, which is ruled out by the deterministic dynamical motion. The Poincaré sections of Figure 4-18 reveal the folded, layered structure of the attractors. Chaotic attractors are fractals, but space does not permit further discussion of this extremely interesting phenomenom.

4.7 MAPPING

If we use n to denote the time sequence of a system and x to denote a physical observable of the system, we can describe the progression of a nonlinear system at a particular moment by investigating how the $(n+1)$th state (or iterate) depends on the nth state. An example of such a simple, nonlinear behavior is $x_{n+1} = (2x_n+3)^2$. This relationship, $x_{n+1} = f(x_n)$, is called **mapping** and is often used to describe the progression of the system. The Poincaré section plots previously discussed are examples of two-dimensional maps. A physical example appropriate for mapping might be the temperature of the space shuttle orbiter tiles while the shuttle descends through the atmosphere. After the orbiter has been on the ground for some time, the temperature T_{n+1} is the same as T_n, but this was not true while the shuttle plummented through the atmosphere from its orbit. Modeling the tile temperatures correctly with a mathematical model is difficult, and linear assumptions are often first assumed in such calculations with nonlinear terms added to make more realistic calculations.

We can write a *difference equation* using $f(\alpha,x_n)$ where x_n is restricted to a real number in the interval $(0,1)$ between 0 and 1, and α is a model-dependent parameter.

$$x_{n+1} = f(\alpha, x_n) \qquad (4.44)$$

The function $f(\alpha,x_n)$ generates the value of x_{n+1} from x_n, and the collection of points generated is said to be a **map** of the function itself. The equations, which are often nonlinear, are amenable to numerical solution by iteration, starting with x_1. We will restrict ourselves here to one-dimensional maps, but two-dimensional (and higher order) equations are possible.

Mapping can best be understood by looking at an example. Let us consider the "logistic" equation, a simple one-dimensional equation given by

$$f(\alpha,x) = \alpha x(1 - x) \qquad (4.45)$$

so that the iterative equation becomes

$$x_{n+1} = \alpha x_n(1 - x_n) \qquad (4.46)$$

We follow the discussion of Bessoir and Wolf (Be91) who use the logistic equation for a biological application example of studying the population growth of fish in a pond, where the pond is well isolated from external effects such as weather. The iterations, or n values, represent the annual fish population, where x_1 is the number of fish in the pond at the beginning of the first year of the study. If x_1 is small, the fish population may grow rapidly in the early years because of available resources, but overpopulation may eventually deplete the number of fish. The population x_n

is scaled so that its value fits in the interval (0, 1) between 0 and 1. The factor α is a model-dependent parameter representing average effects of environmental factors (e.g., fishermen, floods, drought, predators) that may affect the fish. The factor α may be varied as desired in the study, but experience shows that α should be limited in this example to the interval (0, 4) to prevent the fish population from becoming negative or infinite.

The results of the logistic equation are most easily observed by graphical means in a map called the *logistic map*. The iteration x_{n+1} is plotted versus x_n in Figure 4-20a for a value of $\alpha = 2.0$. Starting with an initial value x_1 on the horizontal (x_n) axis, we move up until we intersect with the curve $x_{n+1} = 2x_n(1 - x_n)$, and then we move to the left where we find x_2 on the vertical axis

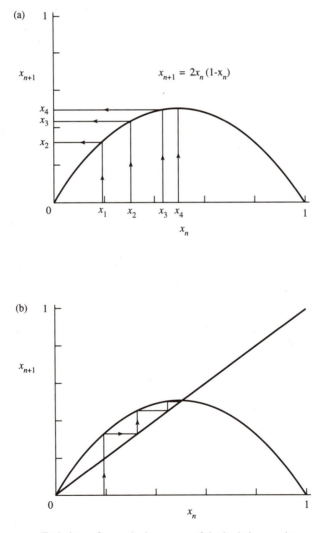

FIGURE 4-20   Techniques for producing a map of the logistic equation.

(x_{n+1}). We then start with this value of x_2 on the horizontal axis and repeat the process to find x_3 on the vertical axis. If we do this for a few iterations, we converge on the value $x = 0.5$, and the fish population stabilizes at half its maximum. We arrive at this result independent of our initial value of x_1 as long as it is not 0 or 1.

An easier way to follow the process is to add the 45° line, $x_{n+1} = x_n$, to the same graph. Then after initially intersecting the curve from x_1 one moves horizontally to intersect with the 45° line to find x_2 and then moves up vertically to find the next iterative value of x_3. This process can go on and reach the same result as in Figure 4-20a. We show the process in Figure 4-20b to indicate that this method is easier to use than the one without the 45° line.

In practice, we want to study the behavior of the system when the model parameter α is varied. In the present case, for values of α less than 3.0, stable populations will result (Figure 4-21a). The solutions follow a square spiral path

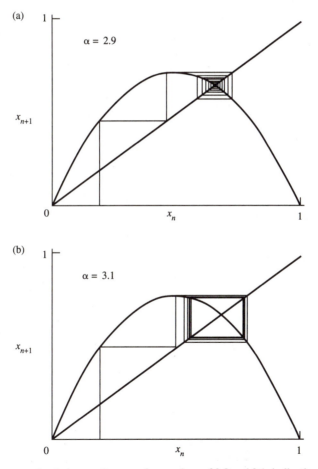

FIGURE 4-21   Logistic equation map for α values of 2.9 and 3.1, indicating the multiple solutions possible for $\alpha > 3.0$.

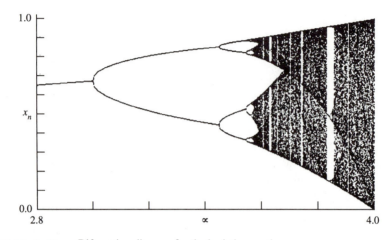

FIGURE 4-22   Bifurcation diagram for the logistic equation map.

to the central, final value. For values of α just above 3.0, more than one solution for the fish population occurs (Figure 4-21b). The solutions follow a path similar to the square spiral, which converges to the two points at which the square intersects the "iteration line," rather than to a single point. Such a change in the number of solutions to an equation, when a parameter such as α is varied, is called a **bifurcation**.

We obtain a more general view of the global picture by plotting a *bifurcation diagram*, which consists of x_n, determined after many iterations to avoid initial effects, plotted as a function of the model parameter α. Many new interesting effects emerge indicating regions and *windows* of stability as well as those of chaotic dynamics. We show the bifurcation diagram in Figure 4-22 for the logistic equation over the range of α values from 2.8 to 4.0. For the value of $\alpha = 2.9$ shown in Figure 4-21a, we observe that after a few iterations, a stable configuration for $x = 0.655$ results. An N cycle is an orbit that returns to its original position after N iterations, that is, $x_{N+i} = x_i$. The period for $\alpha = 2.9$ is then a *one cycle*. For $\alpha = 3.1$ (Figure 4-21b), the value of x oscillates between 0.558 and 0.765 (two cycle) after a few iterations evolve. The bifurcation occuring at 3.0 is called a *pitchfork bifurcation* because of the obvious shape of the diagram caused by the splitting. At $\alpha = 3.1$, the period doubling effect has $x_{n+2} = x_n$. At $\alpha = 3.45$, the two-cycle bifurcation evolves into a four cycle, and the bifurcation and period doubling continues up to an infinite number of cycles near $\alpha = 3.57$. Chaos occurs for many of the α values between 3.57 and 4.0, but there are still windows of periodic motion, with an especially wide window around 3.84. A really interesting behavior occurs for $\alpha = 3.82831$ (Problem 4-11). An apparent periodic cycle of 3 years seems to occur for several periods, but then it suddenly violently changes for a few years, and then returns again to the 3-year cycle. This *intermittent* behavior could certainly prove devastating to a biological study operating over several years that suddenly turns chaotic without apparent reason.

EXAMPLE **4.2** -

Let $\Delta\alpha_n = \alpha_n - \alpha_{n-1}$ be the width between successive period doubling bifurcations of the logistic map that we have been discussing. For example, from Figure 4-22, we let $\alpha_1 = 3.0$ where the first bifurcation occurs and $\alpha_2 = 3.449490$ where the next one occurs. Let δ_n be defined as the ratio

$$\delta_n = \frac{\Delta\alpha_n}{\Delta\alpha_{n+1}} \tag{4.47}$$

and let $\delta_n \to \delta$ as $n \to \infty$. Find δ_n for the first few bifurcations and the limit δ.

Solution: Although we could program this numerical calculation with a computer, we will use one of the commercially available software programs (Be91) to work this example. We make a table of the α_n values using the computer program, find $\Delta\alpha_n$, and then determine a few values of α_n.

n	α_n	$\Delta\alpha$	δ_n
1	3.0		
2	3.449490	0.449490	4.7515
3	3.544090	0.094600	4.6562
4	3.564407	0.020317	4.6684
5	3.568759	0.004352	
∞	3.5699456		4.6692

As α_n approaches the limit 3.5699456, the number of period doublings approaches infinity, and the ratio δ_n, called *Feigenbaum's number*, approaches 4.669202. This result was first found by Mitchell Feigenbaum in the 1970s, and he found that the limit δ was a universal property of the period doubling route to chaos when the function $f(\alpha, x)$ has a quadratic maximum. It is a remarkable fact that this universality is not confined to one-dimensional mappings; it is also true for two-dimensional maps and has been confirmed for several cases. Feigenbaum claims to have found this result using a programmable hand calculator. The calculation obviously has to be carried to many significant figures to establish its accuracy, and such a calculation was not possible before such calculators (or computers) were available.

- ●

4.8 CHAOS IDENTIFICATION

In our driven and damped pendulum, we found that chaotic motion occurs for some values of the parameters, but not for others. What are the characteristics of chaos and how can we identify them? Chaos does not represent periodic motion, and its

limiting motion will not be periodic. Chaos can generally be described as having a sensitive dependence on initial conditions. We can demonstrate this effect by the following example.

E X A M P L E **4.3** -

Consider the nonlinear relation $x_{n+1} = f(\alpha, x_n) = \alpha x_n(1 - x_n^2)$. Let $\alpha = 2.5$ and make two numerical calculations with initial x_1 values of 0.700000000 and 0.700000001. Plot the results and find the interation n where the solutions have clearly diverged.

Solution: The iterative equation that we are considering is

$$x_{n+1} = \alpha x_n(1 - x_n^2) \tag{4.48}$$

We perform a short numeric calculation and plot the results of iterations for the two initial values on the same graph. The result is shown in Figure 4-23 where there is no observed difference for x_{n+1} until n reaches at least 30. By $n = 39$, the difference in the two results is marked, despite the original values differing by only 1 part in 10^8.

If the computations are made without error, and the difference between iterated values doubled on the average for each iteration, then there will be an exponential increase such as

$$2^n = e^{n\ln 2}$$

where n is the number of iterations undergone. For the iterates to be separated by the order of unity (the size of the attractor), we will have

$$2^n 10^{-8} \sim 1$$

which gives $n = 27$. That is, after 27 iterations, the difference between the two

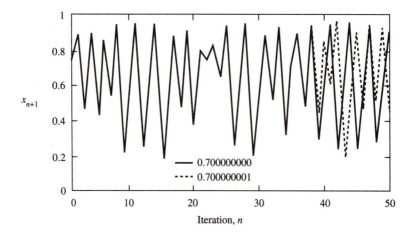

FIGURE 4-23

iterates reaches the full range of x_n. To have the results differ by unity for $n = 40$ iterations, we would have to know the initial values with a precision of 1 part in 10^{12}!

The previous example indicates the sensitive dependence on initial conditions that is characteristic of chaos. The two results can still be determined in this case, but it is rare to know the initial values to a precision of 10^{-8}. If we add another factor of 10 to the precision of x_1, we gain only four interative steps of agreement in the calculation. We must accept the reality that increasing the precision of the initial conditions only gains us a little in the accuracy of the ultimate measurement. This exponential growth of an initial error will ultimately prevent us from predicting the outcome of a measurement.

The effect of sensitive dependence on initial conditions has been called the "butterfly" effect. A butterfly moving slowly through the air may cause an extremely small effect on the airflow that will prevent us from predicting the weather patterns next week. Background noise or thermal effects will usually add uncertainties larger than the ones we have discussed here, and we cannot distinguish these effects from measurement errors. Precise predictive power of many steps is just not possible.

Lyapunov Exponents

One method to quantify the sensitive dependence on initial conditions for chaotic behavior uses the *Lyapunov characteristic exponent*. It is named after the Russian mathematician A. M. Lyapunov (1857–1918). There are as many Lyapunov exponents for a particular system as there are variables. We will limit ourselves at first to considering only one variable and therefore one exponent. Consider a system with two initial states differing by a small amount; we call the initial states x_0 and $x_0 + \varepsilon$. We want to investigate the eventual values of x_n after n iterations from the two initial values. The Lyapunov exponent λ represents the coefficient of the *average* exponential growth per unit time between the two states. After n iterations, the difference d_n between the two x_n values is approximately

$$d_n = \varepsilon e^{n\lambda} \tag{4.49}$$

From this equation, we can see that if λ is negative, the two orbits will eventually converge, but if positive, the nearby trajectories diverge and chaos results.

Let us look at a one-dimensional map described by $x_{n+1} = f(x_n)$. The initial difference between the states is $d_0 = \varepsilon$, and after one iteration, the difference d_1 is

$$d_1 = f(x_0 + \varepsilon) - f(x_0) \simeq \varepsilon \left. \frac{df}{dx} \right|_{x_0}$$

where the last result on the right side occurs because ε is very small. After n iterations, the difference d_n between the two initially nearby states is given by

$$d_n = f^n(x+\varepsilon) - f^n(x_0) = \varepsilon e^{n\lambda} \tag{4.50}$$

where we have indicated the nth iterate of the map $f(x)$ by the superscript n. If we divide by ε and take the logarithm of both sides, we have

$$\ln\left(\frac{f^n(x+\varepsilon) - f^n(x_0)}{\varepsilon}\right) = \ln(e^{n\lambda}) = n\lambda$$

and because ε is very small, we have for λ,

$$\lambda = \frac{1}{n}\ln\left(\frac{f^n(x+\varepsilon) - f^n(x_0)}{\varepsilon}\right) = \frac{1}{n}\ln\left|\frac{df^n(x)}{dx}\right|_{x_0} \qquad (4.51)$$

The value of $f^n(x_0)$ is obtained by iterating the function $f(x_0)$ n times.

$$f^n(x_0) = f(f(\cdots (f(x_0)) \cdots))$$

We use the derivative chain rule of the nth iterate to obtain

$$\frac{df^n(x)}{dx}\bigg|_{x_0} = \frac{df}{dx}\bigg|_{x_{n-1}} \frac{df}{dx}\bigg|_{x_{n-2}} \cdots \frac{df}{dx}\bigg|_{x_0}$$

We take the limit as $n \to \infty$ and finally obtain

$$\lambda = \lim_{n \to \infty} \frac{1}{n} \sum_{i=0}^{n-1} \ln\left|\frac{df(x_i)}{dx}\right| \qquad (4.52)$$

We plot the Lyapunov exponent as a function of α in Figure 4-24 for the logistic map. We note the agreement of the sign of λ with the discussion of chaotic behavior in Section 4.6. The value of λ is zero when bifurcation occurs, because $|df/dx| = 1$, and the solution becomes unstable (see Problem 4-16). A superstable

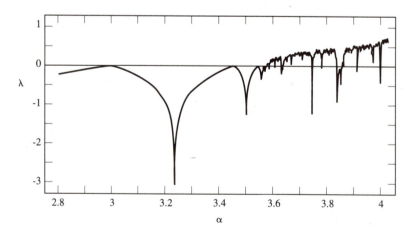

FIGURE 4-24   Lyapunov exponent as a function of α for the logistic equation map. A value of $\lambda > 0$ indicates chaos.

point occurs where $df(x)/dx = 0$, and this implies that $\lambda = -\infty$. From Figure 4-24 as λ goes above 0, we see there are windows where λ returns to $\lambda < 0$ and periodic orbits occur amid the chaotic behavior. The relatively wide window just above 3.8 is apparent.

Remember that for n dimensional maps, there will be n Lyapunov exponents. Only one of them need be positive for chaos to occur. For dissipative systems, the phase space volume will decrease as time passes. This means the sum of the Lyapunov exponents will be negative.

The calculation of Lyapunov exponents for the damped and driven pendulum is difficult, because one has to deal with the solutions of differential equations rather than maps such as those of the logistic equation. Nevertheless, these calculations have been done, and we show in Figure 4-25 the Lyapunov exponents, three of them because of the three dimensions (calculated using Baker's program [Ba90]). The parameters are the same as those discussed in Section 4.6: $c = 0.05$, $\omega = 0.7$, and $F = 0.4$ (periodic) and $F = 0.6$ (chaotic). For both cases, we must make at least several hundred iterations to make sure transient effects have died out. Note that one of the Lyapunov exponents is zero, because it does not contribute to the expansion or contraction of the phase space volume. For the case of $F = 0.4$, none

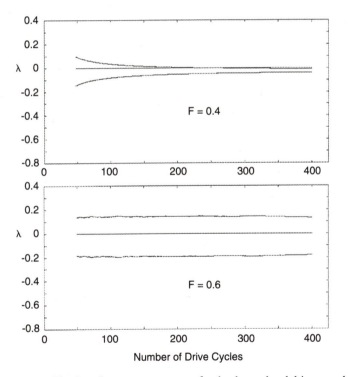

FIGURE 4-25 The three Lyapunov exponents for the damped and driven pendulum. The values of λ are those approached as $t \to \infty$ (large number of cycles).

of the Lyapunov exponents is greater than zero after 350 iterations, but for the $F = 0.6$ driven case, one of the exponents is still well above zero. The motion is chaotic for $F = 0.6$, as we found earlier in Fig. 4-18. However, because the motion described in Fig. 4-25 is damped, the sum of the three Lyapunov exponents is negative for both cases, as it should be.

P R O B L E M S

4-1. Refer to Example 4.1. If each of the springs must be stretched a distance d to attach the particle at the equilibrium position (i.e., in its equilibrium position, the particle is subject to two equal and oppositely directed forces of magnitude kd), then show that the potential in which the particle moves is approximately

$$U(x) \cong (kd/l)x^2 + [k(l - d)/4l^3]x^4$$

4-2. Construct a phase diagram for the potential in Figure 4-1.

4-3. Construct a phase diagram for the potential $U(x) = -(\lambda/3)x^3$.

4-4. Lord Rayleigh used the equation

$$\ddot{x} - (a - b\dot{x}^2)\dot{x} + \omega_0^2 x = 0$$

in his discussion of nonlinear effects in acoustic phenomena.* Show that differentiating this equation with respect to time and making the substitution $y = y_0 \sqrt{3b/a}\, \dot{x}$ results in van der Pol's equation:

$$\ddot{y} - \frac{a}{y_0^2}(y_0^2 - y^2)\dot{y} + \omega_0^2 y = 0$$

4-5. Solve by a successive approximation procedure, and obtain a result accurate to four significant figures:

(a) $x+x^2 + 1 = \tan x$, $0 \le x \le \pi/2$
(b) $x(x + 3) = 10 \sin x$, $x > 0$
(c) $1 + x + \cos x = e^x$, $x > 0$

(It may be profitable to make a crude graph to choose a reasonable first approximation.)

4-6. Derive the expression for the phase paths of the plane pendulum if the total energy is $E > 2mgl$. Note that this is just the case of a particle moving in a periodic potential $U(\theta) = mgl(1 - \cos \theta)$.

4-7. Consider the free motion of a plane pendulum whose amplitude is not small. Show that the *horizontal component* of the motion may be represented by the approximate expres-

---

* J. W. S. Rayleigh, *Phil. Mag.* **15** (April 1883); see also Ra94, Section 68a.

sion (components through the third order are included)

$$\ddot{x} + \omega_0^2(1 + \frac{x_0^2}{l^2})x - \varepsilon x^3 = 0$$

where $\omega_0^2 = g/l$ and $\varepsilon = 3g/2l^3$, with l equal to the length of the suspension.

4-8. A mass m moves in one dimension and is subject to a constant force $+F_0$ when $x < 0$ and to a constant force $-F_0$ when $x > 0$. Describe the motion by constructing a phase diagram. Calculate the period of the motion in terms of m, F_0, and the amplitude A (disregard damping).

4-9. Investigate the motion of an undamped particle subject to a force of the form

$$F(x) = \begin{cases} -kx, & |x| < a \\ -(k + \delta)x + \delta a, & |x| > a \end{cases}$$

where k and δ are positive constants.

4-10. The parameters $F = 0.7$ and $c = 0.05$ are fixed for Equation 4.43 describing the driven, damped pendulum. Determine which of the values for ω (0.1, 0.2, 0.3, . . . , 1.5) produce chaotic motion. Produce a phase plot for $\omega = 0.3$. Do this problem numerically.

4-11. A really interesting situation occurs for the logistic equation, Equation 4.46, when $\alpha = 3.82831$ and $x_1 = 0.51$. Show that a three cycle occurs with the approximate x values 0.16, 0.52, and 0.96 for the first 80 cycles before the behavior apparently turns chaotic. Find for what iteration the next apparently periodic cycle occurs and for how many cycles it stays periodic.

4-12. Let the value of α in the logistic equation, Equation 4.46, be equal to 0.9. Make a map like that in Figure 4-20 when $x_1 = 0.4$. Make the plot for three other values of x_1 for which $0 < x_1 < 1$.

4-13. Perform the numerical calculation done in Example 4.3 and show that the two calculations clearly diverge by $n = 39$. Next, let the second initial value agree to within another factor of 10 (i.e., 0.7000000001), and confirm the statement in the text that only four more iterations are gained in the agreement between the two initial values.

4-14. Use the function described in Example 4.3, $x_{n+1} = \alpha x_n(1 - x_n^2)$ where $\alpha = 2.5$. Consider two starting values of x_1 that are similar, 0.9000000 and 0.9000001. Make a plot of x_n versus n for the two starting values and determine the lowest value of n for which the two values diverge by more than 30%.

4-15. Use direct numerical calculation to show that the map $f(x) = \alpha \sin \pi x$ also leads to the Feigenbaum constant, where x and α are limited to the interval (0, 1).

4-16. The curve $x_{n+1} = f(x_n)$ intersects the curve $x_{n+1} = x_n$ at x_0. The expansion of x_{n+1} about x_0 is $x_{n+1} - x_0 = \beta(x_n - x_0)$ where $\beta = (df/dx)$ at $x = x_0$.
(a) Describe the geometrical sequence that the successive values of $x_{n+1} - x_0$ form.
(b) Show that the intersection is stable when $|\beta| < 1$ and unstable when $|\beta| > 1$.

4-17. The *tent* map is represented by the following iterations:

$$x_{n+1} = 2\alpha x_n \qquad \text{for } 0 < x < 1/2$$
$$x_{n+1} = 2\alpha(1 - x_n) \qquad \text{for } 1/2 < x < 1$$

where $0 < \alpha < 1$. Make a map up to 20 iterations for $\alpha = 0.4$ and 0.7 with $x_1 = 0.2$. Does it appear that either of the maps represent chaotic behavior?

4-18. Plot the bifurcation diagram for the *tent* map of the previous problem. Discuss the results for the various regions.

4-19. Show analytically that the Lyapunov exponent for the *tent* maps is $\lambda = \ln(2\alpha)$. This indicates that chaotic behavior occurs for $\alpha > 1/2$.

4-20. Consider the Henon map described by

$$x_{n+1} = y_n + 1 - ax_n^2$$
$$y_{n+1} = bx_n$$

Let $a = 1.4$ and $b = 0.3$, and use a computer to plot the first 10,000 points (x_n, y_n) starting from the initial values $x_0 = 0$, $y_0 = 0$. Choose the plot region as $-1.5 < x < 1.5$ and $-0.45 < y < 0.45$.

4-21. Make a plot of the Henon map, this time starting from the initial values $x_0 = 0.63135448$, $y_0 = 0.18940634$. Compare the shape of this plot with that obtained in the previous problem. Is the shape of the curves independent of the initial conditions?

4-22. A circuit with a nonlinear inductor can be modeled by the first-order differential equations

$$\frac{dx}{dt} = y$$

$$\frac{dy}{dt} = -ky - x^3 + B \cos t$$

Chaotic oscillations for this situation have been extensively studied. Use a computer to construct the Poincaré section plot for the case $k = 0.1$ and $9.8 \leq B \leq 13.4$. Describe the map.

4-23. The motion of a bouncing ball, on successive bounces, when the floor oscillates sinusoidally can be described by the Chirikov map:

$$p_{n+1} = p_n - K \sin q_n$$

$$q_{n+1} = q_n + p_{n+1}$$

where $-\pi \leq p \leq \pi$ and $-\pi \leq q \leq \pi$. Construct two dimensional maps for $K = 0.8, 3.2$, and 6.4 by starting with random values of p and q and iterating them. Use periodic boundary conditions, which means that if the iterated values of p or q exceed π, a value of 2π is subtracted and whenever they are less than $-\pi$, a value of 2π is added. Examine the maps after thousands of iterations and discuss the differences.

C H A P T E R

5

GRAVITATION

5.1 INTRODUCTION

By 1666, Newton had formulated and numerically checked the gravitation law he eventually published in his book *Principia* in 1687. Newton waited almost 20 years to publish his results because he could not justify his method of numerical calculation in which he considered the Earth and the moon as point masses. With mathematics formulated on calculus (which Newton later invented), we have a much easier time proving the problem Newton found so difficult in the seventeenth century.

Newton's law of universal gravitation states that *each mass particle attracts every other particle in the universe with a force that varies directly as the product of the two masses and inversely as the square of the distance between them.* In mathematical form, we write the law as

$$\mathbf{F} = -G\frac{mM}{r^2}\mathbf{e}_r \tag{5.1}$$

where at a distance r from a particle of mass M a second particle of mass m experiences an attractive force (see Figure 5-1). The unit vector $\mathbf{e}_r$ points from M to m, and the minus sign ensures that the force is attractive—that is, that m is attracted toward M.

A laboratory verification of the law and a determination of the value of G was made in 1798 by the English physicist Henry Cavendish (1731–1810). Cavendish's experiment, described in many elementary physics texts, used a torsion balance

FIGURE 5-1   Particle *m* feels an attractive gravitational force toward *M*.

with two small spheres fixed at the ends of a light rod. The two spheres were attracted to two other large spheres that could be placed on either side of the smaller spheres. The best value for G that has been obtained to date is $6.6726 \pm 0.0008 \times 10^{-11}$ N $\cdot$ m^2/kg^2. Interestingly, although G is perhaps the oldest known of the fundamental constants, we know it with less precision than we know most of the modern fundamental constants such as e, c, and $\hbar$.

In the form of Equation 5.1, the law strictly applies only to *point particles*. If one or both of the particles is replaced by a body with a certain extension, we must make an additional hypothesis before we can calculate the force. We must assume that the gravitational force field is a *linear field*. In other words, we assume that it is possible to calculate the net gravitational force on a particle due to many other particles by simply taking the vector sum of all the individual forces. For a body consisting of a continuous distribution of matter, the sum becomes an integral (Figure 5-2):

$$\mathbf{F} = -Gm \int_V \frac{\rho(\mathbf{r}')\mathbf{e}_r}{r^2}\, dv' \tag{5.2}$$

where $\rho(\mathbf{r}')$ is the mass density and dv' is the element of volume at the position defined by the vector $\mathbf{r}'$ from the (arbitrary) origin to the point within the mass distribution.

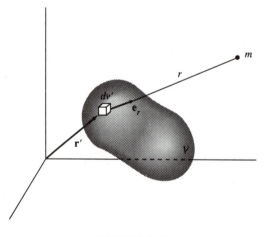

FIGURE 5-2

If both the body of mass M and the body of mass m have finite extension, a second integration over the volume of m will be necessary to compute the total gravitational force.

The **gravitational field vector g** is the vector representing the force per unit mass exerted on a particle in the field of a body of mass M. Thus

$$\mathbf{g} = \frac{\mathbf{F}}{m} = -G\frac{M}{r^2}\mathbf{e}_r \tag{5.3}$$

or

$$\mathbf{g} = -G\int_V \frac{\rho(\mathbf{r}')\mathbf{e}_r}{r^2}\,dv' \tag{5.4}$$

Note that the direction of $\mathbf{e}_r$ varies with r' (in Figure 5-2).

The quantity $\mathbf{g}$ has the dimensions of *force per unit mass*, also equal to *acceleration*. In fact, near the surface of the Earth, the magnitude of $\mathbf{g}$ is just the quantity that we call the **gravitational acceleration constant**. Measurement with a simple pendulum (or some more sophisticated variation) is sufficient to show that $|\mathbf{g}|$ is approximately 9.80 m/s^2 (or 9.80 N/kg) at the surface of the Earth.

5.2 GRAVITATIONAL POTENTIAL

The gravitational field vector $\mathbf{g}$ varies as $1/r^2$ and therefore satisfies the requirement* that permits $\mathbf{g}$ to be represented as the gradient of a scalar function. Hence, we can write

$$\mathbf{g} \equiv -\nabla\Phi \tag{5.5}$$

where Φ is called the **gravitational potential** and has dimensions of (*force per unit mass*) × (*distance*), or *energy per unit mass*.

Because $\mathbf{g}$ has only a radial variation, the potential Φ can have at most a variation with r. Therefore, using Equation 5.3 for $\mathbf{g}$, we have

$$\nabla\Phi = \frac{d\Phi}{dr}\mathbf{e}_r = G\frac{M}{r^2}\mathbf{e}_r$$

Integrating, we obtain

$$\Phi = -G\frac{M}{r} \tag{5.6}$$

---

* That is, $\nabla \times \mathbf{g} \equiv 0$.

The possible constant of integration has been suppressed, because the potential is undetermined to within an additive constant; that is, only *differences* in potential are meaningful, not particular values. We usually remove the ambiguity in the value of the potential by arbitrarily requiring that $\Phi \to 0$ as $r \to \infty$; then Equation 5.6 correctly gives the potential for this condition.

The potential due to a continuous distribution of matter is

$$\Phi = -G\int_V \frac{\rho(\mathbf{r}')}{r}\,dv' \tag{5.7}$$

Similarly, if the mass is distributed only over a thin shell (i.e., a *surface* distribution), then

$$\Phi = -G\int_S \frac{\rho_s}{r}\,da' \tag{5.8}$$

where ρ_s is the surface density of mass (or *areal mass density*).

Finally, if there is a *line source* with linear mass density ρ_l, then

$$\Phi = -G\int_\Gamma \frac{\rho_l}{r}\,ds' \tag{5.9}$$

The physical significance of the gravitational potential function becomes clear if we consider the work per unit mass dW' that must be done by an outside agent on a body in a gravitational field to displace the body a distance $d\mathbf{r}$. In this case, work is equal to the scalar product of the force and the displacement. Thus, for the work done *on* the body per unit mass, we have

$$dW' = -\mathbf{g}\cdot d\mathbf{r} = (\nabla\Phi)\cdot d\mathbf{r}$$

$$= \sum_i \frac{\partial\Phi}{\partial x_i}\,dx_i = d\Phi \tag{5.10}$$

because Φ is a function only of the coordinates of the point at which it is measured: $\Phi = \Phi(x_1, x_2, x_3) = \Phi(x_i)$. Therefore the amount of work per unit mass that must be done on a body to move it from one position to another in a gravitational field is equal to the difference in potential at the two points.

If the final position is farther from the source of mass M than the initial position, work has been done *on* the unit mass. The positions of the two points are arbitrary, and we may take one of them to be at infinity. If we define the potential to be zero at infinity, we may interpret Φ at any point to be the work per unit mass required to bring the body from infinity to that point. The *potential energy* is equal to the mass of the body multiplied by the potential Φ. If U is the potential energy, then

$$U = m\Phi \tag{5.11}$$

and the force on a body is given by the negative of the gradient of the potential energy of that body,

$$\mathbf{F} = -\nabla U \tag{5.12}$$

which is just the expression we have previously used (Equation 2.88).

We note that both the potential and the potential energy *increase* when work is done *on* the body. (The potential, according to our definition, is always negative and only approaches its maximum value, that is, zero, as r tends to infinity.)

A certain potential energy exists whenever a body is placed in the gravitational field of a source mass. This potential energy resides in the *field*,* but it is customary under these circumstances to speak of the potential energy "of the body." We shall continue this practice here. We may also consider the source mass itself to have an intrinsic potential energy. This potential energy is equal to the gravitational energy released when the body was formed or, conversely, is equal to the energy that must be supplied (i.e., the work that must be done) to disperse the mass over the sphere at infinity. For example, when interstellar gas condenses to form a star, the gravitational energy released goes largely into the initial heating of the star. As the temperature increases, energy is radiated away as electromagnetic radiation. In all the problems we treat, the structure of the bodies is considered to remain unchanged during the process we are studying. Thus, there is no change in the intrinsic potential energy, and it may be neglected for the purposes of whatever calculation we are making.

EXAMPLE 5.1 -

What is the gravitational potential both inside and outside a spherical shell of inner radius b and outer radius a?

Solution: One of the important problems of gravitational theory concerns the calculation of the gravitational force due to a homogeneous sphere. This problem is a special case of the more general calculation for a homogeneous spherical shell. A solution to the problem of the shell can be obtained by directly computing the force on an arbitrary object of unit mass brought into the field (see Problem 5-6), but it is easier to use the potential method.

We consider the shell shown in Figure 5-3 and calculate the potential at point P a distance R from the center of the shell. Because the problem has symmetry about the line connecting the center of the sphere and the field point P, the azimuthal angle ϕ is not shown in Figure 5-3 and we can immediately integrate over $d\phi$ in the expression for the potential. Thus,

$$\Phi = -G \int_V \frac{\rho(r')}{r} dv'$$

$$= -2\pi\rho G \int_b^a r'^2 dr' \int_0^\pi \frac{\sin\theta}{r} d\theta \tag{5.13}$$

---

* See, however, the remarks at the end of Section 9.5 regarding the energy in a field.

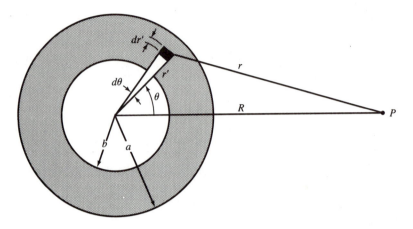

FIGURE 5-3

where we have assumed a homogeneous mass distribution for the shell, $\rho(r') = \rho$. According to the law of cosines,

$$r^2 = r'^2 + R^2 - 2r' R \cos \theta \tag{5.14}$$

Because R is a constant, for a given r' we may differentiate this equation and obtain

$$2r \, dr = 2r'R \sin \theta \, d\theta$$

or

$$\frac{\sin \theta}{r} d\theta = \frac{dr}{r'R} \tag{5.15}$$

Substituting this expression into Equation 5.13, we have

$$\Phi = -\frac{2\pi\rho G}{R} \int_b^a r' \, dr' \int_{r_{min}}^{r_{max}} dr \tag{5.16}$$

The limits on the integral over dr depend on the location of point P. If P is *outside* the shell, then

$$\Phi(R > a) = -\frac{2\pi\rho G}{R} \int_b^a r' \, dr' \int_{R-r'}^{R+r'} dr$$

$$= -\frac{4\pi\rho G}{R} \int_b^a r'^2 \, dr'$$

$$= -\frac{4}{3}\frac{\pi\rho G}{R}(a^3 - b^3) \tag{5.17}$$

But the mass M of the shell is

$$M = \frac{4}{3}\pi\rho(a^3 - b^3) \tag{5.18}$$

so the potential is

$$\boxed{\Phi(R > a) = -\frac{GM}{R}}$$

(5.19)

If the field point lies inside the shell, then

$$\Phi(R < b) = -\frac{2\pi\rho G}{R} \int_b^a r' \, dr' \int_{r'-R}^{r'+R} dr$$

$$= -4\pi\rho G \int_b^a r' \, dr'$$

$$= -2\pi\rho G(a^2 - b^2)$$

(5.20)

The potential is therefore constant and independent of position within the shell.

Finally, if we wish to calculate the potential for points *within* the shell, we need only replace the lower limit of integration in the expression for $\Phi(R < b)$ by the variable R, replace the upper limit of integration in the expression for $\Phi(R > a)$ by R, and add the results. We find

$$\Phi(b < R < a) = -\frac{4\pi\rho G}{3R}(R^3 - b^3) - 2\pi\rho G(a^2 - R^2)$$

$$= -4\pi\rho G\left(\frac{a^2}{2} - \frac{b^3}{3R} - \frac{R^2}{6}\right)$$

(5.21)

We see that if $R \rightarrow a$, then Equation 5.21 yields the same result as Equation 5.19 for the same limit. Similarly, Equations 5.21 and 5.20 produce the same result for the limit $R \rightarrow b$. The potential is therefore *continuous*. If the potential were not continuous at some point, the gradient of the potential—and hence, the force— would be infinite at that point. Because infinite forces do not represent physical reality, we conclude that realistic potential functions must always be continuous.

Note that we treated the mass shell as homogeneous. In order to perform calculations for a solid, massive body like a planet that has a spherically symmetric mass distribution, we could add up a number of shells or, if we choose, we could allow the density to change as a function of radius.

-- ●

The results of Example 5.1 are very important. Equation 5.19 states that the potential at any point outside of a spherically symmetric distribution of matter (shell or solid, because solids are composed of many shells) is independent of the size of the distribution. Therefore, to calculate the external potential (or the force), we consider all the mass to be concentrated at the center. Equation 5.20 indicates that the potential is constant (and the force zero) anywhere inside a spherically symmetric mass shell. And finally, at points within the mass shell, the potential given by Equation 5.21 is consistent with both of the previous results.

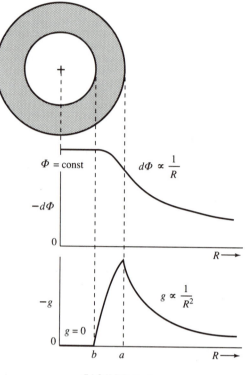

FIGURE 5-4

The magnitude of the field vector **g** may be computed from $g = -d\Phi/dR$ for each of the three regions. The results are

$$
\left.
\begin{aligned}
g(R < b) &= 0 \\
g(b < R < a) &= \frac{4\pi\rho G}{3}\left(\frac{b^3}{R^2} - R\right) \\
g(R > a) &= -\frac{GM}{R^2}
\end{aligned}
\right\}
\tag{5.22}
$$

We see that not only the potential but also the field vector (and hence, the force) are continuous. The *derivative* of the field vector, however, is not continuous across the outer and inner surfaces of the shell.

All these results for the potential and the field vector can be summarized as in Figure 5-4.

EXAMPLE 5.2 --

Consider a thin uniform circular ring of radius a and mass M. A mass m is placed in the plane of the ring. Find a position of equilibrium and determine whether it is stable.

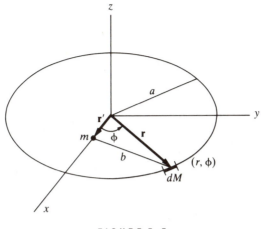

FIGURE 5-5

Solution: From symmetry, we might believe that the mass m placed in the center of the ring (Figure 5-5) should be in equilibrium because it is uniformly surrounded by mass. Put mass m at a distance r' from the center of the ring, and place the x-axis along this direction.

The potential is given by Equation 5.7 where $\rho = M/2\pi a$:

$$d\Phi = -G\frac{dM}{b} = -\frac{Ga\rho}{b}\,d\phi \qquad (5.23)$$

where b is the distance between dM and m, and $dM = \rho a\,d\phi$. Let $\mathbf{r}$ and $\mathbf{r}'$ be the position vectors to dM and m, respectively.

$$b = |\mathbf{r} - \mathbf{r}'| = |a\cos\phi\,\mathbf{e}_1 + a\sin\phi\,\mathbf{e}_2 - r'\mathbf{e}_1|$$

$$= |(a\cos\phi - r')\mathbf{e}_1 + a\sin\phi\,\mathbf{e}_2| = [(a\cos\phi - r')^2 + a^2\sin^2\phi]^{1/2}$$

$$= (a^2 + r'^2 - 2ar'\cos\phi)^{1/2} = a\left[1 + \left(\frac{r'}{a}\right)^2 - \frac{2r'}{a}\cos\phi\right]^{1/2} \qquad (5.24)$$

Integrating Equation 5.23 gives

$$\Phi(r') = -G\int\frac{dM}{b} = -\rho aG\int_0^{2\pi}\frac{d\phi}{b}$$

$$= -\rho G\int_0^{2\pi}\frac{d\phi}{\left[1 + \left(\frac{r'}{a}\right)^2 - \frac{2r'}{a}\cos\phi\right]^{1/2}} \qquad (5.25)$$

The integral in Equation 5.25 is difficult, so let us consider positions close to the equilibrium point, $r' = 0$. If $r' \ll a$, we can expand the denominator in Equation 5.25.

$$\left[1 + \left(\frac{r'}{a}\right)^2 - \frac{2r'}{a}\cos\phi\right]^{-1/2} = 1 - \frac{1}{2}\left[\left(\frac{r'}{a}\right)^2 - \frac{2r'}{a}\cos\phi\right]$$

$$+ \frac{3}{8}\left[\left(\frac{r'}{a}\right)^2 - \frac{2r'}{a}\cos\phi\right]^2 + \cdots$$

$$= 1 + \frac{r'}{a}\cos\phi + \frac{1}{2}\left(\frac{r'}{a}\right)^2(3\cos^2\phi - 1) + \cdots \quad \textbf{(5.26)}$$

Equation 5.25 becomes

$$\Phi(r') = -\rho G \int_0^{2\pi} \left\{1 + \frac{r'}{a}\cos\phi + \frac{1}{2}\left(\frac{r'}{a}\right)^2(3\cos^2\phi - 1) + \cdots\right\} d\phi \quad \textbf{(5.27)}$$

which is easily integrated with the result

$$\Phi(r') = -\frac{MG}{a}\left[1 + \frac{1}{4}\left(\frac{r'}{a}\right)^2 + \cdots\right] \quad \textbf{(5.28)}$$

The potential energy $U(r')$ is from Equation 5.11, simply

$$U(r') = m\Phi(r') = -\frac{mMG}{a}\left[1 + \frac{1}{4}\left(\frac{r'}{a}\right)^2 + \cdots\right] \quad \textbf{(5.29)}$$

The position of equilibrium is found (from Equation 2.100) by

$$\frac{dU(r')}{dr'} = 0 = -\frac{mMG}{a}\frac{1}{2}\frac{r'}{a^2} + \cdots \quad \textbf{(5.30)}$$

so $r' = 0$ is an equilibrium point. We use Equation 2.103 to determine the stability:

$$\frac{d^2U(r')}{dr'^2} = -\frac{mMG}{2a^3} + \cdots < 0 \quad \textbf{(5.31)}$$

so the equilibrium point is unstable.

This last result is not obvious, because we might be led to believe that a small displacement from $r' = 0$ might still be returned to $r' = 0$ by the gravitational forces from all the mass in the ring surrounding it.

--- ●

Poisson's Equation

It is useful to compare these properties of gravitational fields with some of the familiar results from electrostatics that were determined in the formulation of Maxwell's equations. Consider an arbitrary surface as in Figure 5-6 with a mass m placed somewhere inside. Similar to electric flux, let's find the gravitational flux Φ_m emanating from mass m through the arbitrary surface S.

$$\Phi_m = \int_S \mathbf{n} \cdot \mathbf{g}\, da \quad \textbf{(5.32)}$$

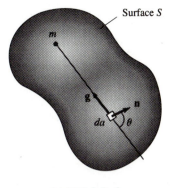

Surface S

FIGURE 5-6

where the integral is over the surface S and the unit vector **n** is normal to the surface at the differential area da. If we substitute **g** from Equation 5.3 for the gravitational field vector for a body of mass m, we have for the scalar product $\mathbf{n} \cdot \mathbf{g}$,

$$\mathbf{n} \cdot \mathbf{g} = -Gm \frac{\cos \theta}{r^2}$$

where θ is the angle between **n** and **g**. We substitute this into Equation 5.32 and obtain

$$\Phi_m = -Gm \int_S \frac{\cos \theta}{r^2} \, da$$

The integral is over the solid angle of the arbitrary surface and has the value 4π radians, which gives for the mass flux

$$\Phi_m = \int_S \mathbf{n} \cdot \mathbf{g} \, da = -4\pi Gm \tag{5.33}$$

Note that it is immaterial where the mass is located inside the surface S. We can generalize this result for many masses m_i inside the surface S by summing over the masses.

$$\int_S \mathbf{n} \cdot \mathbf{g} \, da = -4\pi G \sum_i m_i \tag{5.34}$$

If we change to a continuous mass distribution within surface S, we have

$$\int_S \mathbf{n} \cdot \mathbf{g} \, da = -4\pi G \int_V \rho \, dv \tag{5.35}$$

where the integral on the right-hand side is over the volume V enclosed by S, ρ is the mass density, and dv is the differential volume. We use Gauss's divergence theorem to rewrite this result. Gauss's divergence theorem, Equation 1.130 where

$da = \mathbf{n}\, da$, is

$$\int_S \mathbf{n} \cdot \mathbf{g}\, da = \int_V \mathbf{\nabla} \cdot \mathbf{g}\, dv \tag{5.36}$$

If we set the right-hand sides of Equations 5.35 and 5.36 equal, we have

$$\int_V (-4\pi G)\rho\, dv = \int_V \mathbf{\nabla} \cdot \mathbf{g}\, dv$$

and because the surface S, and its volume V, is completely arbitrary, the two integrands must be equal.

$$\mathbf{\nabla} \cdot \mathbf{g} = -4\pi G\rho \tag{5.37}$$

This result is similar to the differential form of Gauss's law for electric field, $\mathbf{\nabla} \cdot \mathbf{E} = \rho/\varepsilon$, where ρ in this case is the charge density.

We insert $\mathbf{g} = -\mathbf{\nabla}\Phi$ from Equation 5.5 into the left-hand side of Equation 5.37 and obtain $\mathbf{\nabla} \cdot \mathbf{g} = -\mathbf{\nabla} \cdot \mathbf{\nabla}\Phi = -\nabla^2\Phi$. Equation 5.37 becomes

$$\nabla^2\Phi = 4\pi G\rho \tag{5.38}$$

which is known as *Poisson's equation* and is useful in a number of potential theory applications. When the right-hand side of Equation 5.38 is zero, the result $\nabla^2\Phi = 0$ is an even better known equation called *Laplace's equation*. Poisson's equation is useful in developing Green's functions, whereas we often encounter Laplace's equation when dealing with various coordinate systems.

5.3 LINES OF FORCE AND EQUIPOTENTIAL SURFACES

Let us consider a mass that gives rise to a gravitational field that can be described by a field vector $\mathbf{g}$. Let us draw a line outward from the surface of the mass such that the direction of the line at every point is the same as the direction of $\mathbf{g}$ at that point. This line will extend from the surface of the mass to infinity. Such a line is called a **line of force**.

By drawing similar lines from every small increment of surface area of the mass, we can indicate the direction of the force field at any arbitrary point in space. The lines of force for a single point mass are all straight lines extending from the mass to infinity. Defined in this way, the lines of force are related only to the *direction* of the force field at any point. We may consider, however, that the *density* of such lines—that is, the number of lines passing through a unit area oriented perpendicular to the lines—is proportional to the *magnitude* of the force at that area. The lines-of-force picture is thus a convenient way to visualize both the magnitude and the direction (i.e., the *vector* property) of the field.

The potential function is defined at every point in space (except at the position of a point mass). Therefore, the equation

$$\Phi = \Phi(x_1, x_2, x_3) = \text{constant} \tag{5.39}$$

defines a surface on which the potential is constant. Such a surface is called an **equipotential surface**. The field vector **g** is equal to the gradient of Φ, so **g** can have no component *along* an equipotential surface. It therefore follows that every line of force must be normal to every equipotential surface. Thus, the field does no work on a body moving along an equipotential surface. Because the potential function is single valued, no two equipotential surfaces can intersect or touch. The surfaces of equal potential that surround a single, isolated point mass (or any spherically symmetric mass) are all spheres. Consider two point masses M that are separated by a certain distance. If r_1 is the distance from one mass to some point in space and if r_2 is the distance from the other mass to the same point, then

$$\Phi = -GM \left(\frac{1}{r_1} + \frac{1}{r_2} \right) = \text{constant} \qquad (5.40)$$

defines the equipotential surfaces. Several of these surfaces are shown in Figure 5-7 for this two-particle system. In three dimensions, the surfaces are generated by rotating this diagram around the line connecting the two masses.

5.4 WHEN IS THE POTENTIAL CONCEPT USEFUL?

The use of potentials to describe the effects of "action-at-a-distance" forces is an extremely important and powerful technique. We should not, however, lose sight

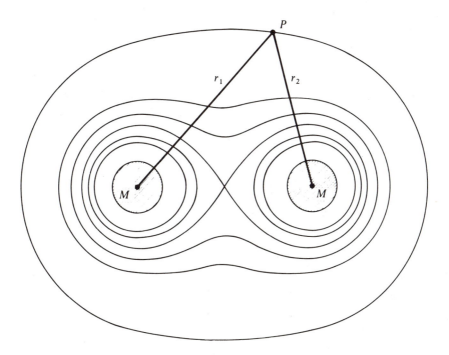

FIGURE 5-7

of the fact that the ultimate justification for using a potential is to provide a convenient means of calculating the force on a body (or the energy for the body in the field)—for it is the *force* (and energy) and not the *potential* that is the physically meaningful quantity. Thus, in some problems it may be easier to calculate the force directly, rather than computing a potential and then taking the gradient. The advantage of using the potential method is that the potential is a *scalar* quantity*: We need not deal with the added complication of sorting out the components of a vector until the gradient operation is performed. In direct calculations of the force, the components must be carried through the entire computation. Some skill, then, is necessary in choosing the particular approach to use. For example, if a problem has a particular symmetry that, from physical considerations, allows us to determine that the force has a certain direction, then the choice of that direction as one of the coordinate directions reduces the vector calculation to a simple scalar calculation. In such a case, the direct calculation of the force may be sufficiently straightforward to obviate the necessity of using the potential method. Every problem requiring a force must be examined to discover the easiest method of computation.

E X A M P L E   5.3   ---

Consider a thin uniform disk of mass M and radius a. Find the force on a mass m located along the axis of the disk.

Solution: We solve this problem by using both the potential and direct force approaches. Consider Figure 5-8. The differential potential $d\Phi$ at a distance z is given by

$$d\Phi = -G\frac{dM}{r} \tag{5.41}$$

The differential mass dM is a thin ring of width dx, because we have azimuthal symmetry.

$$dM = \rho \, dA = \rho 2\pi x \, dx \tag{5.42}$$

$$d\Phi = -2\pi\rho G\frac{x \, dx}{r} = -2\pi\rho G\frac{x \, dx}{(x^2 + z^2)^{1/2}}$$

$$\Phi(z) = -\pi\rho G\int_0^a \frac{2x \, dx}{(x^2 + z^2)^{1/2}} = -2\pi\rho G(x^2 + z^2)^{1/2}\Big|_0^a$$

$$= -2\pi\rho G[(a^2 + z^2)^{1/2} - z] \tag{5.43}$$

---

* We shall see in Chapter 7 another example of a scalar function from which vector results may be obtained. This is the **Lagrangian function**, which, to emphasize the similarity, is sometimes (mostly in older treatments) called the *kinetic potential*.

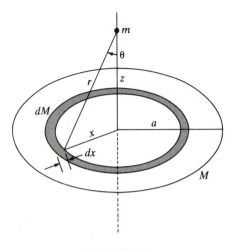

FIGURE 5-8

We find the force from

$$\mathbf{F} = -\nabla U = -m\nabla\Phi \qquad (5.44)$$

From symmetry, we have only a force in the z direction,

$$F_z = -m\frac{\partial\Phi(z)}{\partial z} = +2\pi m\rho G\left[\frac{z}{(a^2 + z^2)^{1/2}} - 1\right] \qquad (5.45)$$

In our second method, we compute the force directly using Equation 4.2:

$$d\mathbf{F} = -Gm\frac{dM'}{r^2}\mathbf{e}_r \qquad (5.46)$$

where dM' refers to the mass of a small differential area more like a square than a thin ring. The vectors complicate matters. How can symmetry help? For every small dM' on one side of the thin ring of width dx, another dM' exists on the other side that exactly cancels the horizontal component of $d\mathbf{F}$ on m. Similarly, all horizontal components cancel, and we need only consider the vertical component of $d\mathbf{F}$ along z.

$$dF_z = \cos\theta|d\mathbf{F}| = -mG\frac{\cos\theta\, dM'}{r^2}$$

and, because $\cos\theta = z/r$,

$$dF_z = -mG\frac{z\, dM'}{r^3}$$

Now we integrate over the mass $dM' = \rho 2\pi x\, dx$ around the ring and obtain

$$dF_z = -mG\rho\frac{2\pi xz\, dx}{r^3}$$

and

$$F_z = -\pi m\rho Gz \int_0^a \frac{2x\,dx}{(z^2 + x^2)^{3/2}}$$

$$= -\pi m\rho Gz \left[\frac{-2}{(z^2 + x^2)^{1/2}} \right]\Big|_0^a$$

$$= 2\pi m\rho G \left[\frac{z}{(a^2 + z^2)^{1/2}} - 1 \right] \tag{5.47}$$

which is identical to Equation 5.45. Notice that the value of F_z is negative, indicating that the force is downward in Figure 5-8 and attractive.

-- ●

5.5 OCEAN TIDES

The ocean tides have long been of interest to humans. Galileo tried unsuccessfully to explain ocean tides but could not account for the timing of the approximately two high tides each day. Newton finally gave an adequate explanation. The tides are caused by the gravitational attraction of the ocean to both the moon and the sun, but there are several complicating factors.

The calculation is complicated by the fact that the surface of the Earth is not an inertial system. The Earth and moon rotate about their center of mass (and move about the sun), so we may regard the water nearest the moon as being pulled away from the Earth, and the Earth as being pulled away from the water farthest from the moon. However, the Earth rotates while the moon rotates about the Earth. Let's first consider only the effect of the moon, adding the effect of the sun later. We will assume a simple model whereby the Earth's surface is completely covered with water, and we shall add the effect of the Earth's rotation at an appropriate time. We set up an inertial frame of reference $x'y'z'$ as shown in Figure 5-9a. We let M_m be the mass of the moon, r the radius of a circular Earth, and D the distance from the center of the moon to the center of the Earth. We consider the effect of both the moon's and Earth's gravitational attraction on a small mass m placed on the surface of the Earth. As displayed in Figure 5-9a, the position vector of the mass m from the moon is $\mathbf{R}$, from the center of the Earth is $\mathbf{r}$, and from our inertial system $\mathbf{r}'_m$. The position vector from the inertial system to the center of the Earth is $\mathbf{r}'_E$. As measured from the inertial system, the force on m, due to the Earth and the moon, is

$$m\ddot{\mathbf{r}}'_m = -\frac{GmM_E}{r^2}\mathbf{e}_r - \frac{GmM_m}{R^2}\mathbf{e}_R \tag{5.48}$$

Similarly, the force on the center of mass of the Earth caused by the moon is

$$M_E\ddot{\mathbf{r}}'_E = -\frac{GM_EM_m}{D^2}\mathbf{e}_D \tag{5.49}$$

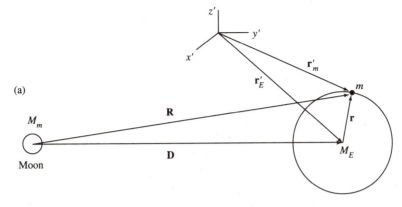

(a)

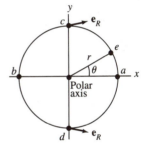

(b)

Moon

FIGURE 5-9

We want to find the acceleration $\ddot{r}$ as measured in the noninertial system placed at the center of the Earth. Therefore, we want

$$\ddot{r} = \ddot{r}'_m - \ddot{r}'_E = \frac{m\ddot{r}'_m}{m} - \frac{M_E\ddot{r}'_E}{M_E}$$

$$= -\frac{GM_E}{r^2}\mathbf{e}_r - \frac{GM_m}{R^2}\mathbf{e}_R + \frac{GM_m}{D^2}\mathbf{e}_D$$

$$= -\frac{GM_E}{r^2}\mathbf{e}_r - GM_m\left(\frac{\mathbf{e}_R}{R^2} - \frac{\mathbf{e}_D}{D^2}\right) \tag{5.50}$$

The first part is due to the Earth, and the second part is the acceleration from the **tidal** force, which is responsible for producing the ocean tides. It is due to the difference between the moon's gravitational pull at the center of the Earth and on the Earth's surface.

We next find the effect of the tidal force at various points on Earth as noted in Figure 5-9b. We show a polar view of the Earth with the polar axis along the

z-axis. The tidal force $\mathbf{F}_T$ on the mass m on the Earth's surface is

$$\mathbf{F}_T = -GmM_m\left(\frac{\mathbf{e}_R}{R^2} - \frac{\mathbf{e}_D}{D^2}\right) \tag{5.51}$$

where we have used only the second part of Equation 5.50. We look first at point a, the farthest point on Earth from the moon. Both unit vectors $\mathbf{e}_R$ and $\mathbf{e}_D$ are pointing in the same direction away from the moon along the x-axis. Because $R > D$, the second term in Equation 5.51 predominates, and the tidal force is along the $+x$-axis as shown in Figure 5-9b. For point b, $R < D$ and the tidal force has the same magnitude as at point a, but is along the $-x$-axis. The magnitude of the tidal force along the x-axis, F_{Tx}, is

$$F_{Tx} = -GmM_m\left(\frac{1}{R^2} - \frac{1}{D^2}\right) = -GmM_m\left(\frac{1}{(D+r)^2} - \frac{1}{D^2}\right)$$
$$= -\frac{GmM_m}{D^2}\left(\frac{1}{\left(1+\dfrac{r}{D}\right)^2} - 1\right)$$

We expand the first term in brackets using the $(1 + x)^{-2}$ expansion in Equation D.9.

$$F_{Tx} = -\frac{GmM_m}{D^2}\left[1 - 2\frac{r}{D} + 3\left(\frac{r}{D}\right)^2 - \cdots - 1\right] = +\frac{2GmM_mr}{D^3} \tag{5.52}$$

where we have kept only the largest nonzero term in the expansion, because $r/D = 0.02$.

For point c, the unit vector $\mathbf{e}_R$ (Figure 5-9b) is not quite exactly along $\mathbf{e}_D$, but the x-axis components approximately cancel, because $R \approx D$ and the x-components of $\mathbf{e}_R$ and $\mathbf{e}_D$ are similar. There will be a small component of $\mathbf{e}_R$ along the y-axis. We approximate the y-component of $\mathbf{e}_R$ by $(r/D)\mathbf{j}$, and the tidal force at point c, call it $\mathbf{F}_{Ty}$, is along the y-axis and has the magnitude

$$F_{Ty} = -GmM_m\left(\frac{1}{D^2}\frac{r}{D}\right) = -\frac{GmM_mr}{D^3} \tag{5.53}$$

Note that this force is along the $-y$-axis toward the center of the Earth at point c. We find similarly at point D the same magnitude, but the component of $\mathbf{e}_R$ will be along the $-y$-axis, so the force itself, with the sign of Equation 5.53, will be along the $+y$-axis toward the center of the Earth. We indicate the tidal forces at points a, b, c, and d on Figure 5-10a.

We determine the force at an arbitrary point e by noting that the x- and y-components of the tidal force can be found by substituting x and y for r in F_{Tx} and F_{Ty}, respectively, in Equations 5.52 and 5.53.

$$F_{Tx} = \frac{2GmM_mx}{D^3}$$

$$F_{Ty} = -\frac{GmM_my}{D^3}$$

(a)

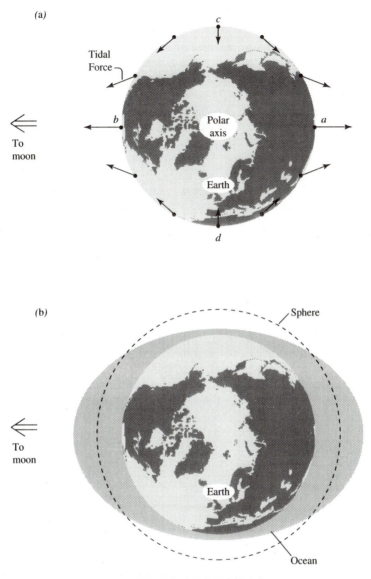

FIGURE 5-10

Then at an arbitrary point such as e, we let $x = r \cos \theta$ and $y = r \sin \theta$, so we have

$$F_{Tx} = \frac{2GmM_mr \cos \theta}{D^3} \qquad (5.54a)$$

$$F_{Ty} = -\frac{GmM_mr \sin \theta}{D^3} \qquad (5.54b)$$

Equations 5.54a and b give the tidal force around the Earth for all angles θ. Note that they give the correct result at points a, b, c, and d.

Figure 5-10a gives a representation of the tidal forces. For our simple model, these forces lead to the water along the y-axis being more shallow than along the x-axis. We show an exaggerated result in Figure 5-10b. As the Earth makes a revolution about its own axis every 24 hours, we will observe two high tides a day.

A quick calculation shows that the sun's gravitational attraction is about 175 times stronger than the moon's on the Earth's surface, so we would expect tidal forces from the sun as well. The tidal force calculation is similar to the one we have just performed for the moon. The result (Problem 5-18) is that the tidal force due to the sun is 0.46 that of the moon, a sizable effect. Despite the stronger attraction due to the sun, the gravitational force gradient over the surface of the Earth is much smaller, because of the much larger distance to the sun.

E X A M P L E 5.4 ---

Calculate the maximum height change in the ocean tides caused by the moon.

Solution: We continue to use our simple model of the ocean surrounding the Earth. Newton proposed a solution to this calculation by imagining that two wells be dug, one along the direction of high tide (our x-axis) and one along the direction of low tide (our y-axis). If the tidal height change we want to determine is h, then the difference in potential energy of mass m due to the height difference is mgh. Let's calculate the difference in work if we move the mass m from point c in Figure 5-11 to the center of the Earth and then to point a. This work W done by gravity must equal the potential energy change mgh. The work W is

$$W = \int_{r+\delta_1}^{0} F_{Ty}\, dy + \int_{0}^{r+\delta_2} F_{Tx}\, dx$$

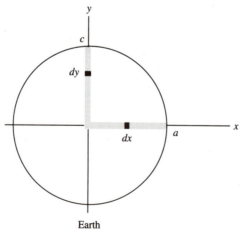

FIGURE 5-11

where we use the tidal forces F_{Ty} and F_{Tx} of Equations 5.54. The small distances δ_1 and δ_2 are to account for the small variations from a spherical Earth, but these values are so small they can be henceforth neglected. The value for W becomes

$$W = \frac{GmM_m}{D^3}\left[\int_r^0 (-y)dy + \int_0^r 2x\,dx\right]$$

$$= \frac{GmM_m}{D^3}\left(\frac{r^2}{2} + r^2\right) = \frac{3GmM_mr^2}{2D^3}$$

Because this work is equal to mgh, we have

$$mgh = \frac{3GmM_mr^2}{2D^3}$$

$$h = \frac{3GM_mr^2}{2gD^3}$$

Note that the mass m cancels, and the value of h does not depend on m. Nor does it depend on the substance, so to the extent the Earth is plastic, similar tidal effects should be (and are) observed for the surface land. If we insert the known values of the constants into Equation 5.55, we find

$$h = \frac{3(6.67 \times 10^{-11}\ \text{m}^3/\text{kg·s}^2)(7.350 \times 10^{22}\ \text{kg})(6.37 \times 10^6\ \text{m})^2}{2(9.80\ \text{m/s}^2)(3.84 \times 10^8\ \text{m})^3} = 0.54\ \text{m}$$

--- ●

The highest tides (called *spring* tides) occur when the Earth, moon, and sun are lined up (new moon and full moon), and the smallest tides (called *neap* tides) occur for the first and third quarters of the moon when the sun and moon are at right angles to each other, partially cancelling their effects. The maximum tide, which occurs every 2 weeks, should be $1.46h = 0.83$ m for the spring tides.

An observer who has spent much time near the ocean has noticed that typical oceanshore tides are greater than those calculated in Example 5.4. Several other effects come into play. The Earth is not covered completely with water, and the continents play a significant role, especially the shelfs and narrow estuaries. Local effects can be dramatic, leading to tidal changes of several meters. The tides in midocean, however, are similar to what we have calculated. Resonances can affect the natural oscillation of the bodies of water and cause tidal changes. Tidal friction between the water and the Earth leads to a significant amount of energy loss on Earth. The Earth is not rigid, and it is also distorted by tidal forces.

In addition to the effects just discussed, remember that as the Earth rotates, the moon is also orbiting the Earth. This leads to the result that there are not quite exactly two high tides per day, because they occur once every 12 h and 26 min (Problem 5-19). The plane of the moon's orbit about Earth is also not perpendicular to the Earth's rotation axis. This causes one high tide each day to be slightly higher than the other. The tidal friction between water and land mentioned previously also results in the Earth "dragging" the ocean with it as the Earth rotates. This causes

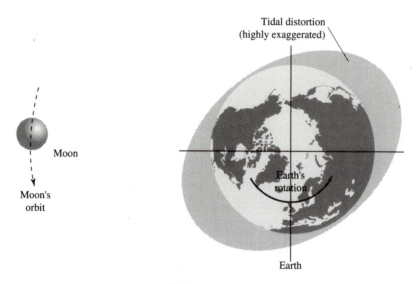

FIGURE 5-12

the high tides to be not quite along the Earth-moon axis, but rather several degrees apart as shown in Figure 5-12.

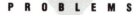

P R O B L E M S

5-1. Sketch the equipotential surfaces and the lines of force for two point masses separated by a certain distance. Next, consider one of the masses to have a fictitious negative mass − *M*. Sketch the equipotential surfaces and lines of force for this case. To what kind of physical situation does this set of equipotentials and field lines apply? (Note that the lines of force have *direction*; indicate this with appropriate arrows.)

5-2. If the field vector is independent of the radial distance within a sphere, find the function describing the density $\rho = \rho(r)$ of the sphere.

5-3. Assuming that air resistance is unimportant, calculate the minimum velocity a particle must have at the surface of the Earth to escape from the Earth's gravitational field. Obtain a numerical value for the result. (This velocity is called the *escape velocity*.)

5-4. A particle is attracted toward a center of force according to the relation $F = -mk^2/x^3$. Show that the time required for the particle to reach the force center from a distance d is d^2/k.

5-5. A particle falls to the Earth starting from rest at a great height. Neglect air resistance and show that the particle requires approximately $\frac{9}{11}$ of the total time of fall to traverse the first half of the distance.

5-6. Compute directly the gravitational force on a unit mass at a point exterior to a homogeneous sphere of matter.

5-7. Calculate the gravitational potential due to a thin rod of length l and mass M at a distance R from the center of the rod and in a direction perpendicular to the rod.

5-8. Calculate the gravitational field vector due to a homogeneous cylinder at exterior points on the axis of the cylinder. Perform the calculation (**a**) by computing the force directly and (**b**) by computing the potential first.

5-9. Calculate the potential due to a thin circular ring of radius a and mass M for points lying in the plane of the ring and exterior to it. The result can be expressed as an elliptic integral.* Assume that the distance from the center of the ring to the field point is large compared with the radius of the ring. Expand the expression for the potential and find the first correction term.

5-10. Find the potential at off-axis points due to a thin circular ring of radius a and mass M. Let R be the distance from the center of the ring to the field point, and let θ be the angle between the line connecting the center of the ring with the field point and the axis of the ring. Assume $R \gg a$ so that terms of order $(a/R)^3$ and higher may be neglected.

5-11. Consider a massive body of arbitrary shape and a spherical surface that is exterior to and does not contain the body. Show that the average value of the potential due to the body taken over the spherical surface is equal to the value of the potential at the center of the sphere.

5-12. In the previous problem, let the massive body be inside the spherical surface. Now show that the average value of the potential over the surface of the sphere is equal to the value of the potential that would exist on the surface of the sphere if all the mass of the body were concentrated at the center of the sphere.

5-13. A planet of density ρ_1 (spherical core, radius R_1) with a thick spherical cloud of dust (density ρ_2, radius R_2) is discovered. What is the force on a particle of mass m placed within the dust cloud?

5-14. Show that the gravitational self-energy (energy of assembly piecewise from infinity) of a uniform sphere of mass M and radius R is

$$U = -\frac{3}{5}\frac{GM^2}{R}$$

5-15. A particle is dropped into a hole drilled straight through the center of the Earth. Neglecting rotational effects, show that the particle's motion is simple harmonic if you assume the Earth has uniform density. Show that the period of the oscillation is about 84 min.

5-16. A uniformly solid sphere of mass M and radius R is fixed a distance h above a thin infinite sheet of mass density ρ_s (mass/area). With what force does the sphere attract the sheet?

5-17. Newton's model of the tidal height, using the two water wells dug to the center of the Earth, used the fact that the pressure at the bottom of the two wells should be the same. Assume water is incompressible and find the tidal height difference h, Equation 5.55, due to the moon using this model. (*Hint:* $\displaystyle\int_0^{y_{max}} \rho g_y \, dy = \int_0^{x_{max}} \rho g_x \, dx$; $h = x_{max} - y_{max}$, where $x_{max} + y_{max} = 2R_{earth}$, and R_{earth} is the Earth's median radius.)

---

* See Appendix B for a list of some elliptic integrals.

5-18. Show that the ratio of maximum tidal heights due to the moon and sun is given by

$$\frac{M_m}{M_s}\left(\frac{R_{Es}}{D}\right)^3$$

and that this value is 2.2. R_{Es} is the distance between the sun and Earth, and M_s is the sun's mass.

5-19. The orbital revolution of the moon about the Earth takes about 27.3 days and is in the same direction as the Earth's rotation (24 h). Use this information to show that high tides occur everywhere on Earth every 12 h and 26 min.

6

SOME METHODS IN THE CALCULUS OF VARIATIONS

6.1 INTRODUCTION

Many problems in Newtonian mechanics are more easily analyzed by means of alternative statements of the laws, including **Lagrange's equation** and **Hamilton's principle**.* As a prelude to these techniques, we consider in this chapter some general principles of the techniques of the calculus of variations.

Emphasis will be placed on those aspects of the theory of variations that have a direct bearing on classical systems, omitting some existence proofs. Our primary interest here is in determining the path that gives extremum solutions, for example, the shortest distance (or time) between two points. A well-known example of the use of the theory of variations is **Fermat's principle**: Light travels by the path that takes the least amount of time (see Problem 6-7).

---

* The development of the calculus of variations was begun by Newton (1686) and was extended by Johann and Jakob Bernoulli (1696) and by Euler (1744). Adrien Legendre (1786), Joseph Lagrange (1788), Hamilton (1833), and Jacobi (1837) all made important contributions. The names of Peter Dirichlet (1805–1859) and Karl Weierstrass (1815–1879) are particularly associated with the establishment of a rigorous mathematical foundation for the subject.

6.2 STATEMENT OF THE PROBLEM

The basic problem of the calculus of variations is to determine the function $y(x)$ such that the integral

$$J = \int_{x_1}^{x_2} f\{y(x), y'(x); x\} \, dx \tag{6.1}$$

is an *extremum* (i.e., either a maximum or a minimum). In Equation 6.1, $y'(x) \equiv dy/dx$, and the semicolon in f separates the independent variable x from the dependent variable $y(x)$ and its derivative $y'(x)$. The functional* f is considered as given, and the limits of integration are fixed.† The function $y(x)$ is then to be varied until an extreme value of J is found. By this we mean that if a function $y = y(x)$ gives the integral J a minimum value, then any *neighboring function*, no matter how close to $y(x)$, must make J increase. The definition of a neighboring function may be made as follows. We give all possible functions y a parametric representation $y = y(\alpha, x)$ such that, for $\alpha = 0$, $y = y(0, x) = y(x)$ is the function that yields an extremum for J. We can then write

$$y(\alpha, x) = y(0, x) + \alpha \eta(x) \tag{6.2}$$

where $\eta(x)$ is some function of x that has a continuous first derivative and that vanishes at x_1 and x_2, because the varied function $y(\alpha, x)$ must be identical with $y(x)$ at the endpoints of the path: $\eta(x_1) = \eta(x_2) = 0$. The situation is depicted schematically in Figure 6-1.

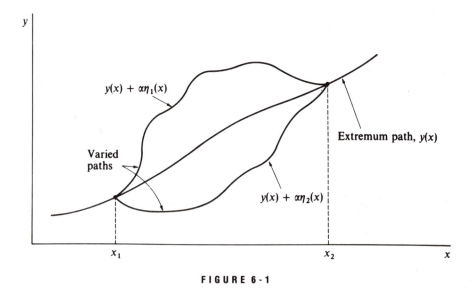

FIGURE 6-1

---

* The quantity f depends on the functional form of the dependent variable $y(x)$ and is called a *functional*.

† It is not necessary that the limits of integration be considered fixed. If they are allowed to vary, the problem increases to finding not only $y(x)$ but also x_1 and x_2 such that J is an extremum.

If functions of the type given by Equation 6.2 are considered, the integral J becomes a function of the parameter α:

$$J(\alpha) = \int_{x_1}^{x_2} f\{y(\alpha, x), y'(\alpha, x); x\} \, dx \tag{6.3}$$

The condition that the integral have a *stationary value* (i.e., that an extremum results) is that J be independent of α in first order along the path giving the extremum ($\alpha = 0$), or, equivalently, that

$$\left. \frac{\partial J}{\partial \alpha} \right|_{\alpha=0} = 0 \tag{6.4}$$

for all functions $\eta(x)$. This is only a *necessary* condition; it is not sufficient.

E X A M P L E 6.1 --

Consider the function $f = (dy/dx)^2$, where $y(x) = x$. Add to $y(x)$ the function $\eta(x) = \sin x$, and find $J(\alpha)$ between the limits of $x = 0$ and $x = 2\pi$. Show that the stationary value of $J(\alpha)$ occurs for $\alpha = 0$.

Solution: We may construct neighboring varied paths by adding to $y(x)$,

$$y(x) = x \tag{6.5}$$

the sinusoidal variation $\alpha \sin x$,

$$y(\alpha, x) = x + \alpha \sin x \tag{6.6}$$

These paths are illustrated in Figure 6-2 for $\alpha = 0$ and for two different nonvanishing values of α. Clearly, the function $\eta(x) = \sin x$ obeys the endpoint conditions, that is, $\eta(0) = 0 = \eta(2\pi)$. To determine $f(y, y'; x)$ we first determine

$$\frac{dy(\alpha, x)}{dx} = 1 + \alpha \cos x \tag{6.7}$$

then

$$f = \left(\frac{dy(\alpha, x)}{dx} \right)^2 = 1 + 2\alpha \cos x + \alpha^2 \cos^2 x \tag{6.8}$$

Equation 6.3 now becomes

$$J(\alpha) = \int_0^{2\pi} (1 + 2\alpha \cos x + \alpha^2 \cos^2 x) \, dx \tag{6.9}$$

$$= 2\pi + \alpha^2 \pi \tag{6.10}$$

Thus we see the value of $J(\alpha)$ is always greater than $J(0)$, no matter what value (positive or negative) we choose for α. The condition of Equation 6.4 is also satisfied.

--- ●

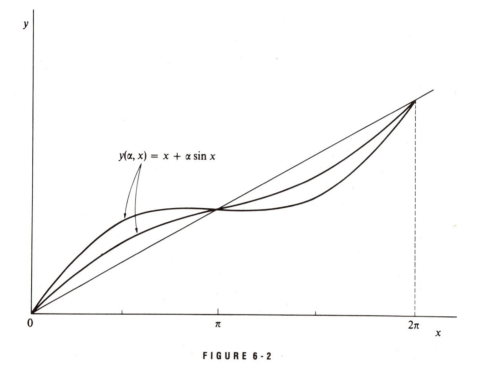

$$y(\alpha, x) = x + \alpha \sin x$$

FIGURE 6-2

6.3 EULER'S EQUATION

To determine the result of the condition expressed by Equation 6.4, we perform the indicated differentiation in Equation 6.3:

$$\frac{\partial J}{\partial \alpha} = \frac{\partial}{\partial \alpha} \int_{x_1}^{x_2} f\{y, y'; x\} \, dx \tag{6.11}$$

Because the limits of integration are fixed, the differential operation affects only the integrand. Hence,

$$\frac{\partial J}{\partial \alpha} = \int_{x_1}^{x_2} \left(\frac{\partial f}{\partial y} \frac{\partial y}{\partial \alpha} + \frac{\partial f}{\partial y'} \frac{\partial y'}{\partial \alpha} \right) dx \tag{6.12}$$

From Equation 6.2, we have

$$\frac{\partial y}{\partial \alpha} = \eta(x); \qquad \frac{\partial y'}{\partial \alpha} = \frac{d\eta}{dx} \tag{6.13}$$

Equation 6.12 becomes

$$\frac{\partial J}{\partial \alpha} = \int_{x_1}^{x_2} \left(\frac{\partial f}{\partial y} \eta(x) + \frac{\partial f}{\partial y'} \frac{d\eta}{dx} \right) dx \tag{6.14}$$

The second term in the integrand can be integrated by parts:

$$\int u \, dv = uv - \int v \, du \tag{6.15}$$

$$\int_{x_1}^{x_2} \frac{\partial f}{\partial y'} \frac{d\eta}{dx} \, dx = \frac{\partial f}{\partial y'} \eta(x) \Big|_{x_1}^{x_2} - \int_{x_1}^{x_2} \frac{d}{dx} \left(\frac{\partial f}{\partial y'} \right) \eta(x) \, dx \tag{6.16}$$

The integrated term vanishes because $\eta(x_1) = \eta(x_2) = 0$. Therefore, Equation 6.12 becomes

$$\frac{\partial J}{\partial \alpha} = \int_{x_1}^{x_2} \left[\frac{\partial f}{\partial y} \eta(x) - \frac{d}{dx} \left(\frac{\partial f}{\partial y'} \right) \eta(x) \right] dx$$

$$= \int_{x_1}^{x_2} \left(\frac{\partial f}{\partial y} - \frac{d}{dx} \frac{\partial f}{\partial y'} \right) \eta(x) \, dx \tag{6.17}$$

The integral in Equation 6.17 now appears to be independent of α. But the functions y and y' with respect to which the derivatives of f are taken are still functions of α. Because $(\partial J/\partial \alpha)|_{\alpha=0}$ must vanish for the extremum value and because $\eta(x)$ is an arbitrary function (subject to the conditions already stated), the integrand in Equation 6.17 must itself vanish for $\alpha = 0$:

$$\boxed{\frac{\partial f}{\partial y} - \frac{d}{dx} \frac{\partial f}{\partial y'} = 0} \quad \text{Euler's equation} \tag{6.18}$$

where now y and y' are the original functions, independent of α. This result is known as **Euler's equation,** * which is a necessary condition for J to have an extremum value.

E X A M P L E   6.2 ---

We can use the calculus of variations to solve a classic problem in the history of physics: the *brachistochrone*.† Consider a particle moving in a constant force field starting at rest from some point (x_1, y_1) to some lower point (x_2, y_2). Find the path that allows the particle to accomplish the transit in the least possible time.

Solution: The coordinate system may be chosen so that the point (x_1, y_1) is at the origin. Further, let the force field be directed along the positive x-axis as in

---

* Derived first by Euler in 1744. When applied to mechanical systems, this is known as the *Euler-Lagrange equation.*

† First solved by Johann Bernoulli (1667–1748) in 1696.

Figure 6-3. Because the force on the particle is constant—and if we ignore the possibility of friction—the field is conservative, and the total energy of the particle is $T + U = $ const. If we measure the potential from the point $x = 0$ [i.e., $U(x = 0) = 0$], then, because the particle starts from rest, $T + U = 0$. The kinetic energy is $T = \frac{1}{2}mv^2$, and the potential energy is $U = -Fx = -mgx$, where g is the acceleration imparted by the force. Thus

$$v = \sqrt{2gx} \tag{6.19}$$

The time required for the particle to make the transit from the origin to (x_2, y_2) is

$$t = \int_{(x_1, y_1)}^{(x_2, y_2)} \frac{ds}{v} = \int \frac{(dx^2 + dy^2)^{1/2}}{(2gx)^{1/2}}$$

$$= \int_{x_1=0}^{x_2} \left(\frac{1 + y'^2}{2gx}\right)^{1/2} dx \tag{6.20}$$

The time of transit is the quantity for which a minimum is desired. Because the constant $(2g)^{-1/2}$ does not affect the final equation, the functional f may be identified as

$$f = \left(\frac{1 + y'^2}{x}\right)^{1/2} \tag{6.21}$$

And, because $\partial f/\partial y = 0$, the Euler equation (Equation 6.18) becomes

$$\frac{d}{dx}\frac{\partial f}{\partial y'} = 0$$

or

$$\frac{\partial f}{\partial y'} = \text{constant} \equiv (2a)^{-1/2}$$

where a is a new constant.

Performing the differentiation $\partial f/\partial y'$ on Equation 6.21 and squaring the result, we have

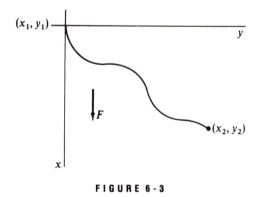

FIGURE 6-3

$$\frac{y'^2}{x(1 + y'^2)} = \frac{1}{2a} \tag{6.22}$$

This may be put in the form

$$y = \int \frac{x \, dx}{(2ax - x^2)^{1/2}} \tag{6.23}$$

We now make the following change of variable:

$$x = a(1 - \cos \theta)$$
$$dx = a \sin \theta \, d\theta \tag{6.24}$$

The integral in Equation 6.23 then becomes

$$y = \int a(1 - \cos \theta) \, d\theta$$

and

$$y = a(\theta - \sin \theta) + \text{constant} \tag{6.25}$$

The parametric equations for a *cycloid** passing through the origin are

$$\left. \begin{array}{l} x = a(1 - \cos \theta) \\ y = a(\theta - \sin \theta) \end{array} \right\} \tag{6.26}$$

which is just the solution found, with the constant of integration set equal to zero to conform with the requirement that (0, 0) is the starting point of the motion. The path is then as shown in Figure 6-4, and the constant a must be adjusted to allow

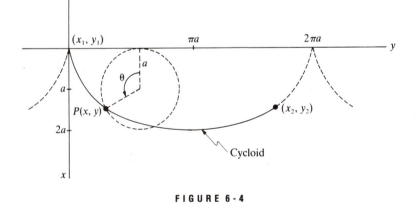

FIGURE 6-4

---

* A cycloid is a curve traced by a point on a circle rolling on a plane along a line in the plane. See the dashed sphere rolling along $x = 0$ in Figure 6-4.

the cycloid to pass through the specified point (x_2, y_2). Solving the problem of the brachistochrone does indeed yield a path the particle traverses in a *minimum* time. But the procedures of variational calculus are designed only to produce an extremum—either a minimum or a maximum. It is almost always the case in dynamics that we desire (and find) a minimum for the problem.

EXAMPLE  6.3

Consider the surface generated by revolving a line connecting two fixed points (x_1, y_1) and (x_2, y_2) about an axis coplanar with the two points. Find the equation of the line connecting the points such that the surface area generated by the revolution (i.e., the area of the surface of revolution) is a minimum.

Solution: We assume that the curve passing through (x_1, y_1) and (x_2, y_2) is revolved about the y-axis, coplanar with the two points. To calculate the total area of the surface of revolution, we first find the area dA of a strip. Refer to Figure 6-5.

$$dA = 2\pi x \, ds = 2\pi x (dx^2 + dy^2)^{1/2} \tag{6.27}$$

$$A = 2\pi \int_{x_1}^{x_2} x(1 + y'^2)^{1/2} \, dx \tag{6.28}$$

where $y' = dy/dx$. To find the extremum value we let

$$f = x(1 + y'^2)^{1/2} \tag{6.29}$$

and insert into Equation 6.18:

$$\frac{\partial f}{\partial y} = 0$$

$$\frac{\partial f}{\partial y'} = \frac{xy'}{(1 + y'^2)^{1/2}}$$

therefore,

$$\frac{d}{dx}\left[\frac{xy'}{(1 + y'^2)^{1/2}} \right] = 0$$

$$\frac{xy'}{(1 + y'^2)^{1/2}} = \text{constant} \equiv a \tag{6.30}$$

From Equation 6.30, we determine

$$y' = \frac{a}{(x^2 - a^2)^{1/2}} \tag{6.31}$$

$$y = \int \frac{a \, dx}{(x^2 - a^2)^{1/2}} \tag{6.32}$$

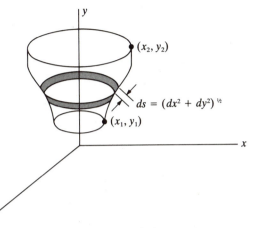

FIGURE 6-5

The solution of this integration is

$$y = a \cosh^{-1}\left(\frac{x}{a}\right) + b \tag{6.33}$$

where a and b are constants of integration determined by requiring the curve to pass through the points (x_1, y_1) and (x_2, y_2). Equation 6.33 can also be written as

$$x = a \cosh\left(\frac{y - b}{a}\right) \tag{6.34}$$

which is more easily recognized as the equation of a *catenary*, the curve of a flexible cord hanging freely between two points of support.

-- ●

Choose two points located at (x_1, y_1) and (x_2, y_2) joined by a curve $y(x)$. We want to find $y(x)$ such that if we revolve the curve around the x-axis, the surface area of the revolution is a minimum. This is the "soap film" problem, because a soap film suspended between two wire circular rings takes this shape (Figure 6-6). We want to minimize the integral of the area $dA = 2\pi y \, ds$ where $ds = \sqrt{1 + y'^2} \, dx$ and $y' = dy/dx$.

$$A = 2\pi \int_{x_1}^{x_2} y\sqrt{1 + y'^2} \, dx \tag{6.35}$$

We find the extremum by setting $f = y\sqrt{1 + y'^2}$ and inserting into Equation 6.18. The derivatives we need are

$$\frac{\partial f}{\partial y} = \sqrt{1 + y'^2}$$

$$\frac{\partial f}{\partial y'} = \frac{yy'}{\sqrt{1 + y'^2}}$$

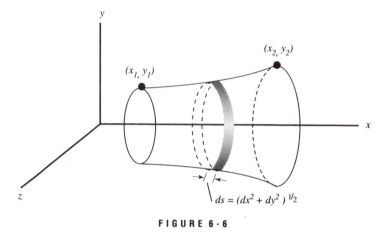

FIGURE 6-6

Equation 6.18 becomes

$$\sqrt{1 + y'^2} = \frac{d}{dx}\left[\frac{yy'}{\sqrt{1 + y'^2}}\right] \tag{6.36}$$

Equation 6.36 does not appear to be a simple equation to solve for $y(x)$. Let's stop and think about whether there might be an easier method of solution. You may have noticed that this problem is just like Example 6.3, but in that case we were minimizing a surface of revolution about the y-axis rather than around the x-axis. The solution to the soap film problem should be identical to Equation 6.34 if we interchange x and y. But how did we end up with such a complicated equation as Equation 6.36? We blindly chose x as the independent variable and decided to find the function $y(x)$. In fact, in general, we can choose the independent variable to be anything we want: x, θ, t, or even y. If we choose y as the independent variable, we would need to interchange x and y in many of the previous equations that led up to Euler's equation (Equation 6.18). It might be easier in the beginning to just interchange the variables that we started with (i.e., call the horizontal axis y in Figure 6-6 and let the independent variable be x). (In a right-handed coordinate system, the x-direction would be down, but that presents no difficulty in this case because of symmetry.) No matter what we do, the solution of our present problem would just parallel Example 6.3. Unfortunately, it is not always possible to look ahead to make the best choice of independent variable. Sometimes we just have to proceed by trial and error.

6.4 THE "SECOND FORM" OF THE EULER EQUATION

A second equation may be derived from Euler's equation that is convenient for functions that do not explicitly depend on x: $\partial f/\partial x = 0$. We first note that for any function $f(y, y'; x)$ the derivative is a sum of terms

$$\frac{df}{dx} = \frac{d}{dx} f\{y, y'; x\} = \frac{\partial f}{\partial y}\frac{dy}{dx} + \frac{\partial f}{\partial y'}\frac{dy'}{dx} + \frac{\partial f}{\partial x}$$

$$= y'\frac{\partial f}{\partial y} + y''\frac{\partial f}{\partial y'} + \frac{\partial f}{\partial x} \tag{6.37}$$

Also

$$\frac{d}{dx}\left(y'\frac{\partial f}{\partial y'}\right) = y''\frac{\partial f}{\partial y'} + y'\frac{d}{dx}\frac{\partial f}{\partial y'}$$

or, substituting from Equation 6.37 for $y''(\partial f/\partial y')$,

$$\frac{d}{dx}\left(y'\frac{\partial f}{\partial y'}\right) = \frac{df}{dx} - \frac{\partial f}{\partial x} - y'\frac{\partial f}{\partial y} + y'\frac{d}{dx}\frac{\partial f}{\partial y'} \tag{6.38}$$

The last two terms in Equation 6.38 may be written as

$$y'\left(\frac{d}{dx}\frac{\partial f}{\partial y'} - \frac{\partial f}{\partial y}\right)$$

which vanishes in view of the Euler equation (Equation 6.18). Therefore,

$$\boxed{\frac{\partial f}{\partial x} - \frac{d}{dx}\left(f - y'\frac{\partial f}{\partial y'}\right) = 0} \tag{6.39}$$

We can use this so-called "second form" of the Euler equation in cases in which f does not depend explicitly on x, and $\partial f/\partial x = 0$. Then,

$$f - y'\frac{\partial f}{\partial y'} = \text{constant} \qquad \left(\text{for } \frac{\partial f}{\partial x} = 0\right) \tag{6.40}$$

E X A M P L E  6.4 –

A *geodesic* **is a line that represents the shortest path between any two points when the path is restricted to a particular surface. Find the geodesic on a sphere.**

Solution: The element of length on the surface of a sphere of radius ρ is given (see Equation F.15 with $dr = 0$) by

$$ds = \rho(d\theta^2 + \sin^2\theta\, d\phi^2)^{1/2} \tag{6.41}$$

The distance s between points 1 and 2 is therefore

$$s = \rho \int_1^2 \left[\left(\frac{d\theta}{d\phi}\right)^2 + \sin^2\theta\right]^{1/2} d\phi \tag{6.42}$$

and, if s is to be a minimum, f is identified as

$$f = (\theta'^2 + \sin^2 \theta)^{1/2} \tag{6.43}$$

where $\theta' \equiv d\theta/d\phi$. Because $\partial f/\partial \phi = 0$, we may use the second form of the Euler equation (Equation 6.40), which yields

$$(\theta'^2 + \sin^2 \theta)^{1/2} - \theta' \cdot \frac{\partial}{\partial \theta'}(\theta'^2 + \sin^2 \theta)^{1/2} = \text{constant} \equiv a \tag{6.44}$$

Differentiating and multiplying through by f, we have

$$\sin^2 \theta = a(\theta'^2 + \sin^2 \theta)^{1/2} \tag{6.45}$$

This may be solved for $d\phi/d\theta = \theta'^{-1}$, with the result

$$\frac{d\phi}{d\theta} = \frac{a \csc^2 \theta}{(1 - a^2 \csc^2 \theta)^{1/2}} \tag{6.46}$$

Solving for ϕ, we obtain

$$\phi = \sin^{-1}\left(\frac{\cot \theta}{\beta}\right) + \alpha \tag{6.47}$$

where α is the constant of integration and $\beta^2 \equiv (1 - a^2)/a^2$. Rewriting Equation 6.47 produces

$$\cot \theta = \beta \sin(\phi - \alpha) \tag{6.48}$$

To interpret this result, we convert the equation to rectangular coordinates by multiplying through by $\rho \sin \theta$ to obtain, on expanding $\sin(\phi - \alpha)$,

$$(\beta \cos \alpha) \rho \sin \theta \sin \phi - (\beta \sin \alpha) \rho \sin \theta \cos \phi = \rho \cos \theta \tag{6.49}$$

Because α and β are constants, we may write them as

$$\beta \cos \alpha \equiv A, \qquad \beta \sin \alpha \equiv B \tag{6.50}$$

Then Equation 6.49 becomes

$$A(\rho \sin \theta \sin \phi) - B(\rho \sin \theta \cos \phi) = (\rho \cos \theta) \tag{6.51}$$

The quantities in the parentheses are just the expressions for y, x, and z, respectively, in spherical coordinates (see Figure F-3, Appendix F); therefore Equation 6.51 may be written as

$$Ay - Bx = z \tag{6.52}$$

which is the equation of a plane passing through the center of the sphere. Hence the geodesic on a sphere is the path that the plane forms at the intersection with the surface of the sphere—a *great circle*. Note that the great circle is the maximum as well as the minimum "straight-line" distance between two points on the surface of a sphere.

6.5 FUNCTIONS WITH SEVERAL DEPENDENT VARIABLES

The Euler equation derived in the preceding section is the solution of the variational problem in which it was desired to find the single function $y(x)$ such that the integral of the functional f was an extremum. The case more commonly encountered in mechanics is that in which f is a functional of several dependent variables:

$$f = f\{y_1(x), y_1'(x), y_2(x), y_2'(x), \cdots; x\} \tag{6.53}$$

or simply

$$f = f\{y_i(x), y_i'(x); x\}, \qquad i = 1, 2, \cdots, n \tag{6.54}$$

In analogy with Equation 6.2, we write

$$y_i(\alpha, x) = y_i(0, x) + \alpha\eta_i(x) \tag{6.55}$$

The development proceeds analogously (cf. Equation 6.17), resulting in

$$\frac{\partial J}{\partial \alpha} = \int_{x_1}^{x_2} \sum_i \left(\frac{\partial f}{\partial y_i} - \frac{d}{dx}\frac{\partial f}{\partial y_i'}\right) \eta_i(x)\, dx \tag{6.56}$$

Because the individual variations—the $\eta_i(x)$—are all independent, the vanishing of Equation 6.56 when evaluated at $\alpha = 0$ requires the separate vanishing of *each* expression in the brackets:

$$\boxed{\frac{\partial f}{\partial y_i} - \frac{d}{dx}\frac{\partial f}{\partial y_i'} = 0, \qquad i = 1, 2, \cdots, n} \tag{6.57}$$

6.6 EULER EQUATIONS WHEN AUXILIARY CONDITIONS ARE IMPOSED

Suppose we want to find, for example, the shortest path between two points on a surface. Then, in addition to the conditions already discussed, there is the condition that the path must satisfy the equation of the surface, say, $g\{y_i; x\} = 0$. Such an equation was implicit in the solution of Example 6.4 for the geodesic on a sphere where the condition was

$$g = \sum_i x_i^2 - \rho^2 = 0 \tag{6.58}$$

that is

$$r = \rho = \text{constant} \tag{6.59}$$

But in the general case, we must make explicit use of the auxiliary equation or equations. These equations are also called **equations of constraint**. Consider the case in which

$$f = f\{y_i, y_i'; x\} = f\{y, y', z, z'; x\} \tag{6.60}$$

The equation corresponding to Equation 6.17 for the case of *two* variables is

$$\frac{\partial J}{\partial \alpha} = \int_{x_1}^{x_2} \left[\left(\frac{\partial f}{\partial y} - \frac{d}{dx} \frac{\partial f}{\partial y'} \right) \frac{\partial y}{\partial \alpha} + \left(\frac{\partial f}{\partial z} - \frac{d}{dx} \frac{\partial f}{\partial z'} \right) \frac{\partial z}{\partial \alpha} \right] dx \qquad (6.61)$$

But now there also exists an equation of constraint of the form

$$g\{y_i; x\} = g\{y, z; x\} = 0 \qquad (6.62)$$

and the variations $\partial y / \partial \alpha$ and $\partial z / \partial \alpha$ are no longer independent, so the expressions in parentheses in Equation 6.61 do not separately vanish at $\alpha = 0$.

Differentiating g from Equation 6.62, we have

$$dg = \left(\frac{\partial g}{\partial y} \frac{\partial y}{\partial \alpha} + \frac{\partial g}{\partial z} \frac{\partial z}{\partial \alpha} \right) d\alpha = 0 \qquad (6.63)$$

where no term in x appears since $\partial x / \partial \alpha = 0$. Now

$$\left. \begin{array}{l} y(\alpha, x) = y(x) + \alpha \eta_1(x) \\[2mm] z(\alpha, x) = z(x) + \alpha \eta_2(x) \end{array} \right\} \qquad (6.64)$$

Therefore, by determining $\partial y / \partial \alpha$ and $\partial z / \partial \alpha$ from Equation 6.64 and inserting into the term in parentheses of Equation 6.63, which, in general, must be zero, we obtain

$$\frac{\partial g}{\partial y} \eta_1(x) = - \frac{\partial g}{\partial z} \eta_2(x) \qquad (6.65)$$

Equation 6.61 becomes

$$\frac{\partial J}{\partial \alpha} = \int_{x_1}^{x_2} \left[\left(\frac{\partial f}{\partial y} - \frac{d}{dx} \frac{\partial f}{\partial y'} \right) \eta_1(x) + \left(\frac{\partial f}{\partial z} - \frac{d}{dx} \frac{\partial f}{\partial z'} \right) \eta_2(x) \right] dx$$

Factoring $\eta_1(x)$ out of the square brackets and writing Equation 6.65 as

$$\frac{\eta_2(x)}{\eta_1(x)} = - \frac{\partial g / \partial y}{\partial g / \partial z}$$

we have

$$\frac{\partial J}{\partial \alpha} = \int_{x_1}^{x_2} \left[\left(\frac{\partial f}{\partial y} - \frac{d}{dx} \frac{\partial f}{\partial y'} \right) - \left(\frac{\partial f}{\partial z} - \frac{d}{dx} \frac{\partial f}{\partial z'} \right) \left(\frac{\partial g / \partial y}{\partial g / \partial z} \right) \right] \eta_1(x) \, dx \qquad (6.66)$$

This latter equation now contains the single arbitrary function $\eta_1(x)$, which is not in any way restricted by Equation 6.64, and on requiring the condition of Equation 6.4, the expression in the brackets must vanish. Thus we have

$$\left(\frac{\partial f}{\partial y} - \frac{d}{dx} \frac{\partial f}{\partial y'} \right) \left(\frac{\partial g}{\partial y} \right)^{-1} = \left(\frac{\partial f}{\partial z} - \frac{d}{dx} \frac{\partial f}{\partial z'} \right) \left(\frac{\partial g}{\partial z} \right)^{-1} \qquad (6.67)$$

The left-hand side of this equation involves only derivatives of f and g with respect to y and y', and the right-hand side involves only derivatives with respect to z and z'. Because y and z are both functions of x, the two sides of Equation 6.67 may be set equal to a function of x, which we write as $-\lambda(x)$:

$$\left.\begin{array}{l} \dfrac{\partial f}{\partial y} - \dfrac{d}{dx}\dfrac{\partial f}{\partial y'} + \lambda(x)\dfrac{\partial g}{\partial y} = 0 \\[3mm] \dfrac{\partial f}{\partial z} - \dfrac{d}{dx}\dfrac{\partial f}{\partial z'} + \lambda(x)\dfrac{\partial g}{\partial z} = 0 \end{array}\right\} \tag{6.68}$$

The complete solution to the problem now depends on finding *three* functions: $y(x)$, $z(x)$, and $\lambda(x)$. But there are *three* relations that may be used: the two equations (Equation 6.68) and the equation of constraint (Equation 6.62). Thus, there is a sufficient number of relations to allow a complete solution. Note that here $\lambda(x)$ is considered to be *undetermined** and is obtained as a part of the solution. The function $\lambda(x)$ is known as a **Lagrange undetermined multiplier**.

For the general case of several dependent variables and several auxiliary conditions, we have the following set of equations:

$$\boxed{\begin{array}{c} \dfrac{\partial f}{\partial y_i} - \dfrac{d}{dx}\dfrac{\partial f}{\partial y'_i} + \displaystyle\sum_j \lambda_j(x)\dfrac{\partial g_j}{\partial y_i} = 0 \qquad\qquad (6.69) \\[4mm] g_j\{y_i; x\} = 0 \qquad\qquad\qquad\qquad\qquad (6.70) \end{array}}$$

If $i = 1, 2, \cdots, m$, and $j = 1, 2, \cdots, n$, Equation 6.69 represents m equations in $m + n$ unknowns, but there are also the n equations of constraint (Equation 6.70). Thus, there are $m + n$ equations in $m + n$ unknowns, and the system is soluble.

Equation 6.70 is equivalent to the set of n differential equations

$$\sum_i \frac{\partial g_j}{\partial y_i}\, dy_i = 0, \qquad \begin{cases} i = 1, 2, \cdots, m \\ j = 1, 2, \cdots, n \end{cases} \tag{6.71}$$

In problems in mechanics, the constraint equations are frequently differential equations rather than algebraic equations. Therefore, equations such as Equation 6.71 are sometimes more useful than the equations represented by Equation 6.70. (See Section 7.5 for an amplification of this point.)

E X A M P L E 6.5 ---

Consider a disk rolling without slipping on an inclined plane (Figure 6-7). Determine the equation of constraint in terms of the "coordinates"[†] y and θ.

Solution: The relation between the coordinates (which are not independent) is

$$y = R\theta \tag{6.72}$$

---

* The function $\lambda(x)$ was introduced in Lagrange's *Mécanique analytique* (Paris, 1788).

† These are actually the *generalized coordinates* discussed in Section 7.3; see also Example 7.9.

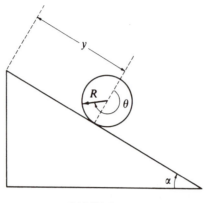

FIGURE 6-7

where R is the radius of the disk. Hence the equation of constraint is

$$g(y, \theta) = y - R\theta = 0 \tag{6.73}$$

and

$$\frac{\partial g}{\partial y} = 1, \qquad \frac{\partial g}{\partial \theta} = -R \tag{6.74}$$

are the quantities associated with λ, the single undetermined multiplier for this case.

6.7 THE δ NOTATION

In analyses that use the calculus of variations, we customarily use a shorthand notation to represent the variation. Thus, Equation 6.17, which can be written as

$$\frac{\partial J}{\partial \alpha} d\alpha = \int_{x_1}^{x_2} \left(\frac{\partial f}{\partial y} - \frac{d}{dx} \frac{\partial f}{\partial y'} \right) \frac{\partial y}{\partial \alpha} d\alpha \, dx \tag{6.75}$$

may be expressed as

$$\delta J = \int_{x_1}^{x_2} \left(\frac{\partial f}{\partial y} - \frac{d}{dx} \frac{\partial f}{\partial y'} \right) \delta y \, dx \tag{6.76}$$

where

$$\left.\begin{array}{l} \dfrac{\partial J}{\partial \alpha} d\alpha \equiv \delta J \\[4mm] \dfrac{\partial y}{\partial \alpha} d\alpha \equiv \delta y \end{array}\right\} \tag{6.77}$$

The condition of extremum then becomes

$$\delta J = \delta \int_{x_1}^{x_2} f\{y, y'; x\}\, dx = 0 \qquad \textbf{(6.78)}$$

Taking the variation symbol δ inside the integral (because, by hypothesis, the limits of integration are not affected by the variation), we have

$$\delta J = \int_{x_1}^{x_2} \delta f\, dx$$

$$= \int_{x_1}^{x_2} \left(\frac{\partial f}{\partial y} \delta y + \frac{\partial f}{\partial y'} \delta y' \right) dx \qquad \textbf{(6.79)}$$

But

$$\delta y' = \delta \left(\frac{dy}{dx} \right) = \frac{d}{dx}(\delta y) \qquad \textbf{(6.80)}$$

so

$$\delta J = \int_{x_1}^{x_2} \left(\frac{\partial f}{\partial y} \delta y + \frac{\partial f}{\partial y'} \frac{d}{dx} \delta y \right) dx \qquad \textbf{(6.81)}$$

Integrating the second term by parts as before, we find

$$\delta J = \int_{x_1}^{x_2} \left(\frac{\partial f}{\partial y} - \frac{d}{dx} \frac{\partial f}{\partial y'} \right) \delta y\, dx \qquad \textbf{(6.82)}$$

Because the variation δy is arbitrary, the extremum condition $\delta J = 0$ requires the integrand to vanish, thereby yielding the Euler equation (Equation 6.18).

Although the δ notation is frequently used, it is important to realize that it is only a shorthand expression of the more precise differential quantities. The varied path δy can be thought of physically as a virtual displacement from the actual path consistent with all the forces and constraints (see Figure 6-8). This variation δy is

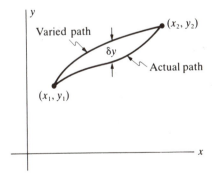

FIGURE 6-8

distinguished from an actual differential displacement dy by the condition that $dt = 0$—that is, that time is fixed. The varied path δy, in fact, need not even correspond to a possible path of motion. The variation must vanish at the endpoints.

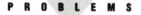

P R O B L E M S

6-1. Consider the line connecting $(x_1, y_1) = (0, 0)$ and $(x_2, y_2) = (1, 1)$. Show explicitly that the function $y(x) = x$ produces a minimum path length by using the varied function $y(\alpha, x) = x + \alpha \sin \pi(1 - x)$. Use the first few terms in the expansion of the resulting elliptic integral to show the equivalent of Equation 6.4.

6-2. Show that the shortest distance between two points on a plane is a straight line.

6-3. Show that the shortest distance between two points in (three-dimensional) space is a straight line.

6-4. Show that the geodesic on the surface of a right circular cylinder is a segment of a helix.

6-5. Consider the surface generated by revolving a line connecting two fixed points (x_1, y_1) and (x_2, y_2) about an axis coplanar with the two points. Find the equation of the line connecting the points such that the surface area generated by the revolution (i.e., the area of the surface of revolution) is a minimum. Obtain the solution by using Equation 6.39.

6-6. Reexamine the problem of the brachistochrone (Example 6.2) and show that the time required for a particle to move (frictionlessly) to the *minimum* point of the cycloid is $\pi\sqrt{a/g}$, *independent* of the starting point.

6-7. Consider light passing from one medium with index of refraction n_1 into another medium with index of refraction n_2 (Figure 6-A). Use Fermat's principle to minimize time, and derive the law of refraction: $n_1 \sin \theta_1 = n_2 \sin \theta_2$.

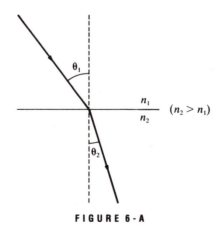

FIGURE 6-A

6-8. Find the dimensions of the parallelepiped of maximum volume circumscribed by (**a**) a sphere of radius R; (**b**) an ellipsoid with semiaxes a, b, c.

6-9. Find an expression involving the function $\phi(x_1, x_2, x_3)$ that has a minimum average value of the square of its gradient within a certain volume V of space.

6-10. Find the ratio of the radius R to the height H of a right-circular cylinder of fixed volume V that minimizes the surface area A.

▷ **6-11.** A disk of radius R rolls without slipping inside the parabola $y = ax^2$. Find the equation of constraint. Express the condition that allows the disk to roll so that it contacts the parabola at one and only one point, independent of its position.

6-12. Repeat Example 6.4, finding the shortest path between any two points on the surface of a sphere, but use the method of the Euler equations with an auxiliary condition imposed.

7

HAMILTON'S PRINCIPLE— LAGRANGIAN AND HAMILTONIAN DYNAMICS

7.1 INTRODUCTION

Experience has shown that a particle's motion in an inertial reference frame is correctly described by the Newtonian equation $\mathbf{F} = \dot{\mathbf{p}}$. If the particle is not required to move in some complicated manner and if rectangular coordinates are used to describe the motion, then usually the equations of motion are relatively simple. But if either of these restrictions is removed, the equations can become quite complex and difficult to manipulate. For example, if a particle is constrained to move on the surface of a sphere, the equations of motion result from the projection of the Newtonian vector equation onto that surface. The representation of the acceleration vector in spherical coordinates is a formidable expression, as the reader who has worked Problem 1-25 can readily testify.

Moreover, if a particle is constrained to move on a given surface, certain forces must exist (called **forces of constraint**) that maintain the particle in contact with the specified surface. For a particle moving on a smooth horizontal surface, the force of constraint is simply $\mathbf{F}_c = -m\mathbf{g}$. But, if the particle is, say, a bead sliding down a curved wire, the force of constraint can be quite complicated. Indeed, in particular situations it may be difficult or even impossible to obtain explicit expressions for the forces of constraint. But in solving a problem by using the Newtonian procedure, we must know *all* the forces, because the quantity $\mathbf{F}$ that appears in the fundamental equation is the *total* force acting on a body.

To circumvent some of the practical difficulties that arise in attempts to apply Newton's equations to particular problems, alternate procedures may be developed.

All such approaches are in essence *a posteriori*, because we know beforehand that a result equivalent to the Newtonian equations must be obtained. Thus, to effect a simplification we need not formulate a *new* theory of mechanics—the Newtonian theory is quite correct—but only devise an alternate method of dealing with complicated problems in a general manner. Such a method is contained in **Hamilton's Principle**, and the equations of motion resulting from the application of this principle are called **Lagrange's equations**.

If Lagrange's equations are to constitute a proper description of the dynamics of particles, they must be equivalent to Newton's equations. On the other hand, Hamilton's Principle can be applied to a wide range of physical phenomena (particularly those involving *fields*) not usually associated with Newton's equations. To be sure, each of the results that can be obtained from Hamilton's Principle was *first* obtained, as were Newton's equations, by the correlation of experimental facts. Hamilton's Principle has not provided us with any new physical theories, but it has allowed a satisfying unification of many individual theories by a single basic postulate. This is not an idle exercise in hindsight, because it is the goal of physical theory not only to give precise mathematical formulation to observed phenomena but also to describe these effects with an economy of fundamental postulates and in the most unified manner possible. Indeed, Hamilton's Principle is one of the most elegant and far-reaching principles of physical theory.

In view of its wide range of applicability (even though this is an after-the-fact discovery), it is not unreasonable to assert that Hamilton's Principle is more "fundamental" than Newton's equations. Therefore, we proceed by first postulating Hamilton's Principle; we then obtain Lagrange's equations and show that these are equivalent to Newton's equations.

Because we have already discussed (in Chapters 2, 3, and 4) dissipative phenomena at some length, we henceforth confine our attention to *conservative* systems. Consequently, we do not discuss the more general set of Lagrange's equations, which take into account the effects of nonconservative forces. The reader is referred to the literature for these details.*

7.2 HAMILTON'S PRINCIPLE

Minimal principles in physics have a long and interesting history. The search for such principles is predicated on the notion that nature always minimizes certain important quantities when a physical process takes place. The first such minimum principles were developed in the field of optics. Hero of Alexandria, in the second century B.C., found that the law governing the reflection of light could be obtained by asserting that a light ray, traveling from one point to another by a reflection from a plane mirror, always takes the shortest possible path. A simple geometric construction verifies that this minimum principle does indeed lead to the equality

---

* See, for example, Goldstein (Go80, Chapter 2) or, for a comprehensive discussion, Whittaker (Wh37, Chapter 8).

of the angles of incidence and reflection for a light ray reflected from a plane mirror. Hero's principle of the *shortest path* cannot, however, yield a correct law for *refraction*. In 1657, Fermat reformulated the principle by postulating that a light ray always travels from one point to another in a medium by a path that requires the least time.* Fermat's principle of *least time* leads immediately, not only to the correct law of reflection, but also to Snell's law of refraction (see Problem 6-7).[†]

Minimum principles continued to be sought, and in the latter part of the seventeenth century the beginnings of the calculus of variations were developed by Newton, Leibniz, and the Bernoullis when such problems as the brachistochrone (see Example 6.2) and the shape of a hanging chain (a catenary) were solved.

The first application of a general minimum principle in mechanics was made in 1747 by Maupertuis, who asserted that dynamical motion takes place with minimum action.[‡] Maupertuis's **principle of least action** was based on theological grounds (action is minimized through the "wisdom of God"), and his concept of "action" was rather vague. (Recall that *action* is a quantity with the dimensions of *length × momentum* or *energy × time*.) Only later was a firm mathematic foundation of the principle given by Lagrange (1760). Although it is a useful form from which to make the transition from classical mechanics to optics and to quantum mechanics, the principle of least action is less general than Hamilton's Principle and, indeed, can be derived from it. We forego a detailed discussion here.[#]

In 1828, Gauss developed a method of treating mechanics by his principle of **least constraint**; a modification was later made by Hertz and embodied in his principle of **least curvature**. These principles[##] are closely related to Hamilton's Principle and add nothing to the content of Hamilton's more general formulation; their mention only emphasizes the continual concern with minimal principles in physics.

In two papers published in 1834 and 1835, Hamilton[§] announced the dynamical principle on which it is possible to base all of mechanics and, indeed, most of classical physics. Hamilton's Principle may be stated as follows[§§]:

Of all the possible paths along which a dynamical system may move from one point to another within a specified time interval (consistent with any constraints), the actual path followed is that which minimizes the time integral of the difference between the kinetic and potential energies.

---

* Pierre de Fermat (1601–1665), a French lawyer, linguist, and amateur mathematician.

[†] In 1661, Fermat correctly deduced the law of refraction, which had been discovered experimentally in about 1621 by Willebrord Snell (1591–1626), a Dutch mathematical prodigy.

[‡] Pierre-Louise-Moreau de Maupertuis (1698–1759), French mathematician and astronomer. The first use to which Maupertuis put the principle of least action was to restate Fermat's derivation of the law of refraction (1744).

[#] See, for example, Goldstein (Go80, pp. 365–371) or Sommerfeld (So50, pp. 204–209).

[##] See, for example, Lindsay and Margenau (Li36, pp. 112–120) or Sommerfeld (So50, pp. 210–214).

[§] Sir William Rowan Hamilton (1805–1865), Scottish mathematician and astronomer, and later, Irish Astronomer Royal.

[§§] The general meaning of "the path of a system" is made clear in Seciton 7.3.

In terms of the calculus of variations, Hamilton's Principle becomes

$$\delta \int_{t_1}^{t_2} (T - U)\, dt = 0 \tag{7.1}$$

where the symbol δ is a shorthand notation to describe the variation discussed in Sections 6.3 and 6.7. This variational statement of the principle requires only that the integral of $T - U$ be an *extremum*, not necessarily a *minimum*. But in almost all important applications in dynamics, the minimum condition occurs.

The kinetic energy of a particle expressed in fixed, rectangular coordinates is a function only of the $\dot{x}_i$, and if the particle moves in a conservative force field, the potential energy is a function only of the x_i:

$$T = T(\dot{x}_i), \quad U = U(x_i)$$

If we define the difference of these quantities to be

$$L \equiv T - U = L(x_i, \dot{x}_i) \tag{7.2}$$

then Equation 7.1 becomes

$$\boxed{\delta \int_{t_1}^{t_2} L(x_i, \dot{x}_i)\, dt = 0} \tag{7.3}$$

The function L appearing in this expression may be identified with the function f of the variational integral (see Section 6.5),

$$\delta \int_{x_1}^{x_2} f\{y_i(x), y_i'(x); x\}\, dx$$

if we make the transformations

$$x \rightarrow t$$
$$y_i(x) \rightarrow x_i(t)$$
$$y_i'(x) \rightarrow \dot{x}_i(t)$$
$$f\{y_i(x), y_i'(x); x\} \rightarrow L(x_i, \dot{x}_i)$$

The Euler-Lagrange equations (Equation 6.57) corresponding to Equation 7.3 are therefore

$$\boxed{\frac{\partial L}{\partial x_i} - \frac{d}{dt}\frac{\partial L}{\partial \dot{x}_i} = 0, \quad i = 1, 2, 3} \qquad \text{Lagrange equations of motion} \tag{7.4}$$

These are the **Lagrange equations of motion** for the particle, and the quantity L is called the **Lagrange function** or **Lagrangian** for the particle.

By way of example, let us obtain the Lagrange equation of motion for the one-dimensional harmonic oscillator. With the usual expressions for the kinetic and potential energies, we have

$$L = T - U = \tfrac{1}{2}m\dot{x}^2 - \tfrac{1}{2}kx^2$$

$$\frac{\partial L}{\partial x} = -kx$$

$$\frac{\partial L}{\partial \dot{x}} = m\dot{x}$$

$$\frac{d}{dt}\left(\frac{\partial L}{\partial \dot{x}}\right) = m\ddot{x}$$

Substituting these results into Equation 7.4 leads to

$$m\ddot{x} + kx = 0$$

which is identical with the equation of motion obtained using Newtonian mechanics.

The Lagrangian procedure seems rather complicated if it can only duplicate the simple results of Newtonian theory. However, let us continue illustrating the method by considering the plane pendulum (see Section 4.4). Using Equation 4.23 for T and U, we have, for the Lagrangian function

$$L = \tfrac{1}{2}ml^2\dot{\theta}^2 - mgl(1 - \cos\theta)$$

We now treat θ *as if it were a rectangular coordinate* and apply the operations specified in Equation 7.4; we obtain

$$\frac{\partial L}{\partial \theta} = -mgl\sin\theta$$

$$\frac{\partial L}{\partial \dot{\theta}} = ml^2\dot{\theta}$$

$$\frac{d}{dt}\left(\frac{\partial L}{\partial \dot{\theta}}\right) = ml^2\ddot{\theta}$$

$$\ddot{\theta} + \frac{g}{l}\sin\theta = 0$$

which again is identical with the Newtonian result (Equation 4.21). This is a remarkable result; it has been obtained by calculating the kinetic and potential energies in terms of θ rather than x and then applying a set of operations designed for use with rectangular rather than angular coordinates. We are therefore led to suspect that the Lagrange equations are more general than the form of Equation 7.4 would indicate. We pursue this matter in Section 7.4.

Another important characteristic of the method used in the two preceding simple examples is that nowhere in the calculations did there enter any statement regarding *force*. The equations of motion were obtained only by specifying certain properties associated *with the particle* (the kinetic and potential energies), and without the necessity of explicitly taking into account the fact that there was an external agency acting *on the particle* (the force). Therefore, insofar as *energy* can be defined independently of Newtonian concepts, Hamilton's Principle allows us to calculate the equations of motion of a body completely without recourse to Newtonian theory. We shall return to this important point in Sections 7.5 and 7.7.

7.3 GENERALIZED COORDINATES

We now seek to take advantage of the flexibility in specifying coordinates that the two examples of the preceding section have suggested is inherent in Lagrange's equations.

We consider a general mechanical system consisting of a collection of n discrete point particles, some of which may be connected to form rigid bodies. We discuss such systems of particles in Chapter 9 and rigid bodies in Chapter 11. To specify the state of such a system at a given time, it is necessary to use n radius vectors. Because each radius vector consists of a triple of numbers (e.g., the rectangular coordinates), $3n$ quantities must be specified to describe the positions of all the particles. If there exist equations of constraint that relate some of these coordinates to others (as would be the case, for example, if some of the particles were connected to form rigid bodies or if the motion were constrained to lie along some path or on some surface), then not all the $3n$ coordinates are independent. In fact, if there are m equations of constraint, then $3n - m$ coordinates are independent, and the system is said to possess $3n - m$ *degrees of freedom.*

It is important to note that if $s = 3n - m$ coordinates are required in a given case, we need not choose s rectangular coordinates or even s curvilinear coordinates (e.g., spherical, cylindrical). We can choose *any* s independent parameters, as long as they completely specify the state of the system. These s quantities need not even have the dimensions of length. Depending on the problem at hand, it may prove more convenient to choose some of the parameters with dimensions of *energy*, some with dimensions of $(length)^2$, some that are *dimensionless*, and so forth. In Example 6.5, we described a disk rolling down an inclined plane in terms of one coordinate that was a length and one that was an angle. We give the name **generalized coordinates** to any set of quantities that completely specifies the state of a system. The generalized coordinates are customarily written as $q_1, q_2, \ldots,$ or simply as the q_j. A set of independent generalized coordinates whose number equals the number s of degrees of freedom of the system and not restricted by the constraints is called a *proper* set of generalized coordinates. In certain instances, it may be advantageous to use generalized coordinates whose number exceeds the number of degrees of freedom and to explicitly take into account the constraint relations through the use of the Lagrange undetermined multipliers. Such would be the case, for example, if we desired to calculate the forces of constraint (see Example 7.8).

The choice of a set of generalized coordinates to describe a system is not unique; there are in general many sets of quantities (in fact, an *infinite* number!) that completely specify the state of a given system. For example, in the problem of the disk rolling down the inclined plane, we might choose as coordinates the height of the center of mass of the disk above some reference level and the distance through which some point on the rim has traveled since the start of the motion. The ultimate test of the "suitability" of a particular set of generalized coordinates is whether the resulting equations of motion are sufficiently simple to allow a straightforward interpretation. Unfortunately, we can state no general rules for selecting the "most suitable" set of generalized coordinates for a given problem. A certain skill must be developed through experience.

In addition to the generalized coordinates, we may define a set of quantities consisting of the time derivatives of q_j: $\dot{q}_1, \dot{q}_2, \ldots$, or simply $\dot{q}_j$. In analogy with the nomenclature for rectangular coordinates, we call $\dot{q}_j$ the **generalized velocities**.

If we allow for the possibility that the equations connecting $x_{\alpha,i}$ and q_j explicitly contain the time, then the set of transformation equations is given by*

$$x_{\alpha,i} = x_{\alpha,i}(q_1, q_2, \ldots, q_s, t), \qquad \begin{cases} \alpha = 1, 2, \ldots, n \\ i = 1, 2, 3 \end{cases}$$

$$= x_{\alpha,i}(q_j, t), \qquad\qquad j = 1, 2, \ldots, s \qquad (7.5)$$

In general, the rectangular components of the velocities depend on the generalized coordinates, the generalized velocities, and the time:

$$\dot{x}_{\alpha,i} = \dot{x}_{\alpha,i}(q_j, \dot{q}_j, t) \qquad (7.6)$$

We may also write the inverse transformations as

$$q_j = q_j(x_{\alpha,i}, t) \qquad (7.7)$$
$$\dot{q}_j = \dot{q}_j(x_{\alpha,i}, \dot{x}_{\alpha,i}, t) \qquad (7.8)$$

Also, there are $m = 3n - s$ equations of constraint of the form

$$f_k(x_{\alpha,i}, t) = 0, \qquad k = 1, 2, \ldots, m \qquad (7.9)$$

E X A M P L E   **7.1**   -

Find a suitable set of generalized coordinates for a point particle moving on the surface of a hemisphere of radius R whose center is at the origin.

Solution: Because the motion always takes place on the surface, we have

$$x^2 + y^2 + z^2 - R^2 = 0, \qquad z \geq 0 \qquad (7.10)$$

Let us choose as our generalized coordinates the cosines of the angles between the x-, y-, and z-axes and the line connecting the particle with the origin. Therefore,

$$q_1 = \frac{x}{R}, \qquad q_2 = \frac{y}{R}, \qquad q_3 = \frac{z}{R} \qquad (7.11)$$

---

* In this chapter, we attempt to simplify the notation by reserving the subscript i to designate rectangular axes; therefore, we always have $i = 1, 2, 3$.

But the sum of the squares of the direction cosines of a line equals unity. Hence,

$$q_1^2 + q_2^2 + q_3^2 = 1 \tag{7.12}$$

This set of q_j does not constitute a proper set of generalized coordinates, because we can write q_3 as a function of q_1 and q_2:

$$q_3 = \sqrt{1 - q_1^2 - q_2^2} \tag{7.13}$$

We may, however, choose $q_1 = x/R$ and $q_2 = y/R$ as proper generalized coordinates, and these quantities, together with the equation of constraint (Equation 7.13)

$$z = \sqrt{R^2 - x^2 - y^2} \tag{7.14}$$

are sufficient to uniquely specify the position of the particle. This should be an obvious result, because only two coordinates (e.g., latitude and longitude) are necessary to specify a point on the surface of a sphere. But the example illustrates the fact that the equations of constraint can always be used to reduce a trial set of coordinates to a proper set of generalized coordinates.

E X A M P L E **7.2** --

Use the (x, y) coordinate system of Figure 7-1 to find the kinetic energy T, potential energy U, and the Lagrangian L for a simple pendulum (length ℓ, mass bob m) moving in the x, y plane. Determine the transformation equations from the (x, y) rectangular system to the coordinate θ. Find the equation of motion.

Solution: We have already examined this general problem in Sections 4.4 and 7.1. When using the Lagrangian method, it is often useful to begin with rectangular coordinates and transform to the most obvious system with the simplest generalized coordinates. In this case, the kinetic and potential energies and the Lagrangian

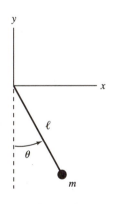

FIGURE 7-1

become

$$T = \tfrac{1}{2}m\dot{x}^2 + \tfrac{1}{2}m\dot{y}^2$$

$$U = mgy$$

$$L = T - U = \tfrac{1}{2}m\dot{x}^2 + \tfrac{1}{2}m\dot{y}^2 - mgy$$

Inspection of Figure 7-1 reveals that the motion can be determined by knowing θ and $\dot{\theta}$. Let's transform x and y into the coordinate θ and then find L in terms of θ.

$$x = \ell \sin \theta$$

$$y = -\ell \cos \theta$$

We now find for $\dot{x}$ and $\dot{y}$

$$\dot{x} = \ell \dot{\theta} \cos \theta$$

$$\dot{y} = \ell \dot{\theta} \sin \theta$$

$$L = \frac{m}{2}\left(\ell^2 \dot{\theta}^2 \cos^2 \theta + \ell^2 \dot{\theta}^2 \sin^2 \theta\right) + mg\ell \cos \theta = \frac{m}{2} \ell^2 \dot{\theta}^2 + mg\ell \cos \theta$$

The only generalized coordinate in the case of the pendulum is the angle θ, and we have expressed the Lagrangian in terms of θ by following a simple procedure of finding L in terms of x and y, finding the transformation equations, and then inserting them into the expression for L. If we do as we did in the previous section and treat θ *as if it were a rectangular coordinate*, we can find the equation of motion as follows:

$$\frac{\partial L}{\partial \theta} = -mg\ell \sin \theta$$

$$\frac{\partial L}{\partial \dot{\theta}} = m\ell^2 \dot{\theta}$$

$$\frac{d}{dt}\left(\frac{\partial L}{\partial \dot{\theta}}\right) = m\ell^2 \ddot{\theta}$$

We insert these relations into Equation 7.4 to find the same equation of motion as found previously.

$$\ddot{\theta} + \frac{g}{\ell} \sin \theta = 0$$

-- ●

The state of a system consisting of n particles and subject to m constraints that connect some of the $3n$ rectangular coordinates is completely specified by $s = 3n - m$ generalized coordinates. We may therefore represent the state of such a system by a point in an s-dimensional space called **configuration space**. Each dimension of this space corresponds to one of the q_j coordinates. We may represent the time history of a system by a curve in configuration space, each point specifying the *configuration* of the system at a particular instant. Through each such point passes an infinity of curves representing possible motions of the system; each curve corresponds to a particular set of initial conditions. We may therefore speak of the "path" of a system as it "moves" through configuration space. But we must be careful not to confuse this terminology with that applied to the motion of a particle along a path in ordinary three-dimensional space.

We should also note that a dynamical path in a configuration space consisting of proper generalized coordinates is automatically consistent with the constraints on the system, because the coordinates are chosen to correspond only to realizable motions of the system.

7.4 LAGRANGE'S EQUATIONS OF MOTION IN GENERALIZED COORDINATES

In view of the definitions in the preceding sections, we may now restate Hamilton's Principle as follows:

Of all the possible paths along which a dynamical system may move from one point to another in configuration space within a specified time interval, the actual path followed is that which minimizes the time integral of the Lagrangian function for the system.

To set up the variational form of Hamilton's Principle in generalized coordinates, we may take advantage of an important property of the Lagrangian we have not so far emphasized. The Lagrangian for a system is defined to be the difference between the kinetic and potential energies. But *energy is a scalar quantity and so the Lagrangian is a scalar function.* Hence the Lagrangian must be *invariant with respect to coordinate transformations.* However, certain transformations that change the Lagrangian but *leave the equations of motion unchanged* are allowed. For example, equations of motion are unchanged if L is replaced by $L + d/dt$ $[f(q_i, t)]$ for a function $f(q_i, t)$ with continuous second partial derivatives. As long as we define the Lagrangian to be the difference between the kinetic and potential energies, we may use different generalized coordinates. (The Lagrangian is, however, indefinite to an additive constant in the potential energy U.) It is therefore immaterial whether we express the Lagrangian in terms of $x_{\alpha,i}$ and $\dot{x}_{\alpha,i}$ or q_j and $\dot{q}_j$:

$$L = T(\dot{x}_{\alpha,i}) - U(x_{\alpha,i})$$

$$= T(q_j, \dot{q}_j, t) - U(q_j, t) \tag{7.15}$$

that is,

$$L = L(q_1, q_2, ..., \dot{q}_s; \dot{q}_1, \dot{q}_2, ..., \dot{q}_s; t)$$

$$= L(q_j, \dot{q}_j, t) \tag{7.16}$$

Thus, Hamilton's Principle becomes

$$\delta \int_{t_1}^{t_2} L(q_j, \dot{q}_j, t)\, dt = 0 \qquad \text{Hamilton's Principle} \qquad (7.17)$$

If we refer to the definitions of the quantities in Section 6.5 and make the identifications

$$x \to t$$

$$y_i(x) \to q_j(t)$$

$$y_i'(x) \to \dot{q}_j(t)$$

$$f\{y_i, y_i'; x\} \to L(q_j, \dot{q}_j, t)$$

then the Euler equations (Equation 6.57) corresponding to the variational problem stated in Equation 7.17 become

$$\frac{\partial L}{\partial q_j} - \frac{d}{dt}\frac{\partial L}{\partial \dot{q}_j} = 0, \qquad j = 1, 2, ..., s \qquad (7.18)$$

These are the Euler-Lagrange equations of motion for the system (usually called simply **Lagrange's equations***). There are s of these equations, and together with the m equations of constraint and the initial conditions that are imposed, they completely describe the motion of the system.[†]

It is important to realize that the validity of Lagrange's equations requires the following two conditions:

1. The forces acting on the system (apart from any forces of constraint) must be derivable from a potential (or several potentials).

2. The equations of constraint must be relations that connect the *coordinates* of the particles and may be functions of the time—that is, we must have constraint relations of the form given by Equation 7.9.

If the constraints can be expressed as in condition 2, they are termed **holonomic** constraints. If the equations do not explicitly contain the time, the constraints are said to be **fixed** or **scleronomic**; moving constraints are **rheonomic**.

---

* First derived for a mechanical system (although not, of course, by using Hamilton's Principle) by Lagrange and presented in his famous treatise *Mécanique analytique* in 1788. In this monumental work, which encompasses all phases of mechanics (statics, dynamics, hydrostatics, and hydrodynamics), Lagrange placed the subject on a firm and unified mathematical foundation. The treatise is mathematical rather than physical; Lagrange was quite proud of the fact that the entire work contains not a single diagram.

[†] Because there are s second-order differential equations, $2s$ initial conditions must be supplied to determine the motion uniquely.

We here consider only the motion of systems subject to conservative forces. Such forces can always be derived from potential functions, so that condition 1 is satisfied. This is not a necessary restriction on either Hamilton's Principle or Lagrange's equations; the theory can readily be extended to include nonconservative forces. Similarly, we can formulate Hamilton's Principle to include certain types of nonholonomic constraints, but the treatment here is confined to holonomic systems.

We now want to work several examples using Lagrange's equations. Experience is the best way to determine a set of generalized coordinates, realize the constraints, and set up the Lagrangian. Once this is done, the remainder of the problem is for the most part mathematical.

E X A M P L E **7.3** ---

Consider the case of projectile motion under gravity in two dimensions as was discussed in Example 2.6. Find the equations of motion in both Cartesian and polar coordinates.

Solution: We use Figure 2-7 to describe the system. In Cartesian coordinates, we use x (horizontal) and y (vertical). In polar coordinates we use r (in radial direction) and θ (elevation angle from horizontal). First, in Cartesian coordinates we have

$$\left. \begin{array}{l} T = \frac{1}{2}m\dot{x}^2 + \frac{1}{2}m\dot{y}^2 \\[2mm] U = mgy \end{array} \right\} \tag{7.19}$$

where $U = 0$ at $y = 0$.

$$L = T - U = \tfrac{1}{2}m\dot{x}^2 + \tfrac{1}{2}m\dot{y}^2 - mgy \tag{7.20}$$

We find the equations of motion by using Equation 7.18:

x:

$$\frac{\partial L}{\partial x} - \frac{d}{dt}\frac{\partial L}{\partial \dot{x}} = 0$$

$$0 - \frac{d}{dt}m\dot{x} = 0$$

$$\ddot{x} = 0 \tag{7.21}$$

y:

$$\frac{\partial L}{\partial y} - \frac{d}{dt}\frac{\partial L}{\partial \dot{y}} = 0$$

$$-mg - \frac{d}{dt}(m\dot{y}) = 0$$

$$\ddot{y} = -g \tag{7.22}$$

By using the initial conditions, Equations 7.21 and 7.22 can be integrated to determine the appropriate equations of motion.

In polar coordinates, we have

$$T = \tfrac{1}{2}m\dot{r}^2 + \tfrac{1}{2}m(r\dot{\theta})^2$$

$$U = mgr \sin \theta$$

where $U = 0$ for $\theta = 0$.

$$L = T - U = \tfrac{1}{2}m\dot{r}^2 + \tfrac{1}{2}mr^2\dot{\theta}^2 - mgr \sin \theta \tag{7.23}$$

r:

$$\frac{\partial L}{\partial r} - \frac{d}{dt}\frac{\partial L}{\partial \dot{r}} = 0$$

$$mr\dot{\theta}^2 - mg \sin \theta - \frac{d}{dt}(m\dot{r}) = 0$$

$$r\dot{\theta}^2 - g \sin \theta - \ddot{r} = 0 \tag{7.24}$$

θ:

$$\frac{\partial L}{\partial \theta} - \frac{d}{dt}\frac{\partial L}{\partial \dot{\theta}} = 0$$

$$- mgr \cos \theta - \frac{d}{dt}(mr^2\dot{\theta}) = 0$$

$$- gr \cos \theta - 2r\dot{r}\dot{\theta} - r^2\ddot{\theta} = 0 \tag{7.25}$$

The equations of motion expressed by Equations 7.21 and 7.22 are clearly simpler than those of Equations 7.24 and 7.25. We should choose Cartesian coordinates as the generalized coordinates to solve this problem. The key in recognizing this was that the potential energy of the system only depended on the y coordinate. In polar coordinates, the potential energy depended on both r and θ.

⬤

EXAMPLE 7.4

A particle of mass m is constrained to move on the inside surface of a smooth cone of half-angle α (see Figure 7-2). The particle is subject to a gravitational force. Determine a set of generalized coordinates and determine the constraints. Find Lagrange's equations of motion, Equation 7.18.

Solution: Let the axis of the cone correspond to the z-axis and let the apex of the cone be located at the origin. Since the problem possesses cylindrical symmetry, we choose r, θ, and z as the generalized coordinates. We have, however, the equation of constraint

$$z = r \cot \alpha \tag{7.26}$$

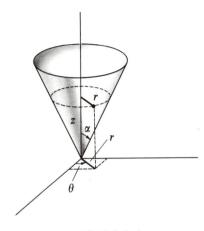

FIGURE 7-2

so there are only two degrees of freedom for the system, and therefore only two proper generalized coordinates. We may use Equation 7.26 to eliminate either the coordinate z or r; we choose to do the former. Then the square of the velocity is

$$v^2 = \dot{r}^2 + r^2\dot{\theta}^2 + \dot{z}^2$$
$$= \dot{r}^2 + r^2\dot{\theta}^2 + \dot{r}^2\cot^2\alpha$$
$$= \dot{r}^2\csc^2\alpha + r^2\dot{\theta}^2 \qquad (7.27)$$

The potential energy (if we choose $U = 0$ at $z = 0$) is

$$U = mgz = mgr\cot\alpha$$

so the Langrangian is

$$L = \tfrac{1}{2}m(\dot{r}^2\csc^2\alpha + r^2\dot{\theta}^2) - mgr\cot\alpha \qquad (7.28)$$

We note first that L does not explicitly contain θ. Therefore $\partial L/\partial\theta = 0$, and the Lagrange equation for the coordinate θ is

$$\frac{d}{dt}\frac{\partial L}{\partial\dot{\theta}} = 0$$

Hence

$$\frac{\partial L}{\partial\dot{\theta}} = mr^2\dot{\theta} = \text{constant} \qquad (7.29)$$

But $mr^2\dot{\theta} = mr^2\omega$ is just the angular momentum about the z-axis. Therefore, Equation 7.29 expresses the conservation of angular momentum about the axis of symmetry of the system.

The Lagrange equation for r is

$$\frac{\partial L}{\partial r} - \frac{d}{dt}\frac{\partial L}{\partial\dot{r}} = 0 \qquad (7.30)$$

Calculating the derivatives, we find

$$\ddot{r} - r\dot{\theta}^2 \sin^2 \alpha + g \sin \alpha \cos \alpha = 0 \tag{7.31}$$

which is the equation of motion for the coordinate r.

We shall return to this example in Section 8.10 and examine the motion in more detail.

--- ●

EXAMPLE 7.5   ---

The point of support of a simple pendulum of length b moves on a massless rim of radius a rotating with constant angular velocity ω. Obtain the expression for the Cartesian components of the velocity and acceleration of the mass m. Obtain also the angular acceleration for the angle θ shown in Figure 7-3.

Solution: We choose the origin of our coordinate system to be at the center of the rotating rim. The Cartesian components of mass m become

$$\left. \begin{array}{l} x = a \cos \omega t + b \sin \theta \\ y = a \sin \omega t - b \cos \theta \end{array} \right\} \tag{7.32}$$

The velocities are

$$\left. \begin{array}{l} \dot{x} = -a\omega \sin \omega t + b\dot{\theta} \cos \theta \\ \dot{y} = a\omega \cos \omega t + b\dot{\theta} \sin \theta \end{array} \right\} \tag{7.33}$$

Taking the time derivative once again gives the acceleration:

$$\ddot{x} = -a\omega^2 \cos \omega t + b(\ddot{\theta} \cos \theta - \dot{\theta}^2 \sin \theta)$$

$$\ddot{y} = -a\omega^2 \sin \omega t + b(\ddot{\theta} \sin \theta + \dot{\theta}^2 \cos \theta)$$

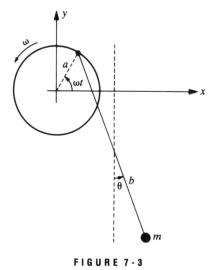

FIGURE 7-3

It should now be clear that the single generalized coordinate is θ. The kinetic and potential energies are

$$T = \tfrac{1}{2}m(\dot{x}^2 + \dot{y}^2)$$

$$U = mgy$$

where $U = 0$ at $y = 0$. The Lagrangian is

$$L = T - U = \frac{m}{2}[a^2\omega^2 + b^2\dot{\theta}^2 + 2b\dot{\theta}a\omega \sin(\theta - \omega t)]$$

$$-mg(a \sin \omega t - b \cos \theta) \qquad (7.34)$$

The derivatives for the Lagrange equation of motion for θ are

$$\frac{d}{dt}\frac{\partial L}{\partial \dot{\theta}} = mb^2\ddot{\theta} + mba\omega(\dot{\theta} - \omega)\cos(\theta - \omega t)$$

$$\frac{\partial L}{\partial \theta} = mb\dot{\theta}a\omega \cos(\theta - \omega t) - mgb \sin \theta$$

which results in the equation of motion (after solving for $\ddot{\theta}$)

$$\ddot{\theta} = \frac{\omega^2 a}{b} \cos(\theta - \omega t) - \frac{g}{b} \sin \theta \qquad (7.35)$$

Notice that this result reduces to the well-known equation of motion for a simple pendulum if $\omega = 0$.

EXAMPLE 7.6

Find the frequency of small oscillations of a simple pendulum placed in a railroad car that has a constant acceleration a in the x-direction.

Solution: A schematic diagram is shown in Figure 7-4a for the pendulum of length ℓ, mass m, and displacement angle θ. We choose a fixed cartesian coordinate system with $x = 0$ and $\dot{x} = v_0$ at $t = 0$. The position and velocity of m become

$$x = v_0 t + \tfrac{1}{2}at^2 + \ell \sin \theta$$

$$y = -\ell \cos \theta$$

$$\dot{x} = v_0 + at + \ell\dot{\theta} \cos \theta$$

$$\dot{y} = \ell\dot{\theta} \sin \theta$$

The kinetic and potential energies are

$$T = \tfrac{1}{2}m(\dot{x}^2 + \dot{y}^2) \qquad U = -mg\ell \cos \theta$$

(a)

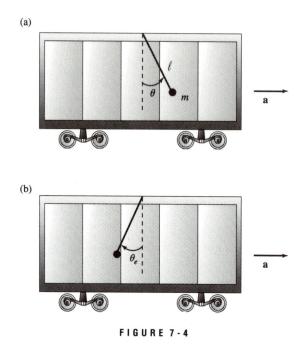

(b)

FIGURE 7-4

and the Lagrangian is

$$L = T - U = \tfrac{1}{2}m(v_0 + at + \ell\dot\theta\cos\theta)^2 + \tfrac{1}{2}m(\ell\dot\theta\sin\theta)^2 + mg\ell\cos\theta$$

The angle θ is the only generalized coordinate, and after taking the derivatives for Lagrange's equations and suitable collection of terms, the equation of motion becomes (Problem 7-2)

$$\ddot\theta = -\frac{g}{\ell}\sin\theta - \frac{a}{\ell}\cos\theta \tag{7.36}$$

We determine the equilibrium angle $\theta = \theta_e$ by setting $\ddot\theta = 0$,

$$0 = g\sin\theta_e + a\cos\theta_e \tag{7.37}$$

The equilibrium angle θ_e, shown in Figure 7-4b, is obtained by

$$\tan\theta_e = -\frac{a}{g} \tag{7.38}$$

Because the oscillations are small and are about the equilibrium angle, let $\theta = \theta_e + \eta$, where η is a small angle.

$$\ddot\theta = \ddot\eta = -\frac{g}{\ell}\sin(\theta_e + \eta) - \frac{a}{\ell}\cos(\theta_e + \eta) \tag{7.39}$$

We expand the sine and cosine terms and use the small angle approximation for $\sin\eta$ and $\cos\eta$, keeping only the first terms in the Taylor series expansions.

$$\ddot{\eta} = -\frac{g}{\ell}(\sin\theta_e\cos\eta + \cos\theta_e\sin\eta) - \frac{a}{\ell}(\cos\theta_e\cos\eta - \sin\theta_e\sin\eta)$$

$$= -\frac{g}{\ell}(\sin\theta_e + \eta\cos\theta_e) - \frac{a}{\ell}(\cos\theta_e - \eta\sin\theta_e)$$

$$= -\frac{1}{\ell}[(g\sin\theta_e + a\cos\theta_e) + \eta(g\cos\theta_e - a\sin\theta_e)]$$

The first term in the brackets is zero because of Equation 7.37, which leaves

$$\ddot{\eta} = -\frac{1}{\ell}(g\cos\theta_e - a\sin\theta_e)\eta \tag{7.40}$$

We use Equation 7.38 to determine $\sin\theta_e$ and $\cos\theta_e$ and after a little manipulation (Problem 7-2), Equation 7.40 becomes

$$\ddot{\eta} = -\frac{\sqrt{a^2 + g^2}}{\ell}\eta \tag{7.41}$$

Because this equation now represents simple harmonic motion, the frequency ω is determined to be

$$\omega^2 = \frac{\sqrt{a^2 + g^2}}{\ell} \tag{7.42}$$

This result seems plausible, because $\omega \to \sqrt{g/\ell}$ for $a = 0$ when the railroad car is at rest.

- ●

E X A M P L E ⬤ 7.7 -

A bead slides along a smooth wire bent in the shape of a parabola $z = cr^2$ (Figure 7-5). The bead rotates in a circle of radius R when the wire is rotating about its vertical symmetry axis with angular velocity ω. Find the value of c.

Solution: Because the problem has cylindrical symmetry, we choose r, θ, and z as the generalized coordinates. The kinetic energy of the bead is

$$T = \frac{m}{2}[\dot{r}^2 + \dot{z}^2 + (r\dot{\theta})^2] \tag{7.43}$$

If we choose $U = 0$ at $z = 0$, the potential energy term is

$$U = mgz \tag{7.44}$$

But r, z, and θ are not independent. The equation of constraint for the parabola is

$$z = cr^2 \tag{7.45}$$

$$\dot{z} = 2c\dot{r}r \tag{7.46}$$

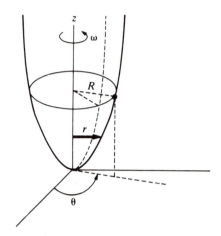

FIGURE 7-5

We also have an explicit time dependence of the angular rotation

$$\theta = \omega t$$

$$\dot{\theta} = \omega \tag{7.47}$$

We can now construct the Lagrangian as being dependent only on r, because there is no direct θ dependence.

$$L = T - U$$

$$= \frac{m}{2}(\dot{r}^2 + 4c^2r^2\dot{r}^2 + r^2\omega^2) - mgcr^2 \tag{7.48}$$

The problem stated that the bead moved in a circle of radius R. The reader might be tempted at this point to let $r = R = $ const. and $\dot{r} = 0$. It would be a mistake to do this now in the Lagrangian. First, we should find the equation of motion for the variable r and then let $r = R$ as a condition of the particular motion. This determines the particular value of c needed for $r = R$.

$$\frac{\partial L}{\partial \dot{r}} = \frac{m}{2}(2\dot{r} + 8c^2r^2\dot{r})$$

$$\frac{d}{dt}\frac{\partial L}{\partial \dot{r}} = \frac{m}{2}(2\ddot{r} + 16c^2r\dot{r}^2 + 8c^2r^2\ddot{r})$$

$$\frac{\partial L}{\partial r} = m(4c^2r\dot{r}^2 + r\omega^2 - 2gcr)$$

Lagrange's equation of motion becomes

$$\ddot{r}(1 + 4c^2r^2) + \dot{r}^2(4c^2r) + r(2gc - \omega^2) = 0 \tag{7.49}$$

which is a complicated result. If, however, the bead rotates with $r = R =$ constant, then $\dot{r} = \ddot{r} = 0$, and Equation 7.49 becomes

$$R(2gc - \omega^2) = 0$$

and

$$c = \frac{\omega^2}{2g} \tag{7.50}$$

is the result we wanted.

--●

EXAMPLE 7.8 ---

Consider the double pulley system shown in Figure 7-6. Use the coordinates indicated and determine the equations of motion.

Solution: Consider the pulleys to be massless, and let l_1 and l_2 be the lengths of rope hanging freely from each of the two pulleys. The distances x and y are measured from the center of the two pulleys.

m_1:

$$v_1 = \dot{x} \tag{7.51}$$

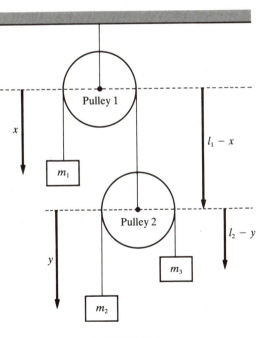

FIGURE 7-6

m_2:

$$v_2 = \frac{d}{dt}(l_1 - x + y) = -\dot{x} + \dot{y} \tag{7.52}$$

m_3:

$$v_3 = \frac{d}{dt}(l_1 - x + l_2 - y) = -\dot{x} - \dot{y} \tag{7.53}$$

$$T = \tfrac{1}{2}m_1 v_1^2 + \tfrac{1}{2}m_2 v_2^2 + \tfrac{1}{2}m_3 v_3^2$$

$$= \tfrac{1}{2}m_1\dot{x}^2 + \tfrac{1}{2}m_2(\dot{y} - \dot{x})^2 + \tfrac{1}{2}m_3(-\dot{x} - \dot{y})^2 \tag{7.54}$$

Let the potential energy $U = 0$ at $x = 0$.

$$U = U_1 + U_2 + U_3$$

$$= -m_1 gx - m_2 g(l_1 - x + y) - m_3 g(l_1 - x + l_2 - y) \tag{7.55}$$

Because T and U have been determined, the equations of motion can be obtained using Equation 7.18. The results are

$$m_1\ddot{x} + m_2(\ddot{x} - \ddot{y}) + m_3(\ddot{x} + \ddot{y}) = (m_1 - m_2 - m_3)g \tag{7.56}$$

$$- m_2(\ddot{x} - \ddot{y}) + m_3(\ddot{x} + \ddot{y}) = (m_2 - m_3)g \tag{7.57}$$

Equations 7.56 and 7.57 can be solved for $\ddot{x}$ and $\ddot{y}$.

-- ●

Examples 7.2–7.8 indicate the ease of using Lagrange's equations. It has been said, probably unfairly, that Lagrangian techniques are simply recipes to follow. The argument is that we lose track of the "physics" by its use. Lagrangian methods, on the contrary, are extremely powerful and allow us to solve problems that otherwise would lead to severe complications using Newtonian methods. Simple problems can perhaps be solved just as easily using Newtonian methods, but the Lagrangian technique can be used to attack a wide range of complex physical situations (including those occurring in quantum mechanics*).

7.5 LAGRANGE'S EQUATIONS WITH UNDETERMINED MULTIPLIERS

Constraints that can be expressed as algebraic relations among the coordinates are holonomic constraints. If a system is subject only to such constraints, we can

---

* See Feynman and Hibbs (Fe65).

always find a proper set of generalized coordinates in terms of which the equations of motion are free from explicit reference to the constraints.

Any constraints that must be expressed in terms of the *velocities* of the particles in the system are of the form

$$f(x_{\alpha,i}, \dot{x}_{\alpha,i}, t) = 0 \tag{7.58}$$

and constitute nonholonomic constraints *unless* the equations can be integrated to yield relations among the coordinates.*

Consider a constraint relation of the form

$$\sum_i A_i \dot{x}_i + B = 0, \qquad i = 1, 2, 3 \tag{7.59}$$

In general, this equation is nonintegrable, and therefore the constraint is nonholonomic. But if A_i and B have the forms

$$A_i = \frac{\partial f}{\partial x_i}, \qquad B = \frac{\partial f}{\partial t}, \qquad f = f(x_i, t) \tag{7.60}$$

then Equation 7.59 may be written as

$$\sum_i \frac{\partial f}{\partial x_i} \frac{dx_i}{dt} + \frac{\partial f}{\partial t} = 0 \tag{7.61}$$

But this is just

$$\frac{df}{dt} = 0$$

which can be integrated to yield

$$f(x_i, t) - \text{constant} = 0 \tag{7.62}$$

so the constraint is actually holonomic.

From the preceding discussion, we conclude that constraints expressible in differential form as

$$\sum_j \frac{\partial f_k}{\partial q_j} dq_j + \frac{\partial f_k}{\partial t} dt = 0 \tag{7.63}$$

are equivalent to those having the form of Equation 7.9.

If the constraint relations for a problem are given in differential form rather than as algebraic expressions, we can incorporate them directly into Lagrange's equations by using the Lagrange undetermined multipliers (see Section 6.6) without first performing the integrations; that is, for constraints expressible as in Equation 6.71,

$$\sum_j \frac{\partial f_k}{\partial q_j} dq_j = 0 \qquad \begin{cases} j = 1, 2, ..., s \\ k = 1, 2, ..., m \end{cases} \tag{7.64}$$

---

* Such constraints are sometimes called "semiholonomic."

the Lagrange equations (Equation 6.69) are

$$\boxed{\frac{\partial L}{\partial q_j} - \frac{d}{dt}\frac{\partial L}{\partial \dot{q}_j} + \sum_k \lambda_k(t)\frac{\partial f_k}{\partial q_j} = 0}$$ (7.65)

In fact, because the variation process involved in Hamilton's Principle holds the time constant at the endpoints, we could add to Equation 7.64 a term $(\partial f_k/\partial t)dt$ without affecting the equations of motion. Thus constraints expressed by Equation 7.63 also lead to the Lagrange equations given in Equation 7.65.

The great advantage of the Lagrangian formulation of mechanics is that the explicit inclusion of the forces of constraint is not necessary; that is, the emphasis is placed on the dynamics of the system rather than the calculation of the forces acting on each component of the system. In certain instances, however, it may be desired to know the forces of constraint. For example, from an engineering standpoint, it would be useful to know the constraint forces for design purposes. It is therefore worth pointing out that in Lagrange's equations expressed as in Equation 7.65, **the undetermined multipliers $\lambda_k(t)$ are just these forces of constraint.**[*] The generalized forces of constraint Q_j are given by

$$Q_j = \sum_k \lambda_k \frac{\partial f_k}{\partial q_j}$$ (7.66)

EXAMPLE 7.9 -

Let us consider again the case of the disk rolling down an inclined plane (see Example 6.5 and Figure 6-7). Find the equations of motion, the force of constraint, and the angular acceleration.

Solution: The kinetic energy may be separated into translational and rotational terms[†]

$$T = \tfrac{1}{2}M\dot{y}^2 + \tfrac{1}{2}I\dot{\theta}^2$$

$$= \tfrac{1}{2}M\dot{y}^2 + \tfrac{1}{4}MR^2\dot{\theta}^2$$

where M is the mass of the disk and R is the radius; $I = \tfrac{1}{2}MR^2$ is the moment of inertia of the disk about a central axis. The potential energy is

$$U = Mg(l - y)\sin\alpha$$ (7.67)

where l is the length of the inclined surface of the plane and where the disk is assumed to have zero potential energy at the bottom of the plane. The Lagrangian

---

[*] See, for example, Goldstein (Go80, p. 47). Explicit calculations of the forces of constraint in some specific problems are carried out by Becker (Be54, Chapters 11 and 13) and by Symon (Sy71, p. 372ff).

[†] We anticipate here a well-known result from rigid-body dynamics discussed in Chapter 11.

is therefore

$$L = T - U$$
$$= \tfrac{1}{2}M\dot{y}^2 + \tfrac{1}{4}MR^2\dot{\theta}^2 + Mg(y - l)\sin\alpha \tag{7.68}$$

The equation of constraint is

$$f(y, \theta) = y - R\theta = 0 \tag{7.69}$$

The system has only one degree of freedom if we insist that the rolling takes place without slipping. We may therefore choose either y or θ as the proper coordinate and use Equation 7.69 to eliminate the other. Alternatively, we may continue to consider *both* y and θ as generalized coordinates and use the method of undetermined multipliers. The Lagrange equations in this case are

$$\left.\begin{array}{l} \dfrac{\partial L}{\partial y} - \dfrac{d}{dt}\dfrac{\partial L}{\partial \dot{y}} + \lambda\dfrac{\partial f}{\partial y} = 0 \\[2mm] \dfrac{\partial L}{\partial \theta} - \dfrac{d}{dt}\dfrac{\partial L}{\partial \dot{\theta}} + \lambda\dfrac{\partial f}{\partial \theta} = 0 \end{array}\right\} \tag{7.70}$$

Performing the differentiations, we obtain, for the equations of motion,

$$Mg\sin\alpha - M\ddot{y} + \lambda = 0 \tag{7.71a}$$

$$-\tfrac{1}{2}MR^2\ddot{\theta} - \lambda R = 0 \tag{7.71b}$$

Also, from the constraint equation, we have

$$y = R\theta \tag{7.72}$$

These equations (Equations 7.71 and 7.72) constitute a soluble system for the three unknowns y, θ, λ. Differentiating the equation of constraint (Equation 7.72), we obtain

$$\ddot{\theta} = \frac{\ddot{y}}{R} \tag{7.73}$$

Combining Equations 7.71b and 7.73, we find

$$\lambda = -\tfrac{1}{2}M\ddot{y} \tag{7.74}$$

and then using this expression in Equation 7.71a there results

$$\ddot{y} = \frac{2g\sin\alpha}{3} \tag{7.75}$$

with

$$\lambda = -\frac{Mg\sin\alpha}{3} \tag{7.76}$$

so that Equation 7.71b yields

$$\ddot{\theta} = \frac{2g \sin \alpha}{3R} \qquad (7.77)$$

Thus, we have three equations for the quantities $\ddot{y}$, $\ddot{\theta}$, and λ that can be immediately integrated.

We note that if the disk were to slide without friction down the plane, we would have $\ddot{y} = g \sin \alpha$. Therefore, the rolling constraint reduces the acceleration to $\frac{2}{3}$ of the value of frictionless sliding. The magnitude of the force of friction producing the constraint is just λ—that is, $(Mg/3) \sin \alpha$.

The generalized forces of constraint, Equations 7.66, are

$$Q_y = \lambda \frac{\partial f}{\partial y} = \lambda = -\frac{Mg \sin \alpha}{3}$$

$$Q_\theta = \lambda \frac{\partial f}{\partial \theta} = -\lambda R = \frac{MgR \sin \alpha}{3}$$

Note that Q_y and Q_θ are a force and a torque, respectively, and they are the generalized forces of constraint required to keep the disk rolling down the plane without slipping.

Note that we may eliminate $\dot{\theta}$ from the Lagrangian by substituting $\dot{\theta} = \dot{y}/R$ from the equation of constraint:

$$L = \tfrac{3}{4}M\dot{y}^2 + Mg(y - l) \sin \alpha \qquad (7.78)$$

The Lagrangian is then expressed in terms of only one proper coordinate, and the single equation of motion is immediately obtained from Equation 7.18:

$$Mg \sin \alpha - \tfrac{3}{2}M\ddot{y} = 0 \qquad (7.79)$$

which is the same as Equation 7.75. Although this procedure is simpler, it cannot be used to obtain the force of constraint.

--- ●

E X A M P L E   7.10

A particle of mass m starts at rest on top of a smooth fixed hemisphere of radius a. Find the force of constraint, and determine the angle at which the particle leaves the hemisphere.

Solution: See Figure 7-7. Because we are considering the possibility of the particle leaving the hemisphere, we choose the generalized coordinates to be r and θ. The constraint equation is

$$f(r, \theta) = r - a = 0 \qquad (7.80)$$

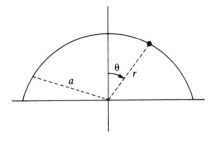

FIGURE 7-7

The Lagrangian is determined from the kinetic and potential energies:

$$T = \frac{m}{2}(\dot{r}^2 + r^2\dot{\theta}^2)$$

$$U = mgr \cos \theta$$

$$L = T - U$$

$$L = \frac{m}{2}(\dot{r}^2 + r^2\dot{\theta}^2) - mgr \cos \theta \qquad (7.81)$$

where the potential energy is zero at the bottom of the hemisphere. The Lagrange equations, Equation 7.65, are

$$\frac{\partial L}{\partial r} - \frac{d}{dt}\frac{\partial L}{\partial \dot{r}} + \lambda \frac{\partial f}{\partial r} = 0 \qquad (7.82)$$

$$\frac{\partial L}{\partial \theta} - \frac{d}{dt}\frac{\partial L}{\partial \dot{\theta}} + \lambda \frac{\partial f}{\partial \theta} = 0 \qquad (7.83)$$

Performing the differentiations on Equation 7.80 gives

$$\frac{\partial f}{\partial r} = 1, \qquad \frac{\partial f}{\partial \theta} = 0 \qquad (7.84)$$

Equations 7.82 and 7.83 become

$$mr\dot{\theta}^2 - mg \cos \theta - m\ddot{r} + \lambda = 0 \qquad (7.85)$$

$$mgr \sin \theta - mr^2\ddot{\theta} - 2mr\dot{r}\dot{\theta} = 0 \qquad (7.86)$$

Next, we apply the constraint $r = a$ to these equations of motion:

$$r = a, \qquad \dot{r} = 0 = \ddot{r}$$

Equations 7.85 and 7.86 then become

$$ma \dot{\theta}^2 - mg \cos \theta + \lambda = 0 \qquad (7.87)$$

$$mga \sin \theta - ma^2\ddot{\theta} = 0 \qquad (7.88)$$

From Equation 7.88, we have

$$\ddot{\theta} = \frac{g}{a} \sin \theta \tag{7.89}$$

We can integrate Equation 7.89 to determine $\dot{\theta}^2$.

$$\ddot{\theta} = \frac{d}{dt} \frac{d\theta}{dt} = \frac{d\dot{\theta}}{dt} = \frac{d\dot{\theta}}{d\theta} \frac{d\theta}{dt} = \dot{\theta} \frac{d\dot{\theta}}{d\theta} \tag{7.90}$$

We integrate Equation 7.89,

$$\int \dot{\theta} \, d\dot{\theta} = \frac{g}{a} \int \sin \theta \, d\theta \tag{7.91}$$

which results in

$$\frac{\dot{\theta}^2}{2} = \frac{-g}{a} \cos \theta + \frac{g}{a} \tag{7.92}$$

where the integration constant is g/a, because $\dot{\theta} = 0$ at $t = 0$ when $\theta = 0$. Substituting $\dot{\theta}^2$ from Equation 7.92 into Equation 7.87 gives, after solving for λ,

$$\lambda = mg(3 \cos \theta - 2) \tag{7.93}$$

which is the force of constraint. The particle falls off the hemisphere at angle θ_0 when $\lambda = 0$.

$$\lambda = 0 = mg(3 \cos \theta_0 - 2) \tag{7.94}$$

$$\theta_0 = \cos^{-1}\left(\frac{2}{3}\right) \tag{7.95}$$

As a quick check, notice that the constraint force is $\lambda = mg$ at $\theta = 0$ when the particle is perched on top of the hemisphere.

--- ●

The usefulness of the method of undetermined multipliers is twofold:

1. The Lagrange multipliers are the forces of constraint that are often needed.

2. When a proper set of generalized coordinates is not desired or too difficult to obtain, the method may be used to increase the number of generalized coordinates by including constraint relations between the coordinates.

7.6 EQUIVALENCE OF LAGRANGE'S AND NEWTON'S EQUATIONS

As we have emphasized from the outset, the Lagrangian and Newtonian formulations of mechanics are equivalent: the viewpoint is different, but the content is the same. We now explicitly demonstrate this equivalence by showing that the two sets of equations of motion are in fact the same.

In Equation 7.18, let us choose the generalized coordinates to be the rectangular coordinates. Lagrange's equations (for a single particle) then become

$$\frac{\partial L}{\partial x_i} - \frac{d}{dt}\frac{\partial L}{\partial \dot{x}_i} = 0, \qquad i = 1, 2, 3 \tag{7.96}$$

or

$$\frac{\partial (T - U)}{\partial x_i} - \frac{d}{dt}\frac{\partial (T - U)}{\partial \dot{x}_i} = 0$$

But in rectangular coordinates and for a conservative system, we have $T = T(\dot{x}_i)$ and $U = U(x_i)$, so

$$\frac{\partial T}{\partial x_i} = 0 \quad \text{and} \quad \frac{\partial U}{\partial \dot{x}_i} = 0$$

Lagrange's equations therefore become

$$-\frac{\partial U}{\partial x_i} = \frac{d}{dt}\frac{\partial T}{\partial \dot{x}_i} \tag{7.97}$$

We also have (for a conservative system)

$$-\frac{\partial U}{\partial x_i} = F_i$$

and

$$\frac{d}{dt}\frac{\partial T}{\partial \dot{x}_i} = \frac{d}{dt}\frac{\partial}{\partial \dot{x}_i}\left(\sum_{j=1}^{3}\frac{1}{2}m\dot{x}_j^2\right) = \frac{d}{dt}(m\dot{x}_i) = \dot{p}_i$$

so Equation 7.97 yields the Newtonian equations, as required:

$$F_i = \dot{p}_i \tag{7.98}$$

Thus, the Lagrangian and Newtonian equations are identical if the generalized coordinates are the rectangular coordinates.

Now let us derive Lagrange's equations of motion using Newtonian concepts. Consider only a single particle for simplicity. We need to transform from the x_i-coordinates to the generalized coordinates q_j. From Equation 7.5, we have

$$x_i = x_i(q_j, t) \tag{7.99}$$

$$\dot{x}_i = \sum_j \frac{\partial x_i}{\partial q_j}\dot{q}_j + \frac{\partial x_i}{\partial t} \tag{7.100}$$

and

$$\frac{\partial \dot{x}_i}{\partial \dot{q}_i} = \frac{\partial x_i}{\partial q_i} \tag{7.101}$$

A generalized momentum p_j associated with q_j is easily determined by

$$p_j = \frac{\partial T}{\partial \dot{q}_j} \tag{7.102}$$

For example, for a particle moving in plane polar coordinates, $T = (\dot{r}^2 + r^2\dot{\theta}^2)m/2$, we have $p_r = m\dot{r}$ for coordinate r and $p_\theta = mr^2\dot{\theta}$ for coordinate θ. Obviously p_r is a linear momentum and p_θ is an angular momentum, so our generalized momentum definition seems consistent with Newtonian concepts.

We can determine a generalized force by considering the *virtual* work δW done by a varied path δx_i as described in Section 6.7.

$$\delta W = \sum_i F_i \delta x_i = \sum_{i,j} F_i \frac{\partial x_i}{\partial q_j} \delta q_j \tag{7.103}$$

$$\equiv \sum_j Q_j \, \delta q_j \tag{7.104}$$

so that the generalized force Q_j associated with q_j is

$$Q_j = \sum_i F_i \frac{\partial x_i}{\partial q_j} \tag{7.105}$$

Just as work is always energy, so is the product of Qq. If q is length, Q is force; if q is an angle, Q is torque. For a conservative system, Q_j is derivable from the potential energy:

$$Q_j = -\frac{\partial U}{\partial q_j} \tag{7.106}$$

Now we are ready to obtain Lagrange's equations:

$$p_j = \frac{\partial T}{\partial \dot{q}_j} = \frac{\partial}{\partial \dot{q}_j}\left(\sum_i \frac{1}{2} m\dot{x}_i^2\right)$$

$$= \sum_i m\dot{x}_i \frac{\partial \dot{x}_i}{\partial \dot{q}_j}$$

$$p_j = \sum_i m\dot{x}_i \frac{\partial x_i}{\partial q_j} \tag{7.107}$$

where we use Equation 7.101 for the last step. Taking the time derivative of Equation 7.107 gives

$$\dot{p}_j = \sum_i \left(m\ddot{x}_i \frac{\partial x_i}{\partial q_j} + m\dot{x}_i \frac{d}{dt}\frac{\partial x_i}{\partial q_j} \right) \tag{7.108}$$

Expanding the last term gives

$$\frac{d}{dt}\frac{\partial x_i}{\partial q_j} = \sum_k \frac{\partial^2 x_i}{\partial q_k \partial q_j}\dot{q}_k + \frac{\partial^2 x_i}{\partial q_j \partial t}$$

and Equation 7.108 becomes

$$\dot{p}_j = \sum_i m\ddot{x}_i \frac{\partial x_i}{\partial q_j} + \sum_{i,k} m\dot{x}_i \frac{\partial^2 x_i}{\partial q_k \partial q_j} \dot{q}_k + \sum_i m\dot{x}_i \frac{\partial^2 x_i}{\partial q_j \partial t} \qquad (7.109)$$

The first term on the right side of Equation 7.109 is just Q_j ($F_i = m\ddot{x}_i$ and Equation 7.105). The sum of the other two terms is $\partial T / \partial q_j$:

$$\frac{\partial T}{\partial q_j} = \sum_i m\dot{x}_i \frac{\partial \dot{x}_i}{\partial q_j}$$

$$= \sum_i m\dot{x}_i \frac{\partial}{\partial q_j} \left(\sum_k \frac{\partial x_i}{\partial q_k} \dot{q}_k + \frac{\partial x_i}{\partial t} \right) \qquad (7.110)$$

where we have used $T = \sum_i 1/2\, m\dot{x}_i^2$ and Equation 7.100.

Equation 7.109 can now be written as

$$\dot{p}_j = Q_j + \frac{\partial T}{\partial q_j} \qquad (7.111)$$

or, using Equations 7.102 and 7.106,

$$\frac{d}{dt} \left(\frac{\partial T}{\partial \dot{q}_j} \right) - \frac{\partial T}{\partial q_j} = Q_j = -\frac{\partial U}{\partial q_j} \qquad (7.112)$$

Because U does not depend on the generalized velocities $\dot{q}_j$, Equation 7.112 can be written

$$\frac{d}{dt} \left[\frac{\partial (T - U)}{\partial \dot{q}_j} \right] - \frac{\partial (T - U)}{\partial q_j} = 0 \qquad (7.113)$$

and using $L = T - U$,

$$\frac{d}{dt} \left(\frac{\partial L}{\partial \dot{q}_j} \right) - \frac{\partial L}{\partial q_j} = 0 \qquad (7.114)$$

which are Lagrange's equations of motion.

7.7 ESSENCE OF LAGRANGIAN DYNAMICS

In the preceding sections, we made several general and important statements concerning the Lagrange formulation of mechanics. Before proceeding further, we should summarize these points to emphasize the differences between the Lagrange and Newtonian viewpoints.

Historically, the Lagrange equations of motion expressed in generalized coordinates were derived before the statement of Hamilton's Principle.* We elected to deduce Lagrange's equations by postulating Hamilton's Principle because this is the

---

* Lagrange's equations, 1788; Hamilton's Principle, 1834.

most straightforward approach and is also the formal method for unifying classical dynamics.

First, we must reiterate that Lagrangian dynamics does not constitute a *new* theory in any sense of the word. The results of a Lagrangian analysis or a Newtonian analysis must be the same for any given mechanical system. The only difference is the method used to obtain these results.

Whereas the Newtonian approach emphasizes an outside agency acting *on* a body (the *force*), the Lagrangian method deals only with quantities associated *with* the body (the kinetic and potential *energies*). In fact, nowhere in the Lagrangian formulation does the concept of *force* enter. This is a particularly important property—and for a variety of reasons. First, because energy is a scalar quantity, the Lagrangian function for a system is invariant to coordinate transformations. Indeed, such transformations are not restricted to be between various orthogonal coordinate systems in ordinary space; they may also be transformations between *ordinary* coordinates and *generalized* coordinates. Thus, it is possible to pass from ordinary space (in which the equations of motion may be quite complicated) to a configuration space that can be chosen to yield maximum simplification for a particular problem. We are accustomed to thinking of mechanical systems in terms of *vector* quantities such as force, velocity, angular momentum, and torque. But in the Lagrangian formulation, the equations of motion are obtained entirely in terms of *scalar* operations in configuration space.

Another important aspect of the force-versus-energy viewpoint is that in certain situations it may not even be possible to state explicitly all the forces acting on a body (as is sometimes the case for forces of constraint), whereas it is still possible to give expressions for the kinetic and potential energies. It is just this fact that makes Hamilton's Principle useful for quantum-mechanical systems where we normally know the energies but not the forces.

The differential statement of mechanics contained in Newton's equations or the integral statement embodied in Hamilton's Principle (and the resulting Lagrangian equations) have been shown to be entirely equivalent. Hence, no distinction exists between these viewpoints, which are based on the description of *physical effects*. But from a philosophical standpoint, we can make a distinction. In the Newtonian formulation, a certain force on a body produces a definite motion—that is, we always associate a definite *effect* with a certain *cause*. According to Hamilton's Principle, however, the motion of a body results from the attempt of nature to achieve a certain *purpose*, namely, to minimize the time integral of the difference between the kinetic and potential energies. The operational solving of problems in mechanics does not depend on adopting one or the other of these views. But historically such considerations have had a profound influence on the development of dynamics (as, for example, in Maupertuis's principle, mentioned in Section 7.2). The interested reader is referred to Margenau's excellent book for a discussion of these matters.*

---

* Margenau (Ma77, Chapter 19).

7.8 A THEOREM CONCERNING THE KINETIC ENERGY

If the kinetic energy is expressed in fixed, rectangular coordinates, the result is a homogeneous quadratic function of $\dot{x}_{\alpha,i}$:

$$T = \frac{1}{2} \sum_{\alpha=1}^{n} \sum_{i=1}^{3} m_\alpha \dot{x}_{\alpha,i}^2 \tag{7.115}$$

We now wish to consider in more detail the dependence of T on the generalized coordinates and velocities. For many particles, Equations 7.99 and 7.100 become

$$x_{\alpha,i} = x_{\alpha,i}(q_j, t), \qquad j = 1, 2, ..., s \tag{7.116}$$

$$\dot{x}_{\alpha,i} = \sum_{j=1}^{s} \frac{\partial x_{\alpha,i}}{\partial q_j} \dot{q}_j + \frac{\partial x_{\alpha,i}}{\partial t} \tag{7.117}$$

Evaluating the square of $\dot{x}_{\alpha,i}$, we obtain

$$\dot{x}_{\alpha,i}^2 = \sum_{j,k} \frac{\partial x_{\alpha,i}}{\partial q_j} \frac{\partial x_{\alpha,i}}{\partial q_k} \dot{q}_j \dot{q}_k + 2 \sum_j \frac{\partial x_{\alpha,i}}{\partial q_j} \frac{\partial x_{\alpha,i}}{\partial t} \dot{q}_j + \left(\frac{\partial x_{\alpha,i}}{\partial t} \right)^2 \tag{7.118}$$

and the kinetic energy becomes

$$T = \sum_\alpha \sum_{i,j,k} \frac{1}{2} m_\alpha \frac{\partial x_{\alpha,i}}{\partial q_j} \frac{\partial x_{\alpha,i}}{\partial q_k} \dot{q}_j \dot{q}_k$$

$$+ \sum_\alpha \sum_{i,j} m_\alpha \frac{\partial x_{\alpha,i}}{\partial q_j} \frac{\partial x_{\alpha,i}}{\partial t} \dot{q}_j$$

$$+ \sum_\alpha \sum_i \frac{1}{2} m_\alpha \left(\frac{\partial x_{\alpha,i}}{\partial t} \right)^2 \tag{7.119}$$

Thus, we have the general result

$$T = \sum_{j,k} a_{jk} \dot{q}_j \dot{q}_k + \sum_j b_j \dot{q}_j + c \tag{7.120}$$

A particularly important case occurs when the system is *scleronomic*, so that the time does not appear explicitly in the equations of transformation (Equation 7.116); then the partial time derivatives vanish:

$$\frac{\partial x_{\alpha,i}}{\partial t} = 0, \qquad b_j = 0, \qquad c = 0$$

Therefore, under these conditions, the kinetic energy is a *homogeneous quadratic function* of the generalized velocities:

$$\boxed{T = \sum_{j,k} a_{jk} \dot{q}_j \dot{q}_k} \tag{7.121}$$

Next, we differentiate Equation 7.121 with respect to $\dot{q}_l$:

$$\frac{\partial T}{\partial \dot{q}_l} = \sum_k a_{lk}\dot{q}_k + \sum_j a_{jl}\dot{q}_j$$

Multiplying this equation by $\dot{q}_l$ and summing over l, we have

$$\sum_l \dot{q}_l \frac{\partial T}{\partial \dot{q}_l} = \sum_{k,l} a_{lk}\dot{q}_k\dot{q}_l + \sum_{j,l} a_{jl}\dot{q}_j\dot{q}_l$$

In this case, *all* the indices are dummies, so both terms on the right-hand side are identical:

$$\boxed{\sum_l \dot{q}_l \frac{\partial T}{\partial \dot{q}_l} = 2 \sum_{j,k} a_{jk}\dot{q}_j\dot{q}_k = 2T} \qquad (7.122)$$

This important result is a special case of *Euler's theorem*, which states that if $f(y_k)$ is a homogeneous function of the y_k that is of degree n, then

$$\sum_k y_k \frac{\partial f}{\partial y_k} = nf \qquad (7.123)$$

7.9 CONSERVATION THEOREMS REVISITED

Conservation of Energy

We saw in our previous arguments* that *time* is homogeneous within an inertial reference frame. Therefore, the Lagrangian that describes a *closed system* (i.e., a system not interacting with anything outside the system) cannot depend explicitly on time,[†] that is,

$$\frac{\partial L}{\partial t} = 0 \qquad (7.124)$$

so that the total derivative of the Lagrangian becomes

$$\frac{dL}{dt} = \sum_j \frac{\partial L}{\partial q_j}\dot{q}_j + \sum_j \frac{\partial L}{\partial \dot{q}_j}\ddot{q}_j \qquad (7.125)$$

where the usual term, $\partial L/\partial t$, does not now appear. But Lagrange's equations are

$$\frac{\partial L}{\partial q_j} = \frac{d}{dt}\frac{\partial L}{\partial \dot{q}_j} \qquad (7.126)$$

---

* See Section 2.3.

[†] The Lagrangian is likewise independent of the time if the system exists in a uniform force field.

Using Equation 7.126 to substitute for $\partial L / \partial q_j$ in Equation 7.125, we have

$$\frac{dL}{dt} = \sum_j \dot{q}_j \frac{d}{dt} \frac{\partial L}{\partial \dot{q}_j} + \sum_j \frac{\partial L}{\partial \dot{q}_j} \ddot{q}_j$$

or

$$\frac{dL}{dt} - \sum_j \frac{d}{dt} \left(\dot{q}_j \frac{\partial L}{\partial \dot{q}_j} \right) = 0$$

so that

$$\frac{d}{dt} \left(L - \sum_j \dot{q}_j \frac{\partial L}{\partial \dot{q}_j} \right) = 0 \qquad (7.127)$$

The quantity in the parentheses is therefore constant in time; denote this constant by $-H$:

$$L - \sum_j \dot{q}_j \frac{\partial L}{\partial \dot{q}_j} = -H = \text{constant} \qquad (7.128)$$

If the potential energy U does not depend explicitly on the velocities $\dot{x}_{\alpha,i}$ or the time t, then $U = U(x_{\alpha,i})$. The relations connecting the rectangular coordinates and the generalized coordinates are of the form $x_{\alpha,i} = x_{\alpha,i}(q_j)$ or $q_j = q_j(x_{\alpha,i})$, where we exclude the possibility of an explicit time dependence in the transformation equations. Therefore, $U = U(q_j)$, and $\partial U / \partial \dot{q}_j = 0$. Thus

$$\frac{\partial L}{\partial \dot{q}_j} = \frac{\partial (T - U)}{\partial \dot{q}_j} = \frac{\partial T}{\partial \dot{q}_j}$$

Equation 7.128 can then be written as

$$(T - U) - \sum_j \dot{q}_j \frac{\partial T}{\partial \dot{q}_j} = -H \qquad (7.129)$$

and, using Equation 7.122, we have

$$(T - U) - 2T = -H$$

or

$$T + U = E = H = \text{constant} \qquad (7.130)$$

The total energy E is a constant of the motion for this case.

The function H, called the **Hamiltonian** of the system, may be defined as in Equation 7.128 (but see Section 7.10). It is important to note that the Hamiltonian H is equal to the total energy E only if the following conditions are met:

1. The equations of the transformation connecting the rectangular and generalized coordinates (Equation 7.116) must be independent of the time, thus ensuring that the kinetic energy is a homogeneous quadratic function of the $\dot{q}_j$.

2.   The potential energy must be velocity independent, thus allowing the elimination of the terms $\partial U/\partial \dot{q}_j$ from the equation for H (Equation 7.129).

The questions "Does $H = E$ for the system?" and "Is energy conserved for the system?", then, pertain to two *different* aspects of the problem, and each question must be examined separately. We may, for example, have cases in which the Hamiltonian does not equal the total energy, but nevertheless, the energy is conserved. Thus, consider a conservative system, and let the description be made in terms of generalized coordinates in motion with respect to fixed, rectangular axes. The transformation equations then contain the time, and the kinetic energy is *not* a homogeneous quadratic function of the generalized velocities. The choice of a mathematically convenient set of generalized coordinates cannot alter the physical fact that energy is conserved. But in the moving coordinate system, the Hamiltonian is no longer equal to the total energy.

Conservation of Linear Momentum

Because space is homogeneous in an inertial reference frame, the Lagrangian of a closed system is unaffected by a translation of the entire system in space. Consider an infinitesimal translation of every radius vector $\mathbf{r}_\alpha$ such that $\mathbf{r}_\alpha \to \mathbf{r}_\alpha + \delta\mathbf{r}$; this amounts to translating the entire system by $\delta\mathbf{r}$. For simplicity, let us examine a system consisting of only a single particle (by including a summation over α we could consider an *n*-particle system in an entirely equivalent manner), and let us write the Lagrangian in terms of rectangular coordinates $L = L(x_i, \dot{x}_i)$. The change in L caused by the infinitesimal displacement $\delta\mathbf{r} = \sum_i \delta x_i \mathbf{e}_i$ is

$$\delta L = \sum_i \frac{\partial L}{\partial x_i} \delta x_i + \sum_i \frac{\partial L}{\partial \dot{x}_i} \delta \dot{x}_i = 0 \tag{7.131}$$

We consider only a varied *displacement*, so that the δx_i are not explicit or implicit functions of the time. Thus,

$$\delta \dot{x}_i = \delta \frac{dx_i}{dt} = \frac{d}{dt} \delta x_i \equiv 0 \tag{7.132}$$

Therefore, δL becomes

$$\partial L = \sum_i \frac{\partial L}{\partial x_i} \delta x_i = 0 \tag{7.133}$$

Because each of the δx_i is an independent displacement, δL vanishes identically only if each of the partial derivatives of L vanishes:

$$\frac{\partial L}{\partial x_i} = 0 \tag{7.134}$$

Then, according to Lagrange's equations,

$$\frac{d}{dt} \frac{\partial L}{\partial \dot{x}_i} = 0 \tag{7.135}$$

and

$$\frac{\partial L}{\partial \dot{x}_i} = \text{constant} \qquad (7.136)$$

or

$$\frac{\partial (T - U)}{\partial \dot{x}_i} = \frac{\partial T}{\partial \dot{x}_i} = \frac{\partial}{\partial \dot{x}_i} \left(\frac{1}{2} m \sum_j \dot{x}_j^2 \right)$$

$$= m\dot{x}_i = p_i = \text{constant} \qquad (7.137)$$

Thus, the homogeneity of space implies that the linear momentum **p** of a closed system is constant in time.

This result may also be interpreted according to the following statement: If the Lagrangian of a system (not necessarily *closed*) is invariant with respect to translation in a certain direction, then the linear momentum of the system in that direction is constant in time.

Conservation of Angular Momentum

We stated in Section 2.3 that one characteristic of an inertial reference frame is that space is *isotropic*—that is, that the mechanical properties of a closed system are unaffected by the orientation of the system. In particular, the Lagrangian of a closed system does not change if the system is rotated through an infinitesimal angle.*

If a system is rotated about a certain axis by an infinitesimal angle $\delta\theta$ (see Figure 7-8), the radius vector **r** to a given point changes to **r** + δ**r**, where (see Equation 1.106)

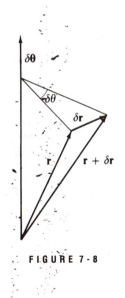

FIGURE 7-8

---

* We limit the rotation to an infinitesimal angle because we wish to be able to represent the rotation by a vector; see Section 1.15.

$$\delta\mathbf{r} = \delta\boldsymbol{\theta} \times \mathbf{r} \tag{7.138}$$

The velocity vectors also change on rotation of the system, and because the transformation equation for all vectors is the same, we have

$$\delta\dot{\mathbf{r}} = \delta\boldsymbol{\theta} \times \dot{\mathbf{r}} \tag{7.139}$$

We consider only a single particle and express the Lagrangian in rectangular coordinates. The change in L caused by the infinitesimal rotation is

$$\delta L = \sum_i \frac{\partial L}{\partial x_i} \delta x_i + \sum_i \frac{\partial L}{\partial \dot{x}_i} \delta \dot{x}_i = 0 \tag{7.140}$$

Equations 7.136 and 7.137 show that the rectangular components of the momentum vector are given by

$$p_i = \frac{\partial L}{\partial \dot{x}_i} \tag{7.141}$$

Lagrange's equations may then be expressed by

$$\dot{p}_i = \frac{\partial L}{\partial x_i} \tag{7.142}$$

Hence, Equation 7.140 becomes

$$\delta L = \sum_i \dot{p}_i \, \delta x_i + \sum_i p_i \, \delta \dot{x}_i = 0 \tag{7.143}$$

or

$$\dot{\mathbf{p}} \cdot \delta\mathbf{r} + \mathbf{p} \cdot \delta\dot{\mathbf{r}} = 0 \tag{7.144}$$

Using Equations 7.138 and 7.139, this equation may be written as

$$\dot{\mathbf{p}} \cdot (\delta\boldsymbol{\theta} \times \mathbf{r}) + \mathbf{p} \cdot (\delta\boldsymbol{\theta} \times \dot{\mathbf{r}}) = 0 \tag{7.145}$$

We may permute in cyclic order the factors of a triple scalar product without altering the value. Thus,

$$\delta\boldsymbol{\theta} \cdot (\mathbf{r} \times \dot{\mathbf{p}}) + \delta\boldsymbol{\theta} \cdot (\dot{\mathbf{r}} \times \mathbf{p}) = 0$$

or

$$\delta\boldsymbol{\theta} \cdot [(\mathbf{r} \times \dot{\mathbf{p}}) + (\dot{\mathbf{r}} \times \mathbf{p})] = 0 \tag{7.146}$$

The terms in the brackets are just the factors that result from the differentiation with respect to time of $\mathbf{r} \times \mathbf{p}$:

$$\delta\boldsymbol{\theta} \cdot \frac{d}{dt} (\mathbf{r} \times \mathbf{p}) = 0 \tag{7.147}$$

Because $\delta\boldsymbol{\theta}$ is arbitrary, we must have

$$\frac{d}{dt} (\mathbf{r} \times \mathbf{p}) = 0 \tag{7.148}$$

so

$$\mathbf{r} \times \mathbf{p} = \text{constant} \qquad (7.149)$$

But $\mathbf{r} \times \mathbf{p} = \mathbf{L}$; the angular momentum of the particle in a closed system is therefore constant in time.

An important corollary of this theorem is the following. Consider a system in an external force field. If the field possesses an axis of symmetry, then the Lagrangian of the system is invariant with respect to rotations about the symmetry axis. Hence, the angular momentum of the system about the axis of symmetry is constant in time. This is exactly the case discussed in Example 7.4; the vertical direction was an axis of symmetry of the system, and the angular momentum about that axis was conserved.

The importance of the connection between *symmetry* properties and the *invariance* of physical quantities can hardly be overemphasized. The association goes beyond momentum conservation—indeed beyond classical systems—and finds wide application in modern theories of field phenomena and elementary particles.

We have derived the conservation theorems for a closed system simply by considering the properties of an inertial reference frame. The results can be summarized as in Table 7-1.

TABLE 7-1		
Characteristic of inertial frame	**Property of Lagrangian**	**Conserved quantity**
Time homogeneous	Not explicit function of time	Total energy
Space homogeneous	Invariant to translation	Linear momentum
Space isotropic	Invariant to rotation	Angular momentum

There are then seven constants (or integrals) of the motion for a closed system: total energy, linear momentum (three components), and angular momentum (three components). These and only these seven integrals have the property that they are *additive* for the particles composing the system; they possess this property whether or not there is an interaction among the particles.

7.10 CANONICAL EQUATIONS OF MOTION— HAMILTONIAN DYNAMICS

In the previous section, we found that if the potential energy of a system is velocity independent, then the linear momentum components in rectangular coordinates are given by

$$p_i = \frac{\partial L}{\partial \dot{x}_i} \qquad (7.150)$$

By analogy, we extend this result to the case in which the Lagrangian is expressed in generalized coordinates and define the **generalized momenta*** according to

$$p_j \equiv \frac{\partial L}{\partial \dot{q}_j} \tag{7.151}$$

(Unfortunately, the customary notations for ordinary momentum and generalized momentum are the same, even though the two quantities may be quite different.) The Lagrange equations of motion are then expressed by

$$\dot{p}_j = \frac{\partial L}{\partial q_j} \tag{7.152}$$

Using the definition of the generalized momenta, Equation 7.128 for the Hamiltonian may be written as

$$H = \sum_j p_j \dot{q}_j - L \tag{7.153}$$

The Lagrangian is considered to be a function of the generalized coordinates, the generalized velocities, and possibly the time. The dependence of L on the time may arise either if the constraints are time dependent or if the transformation equations connecting the rectangular and generalized coordinates explicitly contain the time. (Recall that we do not consider time-dependent potentials.) We may solve Equation 7.151 for the generalized velocities and express them as

$$\dot{q}_j = \dot{q}_j(q_k, p_k, t) \tag{7.154}$$

Thus, in Equation 7.153, we may make a change of variables from the $(q_j, \dot{q}_j, t)$ set to the (q_j, p_j, t) set[†] and express the Hamiltonian as

$$H(q_k, p_k, t) = \sum_j p_j \dot{q}_j - L(q_k, \dot{q}_k, t) \tag{7.155}$$

This equation is written in a manner that stresses the fact that *the Hamiltonian is always considered as a function of the (q_k, p_k, t) set, whereas the Lagrangian is a function of the $(q_k, \dot{q}_k, t)$ set:*

$$H = H(q_k, p_k, t), \qquad L = L(q_k, \dot{q}_k, t) \tag{7.156}$$

---

[*] The terms *generalized coordinates*, *generalized velocities*, and *generalized momenta* were introduced in 1867 by Sir William Thomson (later, Lord Kelvin) and P. G. Tait in their famous treatise *Natural Philosophy*.

[†] This change of variables is similar to that frequently encountered in thermodynamics and falls in the general class of the so-called Legendre transformations (used first by Euler and perhaps even by Leibniz). A general discussion of Legendre transformations with emphasis on their importance in mechanics is given by Lanczos (La49, Chapter 6).

The total differential of H is therefore

$$dH = \sum_k \left(\frac{\partial H}{\partial q_k} dq_k + \frac{\partial H}{\partial p_k} dp_k \right) + \frac{\partial H}{\partial t} dt \qquad (7.157)$$

According to Equation 7.155, we can also write

$$dH = \sum_k \left(\dot{q}_k \, dp_k + p_k \, d\dot{q}_k - \frac{\partial L}{\partial q_k} dq_k - \frac{\partial L}{\partial \dot{q}_k} d\dot{q}_k \right) - \frac{\partial L}{\partial t} dt \qquad (7.158)$$

Using Equations 7.151 and 7.152 to substitute for $\partial L/\partial q_k$ and $\partial L/\partial \dot{q}_k$, the second and fourth terms in the parentheses in Equation 7.158 cancel, and there remains

$$dH = \sum_k (\dot{q}_k \, dp_k - \dot{p}_k \, dq_k) - \frac{\partial L}{\partial t} dt \qquad (7.159)$$

If we identify the coefficients* of dq_k, dp_k, and dt between Equations 7.157 and 7.159, we find

$$\boxed{\begin{aligned} \dot{q}_k &= \frac{\partial H}{\partial p_k} \\[2mm] -\dot{p}_k &= \frac{\partial H}{\partial q_k} \end{aligned}}$$

$$(7.160)$$

Hamilton's equations of motion

$$(7.161)$$

and

$$-\frac{\partial L}{\partial t} = \frac{\partial H}{\partial t} \qquad (7.162)$$

Furthermore, using Equations 7.160 and 7.161 in Equation 7.157, each term in the parentheses vanishes, and it follows that

$$\frac{dH}{dt} = \frac{\partial H}{\partial t} \qquad (7.163)$$

Equations 7.160 and 7.161 are **Hamilton's equations of motion**.[†] Because of their symmetric appearance, they are also known as the **canonical equations of motion**. The description of motion by these equations is termed **Hamiltonian dynamics**.

Equation 7.163 expresses the fact that if H does not explicitly contain the time, then the Hamiltonian is a conserved quantity. We have seen previously (Section

---

* The assumptions implicitly contained in this procedure are examined in the following section.

[†] This set of equations was first obtained by Lagrange in 1809, and Poisson also derived similar equations in the same year. But neither recognized the equations as a basic set of equations of motion; this point was first realized by Cauchy in 1831. Hamilton first derived the equations in 1834 from a fundamental variational principle and made them the basis for a far-reaching theory of dynamics. Thus, the designation "Hamilton's" equations is fully deserved.

7.9) that the Hamiltonian equals the total energy $T + U$ if the potential energy is velocity independent and the transformation equations between $x_{\alpha,i}$ and q_j do not explicitly contain the time. Under these conditions, and if $\partial H/\partial t = 0$, then $H = E =$ constant.

There are $2s$ canonical equations and they replace the s Lagrange equations. (Recall that $s = 3n - m$ is the number of degrees of freedom of the system.) But the canonical equations are *first-order* differential equations, whereas the Lagrange equations are of *second order*.* To use the canonical equations in solving a problem, we must first construct the Hamiltonian as a function of the generalized coordinates and momenta. It may be possible in some instances to do this directly. In more complicated cases, it may be necessary first to set up the Lagrangian and then to calculate the generalized momenta according to Equation 7.151. The equations of motion are then given by the canonical equations.

E X A M P L E   **7.11**   -

Use the Hamiltonian method to find the equations of motion of a particle of mass m constrained to move on the surface of a cylinder defined by $x^2 + y^2 = R^2$. The particle is subject to a force directed toward the origin and proportional to the distance of the particle from the origin: $\mathbf{F} = -k\mathbf{r}$.

Solution: The situation is illustrated in Figure 7-9. The potential corresponding to the force $\mathbf{F}$ is

$$U = \tfrac{1}{2}kr^2 = \tfrac{1}{2}k(x^2 + y^2 + z^2)$$

$$= \tfrac{1}{2}k(R^2 + z^2) \tag{7.164}$$

We can write the square of the velocity in cylindrical coordinates (see Equation 1.101) as

$$v^2 = \dot{R}^2 + R^2\dot{\theta}^2 + \dot{z}^2 \tag{7.165}$$

But in this case, R is a constant, so the kinetic energy is

$$T = \tfrac{1}{2}m(R^2\dot{\theta}^2 + \dot{z}^2) \tag{7.166}$$

We may now write the Lagrangian as

$$L = T - U = \tfrac{1}{2}m(R^2\dot{\theta}^2 + \dot{z}^2) - \tfrac{1}{2}k(R^2 + z^2) \tag{7.167}$$

---

* This is not a special result; any set of s second-order equations can always be replaced by a set of $2s$ first-order equations.

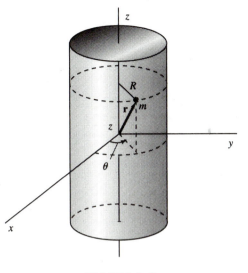

FIGURE 7-9

The generalized coordinates are θ and z, and the generalized momenta are

$$p_\theta = \frac{\partial L}{\partial \dot\theta} = mR^2\dot\theta \qquad (7.168)$$

$$p_z = \frac{\partial L}{\partial \dot z} = m\dot z \qquad (7.169)$$

Because the system is conservative and because the equations of transformation between rectangular and cylindrical coordinates do not explicitly involve the time, the Hamiltonian H is just the total energy expressed in terms of the variables θ, p_θ, z, and p_z. But θ does not occur explicitly, so

$$H(z, p_\theta, p_z) = T + U$$

$$= \frac{p_\theta^2}{2mR^2} + \frac{p_z^2}{2m} + \frac{1}{2}kz^2 \qquad (7.170)$$

where the constant term $\frac{1}{2}kR^2$ has been suppressed. The equations of motion are therefore found from the canonical equations:

$$\dot p_\theta = -\frac{\partial H}{\partial \theta} = 0 \qquad (7.171)$$

$$\dot p_z = -\frac{\partial H}{\partial z} = -kz \qquad (7.172)$$

$$\dot\theta = \frac{\partial H}{\partial p_\theta} = \frac{p_\theta}{mR^2} \qquad (7.173)$$

$$\dot{z} = \frac{\partial H}{\partial p_z} = \frac{p_z}{m} \qquad (7.174)$$

Equations 7.173 and 7.174 just duplicate Equations 7.168 and 7.169. Equations 7.168 and 7.171 give

$$p_\theta = mR^2\dot{\theta} = \text{constant} \qquad (7.175)$$

The angular momentum about the z-axis is thus a constant of the motion. This result is ensured, because the z-axis is the symmetry axis of the problem. Combining Equations 7.169 and 7.172, we find

$$\ddot{z} + \omega_0^2 z = 0 \qquad (7.176)$$

where

$$\omega_0^2 \equiv k/m \qquad (7.177)$$

The motion in the z direction is therefore simple harmonic.

-- ●

The equations of motion for the preceding problem can also be found by the Lagrangian method using the function L defined by Equation 7.167. In this case, the Lagrange equations of motion are easier to obtain than are the canonical equations. In fact, it is quite often true that the Lagrangian method leads more readily to the equations of motion than does the Hamiltonian method. But because we have greater freedom in choosing the variable in the Hamiltonian formulation of a problem (the q_k and the p_k are independent, whereas the q_k and the $\dot{q}_k$ are not), we often gain a certain practical advantage by using the Hamiltonian method. For example, in celestial mechanics—particularly in the event that the motions are subject to perturbations caused by the influence of other bodies—it proves convenient to formulate the problem in terms of Hamiltonian dynamics. Generally speaking, however, the great power of the Hamiltonian approach to dynamics does not manifest itself in simplifying the solutions to mechanics problems; rather, it provides a base we can extend to other fields.

The generalized coordinate q_k and the generalized momentum p_k are **canonically conjugate** quantities. According to Equations 7.160 and 7.161, if q_k does not appear in the Hamiltonian, then $\dot{p}_k = 0$, and the conjugate momentum p_k is a constant of the motion. Coordinates not appearing explicitly in the expressions for T and U are said to be *cyclic*. A coordinate cyclic in H is also cyclic in L. But, even if q_k does not appear in L, the generalized velocity $\dot{q}_k$ related to this coordinate is in general still present. Thus

$$L = L(q_1, ..., q_{k-1}, q_{k+1}, ..., q_s, \dot{q}_1, ..., \dot{q}_s, t)$$

and we accomplish no reduction in the number of degrees of freedom of the system, even though one coordinate is cyclic; there are still s second-order equations to be solved. However, in the canonical formulation, if q_k is cyclic, p_k is constant,

$p_k = \alpha_k$, and

$$H = H(q_1, ..., q_{k-1}, q_{k+1}, ..., q_s, p_1, ..., p_{k-1}, \alpha_k, p_{k+1}, ..., p_s, t)$$

Thus, there are $2s - 2$ first-order equations to be solved, and the problem has, in fact, been reduced in complexity; there are in effect only $s - 1$ degrees of freedom remaining. The coordinate q_k is completely separated, and it is *ignorable* as far as the remainder of the problem is concerned. We calculate the constant α_k by applying the initial conditions, and the equation of motion for the cyclic coordinate is

$$\dot{q}_k = \frac{\partial H}{\partial \alpha_k} \equiv \omega_k \tag{7.178}$$

which can be immediately integrated to yield

$$q_k(t) = \int \omega_k \, dt \tag{7.179}$$

The solution for a cyclic coordinate is therefore trivial to reduce to quadrature. Consequently, the canonical formulation of Hamilton is particularly well suited for dealing with problems in which one or more of the coordinates are cyclic. The simplest possible solution to a problem would result if the problem could be formulated in such a way that *all* the coordinates were cyclic. Then, each coordinate would be described in a trivial manner as in Equation 7.179. It is, in fact, possible to find transformations that render all the coordinates cyclic,* and these procedures lead naturally to a formulation of dynamics particularly useful in constructing modern theories of matter. The general discussion of these topics, however, is beyond the scope of this book.[†]

E X A M P L E 7.12 -

Using the Hamiltonian method, find the equations of motion for a spherical pendulum of mass m and length b (see Figure 7-10).

Solution: The generalized coordinates are θ and ϕ. The kinetic energy is

$$T = \tfrac{1}{2}mb^2\dot{\theta}^2 + \tfrac{1}{2}mb^2 \sin^2\theta\dot{\phi}^2$$

The only force acting on the pendulum (other than at the point of support) is gravity, and we define the potential zero to be at the pendulum's point of attachment.

$$U = -mgb \cos \theta$$

---

* Transformations of this type were derived by Carl Gustav Jacob Jacobi (1804–1851). Jacobi's investigations greatly extended the usefulness of Hamilton's methods, and these developments are known as *Hamilton-Jacobi theory*.

[†] See, for example, Goldstein (Go80, Chapter 10).

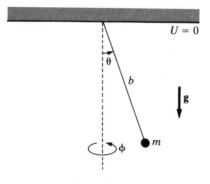

FIGURE 7-10

The generalized momenta are then

$$p_\theta = \frac{\partial L}{\partial \dot\theta} = mb^2 \dot\theta \qquad (7.180)$$

$$p_\phi = \frac{\partial L}{\partial \dot\phi} = mb^2 \sin^2\theta \, \dot\phi \qquad (7.181)$$

We can solve Equations 7.180 and 7.181 for $\dot\theta$ and $\dot\phi$ in terms of p_θ and p_ϕ.

We determine the Hamiltonian from Equation 7.155 or from $H = T + U$ (because the conditions for Equation 7.130 apply).

$$H = T + U$$

$$= \frac{1}{2} mb^2 \frac{p_\theta^2}{(mb^2)^2} + \frac{1}{2} \frac{mb^2 \sin^2\theta p_\phi^2}{(mb^2 \sin^2\theta)^2} - mgb\cos\theta$$

$$= \frac{p_\theta^2}{2mb^2} + \frac{p_\phi^2}{2mb^2 \sin^2\theta} - mgb\cos\theta$$

The equations of motion are

$$\dot\theta = \frac{\partial H}{\partial p_\theta} = \frac{p_\theta}{mb^2}$$

$$\dot\phi = \frac{\partial H}{\partial p_\phi} = \frac{p_\phi}{mb^2 \sin^2\theta}$$

$$\dot p_\theta = -\frac{\partial H}{\partial \theta} = \frac{p_\phi^2 \cos\theta}{mb^2 \sin^3\theta} - mgb\sin\theta$$

$$\dot p_\phi = -\frac{\partial H}{\partial \phi} = 0$$

Because ϕ is cyclic, the momentum p_ϕ about the symmetry axis is constant.

7.11 SOME COMMENTS REGARDING DYNAMICAL VARIABLES AND VARIATIONAL CALCULATIONS IN PHYSICS

We originally obtained Lagrange's equations of motion by stating Hamilton's Principle as a variational integral and then using the results of the preceding chapter on the calculus of variations. Because the method and the application were thereby separated, it is perhaps worthwhile to restate the argument in an orderly but abbreviated way.

Hamilton's Principle is expressed by

$$\delta \int_{t_1}^{t_2} L(q_j, \dot{q}_j, t)\, dt = 0 \qquad (7.182)$$

Applying the variational procedure specified in Section 6.7, we have

$$\int_{t_1}^{t_2} \left(\frac{\partial L}{\partial q_j} \delta q_j + \frac{\partial L}{\partial \dot{q}_j} \delta \dot{q}_j \right) dt = 0$$

Next, we assert that the δq_j and the $\delta \dot{q}_j$ are *not* independent, so the variation operation and the time differentiation can be interchanged:

$$\delta \dot{q}_j = \delta \left(\frac{dq_j}{dt} \right) = \frac{d}{dt} \delta q_j \qquad (7.183)$$

The varied integral becomes (after the integration by parts in which the δq_j are set equal at zero at the endpoints)

$$\int_{t_1}^{t_2} \left(\frac{\partial L}{\partial q_j} - \frac{d}{dt} \frac{\partial L}{\partial \dot{q}_j} \right) \delta q_j\, dt = 0 \qquad (7.184)$$

The requirement that the δq_j be independent variations leads immediately to Lagrange's equations.

In Hamilton's Principle, expressed by the variational integral in Equation 7.182, the Lagrangian is a function of the generalized coordinates and the generalized velocities. But only the q_j are considered as independent variables; the generalized velocities are simply the time derivatives of the q_j. When the integral is reduced to the form given by Equation 7.184, we state that the δq_j are independent variations; thus the integrand must vanish identically, and Lagrange's equations result. We may therefore pose this question: Because the dynamical motion of the system is completely determined by the initial conditions, what is the meaning of the variations δq_j? Perhaps a sufficient answer is that the variables are to be considered geometrically feasible within the limits of the given constraints—although they are not dynamically possible; that is, when using a variational procedure to obtain Lagrange's equations, it is convenient to ignore temporarily the fact that we are dealing with a physical system whose motion is completely determined and subject to no variation and to consider instead only a certain abstract mathematical problem. Indeed, this is the spirit in which any variational calculation relating to a physical process must be carried out. In adopting such a viewpoint, we must not be overly concerned with the fact that the variational procedure may be contrary to certain known physical properties of the system. (For example, energy is generally not conserved in passing from the true path to the varied path.) A variational calculation simply tests various *possible* solutions to a problem and prescribes a method for selecting the *correct* solution.

The canonical equations of motion can also be obtained directly from a variational calculation based on the so-called **modified Hamilton's Principle**. The Lagrangian function can be expressed as (see Equation 7.153):

$$L = \sum_j p_j \dot{q}_j - H(q_j, p_j, t) \tag{7.185}$$

and the statement of Hamilton's Principle contained in Equation 7.182 can be modified to read

$$\delta \int_{t_1}^{t_2} \left(\sum_j p_j \dot{q}_j - H \right) dt = 0 \tag{7.186}$$

Carrying out the variation in the standard manner, we obtain

$$\int_{t_1}^{t_2} \sum_j \left(p_j \delta \dot{q}_j + \dot{q}_j \delta p_j - \frac{\partial H}{\partial q_j} \delta q_j - \frac{\partial H}{\partial p_j} \delta p_j \right) dt = 0 \tag{7.187}$$

In the Hamiltonian formulation, the q_j and the p_j are considered to be independent. The $\dot{q}_j$ are again not independent of the q_j, so Equation 7.183 can be used to express the first term in Equation 7.187 as

$$\int_{t_1}^{t_2} \sum_j p_j \delta \dot{q}_j \, dt = \int_{t_1}^{t_2} \sum_j p_j \frac{d}{dt} \delta q_j \, dt$$

Integrating by parts, the integrated term vanishes, and we have

$$\int_{t_1}^{t_2} \sum_j p_j \delta \dot{q}_j \, dt = - \int_{t_1}^{t_2} \sum_j \dot{p}_j \delta q_j \, dt \tag{7.188}$$

Equation 7.187 then becomes

$$\int_{t_1}^{t_2} \sum_j \left\{ \left(\dot{q}_j - \frac{\partial H}{\partial p_j} \right) \delta p_j - \left(\dot{p}_j + \frac{\partial H}{\partial q_j} \right) \delta q_j \right\} dt = 0 \tag{7.189}$$

If δq_j and δp_j represent *independent variations*, the terms in the parentheses must separately vanish and Hamilton's canonical equations result.

In the preceding section, we obtained the canonical equations by writing two different expressions for the total differential of the Hamiltonian (Equations 7.157 and 7.159) and then equating the coefficients of dq_j and dp_j. Such a procedure is valid if the q_j and the p_j are independent variables. Therefore, both in the previous derivation and in the preceding variational calculation, we obtained the canonical equations by exploring the independent nature of the generalized coordinates and the generalized momenta.

The coordinates and momenta are not actually "independent" in the ultimate sense of the word. For if the time dependence of each of the coordinates is known, $q_j = q_j(t)$, the problem is completely solved. The generalized velocities can be calculated from

$$\dot{q}_j(t) = \frac{d}{dt} q_j(t)$$

and the generalized momenta are

$$p_j = \frac{\partial}{\partial \dot{q}_j} L(q_j, \dot{q}_j, t)$$

The essential point is that, whereas the q_j and the $\dot{q}_j$ are related by a simple time derivative *independent of the manner in which the system behaves*, the connection between the q_j and the p_j are the *equations of motion themselves*. Finding the relations that connect the q_j and the p_j (and thereby eliminating the assumed independence of these quantities) is therefore tantamount to solving the problem.

7.12 PHASE SPACE AND LIOUVILLE'S THEOREM (OPTIONAL)

We pointed out previously that the generalized coordinates q_j can be used to define an *s*-dimensional *configuration space* with every point representing a certain state of the system. Similarly, the *generalized momenta* p_j define an *s*-dimensional *momentum space* with every point representing a certain condition of motion of the system. A given point in configuration space specifies only the position of each of the particles in the system; nothing can be inferred regarding the motion of the particles. The reverse is true for momentum space. In Chapter 3, we found it profitable to represent geometrically the dynamics of simple oscillatory systems by phase diagrams. If we use this concept with more complicated dynamical systems, then a 2*s*-dimensional space consisting of the q_j and the p_j allows us to represent both the positions *and* the momenta of all particles. This generalization is called **Hamiltonian phase space** or, simply, **phase space.***

E X A M P L E ⬤ 7.13 –

Construct the phase diagram for the particle in Example 7.11.

Solution: The particle has two degrees of freedom (θ, z), so the phase space for this example is actually four dimensional: θ, p_θ, z, p_z. But p_θ is constant and therefore may be suppressed. In the z direction, the motion is simple harmonic, and so the projection onto the z-p_z plane of the phase path for any total energy H is just

---

* We previously plotted in the phase diagrams the position versus a quantity proportional to the velocity. In Hamiltonian phase space, this latter quantity becomes the generalized momentum.

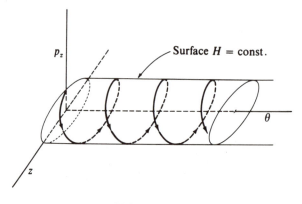

FIGURE 7-11

an ellipse. Because $\dot{\theta}$ = constant, the phase path must represent motion increasing uniformly with θ. Thus, the phase path on any surface H = constant is a **uniform elliptic spiral** (Figure 7-11).

If, at a given time, the position and momenta of all the particles in a system are known, then with these quantities as initial conditions, the subsequent motion of the system is completely determined; that is, starting from a point $q_j(0)$, $p_j(0)$ in phase space, the representative point describing the system moves along a unique phase path. In principle, this procedure can always be followed and a solution obtained. But if the number of degrees of freedom of the system is large, the set of equations of motion may be too complicated to solve in a reasonable time. Moreover, for complex systems, such as a quantity of gas, it is a practical impossibility to determine the initial conditions for each constituent molecule. Because we cannot identify any particular point in phase space as representing the actual conditions at any given time, we must devise some alternative approach to study the dynamics of such systems. We therefore arrive at the point of departure of statistical mechanics. The Hamiltonian formulation of dynamics is ideal for the statistical study of complex systems. We demonstrate this in part by now proving a theorem that is fundamental for such investigations.

For a large collection of particles—say, gas molecules—we are unable to identify the particular point in phase space correctly representing the system. But we may fill the phase space with a collection of points, each representing a *possible* condition of the system; that is, we imagine a large number of systems (each consistent with the known constraints), any of which could conceivably be the actual system. Because we are unable to discuss the details of the particles' motion in the actual system, we substitute a discussion of an *ensemble* of equivalent systems. Each representative point in phase space corresponds to a single system of the ensemble, and the motion of a particular point represents the independent motion of that system. Thus, no two of the phase paths may ever intersect.

We may consider the representative points to be sufficiently numerous that we can define a *density in phase* ρ. The volume elements of the phase space defining the density must be sufficiently large to contain a large number of representative points, but they must also be sufficiently small so that the density varies continuously. The number N of systems whose representative points lie within a volume dv of phase space is

$$N = \rho \, dv \qquad (7.190)$$

where

$$dv = dq_1 \, dq_2 \cdots dq_s \, dp_1 \, dp_2 \cdots dp_s \qquad (7.191)$$

As before, s is the number of degrees of freedom of each system in the ensemble.

Consider an element of area in the $q_k - p_k$ plane in phase space (Figure 7-12). The number of representative points moving across the left-hand edge into the area per unit time is

$$\rho \frac{dq_k}{dt} \, dp_k = \rho \dot{q}_k \, dp_k$$

and the number moving across the lower edge into the area per unit time is

$$\rho \frac{dp_k}{dt} \, dq_k = \rho \dot{p}_k \, dq_k$$

so that the total number of representative points moving *into* the area $dq_k \, dp_k$ per unit time is

$$\rho(\dot{q}_k \, dp_k + \dot{p}_k \, dq_k) \qquad (7.192)$$

By a Taylor series expansion, the number of representative points moving *out of* the area per unit time is (approximately)

$$\left[\rho \dot{q}_k + \frac{\partial}{\partial q_k} (\rho \dot{q}_k) \, dq_k \right] dp_k + \left[\rho \dot{p}_k + \frac{\partial}{\partial p_k} (\rho \dot{p}_k) \, dp_k \right] dq_k \qquad (7.193)$$

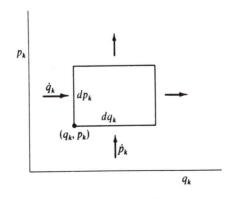

FIGURE 7-12

Thus, the total increase in density in $dq_k \, dp_k$ per unit time is the difference between Equations 7.192 and 7.193:

$$\frac{\partial \rho}{\partial t} dq_k \, dp_k = -\left[\frac{\partial}{\partial q_k} (\rho \dot{q}_k) + \frac{\partial}{\partial p_k} (\rho \dot{p}_k) \right] dq_k \, dp_k \qquad (7.194)$$

After dividing by $dq_k \, dp_k$ and summing this expression over all possible values of k, we find

$$\frac{\partial \rho}{\partial t} + \sum_{k=1}^{s} \left(\frac{\partial \rho}{\partial q_k} \dot{q}_k + \rho \frac{\partial \dot{q}_k}{\partial q_k} + \frac{\partial \rho}{\partial p_k} \dot{p}_k + \rho \frac{\partial \dot{p}_k}{\partial p_k} \right) = 0 \qquad (7.195)$$

From Hamilton's equations (Equations 7.160 and 7.161), we have (if the second partial derivatives of H are continuous)

$$\frac{\partial \dot{q}_k}{\partial q_k} + \frac{\partial \dot{p}_k}{\partial p_k} = 0 \qquad (7.196)$$

so Equation 7.195 becomes

$$\frac{\partial \rho}{\partial t} + \sum_{k} \left(\frac{\partial \rho}{\partial q_k} \frac{dq_k}{dt} + \frac{\partial \rho}{\partial p_k} \frac{dp_k}{dt} \right) = 0 \qquad (7.197)$$

But this is just the total time derivative of ρ, so we conclude that

$$\boxed{\frac{d\rho}{dt} = 0} \qquad (7.198)$$

This important result, known as **Liouville's theorem**,* states that the density of representative points in phase space corresponding to the motion of a system of particles remains constant during the motion. It must be emphasized that we have been able to establish the invariance of the density ρ only because the problem was formulated in *phase space;* an equivalent theorem for configuration space does not exist. Thus, we must use Hamiltonian dynamics (rather than Lagrangian dynamics) to discuss ensembles in statistical mechanics.

Liouville's theorem is important not only for aggregates of microscopic particles, as in the statistical mechanics of gaseous systems and the focusing properties of charged-particle accelerators, but also in certain macroscopic systems. For example, in stellar dynamics, the problem is inverted and by studying the distribution function ρ of stars in the galaxy, the potential U of the galactic gravitational field may be inferred.

7.13 VIRIAL THEOREM (OPTIONAL)

Another important result of a statistical nature is worthy of mention. Consider a collection of particles whose position vectors $\mathbf{r}_\alpha$ and momenta $\mathbf{p}_\alpha$ are both bounded

---

* Published in 1838 by Joseph Liouville (1809–1882).

(i.e., remain finite for all values of the time). Define a quantity

$$S \equiv \sum_{\alpha} \mathbf{p}_{\alpha} \cdot \mathbf{r}_{\alpha} \tag{7.199}$$

The time derivative of S is

$$\frac{dS}{dt} = \sum_{\alpha} (\mathbf{p}_{\alpha} \cdot \dot{\mathbf{r}}_{\alpha} + \dot{\mathbf{p}}_{\alpha} \cdot \mathbf{r}_{\alpha}) \tag{7.200}$$

If we calculate the average value of dS/dt over a time interval τ, we find

$$\left\langle \frac{dS}{dt} \right\rangle = \frac{1}{\tau} \int_0^{\tau} \frac{dS}{dt} \, dt = \frac{S(\tau) - S(0)}{\tau} \tag{7.201}$$

If the system's motion is periodic—and if τ is some integer multiple of the period—then $S(\tau) = S(0)$, and $\langle \dot{S} \rangle$ vanishes. But even if the system does not exhibit any periodicity, then—because S is by hypothesis a bounded function—we can make $\langle \dot{S} \rangle$ as small as desired by allowing the time τ to become sufficiently long. Therefore, the time average of the right-hand side of Equation 7.201 can always be made to vanish (or at least to approach zero). Thus, in this limit, we have

$$\left\langle \sum_{\alpha} \mathbf{p}_{\alpha} \cdot \dot{\mathbf{r}}_{\alpha} \right\rangle = - \left\langle \sum_{\alpha} \dot{\mathbf{p}}_{\alpha} \cdot \mathbf{r}_{\alpha} \right\rangle \tag{7.202}$$

On the left-hand side of this equation, $\mathbf{p}_{\alpha} \cdot \dot{\mathbf{r}}_{\alpha}$ is twice the kinetic energy. On the right-hand side, $\dot{\mathbf{p}}_{\alpha}$ is just the force $\mathbf{F}_{\alpha}$ on the αth particle. Hence,

$$\left\langle 2 \sum_{\alpha} T_{\alpha} \right\rangle = - \left\langle \sum_{\alpha} \mathbf{F}_{\alpha} \cdot \mathbf{r}_{\alpha} \right\rangle \tag{7.203}$$

The sum over T_{α} is the total kinetic energy T of the system, so we have the general result

$$\boxed{\langle T \rangle = - \frac{1}{2} \left\langle \sum_{\alpha} \mathbf{F}_{\alpha} \cdot \mathbf{r}_{\alpha} \right\rangle} \tag{7.204}$$

The right-hand side of this equation was called by Clausius* the **virial** of the system, and the **virial theorem** states that *the average kinetic energy of a system of particles is equal to its virial.*

E X A M P L E 7.14 ---

Consider an ideal gas containing N atoms in a container of volume V, pressure P, and absolute temperature T_1 (not to be confused with the kinetic energy T). Use the virial theorem to derive the equation of state for a perfect gas.

---

*Rudolph Julius Emmanuel Clausius (1822–1888), a German physicist and one of the founders of thermodynamics.

Solution: According to the equipartition theorem, the average kinetic energy of each atom in the ideal gas is $3/2 \, kT_1$, where k is the Boltzmann constant. The total average kinetic energy becomes

$$\langle T \rangle = \frac{3}{2} NkT_1 \tag{7.205}$$

The right-hand side of the virial theorem (Equation 7.204) contains the forces $\mathbf{F}_\alpha$. For an ideal perfect gas, no force of interaction occurs between atoms. The only force is represented by the force of constraint of the walls. The atoms bounce elastically off the walls, which are exerting a pressure on the atoms.

Because the pressure is force per unit area, we find the instantaneous differential force over a differential area to be

$$d\mathbf{F}_\alpha = -\mathbf{n}P \, dA \tag{7.206}$$

where $\mathbf{n}$ is a unit vector normal to the surface dA and pointing outward. The right-hand side of the virial theorem becomes

$$-\frac{1}{2} \left\langle \sum_\alpha \mathbf{F}_\alpha \cdot \mathbf{r}_\alpha \right\rangle = \frac{P}{2} \int \mathbf{n} \cdot \mathbf{r} \, dA \tag{7.207}$$

We use the divergence theorem to relate the surface integral to a volume integral.

$$\int \mathbf{n} \cdot \mathbf{r} \, dA = \int \nabla \cdot \mathbf{r} \, dV = 3 \int dV = 3V \tag{7.208}$$

The virial theorem result is

$$\frac{3}{2} NkT = \frac{3PV}{2}$$

$$NkT = PV \tag{7.209}$$

which is the ideal gas law.

- ●

If the forces $\mathbf{F}_\alpha$ can be derived from potentials U_α, Equation 7.204 may be rewritten as

$$\langle T \rangle = \frac{1}{2} \left\langle \sum_\alpha \mathbf{r}_\alpha \cdot \nabla U_\alpha \right\rangle \tag{7.210}$$

Of particular interest is the case of two particles that interact according to a central power-law force: $F \propto r^n$. Then, the potential is of the form

$$U = kr^{n+1} \tag{7.211}$$

Therefore

$$\mathbf{r} \cdot \nabla U = r \frac{dU}{dr} = k(n+1)r^{n+1} = (n+1)U \tag{7.212}$$

and the virial theorem becomes

$$\boxed{\langle T \rangle = \frac{n+1}{2} \langle U \rangle}$$

(7.213)

If the particles have a gravitational interaction, then $n = -2$, and

$$\langle T \rangle = -\tfrac{1}{2}\langle U \rangle, \qquad n = -2$$

This relation is useful in calculating, for example, the energetics in planetary motion.

P R O B L E M S

7-1. A disk rolls without slipping across a horizontal plane. The plane of the disk remains vertical, but it is free to rotate about a vertical axis. What generalized coordinates may be used to describe the motion? Write a differential equation describing the rolling constraint. Is this equation integrable? Justify your answer by a physical argument. Is the constraint holonomic?

7-2. Work out Example 7.6 showing all the steps, in particular those leading to Equations 7.36 and 7.41. Explain why the sign of the acceleration a cannot affect the frequency ω. Give an argument why the signs of a^2 and g^2 in the solution of ω^2 in Equation 7.42 are the same.

7-3. A sphere of radius ρ is constrained to roll without slipping on the lower half of the inner surface of a hollow cylinder of inside radius R. Determine the Lagrangian function, the equation of constraint, and Lagrange's equations of motion. Find the frequency of small oscillations.

▷ **7-4.** A particle moves in a plane under the influence of a force $f = -Ar^{\alpha-1}$ directed toward the origin; A and α (> 0) are constants. Choose appropriate generalized coordinates, and let the potential energy be zero at the origin. Find the Lagrangian equations of motion. Is the angular momentum about the origin conserved? Is the total energy conserved?

7-5. Consider a vertical plane in a constant gravitational field. Let the origin of a coordinate system be located at some point in this plane. A particle of mass m moves in the vertical plane under the influence of gravity and under the influence of an additional force $f = -Ar^{\alpha-1}$ directed toward the origin (r is the distance from the origin; A and α [$\neq 0$ or 1] are constants). Choose appropriate generalized coordinates, and find the Lagrangian equations of motion. Is the angular momentum about the origin conserved? Explain.

7-6. A hoop of mass m and radius R rolls without slipping down an inclined plane of mass M, which makes an angle α with the horizontal. Find the Lagrange equations and the integrals of the motion if the plane can slide without friction along a horizontal surface.

7-7. A double pendulum consists of two simple pendula, with one pendulum suspended from the bob of the other. If the two pendula have equal lengths and have bobs of equal mass and if both pendula are confined to move in the same plane, find Lagrange's equations of motion for the system. Do not assume small angles.

7-8. Consider a region of space divided by a plane. The potential energy of a particle in region 1 is U_1 and in region 2 it is U_2. If a particle of mass m and with speed v_1 in region 1 passes from region 1 to region 2 such that its path in region 1 makes an angle θ_1 with the normal to the plane of separation and an angle θ_2 with the normal when in region 2, show that

$$\frac{\sin \theta_1}{\sin \theta_2} = \left(1 + \frac{U_1 - U_2}{T_1}\right)^{1/2}$$

where $T_1 = \frac{1}{2}mv_1^2$. What is the optical analog of this problem?

7-9. A disk of mass M and radius R rolls without slipping down a plane inclined from the horizontal by an angle α. The disk has a short weightless axle of negligible radius. From this axis is suspended a simple pendulum of length $l < R$ and whose bob has a mass m. Consider that the motion of the pendulum takes place in the plane of the disk, and find Lagrange's equations for the system.

7-10. Two blocks, each of mass M, are connected by an extensionless, uniform string of length l. One block is placed on a smooth horizontal surface, and the other block hangs over the side, the string passing over a frictionless pulley. Describe the motion of the system **(a)** when the mass of the string is negligible and **(b)** when the string has a mass m.

7-11. A particle of mass m is constrained to move on a circle of radius R. The circle rotates in space about one point on the circle, which is fixed. The rotation takes place in the plane of the circle and with constant angular speed ω. In the absence of a gravitational force, show that the particle's motion about one end of a diameter passing through the pivot point and the center of the circle is the same as that of a plane pendulum in a uniform gravitational field. Explain why this is a reasonable result.

7-12. A particle of mass m rests on a smooth plane. The plane is raised to an inclination angle θ at a constant rate α ($\theta = 0$ at $t = 0$), causing the particle to move down the plane. Determine the motion of the particle.

7-13. A simple pendulum of length b and bob with mass m is attached to a massless support moving horizontally with constant acceleration a. Determine **(a)** the equations of motion and **(b)** the period for small oscillations.

7-14. A simple pendulum of length b and bob with mass m is attached to a massless support moving vertically upward with constant acceleration a. Determine **(a)** the equations of motion and **(b)** the period for small oscillations.

7-15. A pendulum consists of a mass m suspended by a massless spring with unextended length b and spring constant k. Find Lagrange's equations of motion.

7-16. The point of support of a simple pendulum of mass m and length b is driven horizontally by $x = a \sin \omega t$. Find the pendulum's equation of motion.

7-17. A particle of mass m can slide freely along a wire AB whose perpendicular distance to the origin O is h (see Figure 7-A, page 287). The line OC rotates about the origin at a constant angular velocity $\dot{\theta} = \omega$. The position of the particle can be described in terms of the angle θ and the distance q to the point C. If the particle is subject to a gravitational force, and if the initial conditions are

$$\theta(0) = 0, \qquad q(0) = 0, \qquad \dot{q}(0) = 0$$

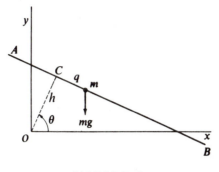

FIGURE 7-A

show that the time dependence of the coordinate q is

$$q(t) = \frac{g}{2\omega^2} (\cosh \omega t - \cos \omega t)$$

Sketch this result. Compute the Hamiltonian for the system, and compare with the total energy. Is the total energy conserved?

7-18. A pendulum is constructed by attaching a mass m to an extensionless string of length l. The upper end of the string is connected to the uppermost point on a vertical disk of radius R $(R < l/\pi)$ as in Figure 7-B. Obtain the pendulum's equation of motion, and find the frequency of small oscillations. Find the line about which the angular motion extends equally in either direction (i.e., $\theta_1 = \theta_2$).

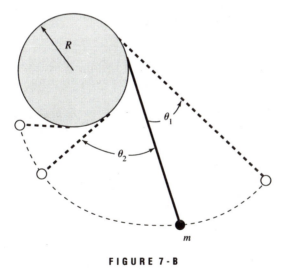

FIGURE 7-B

7-19. Two masses m_1 and m_2 $(m_1 \neq m_2)$ are connected by a rigid rod of length d and of negligible mass. An extensionless string of length l_1 is attached to m_1 and connected to a fixed point of support P. Similarly, a string of length l_2 ($l_1 \neq l_2$) connects m_2 and P. Obtain the equation describing the motion in the plane of m_1, m_2, and P, and find the frequency of small oscillations around the equilibrium position.

7-20. A circular hoop is suspended in a horizontal plane by three strings, each of length l, which are attached symmetrically to the hoop and are connected to fixed points lying in a plane above the hoop. At equilibrium, each string is vertical. Show that the frequency of small rotational oscillations about the vertical through the center of the hoop is the same as that for a simple pendulum of length l.

7-21. A particle is constrained to move (without friction) on a circular wire rotating with constant angular speed ω about a vertical diameter. Find the equilibrium position of the particle, and calculate the frequency of small oscillations around this position. Find and interpret physically a critical angular velocity $\omega = \omega_c$ that divides the particle's motion into two distinct types. Construct phase diagrams for the two cases $\omega < \omega_c$ and $\omega > \omega_c$.

7-22. A particle of mass m moves in one dimension under the influence of a force

$$F(x,t) = \frac{k}{x^2} e^{-(t/\tau)}$$

where k and τ are positive constants. Compute the Lagrangian and Hamiltonian functions. Compare the Hamiltonian and the total energy, and discuss the conservation of energy for the system.

7-23. Consider a particle of mass m moving freely in a conservative force field whose potential function is U. Find the Hamiltonian function, and show that the canonical equations of motion reduce to Newton's equations. (Use rectangular coordinates.)

▷ **7-24.** Consider a simple plane pendulum consisting of a mass m attached to a string of length l. After the pendulum is set into motion, the length of the string is shortened at a constant rate

$$\frac{dl}{dt} = -\alpha = \text{constant}$$

The suspension point remains fixed. Compute the Lagrangian and Hamiltonian functions. Compare the Hamiltonian and the total energy, and discuss the conservation of energy for the system.

7-25. A particle of mass m moves under the influence of gravity along the spiral $z = k\theta$, $r = \text{constant}$, where k is a constant and z is vertical. Obtain the Hamiltonian equations of motion.

▷ **7-26.** Determine the Hamiltonian and Hamilton's equations of motion for **(a)** a simple pendulum and **(b)** a simple Atwood machine (single pulley).   I FOR PULLEY

7-27. A massless spring of length b and spring constant k connects two particles of masses m_1 and m_2. The system rests on a smooth table and may oscillate and rotate.
(a) Determine Lagrange's equations of motion.
(b) What are the generalized momenta associated with any cyclic coordinates?
(c) Determine Hamilton's equations of motion.

7-28. A particle of mass m is attracted to a force center with the force of magnitude k/r^2. Use plane polar coordinates and find Hamilton's equations of motion.

7-29. Consider the pendulum described in Problem 7-15. The pendulum's point of support rises vertically with constant acceleration a.
(a) Using the Lagrangian method find the equations of motion.

(b) Determine the Hamiltonian and Hamilton's equations of motion.

(c) What is the period of small oscillations?

7-30. Consider any two continuous functions of the generalized coordinates and momenta $g(q_k, p_k)$ and $h(q_k, p_k)$. The **Poisson brackets** are defined by

$$[g, h] \equiv \sum_k \left(\frac{\partial g}{\partial q_k} \frac{\partial h}{\partial p_k} - \frac{\partial g}{\partial p_k} \frac{\partial h}{\partial q_k} \right)$$

Verify the following properties of the Poisson brackets:

(a) $\dfrac{dg}{dt} = [g, H] + \dfrac{\partial g}{\partial t}$ **(b)** $\dot{q}_j = [q_j, H], \quad \dot{p}_j = [p_j, H]$

(c) $[p_k, p_j] = 0, \ [q_k, q_j] = 0$ **(d)** $[q_k, p_j] = \delta_{kj}$

where H is the Hamiltonian. If the Poisson bracket of two quantities vanishes, the quantities are said to *commute*. If the Poisson bracket of two quantities equals unity, the quantities are said to be *canonically conjugate*. Show that any quantity that does not depend explicitly on the time and that commutes with the Hamiltonian is a constant of the motion of the system. Poisson-bracket formalism is of considerable importance in quantum mechanics.

7-31. A spherical pendulum consists of a bob of mass m attached to a weightless, extensionless rod of length l. The end of the rod opposite the bob pivots freely (in all directions) about some fixed point. Set up the Hamiltonian function in spherical coordinates. (If $p_\phi = 0$, the result is the same as that for the plane pendulum.) Combine the term that depends on p_ϕ with the ordinary potential energy term to define an *effective* potential $V(\theta, p_\phi)$. Sketch V as a function of θ for several values of p_ϕ, including $p_\phi = 0$. Discuss the features of the motion, pointing out the differences between $p_\phi = 0$ and $p_\phi \neq 0$. Discuss the limiting case of the conical pendulum ($\theta = $ constant) with reference to the V-θ diagram.

7-32. A particle moves in a spherically symmetric force field with potential energy given by $U(r) = -k/r$. Calculate the Hamiltonian function in spherical coordinates, and obtain the canonical equations of motion. Sketch the path that a representative point for the system would follow on a surface $H = $ constant in phase space. Begin by showing that the motion must lie in a plane so that the phase space is four dimensional (r, θ, p_r, p_θ but only the first three are nontrivial). Calculate the projection of the phase path on the r-p_r plane, then take into account the variation with θ.

7-33. Determine the Hamiltonian and Hamilton's equations of motion for the double Atwood machine of Example 7.8.

7-34. A particle of mass m slides down a smooth circular wedge of mass M as shown in Figure 7-C. The wedge rests on a smooth horizontal table. Find **(a)** the equation of motion of m and M and **(b)** the reaction of the wedge on m.

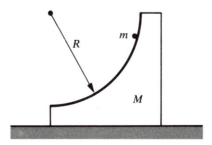

FIGURE 7-C

7-35. Four particles are directed upward in a uniform gravitational field with the following initial conditions:

$$\textbf{(1) } z(0) = z_0; \qquad p_z(0) = p_0$$
$$\textbf{(2) } z(0) = z_0 + \Delta z_0; \qquad p_z(0) = p_0$$
$$\textbf{(3) } z(0) = z_0; \qquad p_z(0) = p_0 + \Delta p_0$$
$$\textbf{(4) } z(0) = z_0 + \Delta z_0; \qquad p_z(0) = p_0 + \Delta p_0$$

Show by direct calculation that the representative points corresponding to these particles always define an area in phase space equal to $\Delta z_0 \, \Delta p_0$. Sketch the phase paths, and show for several times $t > 0$ the shape of the region whose area remains constant.

7-36. Discuss the implications of Liouville's theorem on the focusing of beams of charged particles by considering the following simple case. An electron beam of circular cross section (radius R_0) is directed along the z-axis. The density of electrons across the beam is constant, but the momentum components transverse to the beam (p_x and p_y) are distributed uniformly over a circle of radius p_0 in momentum space. If some focusing system reduces the beam radius from R_0 to R_1, find the resulting distribution of the transverse momentum components. What is the physical meaning of this result? (Consider the angular divergence of the beam.)

8

CENTRAL-FORCE MOTION

8.1 INTRODUCTION

The motion of a system consisting of two bodies affected by a force directed along the line connecting the centers of the two bodies (i.e., a *central force*) is an extremely important physical problem—one we can solve completely. The importance of such a problem lies in large measure in two quite different realms of physics: the motion of celestial bodies—planets, moons, comets, double stars, and the like—and certain two-body nuclear interactions, such as the scattering of α particles by nuclei. In the prequantum-mechanics days, physicists also described the hydrogen atom in terms of a classical two-body central force. Although such a description is still useful in a qualitative sense, the quantum-theoretical approach must be used for a detailed description. In addition to some general considerations regarding motion in central-force fields, we discuss in this and the following chapter several of the problems of two bodies encountered in celestial mechanics and in nuclear and particle physics.

8.2 REDUCED MASS

Describing a system consisting of two particles requires the specification of six quantities; for example, the three components of each of the two vectors $\mathbf{r}_1$ and $\mathbf{r}_2$

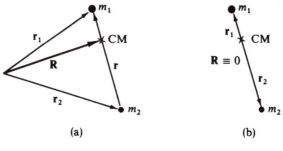

FIGURE 8-1

for the particles.* Alternatively, we may choose the three components of the center-of-mass vector $\mathbf{R}$ and the three components of $\mathbf{r} \equiv \mathbf{r}_1 - \mathbf{r}_2$ (see Figure 8-1a). Here, we restrict our attention to systems without frictional losses and for which the potential energy is a function only of $r = |\mathbf{r}_1 - \mathbf{r}_2|$. The Lagrangian for such a system may be written as

$$L = \tfrac{1}{2}m_1|\dot{\mathbf{r}}_1|^2 + \tfrac{1}{2}m_2|\dot{\mathbf{r}}_2|^2 - U(r) \tag{8.1}$$

Because translational motion of the system as a whole is uninteresting from the standpoint of the particle orbits with respect to one another, we may choose the origin for the coordinate system to be the particles' center of mass—that is, $\mathbf{R} \equiv 0$ (see Figure 8-lb). Then (see Section 9.2)

$$m_1\mathbf{r}_1 + m_2\mathbf{r}_2 = 0 \tag{8.2}$$

This equation, combined with $\mathbf{r} = \mathbf{r}_1 - \mathbf{r}_2$, yields

$$\left.\begin{array}{l} \mathbf{r}_1 = \dfrac{m_2}{m_1 + m_2}\,\mathbf{r} \\[4mm] \mathbf{r}_2 = -\dfrac{m_1}{m_1 + m_2}\,\mathbf{r} \end{array}\right\} \tag{8.3}$$

Substituting Equation 8.3 into the expression for the Lagrangian gives

$$\boxed{L = \tfrac{1}{2}\mu|\dot{\mathbf{r}}|^2 - U(r)} \tag{8.4}$$

where μ is the **reduced mass**,

$$\boxed{\mu \equiv \dfrac{m_1 m_2}{m_1 + m_2}} \tag{8.5}$$

---

* The orientation of the particles is assumed to be unimportant; that is, they are spherically symmetric (or are point particles).

We have therefore formally reduced the problem of the motion of two bodies to an *equivalent one-body problem* in which we must determine only the motion of a "particle" of mass μ in the central field described by the potential function $U(r)$. Once we obtain the solution for $\mathbf{r}(t)$ by applying the Lagrange equations to Equation 8.4, we can find the individual motions of the particles, $\mathbf{r}_1(t)$ and $\mathbf{r}_2(t)$, by using Equation 8.3. This latter step is not necessary if only the orbits relative to one another are required.

8.3 CONSERVATION THEOREMS— FIRST INTEGRALS OF THE MOTION

The system we wish to discuss consists of a particle of mass μ moving in a central-force field described by the potential function $U(r)$. Because the potential energy depends only on the distance of the particle from the force center and not on the orientation, the system possesses **spherical symmetry**; that is, the system's rotation about any fixed axis through the center of force cannot affect the equations of motion. We have already shown (see Section 7.9) that under such conditions the angular momentum of the system is conserved:

$$\mathbf{L} = \mathbf{r} \times \mathbf{p} = \text{constant} \qquad (8.6)$$

From this relation, it should be clear that both the radius vector and the linear momentum vector of the particle lie always in a plane normal to the angular momentum vector $\mathbf{L}$, which is fixed in space (see Figure 8-2). Therefore, we have only a two-dimensional problem, and the Lagrangian may then be conveniently expressed in plane polar coordinates:

$$\boxed{L = \tfrac{1}{2}\mu(\dot{r}^2 + r^2\dot{\theta}^2) - U(r)} \qquad (8.7)$$

Because the Lagrangian is cyclic in θ, the angular momentum conjugate to the coordinate θ is conserved:

$$\dot{p}_\theta = \frac{\partial L}{\partial \theta} = 0 = \frac{d}{dt}\frac{\partial L}{\partial \dot{\theta}} \qquad (8.8)$$

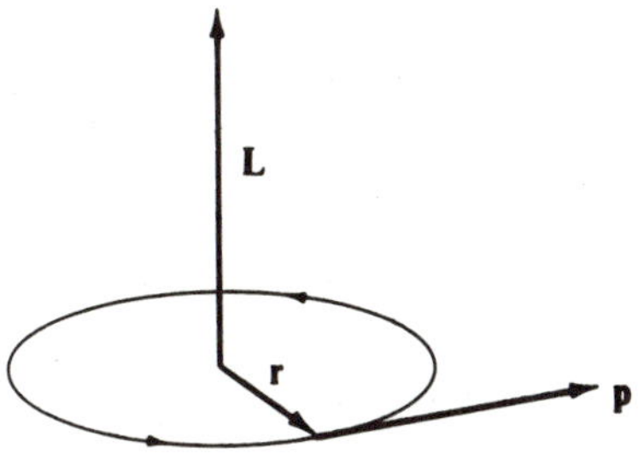

FIGURE 8-2

or

$$p_\theta \equiv \frac{\partial L}{\partial \dot{\theta}} = \mu r^2 \dot{\theta} = \text{constant} \tag{8.9}$$

The system's symmetry has therefore permitted us to integrate immediately one of the equations of motion. The quantity p_θ is a *first integral* of the motion, and we denote its constant value by the symbol l:

$$\boxed{l \equiv \mu r^2 \dot{\theta} = \text{constant}} \tag{8.10}$$

That l is constant has a simple geometric interpretation. Referring to Figure 8-3, we see that in describing the path $\mathbf{r}(t)$, the radius vector sweeps out an area $\frac{1}{2}r^2 d\theta$ in a time interval dt:

$$dA = \frac{1}{2}r^2 d\theta \tag{8.11}$$

On dividing by the time interval, the **areal velocity** is shown to be

$$\frac{dA}{dt} = \frac{1}{2}r^2 \frac{d\theta}{dt} = \frac{1}{2}r^2 \dot{\theta}$$

$$= \frac{l}{2\mu} = \text{constant} \tag{8.12}$$

Thus, the areal velocity is constant in time. This result was obtained empirically by Kepler for planetary motion, and it is known as **Kepler's Second Law.*** It is important to note that the conservation of the areal velocity is not limited to an inverse-square-law force (the case for planetary motion) but is a general result for central-force motion.

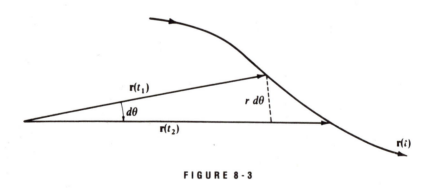

r(t₁)

dθ

r dθ

r(t₂)

r(t)

FIGURE 8-3

---

* Published by Johannes Kepler (1571–1630) in 1609 after an exhaustive study of the compilations made by Tycho Brahe (1546–1601) of the positions of the planet Mars. Kepler's First Law deals with the shape of planetary orbits (see Section 8.7).

Because we have eliminated from consideration the uninteresting uniform motion of the system's center of mass, the conservation of linear momentum adds nothing new to the description of the motion. The conservation of energy is thus the only remaining first integral of the problem. The conservation of the total energy E is automatically ensured because we have limited the discussion to non-dissipative systems. Thus,

$$T + U = E = \text{constant} \qquad (8.13)$$

and

$$E = \tfrac{1}{2}\mu(\dot{r}^2 + r^2\dot{\theta}^2) + U(r)$$

or

$$E = \tfrac{1}{2}\mu\dot{r}^2 + \frac{1}{2}\frac{l^2}{\mu r^2} + U(r) \qquad (8.14)$$

8.4 EQUATIONS OF MOTION

When $U(r)$ is specified, Equation 8.14 completely describes the system, and the integration of this equation gives the general solution of the problem in terms of the parameters E and l. Solving Equation 8.14 for $\dot{r}$, we have

$$\dot{r} = \frac{dr}{dt} = \pm\sqrt{\frac{2}{\mu}(E - U) - \frac{l^2}{\mu^2 r^2}} \qquad (8.15)$$

This equation can be solved for dt and integrated to yield the solution $t = t(r)$. An inversion of this result then gives the equation of motion in the standard form $r = r(t)$. At present, however, we are interested in the equation of the path in terms of r and θ. We can write

$$d\theta = \frac{d\theta}{dt}\frac{dt}{dr}\,dr = \frac{\dot{\theta}}{\dot{r}}\,dr \qquad (8.16)$$

Into this relation, we can substitute $\dot{\theta} = l/\mu r^2$ (Equation 8.10) and the expression for $\dot{r}$ from Equation 8.15. Integrating, we have

$$\theta(r) = \int \frac{\pm(l/r^2)\,dr}{\sqrt{2\mu\left(E - U - \dfrac{l^2}{2\mu r^2}\right)}} \qquad (8.17)$$

Furthermore, because l is constant in time, $\dot{\theta}$ cannot change sign and therefore $\theta(t)$ must increase monotonically with time.

Although we have reduced the problem to the formal evaluation of an integral, the actual solution can be obtained only for certain specific forms of the force law.

If the force is proportional to some power of the radial distance, $F(r) \propto r^n$, then the solution can be expressed in terms of elliptic integrals for certain integer and fractional values of n. Only for $n = 1$, -2, and -3 are the solutions expressible in terms of circular functions.* The case $n = 1$ is just that of the harmonic oscillator (see Chapter 3), and the case $n = -2$ is the important inverse-square-law force treated in Sections 8.6 and 8.7. These two cases, $n = 1$, -2, are of prime importance in physical situations. Details of some other cases of interest will be found in the problems at the end of this chapter.

We have therefore solved the problem in a formal way by combining the equations that express the conservation of energy and angular momentum into a single result, which gives the equation of the orbit $\theta = \theta(r)$. We can also attack the problem using Lagrange's equation for the coordinate r:

$$\frac{\partial L}{\partial r} - \frac{d}{dt}\frac{\partial L}{\partial \dot{r}} = 0$$

Using Equation 8.7 for L, we find

$$\mu(\ddot{r} - r\dot{\theta}^2) = -\frac{\partial U}{\partial r} = F(r) \tag{8.18}$$

Equation 8.18 can be cast in a form more suitable for certain types of calculations by making a simple change of variable:

$$u \equiv \frac{1}{r}$$

First, we compute

$$\frac{du}{d\theta} = -\frac{1}{r^2}\frac{dr}{d\theta} = -\frac{1}{r^2}\frac{dr}{dt}\frac{dt}{d\theta} = -\frac{1}{r^2}\frac{\dot{r}}{\dot{\theta}}$$

But from Equation 8.10, $\dot{\theta} = l/\mu r^2$, so

$$\frac{du}{d\theta} = -\frac{\mu}{l}\dot{r}$$

Next, we write

$$\frac{d^2u}{d\theta^2} = \frac{d}{d\theta}\left(-\frac{\mu}{l}\dot{r}\right) = \frac{dt}{d\theta}\frac{d}{dt}\left(-\frac{\mu}{l}\dot{r}\right) = -\frac{\mu}{l\dot{\theta}}\ddot{r}$$

and with the same substitution for $\dot{\theta}$, we have

$$\frac{d^2u}{d\theta^2} = -\frac{\mu^2}{l^2}r^2\ddot{r}$$

---

* See, for example, Goldstein (Go80, pp. 88–90).

Therefore, solving for $\ddot{r}$ and $r\dot{\theta}^2$ in terms of u, we find

$$\left.\begin{aligned} \ddot{r} &= -\frac{l^2}{\mu^2} u^2 \frac{d^2u}{d\theta^2} \\ r\dot{\theta}^2 &= \frac{l^2}{\mu^2} u^3 \end{aligned}\right\} \tag{8.19}$$

Substituting Equation 8.19 into Equation 8.18, we obtain the transformed equation of motion:

$$\frac{d^2u}{d\theta^2} + u = -\frac{\mu}{l^2}\frac{1}{u^2} F(1/u) \tag{8.20}$$

which we may also write as

$$\boxed{\frac{d^2}{d\theta^2}\left(\frac{1}{r}\right) + \frac{1}{r} = -\frac{\mu r^2}{l^2} F(r)} \tag{8.21}$$

This form of the equation of motion is particularly useful if we wish to find the force law that gives a particular known orbit $r = r(\theta)$.

EXAMPLE 8.1 --

Find the force law for a central-force field that allows a particle to move in a logarithmic spiral orbit given by $r = ke^{\alpha\theta}$, where k and α are constants.

Solution: We use Equation 8.21 to determine the force law $F(r)$. First, we determine

$$\frac{d}{d\theta}\left(\frac{1}{r}\right) = \frac{d}{d\theta}\left(\frac{e^{-\alpha\theta}}{k}\right) = \frac{-\alpha e^{-\alpha\theta}}{k}$$

$$\frac{d^2}{d\theta^2}\left(\frac{1}{r}\right) = \frac{\alpha^2 e^{-\alpha\theta}}{k} = \frac{\alpha^2}{r}$$

From Equation 8.21, we now determine $F(r)$.

$$F(r) = \frac{-l^2}{\mu r^2}\left(\frac{\alpha^2}{r} + \frac{1}{r}\right)$$

$$F(r) = \frac{-l^2}{\mu r^3}(\alpha^2 + 1) \tag{8.22}$$

Thus, the force law is an attractive inverse cube.

--- ●

EXAMPLE **8.2** -

Determine $r(t)$ and $\theta(t)$ for the problem in Example 8.1.

Solution: From Equation 8.10, we find

$$\dot\theta = \frac{l}{\mu r^2} = \frac{l}{\mu k^2 e^{2\alpha\theta}} \tag{8.23}$$

Rearranging Equation 8.23 gives

$$e^{2\alpha\theta}\, d\theta = \frac{l}{\mu k^2}\, dt$$

and integrating gives

$$\frac{e^{2\alpha\theta}}{2\alpha} = \frac{lt}{\mu k^2} + C'$$

where C' is an integration constant. Multiplying by 2α and letting $C = 2\alpha C'$ gives

$$e^{2\alpha\theta} = \frac{2\alpha lt}{\mu k^2} + C \tag{8.24}$$

We solve for $\theta(t)$ by taking the natural logarithm of Equation 8.24:

$$\theta(t) = \frac{1}{2\alpha} \ln\left(\frac{2\alpha lt}{\mu k^2} + C\right) \tag{8.25}$$

We can similarly solve for $r(t)$ by examining Equations 8.23 and 8.24:

$$\frac{r^2}{k^2} = e^{2\alpha\theta} = \frac{2\alpha lt}{\mu k^2} + C$$

$$r(t) = \left[\frac{2\alpha l}{\mu}t + k^2 C\right]^{1/2} \tag{8.26}$$

The integration constant C and angular momentum l needed for Equations 8.25 and 8.26 are determined from the initial conditions.

- ●

EXAMPLE **8.3** -

What is the total energy of the orbit of the previous two examples?

Solution: The energy is found from Equation 8.14. In particular, we need $\dot r$ and $U(r)$.

$$U(r) = -\int F\, dr = \frac{+l^2}{\mu}(\alpha^2 + 1)\int r^{-3}\, dr$$

$$U(r) = -\frac{l^2(\alpha^2 + 1)}{2\mu}\frac{1}{r^2} \tag{8.27}$$

where we have let $U(\infty) = 0$.

We rewrite Equation 8.10 to determine $\dot{r}$:

$$\dot{\theta} = \frac{d\theta}{dt} = \frac{d\theta}{dr}\frac{dr}{dt} = \frac{l}{\mu r^2}$$

$$\dot{r} = \frac{dr}{d\theta}\frac{l}{\mu r^2} = \alpha k e^{\alpha\theta}\frac{l}{\mu r^2} = \frac{\alpha l}{\mu r} \tag{8.28}$$

Substituting Equations 8.27 and 8.28 into Equation 8.14 gives

$$E = \frac{1}{2}\mu\left(\frac{\alpha l}{\mu r}\right)^2 + \frac{l^2}{2\mu r^2} - \frac{l^2(\alpha^2 + 1)}{2\mu r^2}$$

$$E = 0 \tag{8.29}$$

The total energy of the orbit is zero if $U(r = \infty) = 0$.

-- ●

8.5 ORBITS IN A CENTRAL FIELD

The radial velocity of a particle moving in a central field is given by Equation 8.15. This equation indicates that $\dot{r}$ vanishes at the roots of the radical, that is, at points for which

$$E - U(r) - \frac{l^2}{2\mu r^2} = 0 \tag{8.30}$$

The vanishing of $\dot{r}$ implies that a *turning point* in the motion has been reached (see Section 2.6). In general, Equation 8.30 possesses two roots: r_{max} and r_{min}. The motion of the particle is therefore confined to the annular region specified by $r_{max} \geq r \geq r_{min}$. Certain combinations of the potential function $U(r)$ and the parameters E and l produce only a single root for Equation 8.30. In such a case, $\dot{r} = 0$ for all values of the time; hence, $r = $ constant, and the orbit is circular.

If the motion of a particle in the potential $U(r)$ is periodic, then the orbit is *closed*; that is, after a finite number of excursions between the radial limits r_{min} and r_{max}, the motion exactly repeats itself. But if the orbit does not close on itself after a finite number of oscillations, the orbit is said to be *open* (Figure 8-4). From Equation 8.17, we can compute the change in the angle θ that results from one complete transit of r from r_{min} to r_{max} and back to r_{min}. Because the motion is symmetric in time, this angular change is twice that which would result from the

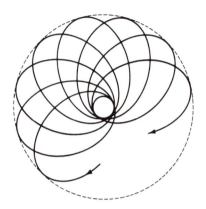

FIGURE 8-4

passage from r_{min} to r_{max}; thus

$$\Delta\theta = 2\int_{r_{min}}^{r_{max}} \frac{(l/r^2)\, dr}{\sqrt{2\mu\left(E - U - \dfrac{l^2}{2\mu r^2}\right)}} \tag{8.31}$$

The path is closed only if $\Delta\theta$ is a rational fraction of 2π—that is, if $\Delta\theta = 2\pi \cdot (a/b)$, where a and b are integers. Under these conditions, after b periods the radius vector of the particle will have made a complete revolutions and will have returned to its original position. We can show (see Problem 8-35) that if the potential varies with some integer power of the radial distance, $U(r) \propto r^{n+1}$, then a closed noncircular path can result *only** if $n = -2$ or $+1$. The case $n = -2$ corresponds to an inverse-square-law force—for example, the gravitational or electrostatic force. The $n = +1$ case corresponds to the harmonic oscillator potential. For the two-dimensional case discussed in Section 3.4, we found that a closed path for the motion resulted if the ratio of the angular frequencies for the x and y motions were rational.

8.6 CENTRIFUGAL ENERGY AND THE EFFECTIVE POTENTIAL

In the preceding expressions for $\dot{r}$, $\Delta\theta$, and so forth, a common term is the radical

$$\sqrt{E - U - \frac{l^2}{2\mu r^2}}$$

---

* Certain fractional values of n also lead to closed orbits, but in general these cases are uninteresting from a physical standpoint.

The last term in the radical has the dimensions of energy and, according to Equation 8.10, can also be written as

$$\frac{l^2}{2\mu r^2} = \frac{1}{2}\mu r^2 \dot{\theta}^2$$

If we interpret this quantity as a "potential energy,"

$$U_c \equiv \frac{l^2}{2\mu r^2} \tag{8.32}$$

then the "force" that must be associated with U_c is

$$F_c = -\frac{\partial U_c}{\partial r} = \frac{l^2}{\mu r^3} = \mu r \dot{\theta}^2 \tag{8.33}$$

This quantity is traditionally called the **centrifugal force**,* although it is not a force in the ordinary sense of the word.[†] We shall, however, continue to use this unfortunate terminology, because it is customary and convenient.

We see that the term $l^2/2\mu r^2$ can be interpreted as the *centrifugal potential energy* of the particle and, as such, can be included with $U(r)$ in an *effective potential energy* defined by

$$\boxed{V(r) \equiv U(r) + \frac{l^2}{2\mu r^2}} \tag{8.34}$$

$V(r)$ is therefore a *fictitious* potential that combines the real potential function $U(r)$ with the energy term associated with the angular motion about the center of force. For the case of inverse-square-law central-force motion, the force is given by

$$F(r) = -\frac{k}{r^2} \tag{8.35}$$

from which

$$U(r) = -\int F(r)\, dr = -\frac{k}{r} \tag{8.36}$$

The effective potential function for gravitational attraction is therefore

$$V(r) = -\frac{k}{r} + \frac{l^2}{2\mu r^2} \tag{8.37}$$

This effective potential and its components are shown in Figure 8-5. The value of

---

* The expression is more readily recognized in the form $F_c = mr\omega^2$. The first real appreciation of centrifugal force was by Huygens, who made a detailed examination in his study of the conical pendulum in 1659.

[†] See Section 10.3 for a more critical discussion of centrifugal force.

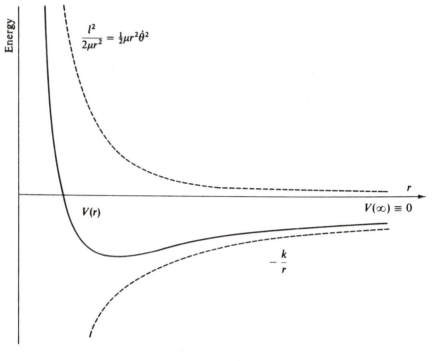

FIGURE 8-5

the potential is arbitrarily taken to be zero at $r = \infty$. (This is implicit in Equation 8.36, where we omitted the constant of integration.)

We may now draw conclusions similar to those in Section 2.6 on the motion of a particle in an arbitrary potential well. If we plot the total energy E of the particle on a diagram similar to Figure 8-5, we may identify three regions of interest (see Figure 8-6). If the total energy is positive or zero (e.g., $E_1 \geq 0$), then the motion is unbounded; the particle moves toward the force center (located at $r = 0$) from infinitely far away until it "strikes" the potential barrier at the *turning point* $r = r_1$ and is reflected back toward infinitely large r. Note that the height of the constant total energy line above $V(r)$ at any r, such as r_5 in Figure 8-6, is equal to $\frac{1}{2}\mu\dot{r}^2$. Thus the radial velocity $\dot{r}$ vanishes and changes sign at the turning point (or points).

If the total energy is negative* and lies between zero and the minimum value of $V(r)$, as does E_2, then the motion is bounded, with $r_2 \leq r \leq r_4$. The values r_2 and r_4 are the turning points, or the **apsidal distances**, of the orbit. If E equals the minimum value of the effective potential energy (see E_3 in Figure 8-6), then the

---

* Note that negative values of the total energy arise only because of the arbitrary choice of $V(r) = 0$ at $r = \infty$.

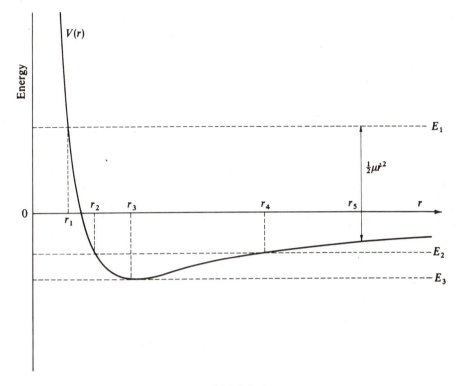

FIGURE 8-6

radius of the particle's path is limited to the single value r_3, and then $\dot{r} = 0$ for all values of the time; hence the motion is circular.

Values of E less than $V_{\min} = -(\mu k^2/2l^2)$ do not result in physically real motion; for such cases $\dot{r}^2 < 0$ and the velocity is imaginary.

The methods discussed in this section are often used in present-day research in general fields, especially atomic, molecular, and nuclear physics. For example, Figure 8-7 shows effective total nucleus-nucleus potentials for the scattering of ^{28}Si and ^{12}C. The total potential includes the coulomb, nuclear, and the centrifugal contributions. The potential for $l = 0\hbar$ indicates the potential with no centrifugal term. For a relative angular momentum value of $l = 20\hbar$, a "pocket" exists where the two scattering nuclei may be bound together (even if only for a short time). For $l = 25\hbar$, the centrifugal "barrier" dominates, and the nuclei are not attracted to each other.

8.7   PLANETARY MOTION—KEPLER'S PROBLEM

The equation for the path of a particle moving under the influence of a central force whose magnitude is inversely proportional to the square of the distance between

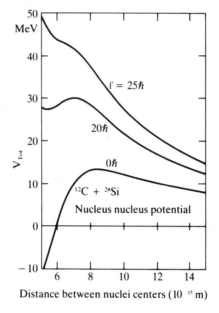

FIGURE 8-7

the particle and the force center can be obtained (see Equation 8.17) from

$$\theta(r) = \int \frac{(l/r^2)\, dr}{\sqrt{2\mu\left(E + \dfrac{k}{r} - \dfrac{l^2}{2\mu r^2}\right)}} + \text{constant} \qquad (8.38)$$

The integral can be evaluated if the variable is changed to $u \equiv l/r$ (see Problem 8-2). If we define the origin of θ so that the minimum value of r is at $\theta = 0$, we find

$$\cos \theta = \frac{\dfrac{l^2}{\mu k} \cdot \dfrac{1}{r} - 1}{\sqrt{1 + \dfrac{2El^2}{\mu k^2}}} \qquad (8.39)$$

Let us now define the following constants:

$$\left.\begin{aligned} \alpha &\equiv \frac{l^2}{\mu k} \\[2mm] \varepsilon &\equiv \sqrt{1 + \frac{2El^2}{\mu k^2}} \end{aligned}\right\} \qquad (8.40)$$

Equation 8.39 can thus be written as

$$\frac{\alpha}{r} = 1 + \varepsilon \cos \theta \qquad \textbf{(8.41)}$$

This is the equation of a conic section with one focus at the origin.* The quantity ε is called the **eccentricity**, and 2α is termed the **latus rectum** of the orbit. Conic sections are formed by the intersection of a plane and a cone. A conic section is formed by the loci of points (formed in a plane) where the ratio of the distance from a fixed point (the focus) to a fixed line (called the directrix) is a constant. The directrix for the parabola is shown in Figure 8-8 by the vertical dashed line, drawn so that $r/r' = 1$.

The minimum value for r in Equation 8.41 occurs when $\theta = 0$, or when $\cos \theta$ is a maximum. Thus the choice of the integration constant in Equation 8.38 corresponds to measuring θ from $r_{\min}$, which position is called the **pericenter**; $r_{\max}$

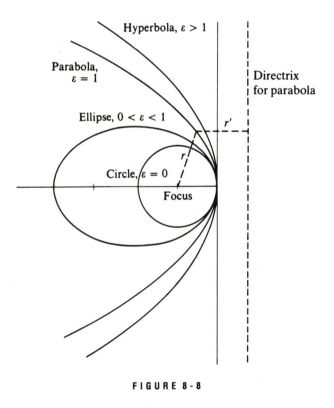

FIGURE 8-8

---

* Johann Bernoulli (1667–1748) appears to have been the first to prove that *all* possible orbits of a body moving in a potential proportional to $1/r$ are conic sections (1710).

corresponds to the **apocenter**. The general term for turning points is **apsides**. The corresponding terms for motion about the sun are *perihelion* and *aphelion*, and for motion about the Earth, *perigee* and *apogee*.

Various values of the eccentricity (and hence of the energy E) classify the orbits according to different conic sections (see Figure 8-8):

$$\varepsilon > 1, \qquad E > 0 \qquad\qquad \text{Hyperbola}$$

$$\varepsilon = 1, \qquad E = 0 \qquad\qquad \text{Parabola}$$

$$0 < \varepsilon < 1, \qquad V_{min} < E < 0 \qquad \text{Ellipse}$$

$$\varepsilon = 0, \qquad E = V_{min} \qquad\quad \text{Circle}$$

$$\varepsilon < 0, \qquad E < V_{min} \qquad\quad \text{Not allowed}$$

For planetary motion, the orbits are ellipses with major and minor axes (equal to $2a$ and $2b$, respectively) given by

$$a = \frac{\alpha}{1 - \varepsilon^2} = \frac{k}{2|E|} \tag{8.42}$$

$$b = \frac{\alpha}{\sqrt{1 - \varepsilon^2}} = \frac{l}{\sqrt{2\mu|E|}} \tag{8.43}$$

Thus, the major axis depends only on the energy of the particle, whereas the minor axis is a function of both first integrals of the motion, E and l. The geometry of elliptic orbits in terms of the parameters α, ε, a, and b is shown in Figure 8-9; P and P' are the foci. From this diagram, we see that the apsidal distances (r_{min} and r_{max} as measured from the foci to the orbit) are given by

$$\left.\begin{aligned} r_{min} = a(1 - \varepsilon) = \frac{\alpha}{1 + \varepsilon} \\[2mm] r_{max} = a(1 + \varepsilon) = \frac{\alpha}{1 - \varepsilon} \end{aligned}\right\} \tag{8.44}$$

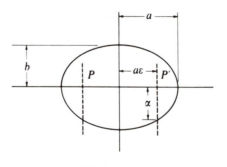

FIGURE 8-9

To find the period for elliptic motion, we rewrite Equation 8.12 for the areal velocity as

$$dt = \frac{2\mu}{l} dA$$

Because the entire area A of the ellipse is swept out in one complete period τ,

$$\int_0^\tau dt = \frac{2\mu}{l} \int_0^A dA$$

$$\tau = \frac{2\mu}{l} A \qquad \textbf{(8.45)}$$

The area of an ellipse is given by $A = \pi ab$, and using a and b from Equations 8.42 and 8.43, we find

$$\tau = \frac{2\mu}{l} \cdot \pi ab = \frac{2\mu}{l} \cdot \pi \cdot \frac{k}{2|E|} \cdot \frac{l}{\sqrt{2\mu|E|}}$$

$$= \pi k \sqrt{\frac{\mu}{2}} \cdot |E|^{-\frac{3}{2}} \qquad \textbf{(8.46)}$$

We also note from Equations 8.42 and 8.43 that the semiminor axis* can be written as

$$b = \sqrt{\alpha a} \qquad \textbf{(8.47)}$$

Therefore, because $\alpha = l^2/\mu k$, the period τ can also be expressed as

$$\boxed{\tau^2 = \frac{4\pi^2 \mu}{k} a^3} \qquad \textbf{(8.48)}$$

This result, that the square of the period is proportional to the cube of the semimajor axis of the elliptic orbit, is known as **Kepler's Third Law.**[†] Note that this result is concerned with the equivalent one-body problem, so account must be taken of the fact that it is the *reduced* mass μ that occurs in Equation 8.48. Kepler actually concluded that the squares of the periods of the planets were proportional to the cubes of the major axes of their orbits—with the same proportionality constant for all planets. In this sense, the statement is only approximately correct, because the reduced mass is different for each planet. In particular, because the

---

* The quantities a and b are called *semimajor* and *semiminor axes*, respectively.

[†] Published by Kepler in 1619. Kepler's Second Law is stated in Section 8.3. The First Law (1609) dictates that the planets move in elliptical orbits with the sun at one focus. Kepler's work preceded by almost 80 years Newton's enunciation of his general laws of motion. Indeed, Newton's conclusions were based to a great extent on Kepler's pioneering studies (and on those of Galileo and Huygens).

TABLE 8-1

SOME PROPERTIES OF THE PRINCIPAL OBJECTS IN THE SOLAR SYSTEM

Name	Semimajor axis of orbit (in astronomical units[a])	Period (yr)	Eccentricity	Mass (in units of the mass of the Earth[b])
Sun	—	—	—	333,480
Mercury	0.3871	0.2408	0.2056	0.0553
Venus	0.7233	0.6152	0.0068	0.8150
Earth	1.0000	1.0000	0.0167	1.000
Eros (asteroid)	1.4583	1.7610	0.2230	2×10^{-9} (?)
Mars	1.5237	1.8809	0.0934	0.1074
Ceres (asteroid)	c	4.6035	0.0765	1/8000 (?)
Jupiter	5.2028	c	0.0483	317.89
Saturn	9.5388	29.456	0.0560	c
Uranus	19.191	84.07	0.0461	14.56
Neptune	30.061	164.81	0.0100	17.15
Pluto	39.529	248.53	0.2484	0.002
Halley (comet)	18	76	0.967	$\sim 10^{-10}$

[a] One astronomical unit (A.U.) is the length of the semimajor axis of the Earth's orbit. One A.U. $\cong$ 1.495×10^{11} m $\cong 93 \times 10^{6}$ miles.

[b] The mass of the Earth is approximately 5.976×10^{24} kg.

[c] See Problem 8-19.

gravitational force is given by

$$F(r) = -\frac{Gm_1 m_2}{r^2} = -\frac{k}{r^2}$$

we identify $k = Gm_1 m_2$. The expression for the square of the period therefore becomes

$$\tau^2 = \frac{4\pi^2 a^3}{G(m_1 + m_2)} \cong \frac{4\pi^2 a^3}{Gm_2}, \qquad m_1 \ll m_2 \qquad \textbf{(8.49)}$$

and Kepler's statement is correct only if the mass m_1 of a planet can be neglected with respect to the mass m_2 of the sun. (But note, for example, that the mass of Jupiter is about $1/1000$ of the mass of the sun, so the departure from the approximate law is not difficult to observe in this case.)

Kepler's laws can now be summarized:

I. *Planets move in elliptical orbits about the sun with the sun at one focus.*

II. *The area per unit time swept out by a radius vector from the sun to a planet is constant.*

III. *The square of a planet's period is proportional to the cube of the major axis of the planet's orbit.*

See Table 8-1 for some properties of the principal objects in the solar system.

EXAMPLE 8.4 --

Halley's comet, which passed around the sun early in 1986, moves in a highly
elliptical orbit with an eccentricity of 0.967 and a period of 76 years. Calculate
its minimum and maximum distances from the sun.

Solution: Equation 8.49 relates the period of motion with the semimajor axes.
Because m (Halley's comet) $\ll m_{sun}$,

$$\alpha = \left(\frac{Gm_{sun}\tau^2}{4\pi^2}\right)^{\frac{1}{3}}$$

$$= \left[\frac{\left(6.67 \times 10^{-11}\,\frac{Nm^2}{kg^2}\right)(1.99 \times 10^{30}\,kg)\left(76\,yr\,\frac{365\,day}{yr}\,\frac{24\,hr}{day}\,\frac{3600\,s}{hr}\right)^2}{4\pi^2}\right]^{\frac{1}{3}}$$

$$\alpha = 2.68 \times 10^{12}\,m$$

Using Equation 8.44, we can determine r_{min} and r_{max}.

$$r_{min} = 2.68 \times 10^{12}\,m(1 - 0.967) = 8.8 \times 10^{10}\,m$$

$$r_{max} = 2.68 \times 10^{12}\,m(1 + 0.967) = 5.27 \times 10^{12}\,m$$

This orbit takes the comet inside the path of Venus, almost to Mercury's orbit, and
out past even the orbit of Neptune and sometimes even to the moderately eccentric
orbit of Pluto. Edmond Halley is generally given the credit for bringing Newton's
work on gravitational and central forces to the attention of the world. After observ-
ing the comet personally in 1682, Halley became interested. Partly as a result of a
bet between Christopher Wren and Robert Hooke, Halley asked Newton in 1684
what paths the planets must follow if the sun pulled them with a force inversely
proportional to the square of their distances. To the astonishment of Halley, Newton
replied, "Why, in ellipses, of course." Newton had worked it out 20 years previously but had not published the result. With painstaking effort, Halley was able in
1705 to predict the next occurrence of the comet, now bearing his name, to be in
1758.

8.8 ORBITAL DYNAMICS

The use of central-force motion is nowhere more useful, important, and interesting
than in space dynamics. Although space dynamics is actually quite complex
because of the gravitational attraction of a spacecraft to various bodies and the
orbital motion involved, we examine two rather simple aspects: a proposed trip to
Mars and flybys past comets and planets.

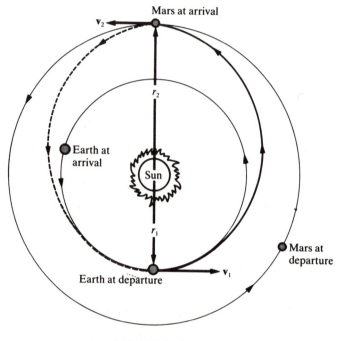

FIGURE 8-10

Orbits are changed by single or multiple thrusts of the rocket engines. The simplest maneuver is a single thrust applied in the orbital plane that does not change the direction of the angular momentum but does change the eccentricity and energy simultaneously. The most economical method of interplanetary transfer consists of moving from one circular heliocentric (sun-oriented motion) orbit to another in the same plane. Earth and Mars represent such a system reasonably well, and a Hohmann transfer (Figure 8-10) represents the path of minimum total energy expenditure.* Two engine burns are required: (1) the first burn injects the spacecraft from the circular Earth orbit to an elliptical transfer orbit that intersects Mars' orbit; (2) the second burn transfers the spacecraft from the elliptical orbit into Mars' orbit.

We can calculate the velocity changes needed for a Hohmann transfer by calculating the velocity of a spacecraft moving in the orbit of the Earth around the sun (r_1 in Figure 8-10) and the velocity needed to "kick" it into an elliptical transfer orbit that can reach Mars' orbit. We are considering only the gravitational attraction of the sun and not that of Earth and Mars.

---

* See Kaplan (Ka76, Chapter 3) for the proof. Walter Hohmann, a German pioneer in space travel research, proposed in 1925 the most energy-efficient method of transferring between elliptical (planetary) orbits in the same plane using only two velocity changes.

For circles and ellipses we have, from Equation 8.42,

$$E = \frac{-k}{2a}$$

For a circular path around the sun, this becomes

$$E = \frac{-k}{2r_1} = \frac{1}{2}mv_1^2 - \frac{k}{r_1} \tag{8.50}$$

where we have $E = T + U$. We solve Equation 8.50 for v_1:

$$v_1 = \sqrt{\frac{k}{mr_1}} \tag{8.51}$$

We denote the semimajor axis of the transfer ellipse by a_t:

$$2a_t = r_1 + r_2$$

If we calculate the energy at the perihelion for the transfer ellipse, we have

$$E_t = \frac{-k}{r_1 + r_2} = \frac{1}{2}mv_{t1}^2 - \frac{k}{r_1} \tag{8.52}$$

where v_{t1} is the perihelion transfer speed. The direction of v_{t1} is along $\mathbf{v}_1$ in Figure 8-10. Solving Equation 8.52 for v_{t1} gives

$$v_{t1} = \sqrt{\frac{2k}{mr_1}\left(\frac{r_2}{r_1 + r_2}\right)} \tag{8.53}$$

The speed transfer Δv_1 needed is just

$$\Delta v_1 = v_{t1} - v_1 \tag{8.54}$$

Similarly, for the transfer from the ellipse to the circular orbit of radius r_2, we have

$$\Delta v_2 = v_2 - v_{t2} \tag{8.55}$$

where

$$v_2 = \sqrt{\frac{k}{mr_2}} \tag{8.56}$$

and

$$\left.\begin{aligned} v_{t2} &= \sqrt{\frac{2}{m}\left(E_t + \frac{k}{r_2}\right)} \\[2mm] v_{t2} &= \sqrt{\frac{2k}{mr_2}\left(\frac{r_1}{r_1 + r_2}\right)} \end{aligned}\right\} \tag{8.57}$$

The direction of v_{t2} is along $\mathbf{v}_2$ in Figure 8-10. The total speed increment can be determined by adding the speed changes, $\Delta v = \Delta v_1 + \Delta v_2$.

The total time required to make the transfer T_t is a half-period of the transfer orbit. From Equation 8.48, we have

$$T_t = \frac{\tau_t}{2}$$

$$T_t = \pi \sqrt{\frac{m}{k}} \, a^{\frac{3}{2}}_t \tag{8.58}$$

EXAMPLE ● 8.5 —

Calculate the time needed for a spacecraft to make a Hohmann transfer from Earth to Mars and the heliocentric transfer speed required assuming both planets are in coplanar orbits.

Solution: We need to insert the appropriate constants in Equation 8.58.

$$\frac{m}{k} = \frac{m}{GmM_{sun}} = \frac{1}{GM_{sun}}$$

$$= \frac{1}{(6.67 \times 10^{-11} \text{ m}^3/\text{s}^2 \cdot \text{kg})(1.99 \times 10^{30} \text{ kg})}$$

$$= 7.53 \times 10^{-21} \text{ s}^2/\text{m}^3 \tag{8.59}$$

Because k/m occurs so often in solar system calculations, we write it as well.

$$\frac{k}{m} = 1.33 \times 10^{20} \text{ m}^3/\text{s}^2$$

$$a_t = \frac{1}{2}(r_{Earth - sun} + r_{Mars - sun})$$

$$= \frac{1}{2}(1.50 \times 10^{11} \text{ m} + 2.28 \times 10^{11} \text{ m})$$

$$= 1.89 \times 10^{11} \text{ m}$$

$$T_t = \pi (7.53 \times 10^{-21} \text{ s}^2/\text{m}^3)^{\frac{1}{2}}(1.89 \times 10^{11} \text{ m})^{\frac{3}{2}}$$

$$= 2.24 \times 10^7 \text{ s}$$

$$= 259 \text{ days} \tag{8.60}$$

The heliocentric speed needed for the transfer is given in Equation 8.53.

$$v_{t1} = \left[\frac{2(1.33 \times 10^{20} \text{ m}^3/\text{s}^2)(2.28 \times 10^{11} \text{ m})}{(1.50 \times 10^{11} \text{ m})(3.78 \times 10^{11} \text{ m})} \right]^{\frac{1}{2}}$$

$$= 3.27 \times 10^4 \text{ m/s} = 32.7 \text{ km/s}$$

We can compare v_{t1} with the orbital speed of the Earth (Equation 8.51).

$$v_1 = \left[\frac{1.33 \times 10^{20} \text{ m}^3/\text{s}^2}{1.50 \times 10^{11} \text{ m}} \right]^{\frac{1}{2}} = 29.8 \text{ km/s}$$

For transfers to the outer planets, the spacecraft should be launched in the direction of the Earth's orbit in order to gain the Earth's orbital velocity. To transfer to the inner planets (e.g., to Venus), the spacecraft should be launched opposite the Earth's motion. In each case, it is the relative velocity Δv_1 that is important to the spacecraft (i.e., relative to the Earth).

Although the Hohmann transfer path represents the least energy expenditure, it does not represent the shortest time. For a round trip from the Earth to Mars, the spacecraft would have to remain on Mars for 460 days until the Earth and Mars were positioned correctly for the return trip (see Figure 8-11a). The total trip

1. Earth departure
2. Mars arrival
3. Mars departure
4. Earth arrival

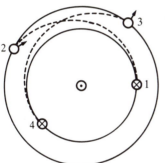

(a) Minimum energy mission requires long stayover on Mars before returning to Earth.

1. Earth departure
2. Mars arrival
3. Mars departure
4. Earth arrival

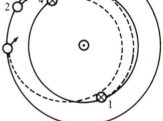

(b) Shorter mission requires more fuel and a closer orbit to the sun.

1. Earth departure
2. Mars arrival
3. Mars departure
4. Venus passage
5. Earth arrival

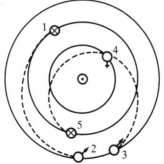

(c) The shorter mission of (b) can be further improved if Venus is positioned for a gravity assist during flyby.

FIGURE 8-11

(259 + 460 + 259 = 978 days = 2.7 yr) would probably be too long. Other schemes either use more fuel to gain speed (Figure 8-11b) or use the slingshot effect of flybys. Such a flyby mission past Venus (see Figure 8-11c) could be done in less than 2 years with only a few weeks near (or on) Mars.

Several spacecraft in recent years have escaped the Earth's gravitational attraction to explore our solar system. Such **interplanetary transfer** can be divided into three segments: (1) the escape from the Earth, (2) a heliocentric transfer to the area of interest, and (3) an encounter with another body—so far, either a planet or a comet. The spacecraft fuel required for such missions can be enormous, but a clever trick has been designed to "steal" energy from other solar system bodies. Because the mass of a spacecraft is so much smaller than the planets (or their moons), the energy loss of the heavenly body is negligible.

We examine a simple version of this flyby or slingshot effect that utilizes gravity assist. A spacecraft coming from infinity approaches a body (labeled B), interacts with B, and recedes. The path is a hyperbola (Figure 8-12). The initial and final velocities, *with respect to B*, are denoted by v_i' and v_f', respectively. The net effect on the spacecraft is a deflection angle of δ with respect to B.

If we examine the system in some inertial frame in which the motion of B occurs, the velocities of the spacecraft can be quite different *because of the motion of B*. The initial velocity v_i is shown in Figure 8-13a, and both v_i and v_f are shown in Figure 8-13b. Notice that the spacecraft has increased its speed as well as changed its direction. An increase in velocity occurs when the spacecraft passes *behind* B's direction of motion. Similarly, a decrease in velocity occurs when the spacecraft passes in *front* of B's motion.

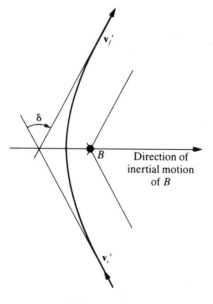

FIGURE 8-12

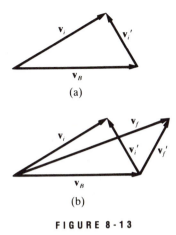

FIGURE 8-13

During the 1970s, scientists at the Jet Propulsion Laboratory of the National Aeronautics and Space Administration (NASA) realized that the four largest planets of our solar system would be in a fortuitous position to allow a spacecraft to fly past them and many of their 32 known moons in a single, relatively short "Grand Tour" mission using the gravity-assist method just discussed. This opportunity of the planets' alignment would not occur again for 175 years. Because of budget constraints, there was not time to develop the new technology needed, and a mission to last only 4 years to visit just Jupiter and Saturn was approved and planned. No special equipment was put on board the twin Voyager spacecrafts for an encounter with Uranus and Neptune. Voyagers 1 and 2 were launched in 1977 for visits to Jupiter in 1979 and Saturn in 1980 (Voyager 1) and 1981 (Voyager 2). Because of the success of these visits to Jupiter and Saturn, funding was later approved to extend Voyager 2's mission to include Uranus and Neptune. The Voyagers are now on their way out of our solar system.

The path of Voyager 2 is shown in Figure 8-14. The slingshot effect of gravity allowed the path of Voyager 2 to be redirected, for example, toward Uranus as it passed Saturn by the method shown in Figure 8-12. The gravitational attraction from Saturn was used to pull the spacecraft off its straight path and redirect it at a different angle. The effect of the orbital motion of Saturn allows an increase in the spacecraft's speed. It was only by using this gravity-assist technique that the spectacular mission of Voyager 2 was made possible in only a brief 12-year period. Voyager 2 passed Uranus in 1986 and Neptune in 1989 before proceeding into interstellar space in one of the most successful space missions ever undertaken. Most planetary missions now take advantage of gravitational assists; for example, the *Galileo* satellite, which photographed the spectacular collisions of the Shoemaker-Levy comet with Jupiter in 1994 and reached Jupiter in 1995, was launched in 1989 but went by Earth twice (1990 and 1992) as well as Venus (1990) to gain speed and redirection.

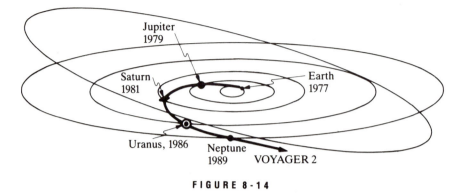

FIGURE 8-14

A spectacular display of flybys occurred in the years 1982–1985 by a spacecraft initially called the International Sun-Earth Explorer 3 (ISEE-3). Launched in 1978, its mission was to monitor the solar wind between the sun and the Earth. For 4 years, the spacecraft circled in the ecliptical plane about 2 million miles from Earth. In 1982—because the United States had decided not to participate in a joint European, Japanese, and Soviet spacecraft investigation of Halley's comet in 1986—NASA decided to reprogram the ISEE-3, renamed it the International Cometary Explorer (ICE), and sent it through the Giacobini-Zinner comet in September 1985, some 6 months before the flybys of other spacecraft with Halley's comet.

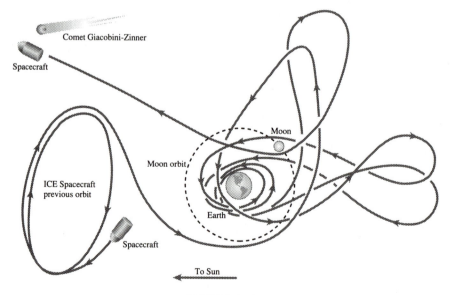

FIGURE 8-15

The subsequent 3-year journey of ICE was spectacular (Figure 8-15). The path of ICE included two close trips to Earth and five flybys of the moon along its billion-mile trip to the comet. During one flyby, the satellite came within 75 miles of the lunar surface. The entire path could be planned precisely because the force law is very well known. The eventual interaction with the comet, some 44 million miles from Earth, included a 20-minute trip through the comet—about 5,000 miles behind the comet's nucleus.

8.9 APSIDAL ANGLES AND PRECESSION (OPTIONAL)

If a particle executes bounded, noncircular motion in a central-force field, then the radial distance from the force center to the particle must always be in the range $r_{max} \geq r \geq r_{min}$; that is, r must be bounded by the apsidal distances. Figure 8-6 indicates that only *two* apsidal distances exist for bounded, noncircular motion. But in executing one complete revolution in θ, the particle may not return to its original position (see Figure 8-4). The angular separation between two successive values of $r = r_{max}$ depends on the exact nature of the force. The angle between any two consecutive apsides is called the **apsidal angle**, and because a closed orbit must be symmetric about any apsis, it follows that all apsidal angles for such motion must be equal. The apsidal angle for elliptical motion, for example, is just π. If the orbit is not closed, the particle reaches the apsidal distances at different points in each revolution; the apsidal angle is not then a rational fraction of 2π, as is required for a closed orbit. If the orbit is *almost* closed, the apsides *precess*, or rotate slowly in the plane of the motion. This effect is exactly analogous to the slow rotation of the elliptical motion of a two-dimensional harmonic oscillator whose natural frequencies for the x and y motions are almost equal (see Section 3.4).

Because an inverse-square-law force requires that all elliptical orbits be exactly closed, the apsides must stay fixed in space for all time. If the apsides move with time, however slowly, this indicates that the force law under which the body moves does not vary exactly as the inverse square of the distance. This important fact was realized by Newton, who pointed out that any advance or regression of a planet's perihelion would require the radial dependence of the force law to be slightly different from $1/r^2$. Thus, Newton argued, the observation of the time dependence of the perihelia of the planets would be a sensitive test of the validity of the form of the universal gravitation law.

In point of fact, for planetary motion within the solar system, one expects that, because of the perturbations introduced by the existence of all the other planets, the force experienced by any planet does not vary exactly as $1/r^2$, if r is measured from the sun. This effect is small, however, and only slight variations of planetary perihelia have been observed. The perihelion of Mercury, for example, which shows the largest effect, advances only about 574" of arc per century.* Detailed calculations of the influence of the other planets on the motion of Mercury predict that

---

* This precession is in addition to the general precession of the equinox with respect to the "fixed" stars, which amounts to 5025.645" ± 0.050" per century.

the rate of advance of the perihelion should be approximately 531″ per century. The uncertainties in this calculation are considerably less than the difference of 43″ between observation and calculation,* and for a considerable time, this discrepancy was the outstanding unresolved difficulty in the Newtonian theory. We now know that the modification introduced into the equation of motion of a planet by the general theory of relativity almost exactly accounts for the difference of 43″. This result is one of the major triumphs of relativity theory.

We next indicate the way the advance of the perihelion can be calculated from the modified equation of motion. To perform this calculation, it is convenient to use the equation of motion in the form of Equation 8.20. If we use the universal gravitational law for $F(r)$, we can write

$$
\frac{d^2u}{d\theta^2} + u = -\frac{m}{l^2}\frac{1}{u^2}F(1/u)
$$

$$
= \frac{Gm^2M}{l^2} \tag{8.61}
$$

where we consider the motion of a body of mass m in the gravitational field of a body of mass M. The quantity u is therefore the reciprocal of the distance between m and M.

The modification of the gravitational force law required by the general theory of relativity introduces into the force a small component that varies as $1/r^4(= u^4)$. Thus, we have

$$
\frac{d^2u}{d\theta^2} + u = \frac{Gm^2M}{l^2} + \frac{3GM}{c^2}u^2 \tag{8.62}
$$

where c is the velocity of propagation of the gravitational interaction and is identified with the velocity of light.[†] To simplify the notation, we define

$$
\left.\begin{array}{c} \dfrac{1}{\alpha} \equiv \dfrac{Gm^2M}{l^2} \\[2ex] \delta \equiv \dfrac{3GM}{c^2} \end{array}\right\} \tag{8.63}
$$

---

* In 1845, the French astronomer Urbain Jean Joseph Le Verrier (1811–1877) first called attention to the irregularity in the motion of Mercury. Similar studies by Le Verrier and by the English astronomer John Couch Adams of irregularities in the motion of Uranus led to the discovery of the planet Neptune in 1846. An interesting account of this episode is given by Turner (Tu04, Chapter 2). We must note, in this regard, that perturbations may be either *periodic* or *secular* (i.e., ever increasing with time). Laplace showed in 1773 (published, 1776) that any perturbation of a planet's mean motion that is caused by the attraction of another planet must be periodic, although the period may be extremely long. This is the case for Mercury; the precession of 531″ per century is periodic, but the period is so long that the change from century to century is small compared with the residual effect of 43″.

[†] One half of the relativistic term results from effects understandable in terms of special relativity, viz., time dilation (1/3) and the relativistic momentum effect (1/6); the velocity is greatest at perihelion and least at aphelion (see Chapter 14). The other half of the term arises from general relativistic effects and is associated with the finite propagation time of gravitational interactions. Thus, the agreement between theory and experiment confirms the prediction that the gravitational propagation velocity is the same as that for light.

and we can write Equation 8.62 as

$$\frac{d^2u}{d\theta^2} + u = \frac{1}{\alpha} + \delta u^2 \tag{8.64}$$

This is a nonlinear equation, and we use a successive approximation procedure to obtain a solution. We choose the first solution to be the solution of Equation 8.64 in the case that the term δu^2 is neglected*:

$$u_1 = \frac{1}{\alpha}(1 + \varepsilon \cos \theta) \tag{8.65}$$

This is the familiar result for the pure inverse-square-law force (see Equation 8.41). Note that α is here the same as that defined in Equation 8.40 except that μ has been replaced by m. If we substitute this expression into the right-hand side of Equation 8.64, we find

$$\frac{d^2u}{d\theta^2} + u = \frac{1}{\alpha} + \frac{\delta}{\alpha^2}[1 + 2\varepsilon \cos \theta + \varepsilon^2 \cos^2 \theta]$$

$$= \frac{1}{\alpha} + \frac{\delta}{\alpha^2}\left[1 + 2\varepsilon \cos \theta + \frac{\varepsilon^2}{2}(1 + \cos 2\theta)\right] \tag{8.66}$$

where $\cos^2\theta$ has been expanded in terms of $\cos 2\theta$. The first trial function u_1, when substituted into the left-hand side of Equation 8.64, reproduces only the first term on the right-hand side: $1/\alpha$. We can therefore construct a second trial function by adding to u_1 a term that reproduces the remainder of the right-hand side (in Equation 8.66). We can verify that such a particle integral is

$$u_p = \frac{\delta}{\alpha^2}\left[\left(1 + \frac{\varepsilon^2}{2}\right) + \varepsilon\theta \sin \theta - \frac{\varepsilon^2}{6}\cos 2\theta\right] \tag{8.67}$$

The second trial function is therefore

$$u_2 = u_1 + u_p$$

If we stop the approximation procedure at this point, we have

$$u \cong u_2 = u_1 + u_p$$

$$= \left[\frac{1}{\alpha}(1 + \varepsilon \cos \theta) + \frac{\delta\varepsilon}{\alpha^2}\theta \sin \theta\right]$$

$$+ \left[\frac{\delta}{\alpha^2}\left(1 + \frac{\varepsilon^2}{2}\right) - \frac{\delta\varepsilon^2}{6\alpha^2}\cos 2\theta\right] \tag{8.68}$$

where we have regrouped the terms in u_1 and u_p.

Consider the terms in the second set of brackets in Equation 8.68: the first of these is just a constant, and the second is only a small and periodic disturbance of

---

* We eliminate the necessity of introducing an arbitrary phase into the argument of the cosine term by choosing to measure θ from the position of perihelion; i.e., u_1 is a maximum (and hence r_1 is a minimum) at $\theta = 0$.

the normal Keplerian motion. Therefore, on a long time scale neither of these terms contributes, on the average, to any change in the positions of the apsides. But in the first set of brackets, the term proportional to θ produces secular and therefore observable effects. Let us consider the first set of brackets:

$$u_{\text{secular}} = \frac{1}{\alpha}\left[1 + \varepsilon \cos\theta + \frac{\delta\varepsilon}{\alpha}\theta\sin\theta\right] \tag{8.69}$$

Next, we can expand the quantity

$$1 + \varepsilon \cos\left(\theta - \frac{\delta}{\alpha}\theta\right) = 1 + \varepsilon\left(\cos\theta\cos\frac{\delta}{\alpha}\theta + \sin\theta\sin\frac{\delta}{\alpha}\theta\right)$$

$$\cong 1 + \varepsilon\cos\theta + \frac{\delta\varepsilon}{\alpha}\theta\sin\theta \tag{8.70}$$

where we have used the fact that δ is small to approximate

$$\cos\frac{\delta}{\alpha}\theta \cong 1, \qquad \sin\frac{\delta}{\alpha}\theta \cong \frac{\delta}{\alpha}\theta$$

Hence, we can write u_{secular} as

$$u_{\text{secular}} \cong \frac{1}{\alpha}\left[1 + \varepsilon\cos\left(\theta - \frac{\delta}{\alpha}\theta\right)\right] \tag{8.71}$$

We have chosen to measure θ from the position of perihelion at $t = 0$. Successive appearances at perihelion result when the argument of the cosine term in u_{secular} increases to $2\pi, 4\pi, \cdots$, and so forth. But an increase of the argument by 2π requires that

$$\theta - \frac{\delta}{\alpha}\theta = 2\pi$$

or

$$\theta = \frac{2\pi}{1 - (\delta/\alpha)} \cong 2\pi\left(1 + \frac{\delta}{\alpha}\right)$$

Therefore, the effect of the relativistic term in the force law is to displace the perihelion in each revolution by an amount

$$\Delta \cong \frac{2\pi\delta}{\alpha} \tag{8.72a}$$

that is, the apsides rotate slowly in space. If we refer to the definitions of α and δ (Equations 8.63), we find

$$\Delta \cong 6\pi\left(\frac{GmM}{cl}\right)^2 \tag{8.72b}$$

From Equations 8.40 and 8.42, we can write $l^2 = \mu k a(1 - \varepsilon^2)$; then, because $k = GmM$ and $\mu \cong m$, we have

$$\Delta \cong \frac{6\pi GM}{ac^2(1 - \varepsilon^2)} \qquad (8.72c)$$

We see therefore that the effect is enhanced if the semimajor axis a is small and if the eccentricity is large. Mercury, which is the planet nearest the sun and which has the most eccentric orbit of any planet (except Pluto), provides the most sensitive test of the theory.* The calculated value of the precessional rate for Mercury is 43.03" ± 0.03" of arc per century. The observed value (corrected for the influence of the other planets) is 43.11" ± 0.45",[†] so the prediction of relativity theory is confirmed in striking fashion. The precessional rates for some of the planets are given in Table 8-2.

TABLE 8-2

PRECESSIONAL RATES FOR THE PERIHELIA OF SOME PLANETS

Planet	Precessional rate (seconds of arc/century)	
	Calculated	Observed
Mercury	43.03 ± 0.03	43.11 ± 0.45
Venus	8.63	8.4 ± 4.8
Earth	3.84	5.0 ± 1.2
Mars	1.35	—
Jupiter	0.06	—

8.10 STABILITY OF CIRCULAR ORBITS (OPTIONAL)

In Section 8.6, we pointed out that the orbit is circular if the total energy equals the minimum value of the effective potential energy, $E = V_{min}$. More generally, however, a circular orbit is allowed for *any* attractive potential, because the attractive force can *always* be made to just balance the centrifugal force by the proper choice of radial velocity. Although circular orbits are therefore always possible in a central, attractive force field, such orbits are not necessarily stable. A circular orbit at $r = \rho$ exists if $\dot{r}\,|_{r=\rho} = 0$ for all t; this is possible if $(\partial V/\partial r)\,|_{r=\rho} = 0$. But only if the effective potential has a *true minimum* does stability result. All other equilibrium circular orbits are unstable.

---

* Alternatively, we can say that the relativistic advance of the perihelion is a maximum for Mercury because the orbital velocity is greatest for Mercury and the relativistic parameter v/c largest.

† R. L. Duncombe, *Astron. J.* **61**, 174 (1956); see also G. M. Clemence, *Rev. Mod. Phys.* **19**, 361 (1947).

Let us consider an attractive central force with the form

$$F(r) = -\frac{k}{r^n} \tag{8.73}$$

The potential for such a force is

$$U(r) = -\frac{k}{n-1} \cdot \frac{1}{r^{(n-1)}} \tag{8.74}$$

and the effective potential function is

$$V(r) = -\frac{k}{n-1} \cdot \frac{1}{r^{(n-1)}} + \frac{l^2}{2\mu r^2} \tag{8.75}$$

The conditions for a minimum of $V(r)$ and hence for a stable circular orbit with a radius ρ are

$$\left. \frac{\partial V}{\partial r} \right|_{r=\rho} = 0 \quad \text{and} \quad \left. \frac{\partial^2 V}{\partial r^2} \right|_{r=\rho} > 0 \tag{8.76}$$

Applying these criteria to the effective potential of Equation 8.75, we have

$$\left. \frac{\partial V}{\partial r} \right|_{r=\rho} = \frac{k}{\rho^n} - \frac{l^2}{\mu \rho^3} = 0$$

or

$$\rho^{(n-3)} = \frac{\mu k}{l^2} \tag{8.77}$$

and

$$\left. \frac{\partial^2 V}{\partial r^2} \right|_{r=\rho} = -\frac{nk}{\rho^{(n+1)}} + \frac{3l^2}{\mu \rho^4} > 0$$

so

$$-\frac{nk}{\rho^{(n-3)}} + \frac{3l^2}{\mu} > 0 \tag{8.78}$$

Substituting $\rho^{(n-3)}$ from Equation 8.77 into Equation 8.78, we have

$$(3-n)\frac{l^2}{\mu} > 0 \tag{8.79}$$

The condition that a stable circular orbit exist is thus $n < 3$.

Next, we apply a more general procedure and inquire about the frequency of oscillation about a circular orbit in a general force field. We write the force as

$$F(r) = -\mu g(r) = -\frac{\partial U}{\partial r} \tag{8.80}$$

Equation 8.18 can now be written as

$$\ddot{r} - r\dot{\theta}^2 = -g(r) \tag{8.81}$$

Substituting for $\dot{\theta}$ from Equation 8.10,

$$\ddot{r} - \frac{l^2}{\mu^2 r^3} = -g(r) \tag{8.82}$$

We now consider the particle to be initially in a circular orbit with radius ρ and apply a perturbation of the form $r \to \rho + x$, where x is small. Because ρ = constant, we also have $\ddot{r} \to \ddot{x}$. Thus

$$\ddot{x} - \frac{l^2}{\mu^2 \rho^3 [1 + (x/\rho)]^3} = -g(\rho + x) \tag{8.83}$$

But by hypothesis $(x/\rho) \ll 1$, so we can expand the quantity:

$$[1 + (x/\rho)]^{-3} = 1 - 3(x/\rho) + \cdots \tag{8.84}$$

We also assume that $g(r) = g(\rho + x)$ can be expanded in a Taylor series about the point $r = \rho$:

$$g(\rho + x) = g(\rho) + xg'(\rho) + \cdots \tag{8.85}$$

where

$$g'(\rho) \equiv \frac{dg}{dr}\bigg|_{r=\rho}$$

If we neglect all terms in x^2 and higher powers, then the substitution of Equations 8.84 and 8.85 into Equation 8.83 yields

$$\ddot{x} - \frac{l^2}{\mu^2 \rho^3}[1 - 3(x/\rho)] \cong -[g(\rho) + xg'(\rho)] \tag{8.86}$$

Recall that we assumed the particle to be initially in a circular orbit with $r = \rho$. Under such a condition, no radial motion occurs—that is, $\dot{r}\,|_{r=\rho} = 0$. Then, also, $\ddot{r}\,|_{r=\rho} = 0$. Therefore, evaluating Equation 8.82 at $r = \rho$, we have

$$g(\rho) = \frac{l^2}{\mu^2 \rho^3} \tag{8.87}$$

Substituting this relation into Equation 8.86, we have, approximately,

$$\ddot{x} - g(\rho)[1 - 3(x/\rho)] \cong -[g(\rho) + xg'(\rho)]$$

or

$$\ddot{x} + \left[\frac{3g(\rho)}{\rho} + g'(\rho)\right]x \cong 0 \tag{8.88}$$

If we define

$$\omega_0^2 \equiv \frac{3g(\rho)}{\rho} + g'(\rho) \tag{8.89}$$

then Equation 8.88 becomes the familiar equation for the undamped harmonic oscillator:

$$\ddot{x} + \omega_0^2 x = 0 \tag{8.90}$$

The solution to this equation is

$$x(t) = Ae^{+i\omega_0 t} + Be^{-i\omega_0 t} \tag{8.91}$$

If $\omega_0^2 < 0$, so that ω_0 is imaginary, then the second term becomes $B \exp(|\omega_0| t)$, which clearly increases without limit as time increases. The condition for oscillation is therefore $\omega_0^2 > 0$, or

$$\frac{3g(\rho)}{\rho} + g'(\rho) > 0 \tag{8.92a}$$

Because $g(\rho) > 0$ (see Equation 8.87), we can divide through by $g(\rho)$ and write this inequality as

$$\frac{g'(\rho)}{g(\rho)} + \frac{3}{\rho} > 0 \tag{8.92b}$$

or, because $g(r)$ and $F(r)$ are related by a constant multiplicative factor, stability results if

$$\boxed{\frac{F'(\rho)}{F(\rho)} + \frac{3}{\rho} > 0} \tag{8.93}$$

We now compare the condition on the force law imposed by Equation 8.93 with that previously obtained for a power-law force:

$$F(r) = -\frac{k}{r^n} \tag{8.94}$$

Equation 8.93 becomes

$$\frac{nk\rho^{-(n+1)}}{-k\rho^{-n}} + \frac{3}{\rho} > 0$$

or

$$(3 - n) \cdot \frac{1}{\rho} > 0 \tag{8.95}$$

and we are led to the same condition as before—that is, $n < 3$. (We must note, however, that the case $n = 3$ needs further examination; see Problem 8-22.)

EXAMPLE **8.6** -

Investigate the stability of circular orbits in a force field described by the potential function

$$U(r) = \frac{-k}{r} e^{-(r/a)} \tag{8.96}$$

where $k > 0$ and $a > 0$.

Solution: This potential is called the **screened Coulomb potential** (when $k = Ze^2/4\pi\varepsilon_0$, where Z is the atomic number and e is the electron charge) because it falls off with distance more rapidly than $1/r$ and hence approximates the electrostatic potential of the atomic nucleus in the vicinity of the nucleus by taking into account the partial "cancellation" or "screening" of the nuclear charge by the atomic electrons. The force is found from

$$F(r) = -\frac{\partial U}{\partial r} = -k\left(\frac{1}{ar} + \frac{1}{r^2}\right)e^{-(r/a)}$$

and

$$\frac{\partial F}{\partial r} = k\left(\frac{1}{a^2 r} + \frac{2}{ar^2} + \frac{2}{r^3}\right)e^{-(r/a)}$$

The condition for stability (see Equation 8.93) is

$$3 + \rho\frac{F'(\rho)}{F(\rho)} > 0$$

Therefore

$$3 + \frac{\rho k\left(\dfrac{1}{a^2\rho} + \dfrac{2}{a\rho^2} + \dfrac{2}{\rho^3}\right)}{-k\left(\dfrac{1}{a\rho} + \dfrac{1}{\rho^2}\right)} > 0$$

which simplifies to

$$a^2 + a\rho - \rho^2 > 0$$

We may write this as

$$\frac{a^2}{\rho^2} + \frac{a}{\rho} - 1 > 0$$

Stability thus results for all $q \equiv a/\rho$ that exceed the value satisfying the equation

$$q^2 + q - 1 = 0$$

The positive (and therefore the only physically meaningful) solution is

$$q = \tfrac{1}{2}(\sqrt{5} - 1) \cong 0.62$$

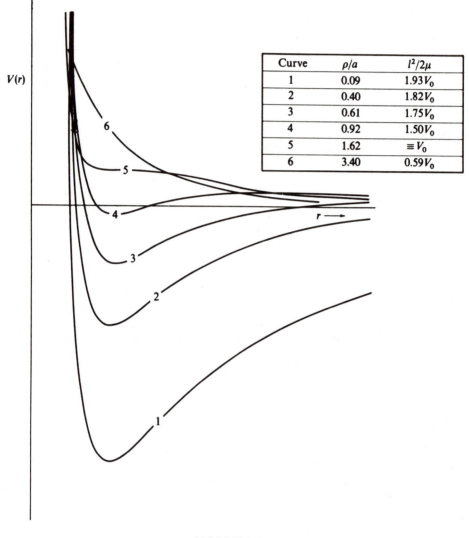

$V(r)$

Curve	ρ/a	$l^2/2\mu$
1	0.09	$1.93V_0$
2	0.40	$1.82V_0$
3	0.61	$1.75V_0$
4	0.92	$1.50V_0$
5	1.62	$\equiv V_0$
6	3.40	$0.59V_0$

$r \longrightarrow$

FIGURE 8-16

If, then, the angular momentum and energy allow a circular orbit at $r = \rho$, the motion is stable if

$$\frac{a}{\rho} \gtrsim 0.62$$

or

$$\rho \lesssim 1.62a \qquad (8.97)$$

The stability condition for orbits in a screened potential is illustrated graphically in Figure 8-16, which shows the potential $V(r)$ for various values of ρ/a.

The force constant k is the same for all the curves, but $l^2/2\mu$ has been adjusted to maintain the minimum of the potential at the same value of the radius as a is changed. For $\rho/a < 1.62$, a true minimum exists for the potential, indicating that the circular orbit is stable with respect to small oscillations. For $\rho/a > 1.62$, there is no minimum, so circular orbits cannot exist. For $\rho/a = 1.62$, the potential has zero slope at the position that a circular orbit would occupy. The orbit is unstable at this position, because ω_0^2 is zero in Equation 8.90 and the displacement x increases linearly with time.

An interesting feature of this potential function is that under certain conditions there can exist bound orbits for which the total energy is positive (see, for example, curve 4 in Figure 8-16).

E X A M P L E   8.7   --

Determine whether a particle moving on the surface of a cone (see Example 7.4) can have a stable circular orbit.

Solution: In Example 7.4, we found that the angular momentum about the z-axis was a constant of the motion:

$$l = mr^2\dot{\theta} = \text{constant}$$

We also found the equation of motion for the coordinate r:

$$\ddot{r} - r\dot{\theta}^2 \sin^2 \alpha + g \sin \alpha \cos \alpha = 0 \qquad \textbf{(8.98)}$$

If the initial conditions are appropriately selected, the particle can move in a circular orbit about the vertical axis with the plane of the orbit at a constant height z_0 above the horizontal plane passing through the apex of the cone. Although this problem does not involve a central force, certain aspects of the motion are the same as for the central-force case. Thus we may discuss, for example, the stability of circular orbits for the particle. To do this, we perform a perturbation calculation.

First, we assume that a circular orbit exists for $r = \rho$. Then, we apply the perturbation $r \to \rho + x$. The quantity $r\dot{\theta}^2$ in Equation 8.98 can be expressed as

$$r\dot{\theta}^2 = r \cdot \frac{l^2}{m^2 r^4} = \frac{l^2}{m^2 r^3}$$

$$= \frac{l^2}{m^2}(\rho + x)^{-3} = \frac{l^2}{m^2 \rho^3}\left(1 + \frac{x}{\rho}\right)^{-3}$$

$$\cong \frac{l^2}{m^2 \rho^3}\left(1 - 3\frac{x}{\rho}\right)$$

where we have retained only the first term in the expansion, because x/ρ is by hypothesis a small quantity.

Then, because $\ddot{p} = 0$, Equation 8.98 becomes, approximately,

$$\ddot{x} - \frac{l^2 \sin^2 \alpha}{m^2 \rho^3} \left(1 - 3\frac{x}{\rho}\right) + g \sin \alpha \cos \alpha = 0$$

or

$$\ddot{x} + \left(\frac{3l^2 \sin^2 \alpha}{m^2 \rho^4}\right) x - \frac{l^2 \sin \alpha}{m^2 \rho^3} + g \sin \alpha \cos \alpha = 0 \qquad (8.99)$$

If we evaluate Equation 8.98 at $r = \rho$, then $\ddot{r} = 0$, and we have

$$g \sin \alpha \cos \alpha = \rho \dot{\theta}^2 \sin^2 \alpha$$

$$= \frac{l^2}{m^2 \rho^3} \sin^2 \alpha$$

In view of this result, the last two terms in Equation 8.99 cancel, and there remains

$$\ddot{x} + \left(\frac{3l^2 \sin^2 \alpha}{m^2 \rho^4}\right) x = 0 \qquad (8.100)$$

The solution to this equation is just a harmonic oscillation with a frequency ω, where

$$\omega = \frac{\sqrt{3}\, l}{m\rho^2} \sin \alpha \qquad (8.101)$$

Thus, the circular orbit is stable.

--- ●

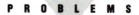

PROBLEMS

8-1. In Section 8.2, we showed that the motion of two bodies interacting only with each other by central forces could be reduced to an equivalent one-body problem. Show by explicit calculation that such a reduction is also possible for bodies moving in an external uniform gravitational field.

8-2. Perform the integration of Equation 8.38 to obtain Equation 8.39.

8-3. A particle moves in a circular orbit in a force field given by

$$F(r) = -k/r^2$$

Show that, if k suddenly decreases to half its original value, the particle's orbit becomes parabolic.

8-4. Perform an explicit calculation of the time average (i.e., the average over one complete period) of the potential energy for a particle moving in an elliptical orbit in a central inverse-square-law force field. Express the result in terms of the force constant of the field and the semimajor axis of the ellipse. Perform a similar calculation for the kinetic energy. Compare the results and thereby verify the virial theorem for this case.

8-5. Two particles moving under the influence of their mutual gravitational force describe circular orbits about one another with a period τ. If they are suddenly stopped in their orbits and allowed to gravitate toward each other, show that they will collide after a time $\tau/4\sqrt{2}$.

8-6. Two gravitating masses m_1 and $m_2(m_1 + m_2 = M)$ are separated by a distance r_0 and released from rest. Show that when the separation is $r(<r_0)$, the speeds are

$$v_1 = m_2\sqrt{\frac{2G}{M}\left(\frac{1}{r} - \frac{1}{r_0}\right)}, \qquad v_2 = m_1\sqrt{\frac{2G}{M}\left(\frac{1}{r} - \frac{1}{r_0}\right)}$$

8-7. Show that the areal velocity is constant for a particle moving under the influence of an attractive force given by $F(r) = -kr$. Calculate the time averages of the kinetic and potential energies and compare with the results of the virial theorem.

8-8. Investigate the motion of a particle *repelled* by a force center according to the law $F(r) = kr$. Show that the orbit can only be hyperbolic.

8-9. A communications satellite is in a circular orbit around the Earth at radius R and velocity v. A rocket accidentally fires quite suddenly, giving the rocket an outward radial velocity v in addition to its original velocity.
(a) Calculate the ratio of the new energy and angular momentum to the old.
(b) Describe the subsequent motion of the satellite and plot $T(r)$, $V(r)$, $U(r)$, and $E(r)$ after the rocket fires.

8-10. Assume the Earth's orbit to be circular and that the sun's mass suddenly decreases by half. What orbit does the Earth then have? Will the Earth escape the solar system?

8-11. A particle moves under the influence of a central force given by $F(r) = -k/r^n$. If the particle's orbit is circular and passes through the force center, show that $n = 5$.

8-12. Consider a comet moving in a parabolic orbit in the plane of the Earth's orbit. If the distance of closest approach of the comet to the sun is βr_E, where r_E is the radius of the Earth's (assumed) circular orbit and where $\beta < 1$, show that the time the comet spends within the orbit of the Earth is given by

$$\sqrt{2(1 - \beta)} \cdot (1 + 2\beta)/3\pi \times 1 \text{ year}$$

If the comet approaches the sun to the distance of the perihelion of Mercury, how many days is it within the Earth's orbit?

8-13. Discuss the motion of a particle in a central inverse-square-law force field for a super-imposed force whose magnitude is inversely proportional to the cube of the distance from the particle to the force center; that is,

$$F(r) = -\frac{k}{r^2} - \frac{\lambda}{r^3}, \qquad k, \lambda > 0$$

Show that the motion is described by a precessing ellipse. Consider the cases $\lambda < l^2/\mu$, $\lambda = l^2/\mu$, and $\lambda > l^2/\mu$.

8-14. Find the force law for a central-force field that allows a particle to move in a spiral orbit given by $r = k\theta^2$, where k is a constant.

8-15. A particle of unit mass moves from infinity along a straight line that, if continued, would allow it to pass a distance $b\sqrt{2}$ from a point P. If the particle is attracted toward P with a force varying as k/r^5, and if the angular momentum about the point P is $\sqrt{k}/b$, show that the trajectory is given by

$$r = b \coth(\theta/\sqrt{2})$$

8-16. A particle executes elliptical (but almost circular) motion about a force center. At some point in the orbit a *tangential* impulse is applied to the particle, changing the velocity from v to $v + \delta v$. Show that the resulting relative change in the major and minor axes of the orbit is twice the relative change in the velocity and that the axes are *increased* if $\delta v > 0$.

8-17. A particle moves in an elliptical orbit in an inverse-square-law central-force field. If the ratio of the maximum angular velocity to the minimum angular velocity of the particle in its orbit is n, then show that the eccentricity of the orbit is

$$\varepsilon = \frac{\sqrt{n} - 1}{\sqrt{n} + 1}$$

8-18. Use Kepler's results (i.e., his first and second laws) to show that the gravitational force must be central and that the radial dependence must be $1/r^2$. Thus, perform an inductive derivation of the gravitational force law.

8-19 Calculate the missing entries denoted by c in Table 8-1.

8-20. For a particle moving in an elliptical orbit with semimajor axis a and eccentricity ε, show that

$$\langle (a/r)^4 \cos \theta \rangle = \varepsilon/(1 - \varepsilon^2)^{5/2}$$

where the slanted brackets denote a time average over one complete period.

8-21. Consider the family of orbits in a central potential for which the total energy is a constant. Show that if a stable circular orbit exists, the angular momentum associated with this orbit is larger than that for any other orbit of the family.

8-22. Discuss the motion of a particle moving in an attractive central-force field described by $F(r) = -k/r^3$.* Sketch some of the orbits for different values of the total energy. Can a circular orbit be stable in such a force field?

8-23. An Earth satellite moves in an elliptical orbit with a period τ, eccentricity ε, and semimajor axis a. Show that the maximum radial velocity of the satellite is $2\pi a\varepsilon/(\tau\sqrt{1 - \varepsilon^2})$.

8-24. An Earth satellite has a perigee of 300 km and an apogee of 3,500 km above the Earth's surface. How far is the satellite above Earth when **(a)** it has rotated 90° around the Earth from perigee and **(b)** it has moved halfway from perigee to apogee?

8-25. An Earth satellite has a speed of 28,070 km/hr when it is at its perigee of 220 km above the Earth's surface. Find the apogee distance, its speed at apogee, and its period of revolution.

8-26. Show that the most efficient way to change the energy of an elliptical orbit for a single short engine thrust is by firing the rocket along the direction of travel at perigee.

---

* This particular force law was extensively investigated by Roger Cotes (1682–1716), and the orbits are known as **Cotes' spirals**.

8-27. A spacecraft in an orbit about Earth has the speed of 10,160 m/s at a perigee of 6,680 km above the Earth's surface. What speed does the spacecraft have at apogee of 42,200 km?

8-28. What is the minimum escape velocity of a spacecraft from the moon?

8-29. The minimum and maximum velocities of a moon rotating around Uranus are $v_{min} = v - v_0$ and $v_{max} = v + v_0$. Find the eccentricity in terms of v and v_0.

8-30. A spacecraft is placed in orbit 200 km above Earth in a circular orbit. Calculate the minimum escape speed from Earth. Sketch the escape trajectory, showing the Earth and the circular orbit. What is the spacecraft's trajectory with respect to Earth?

8-31. Consider a force law of the form

$$F(r) = -\frac{k}{r^2} - \frac{k'}{r^4}$$

Show that if $\rho^2 k > k'$, then a particle can move in a stable circular orbit at $r = \rho$.

8-32. Consider a force law of the form $F(r) = -(k/r^2)\exp(-r/a)$. Investigate the stability of circular orbits in this force field.

8-33. Consider a particle of mass m constrained to move on the surface of a paraboloid whose equation (in cylindrical coordinates) is $r^2 = 4az$. If the particle is subject to a gravitational force, show that the frequency of small oscillations about a circular orbit with radius $\rho = \sqrt{4az_0}$ is

$$\omega = \sqrt{\frac{2g}{a + z_0}}$$

8-34. Consider the problem of the particle moving on the surface of a cone, as discussed in Examples 7.4 and 8.7. Show that the effective potential is

$$V(r) = \frac{l^2}{2mr^2} + mgr \cot \alpha$$

(Note that here r is the radial distance in cylindrical coordinates, not spherical coordinates; see Figure 7-2.) Show that the turning points of the motion can be found from the solution of a cubic equation in r. Show further that only two of the roots are physically meaningful, so that the motion is confined to lie within two horizontal planes that cut the cone.

8-35. An almost circular orbit (i.e., $\varepsilon \ll 1$) can be considered to be a circular orbit to which a small perturbation has been applied. Then, the frequency of the radial motion is given by Equation 8.89. Consider a case in which the force law is $F(r) = -k/r^n$ (where n is an integer), and show that the apsidal angle is $\pi/\sqrt{3 - n}$. Thus, show that a closed orbit generally results only for the harmonic oscillator force and the inverse-square-law force (if values of n equal to or smaller than -6 are excluded).

8-36. A particle moves in an almost circular orbit in a force field described by $F(r) = -(k/r^2)\exp(-r/a)$. Show that the apsides advance by an amount approximately equal to $\pi\rho/a$ in each revolution, where ρ is the radius of the circular orbit and where $\rho \ll a$.

8-37. A communications satellite is in a circular orbit around the Earth at a distance above the Earth equal to the Earth's radius. Find the minimum velocity Δv required to double the height of the satellite and put it in another circular orbit.

8-38. Calculate the minimum Δv required to place a satellite already in the Earth's heliocentric orbit (assumed circular) into the orbit of Venus (also assumed circular and coplanar with the Earth). Consider only the gravitational attraction of the sun. What time of flight would such a trip take?

8-39. Assuming a rocket engine can be fired only once from a low Earth orbit, does a Mars flyby or a Venus flyby require a larger Δv? Explain.

8-40. A spacecraft is being designed to dispose of nuclear waste either by carrying it out of the solar system or crashing into the sun. Assume that no planetary flybys are permitted and that thrusts occur only in the orbital plane. Which mission requires the least energy? Explain.

8-41. A spacecraft is parked in a circular orbit 200 km above the Earth's surface. We want to use a Hohmann transfer to send the spacecraft to the moon's orbit. What are the total Δv and the transfer time required?

8-42. A spacecraft of mass 10,000 kg is parked in a circular orbit 200 km above the Earth's surface. What is the minimum energy required (neglect the fuel mass burned) to place the satellite in a synchronous orbit (i.e., $\tau = 24$ hr)?

9

DYNAMICS OF A
SYSTEM OF PARTICLES

9.1 INTRODUCTION

Thus far, we have treated our dynamical problems primarily in terms of single particles. Even though we have considered extended objects such as projectiles, rockets, and planets, we have been able to treat them as single particles. Generally, we have not had to deal with the internal interactions between the many particles that make up the extended body.

Later, when we treat the dynamics of rigid bodies, we must describe rotational as well as translational motion. We need to prepare the techniques that will allow us to do this.

We first extend our discussion to describe the system of n particles. These particles may form a loose aggregate—such as a pile of rocks or a volume of gas molecules—or form a rigid body in which the constituent particles are restrained from moving relative to one another. We devote the latter part of the chapter to a study of the interaction of two particles ($n = 2$). For the three-body problem ($n = 3$), the solutions become formidable. Perturbation techniques often are used, although great progress has been made through the use of numerical methods with high-speed computers.

Newton's Third Law plays a prominent role in the dynamics of a system of particles because of the internal forces between the particles in the system. We need to make two assumptions concerning the internal forces:

1. The forces exerted by two particles α and β on each other are equal in magnitude and opposite in direction. Let $\mathbf{f}_{\alpha\beta}$ represent the force on the αth

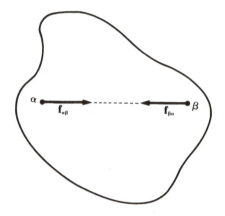

FIGURE 9-1   Example of the strong form of Newton's Third law, where the equal and opposite forces between two particles must lie along a straight line joining the two particles. The force is attractive, as in the molecular attraction in a solid.

particle due to the βth particle. The so-called "weak" form of Newton's Third Law is

$$\mathbf{f}_{\alpha\beta} = -\mathbf{f}_{\beta\alpha} \tag{9.1}$$

2. The forces exerted by two particles α and β on each other, in addition to being equal and opposite, must lie on the straight line joining the two particles. This more restrictive form of Newton's Third Law, often called the "strong" form, is displayed in Figure 9-1.

We must be careful to remember when each form of Newton's Third Law applies. We recall from Section 2.2 that the Third Law is not always valid for moving charged particles; electromagnetic forces are *velocity dependent*. For example, magnetic forces, those forces exerted on a moving charge q in a magnetic field $\mathbf{B}$ ($\mathbf{F} = q\mathbf{v} \times \mathbf{B}$), obey the weak form, but not the strong form, of the Third Law.

9.2   CENTER OF MASS

Consider a system composed of n particles, with each particle's mass described by m_α, where α is an index from $\alpha = 1$ to $\alpha = n$. The total mass of the system is denoted by M,

$$M = \sum_\alpha m_\alpha \tag{9.2}$$

where the summation over α (as in all summations carried out over Greek indices) runs from $\alpha = 1$ to $\alpha = n$. Such a system is displayed in Figure 9-2.

If the vector connecting the origin with the αth particle is $\mathbf{r}_\alpha$, then the vector defining the position of the system's center of mass is

$$\mathbf{R} = \frac{1}{M} \sum_\alpha m_\alpha \mathbf{r}_\alpha \tag{9.3}$$

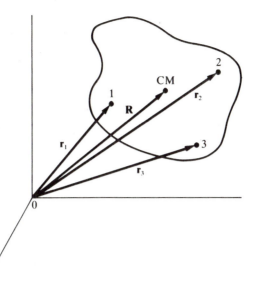

FIGURE 9-2 The position vectors to particles 1, 2, and 3 in the body are indicated, along with the center of mass position vector **R**.

For a continuous distribution of mass, the summation is replaced by an integral,

$$\mathbf{R} = \frac{1}{M} \int \mathbf{r} \, dm \qquad\qquad (9.4)$$

The location of the center of mass of a body is uniquely defined, but the position vector **R** depends on the coordinate system chosen. If the origin in Figure 9-2 were chosen elsewhere, the vector **R** would be different.

EXAMPLE **9.1** –

Find the center of mass of a solid hemisphere of constant density.

Solution: Let the density be ρ, the hemispherical mass be M, and the radius be a.

$$\rho = \frac{M}{\frac{2}{3}\pi a^3}$$

We want to choose the origin of our coordinate system carefully (Figure 9-3) to make the problem as simple as possible. The position coordinates of R are (X, Y, Z). From symmetry, $X = 0$, $Z = 0$. This should be obvious from Equation 9.4,

$$X = \frac{1}{M} \int_{-a}^{a} x \, dm$$

$$Z = \frac{1}{M} \int_{-a}^{a} z \, dm$$

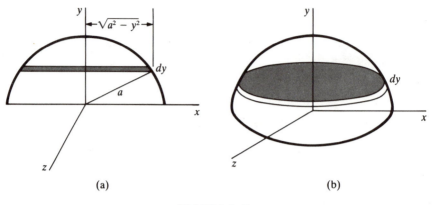

FIGURE 9-3

because we are integrating over an odd power of a variable with symmetric limits. For Y, however, the limits are asymmetric.

$$Y = \frac{1}{M} \int_0^a y \, dm$$

Construct dm so it is placed at a constant value of y. A circular slice perpendicular to the y-axis suffices (see Figure 9-3).

$$dm = \rho dV = \rho \pi (a^2 - y^2) \, dy$$

$$Y = \frac{1}{M} \int_0^a \rho \pi y (a^2 - y^2) \, dy$$

$$Y = \frac{\pi \rho a^4}{4M} = \frac{3a}{8}$$

The position of the center of mass is $(0, 3a/8, 0)$.

-- ●

9.3 LINEAR MOMENTUM OF THE SYSTEM

If a certain group of particles constitutes a *system*, then the resultant force acting on a particle within the system (say, the αth particle) is in general composed of two parts. One part is the resultant of all forces whose origin lies outside of the system; this is called the **external force, $\mathbf{F}_\alpha^{(e)}$**. The other part is the resultant of the forces arising from the interaction of all of the other $n - 1$ particles with the αth particle; this is called the **internal force, $\mathbf{f}_\alpha$**. Force $\mathbf{f}_\alpha$ is given by the vector sum of all the individual forces $\mathbf{f}_{\alpha\beta}$,

$$\mathbf{f}_\alpha = \sum_\beta \mathbf{f}_{\alpha\beta} \tag{9.5}$$

where $\mathbf{f}_{\alpha\beta}$ represents the force on the αth particle due to the βth particle. The total force acting on the αth particle is therefore

$$\mathbf{F}_\alpha = \mathbf{F}_\alpha^{(e)} + \mathbf{f}_\alpha \tag{9.6}$$

Also, according to the weak statement of Newton's Third Law, we have

$$\mathbf{f}_{\alpha\beta} = -\mathbf{f}_{\beta\alpha} \tag{9.1}$$

Newton's Second Law for the αth particle can be written as

$$\dot{\mathbf{p}}_\alpha = m_\alpha \ddot{\mathbf{r}}_\alpha = \mathbf{F}_\alpha^{(e)} + \mathbf{f}_\alpha \tag{9.7}$$

or

$$\frac{d^2}{dt^2}(m_\alpha \mathbf{r}_\alpha) = \mathbf{F}_\alpha^{(e)} + \sum_\beta \mathbf{f}_{\alpha\beta} \tag{9.8}$$

Summing this expression over α, we have

$$\frac{d^2}{dt^2}\sum_\alpha m_\alpha \mathbf{r}_\alpha = \sum_\alpha \mathbf{F}_\alpha^{(e)} + \sum_\alpha \sum_{\substack{\beta \\ \alpha \neq \beta}} \mathbf{f}_{\alpha\beta} \tag{9.9}$$

where the terms $\alpha = \beta$ do not enter in the second sum on the right-hand side, because $\mathbf{f}_{\alpha\alpha} \equiv 0$. The summation on the left-hand side just yields $M\mathbf{R}$ (see Equation 9.3), and the second time derivative is $M\ddot{\mathbf{R}}$. The first term on the right-hand side is the sum of all the external forces and can be written as

$$\sum_\alpha \mathbf{F}_\alpha^{(e)} \equiv \mathbf{F} \tag{9.10}$$

The second term on the right-hand side in Equation 9.9 can be expressed* as

$$\sum_\alpha \sum_{\substack{\beta \\ \alpha \neq \beta}} \mathbf{f}_{\alpha\beta} \equiv \sum_{\alpha,\,\beta \neq \alpha} \mathbf{f}_{\alpha\beta} = \sum_{\alpha < \beta} (\mathbf{f}_{\alpha\beta} + \mathbf{f}_{\beta\alpha})$$

which vanishes[†] according to Equation 9.1. Thus, we have the first important result

$$M\ddot{\mathbf{R}} = \mathbf{F} \tag{9.11}$$

---

* This equation can be verified by explicitly calculating both sides for a single case (e.g., $n = 3$).

[†] The last summation symbol means "sum over all α and β subject to the restrictions $\alpha < \beta$." Note that we can prove the vanishing of

$$\sum_\alpha \sum_{\substack{\beta \\ \alpha \neq \beta}} \mathbf{f}_{\alpha\beta}$$

by appealing to the following argument. Because the summations are carried out over both α and β, these indices are dummies; in particular, we may interchange α and β without affecting the sum. Using the more compact notation, we have

$$\sum_{\alpha,\,\beta \neq \alpha} \mathbf{f}_{\alpha\beta} = \sum_{\beta,\,\alpha \neq \beta} \mathbf{f}_{\beta\alpha}$$

But, by hypothesis, $\mathbf{f}_{\alpha\beta} = -\mathbf{f}_{\beta\alpha}$, so

$$\sum_{\alpha,\,\beta \neq \alpha} \mathbf{f}_{\alpha\beta} = -\sum_{\alpha,\,\beta \neq \alpha} \mathbf{f}_{\alpha\beta}$$

and if a quantity is equal to its negative, it must vanish identically.

which we can express as follows:

I. *The center of mass of a system moves as if it were a single particle, of mass equal to the total mass of the system, acted on by the total external force, and independent of the nature of the internal forces (as long as they follow* $\mathbf{f}_{\alpha\beta} = -\mathbf{f}_{\beta\alpha}$, *the weak form of Newton's Third Law).*

The total linear momentum of the system is

$$\mathbf{P} = \sum_{\alpha} m_{\alpha}\dot{\mathbf{r}}_{\alpha} = \frac{d}{dt}\sum_{\alpha} m_{\alpha}\mathbf{r}_{\alpha} = \frac{d}{dt}(M\mathbf{R}) = M\dot{\mathbf{R}} \qquad (9.12)$$

and

$$\dot{\mathbf{P}} = M\ddot{\mathbf{R}} = \mathbf{F} \qquad (9.13)$$

Thus, the total linear momentum of the system is conserved if there is no external force. From Equations 9.12 and 9.13, we note our second and third important results:

II. *The linear momentum of the system is the same as if a single particle of mass* M *were located at the position of the center of mass and moving in the manner the center of mass moves.*

III. *The total linear momentum for a system free of external forces is constant and equal to the linear momentum of the center of mass (the law of conservation of linear momentum for a system).*

All measurements must be made in an inertial reference system. An example of the linear momentum of a system is given by the explosion of an artillery shell above ground. Because the explosion is an internal effect, the only external force affecting the center of mass velocity is gravity. The center of mass of the artillery shell fragments immediately after the explosion must continue with the velocity of the shell just before the explosion.

EXAMPLE 9.2 ---

A chain of uniform mass density ρ, length b, and mass M ($\rho = M/b$) hangs as shown in Figure 9-4. At time $t = 0$, the ends A and B are adjacent, but end B is released. Find the tension in the chain at point A after end B has fallen a distance x by (a) assuming free fall and (b) by using energy conservation.

Solution: (a) In the case of free fall, let's assume the only forces acting on the system at time t are the tension T acting vertically upward at point A and the gravitational force Mg pulling the chain down. The center of mass momentum reacts to these forces such that

$$\dot{P} = Mg - T \qquad (9.14)$$

The right side of the chain, with mass $\rho(b - x)/2$, is moving at the speed $\dot{x}$, and the left side of the chain is not moving. The total momentum of the system is

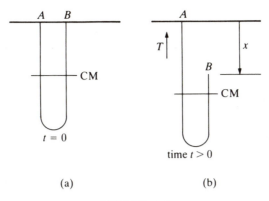

FIGURE 9-4

therefore

$$P = \rho\left(\frac{b-x}{2}\right)\dot{x}$$

and

$$\dot{P} = \frac{\rho}{2}\left[-\dot{x}^2 + \ddot{x}(b-x)\right] \qquad\qquad \textbf{(9.15)}$$

For free fall, we have $x = gt^2/2$, so that

$$\dot{x} = gt = \sqrt{2gx}$$

$$\ddot{x} = g$$

and

$$\dot{P} = \frac{\rho}{2}(gb - 3gx) = Mg - T$$

and finally,

$$T = \frac{Mg}{2}\left(\frac{3x}{b} + 1\right) \qquad\qquad \textbf{(9.16)}$$

(b) Calkin and March (*Am. J. Phys.* **57**, 154[1989]) have found that chains act much like a perfectly flexible, inextensible rope that conserves energy when it falls, with no dissipative mechanisms. We treat the chain as one-dimensional motion, ignoring the small horizontal motion. Let the potential energy U be measured relative to the fixed end of the chain, so that the initial potential energy $U(t=0) = U_0 = -\rho g b^2/4$. A careful geometric construction shows that the potential energy after the chain has dropped a distance x is

$$U = -\frac{1}{4}\rho g(b^2 + 2bx - x^2)$$

The kinetic energy (where we use K instead of T to avoid confusion with tension) is determined from the speed $\dot{x}$ of the right side of the chain, so that

$$K = \frac{\rho}{4}(b - x)\dot{x}^2$$

Because energy is conserved, we must have $K + U = U_0$.

$$\frac{\rho}{4}(b - x)\dot{x}^2 - \frac{1}{4}\rho g\,(b^2 + 2bx - x^2) = -\frac{1}{4}\rho g b^2$$

We solve for $\dot{x}^2$ to obtain

$$\dot{x}^2 = \frac{g(2bx - x^2)}{b - x} \tag{9.17}$$

To find the tension from Equations 9.14 and 9.15, we need to determine $\ddot{x}$. We take the derivative of Equation 9.17 and find

$$\ddot{x} = g + \frac{g(2bx - x^2)}{2(b - x)^2}$$

We now insert $\dot{x}^2$ and $\ddot{x}$ from the two previous equations into Equation 9.15 to determine $\dot{P}$ and insert this value of $\dot{P}$ into Equation 9.14. After collection of terms and solving for T, we obtain

$$T = \frac{Mg}{4b}\frac{1}{(b - x)}(2b^2 + 2bx - 3x^2) \tag{9.18}$$

Note the difference between the two results, Equations 9.16 and 9.18, for the free fall and energy conserving methods. It should be rather easy by experimentation to determine which is correct, because the latter result has the tension rising dramatically ($T \to \infty$) at the end when $x \to b$. Experiments by Calkin and March confirm that the tension does increase rapidly at the end to a maximum of about 25 times the chain's weight, and the observations as a function of x agree well with the calculations. Real chains cannot have an infinite tension.

For the free fall case, the tension in the chain is discontinuous on either side of the bottom bend; the tension is $T_1 = \rho\dot{x}^2/2$ on the fixed side and $T_2 = 0$ on the free side. For the energy conserving case, the tension T_2 on the free side is not zero, and this tension helps gravity pull the chain down. The result is that the chain falls about 15% faster than calculated for the free fall case. For energy-conserving chains, the tension is continuous: $T_1 = T_2 = \rho\dot{x}^2/4$. We examine further properties of the falling chain in the problems.

9.4 ANGULAR MOMENTUM OF THE SYSTEM

It is often more convenient to describe a system by a position vector with respect to the center of mass. The position vector $\mathbf{r}_\alpha$ in the inertial reference system

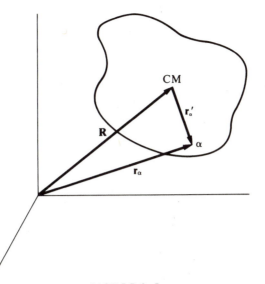

FIGURE 9-5

(see Figure 9-5) becomes

$$\mathbf{r}_\alpha = \mathbf{R} + \mathbf{r}'_\alpha \tag{9.19}$$

where $\mathbf{r}'_\alpha$ is the position vector of the particle α with respect to the center of mass. The angular momentum of the αth particle about the origin is given by Equation 2.81:

$$\mathbf{L}_\alpha = \mathbf{r}_\alpha \times \mathbf{p}_\alpha \tag{9.20}$$

Summing this expression over α, and using Equation 9.19, we have

$$\mathbf{L} = \sum_\alpha \mathbf{L}_\alpha = \sum_\alpha (\mathbf{r}_\alpha \times \mathbf{p}_\alpha) = \sum_\alpha (\mathbf{r}_\alpha \times m_\alpha \dot{\mathbf{r}}_\alpha)$$

$$= \sum_\alpha (\mathbf{r}'_\alpha + \mathbf{R}) \times m_\alpha (\dot{\mathbf{r}}'_\alpha + \dot{\mathbf{R}})$$

$$= \sum_\alpha m_\alpha [(\mathbf{r}'_\alpha \times \dot{\mathbf{r}}'_\alpha) + (\mathbf{r}'_\alpha \times \dot{\mathbf{R}}) + (\mathbf{R} \times \dot{\mathbf{r}}'_\alpha) + (\mathbf{R} \times \dot{\mathbf{R}})] \tag{9.21}$$

The middle two terms can be written as

$$\left(\sum_\alpha m_\alpha \mathbf{r}'_\alpha \right) \times \dot{\mathbf{R}} + \mathbf{R} \times \frac{d}{dt} \left(\sum_\alpha m_\alpha \mathbf{r}'_\alpha \right)$$

which vanishes because

$$\sum_\alpha m_\alpha \mathbf{r}'_\alpha = \sum_\alpha m_\alpha (\mathbf{r}_\alpha - \mathbf{R}) = \sum_\alpha m_\alpha \mathbf{r}_\alpha - \mathbf{R} \sum_\alpha m_\alpha$$

$$\sum_\alpha m_\alpha \mathbf{r}'_\alpha = M\mathbf{R} - M\mathbf{R} \equiv 0 \tag{9.22}$$

This indicates that $\sum_\alpha m_\alpha \mathbf{r}'_\alpha$ specifies the position of the center of mass in the center-of-mass coordinate system and is therefore a null vector. Thus, Equation 9.21 becomes

$$\mathbf{L} = M\mathbf{R} \times \dot{\mathbf{R}} + \sum_\alpha \mathbf{r}'_\alpha \times \mathbf{p}'_\alpha = \mathbf{R} \times \mathbf{P} + \sum_\alpha \mathbf{r}'_\alpha \times \mathbf{p}'_\alpha \qquad (9.23)$$

Our fourth important result is

IV. *The total angular momentum about an origin is the sum of the angular momentum of the center of mass about that origin and the angular momentum of the system about the position of the center of mass.*

The time derivative of the angular momentum of the αth particle is, from Equation 2.83,

$$\dot{\mathbf{L}}_\alpha = \mathbf{r}_\alpha \times \dot{\mathbf{p}}_\alpha \qquad (9.24)$$

and, using Equation 9.13, we have

$$\dot{\mathbf{L}}_\alpha = \mathbf{r}_\alpha \times \left(\mathbf{F}^{(e)}_\alpha + \sum_\beta \mathbf{f}_{\alpha\beta} \right) \qquad (9.25)$$

Summing this expression over α, we have

$$\dot{\mathbf{L}} = \sum_\alpha \dot{\mathbf{L}}_\alpha = \sum_\alpha (\mathbf{r}_\alpha \times \mathbf{F}^{(e)}_\alpha) + \sum_{\alpha, \beta \neq \alpha} (\mathbf{r}_\alpha \times \mathbf{f}_{\alpha\beta}) \qquad (9.26)$$

The last term may be written as

$$\sum_{\alpha, \beta \neq \alpha} (\mathbf{r}_\alpha \times \mathbf{f}_{\alpha\beta}) = \sum_{\alpha < \beta} [(\mathbf{r}_\alpha \times \mathbf{f}_{\alpha\beta}) + (\mathbf{r}_\beta \times \mathbf{f}_{\beta\alpha})]$$

The vector connecting the αth and βth particles (see Figure 9-6) is defined to be

$$\mathbf{r}_{\alpha\beta} \equiv \mathbf{r}_\alpha - \mathbf{r}_\beta \qquad (9.27)$$

and then, because $\mathbf{f}_{\alpha\beta} = -\mathbf{f}_{\beta\alpha}$, we have

$$\sum_{\alpha, \beta \neq \alpha} (\mathbf{r}_\alpha \times \mathbf{f}_{\alpha\beta}) = \sum_{\alpha < \beta} (\mathbf{r}_\alpha - \mathbf{r}_\beta) \times \mathbf{f}_{\alpha\beta}$$

$$= \sum_{\alpha < \beta} (\mathbf{r}_{\alpha\beta} \times \mathbf{f}_{\alpha\beta}) \qquad (9.28)$$

Now we want to limit the discussion to central internal forces and apply the "strong" version of Newton's Third Law. Hence $\mathbf{f}_{\alpha\beta}$ is along the same direction as $\pm\mathbf{r}_{\alpha\beta}$ and

$$\mathbf{r}_{\alpha\beta} \times \mathbf{f}_{\alpha\beta} \equiv 0 \qquad (9.29)$$

and

$$\dot{\mathbf{L}} = \sum_\alpha [\mathbf{r}_\alpha \times \mathbf{F}^{(e)}_\alpha] \qquad (9.30)$$

The right-hand side of this expression is just the sum of all the external torques:

$$\dot{\mathbf{L}} = \sum_\alpha \mathbf{N}^{(e)}_\alpha = \mathbf{N}^{(e)} \qquad (9.31)$$

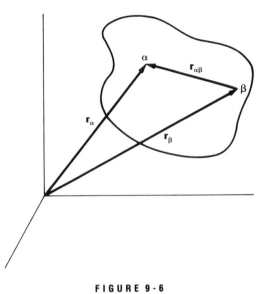

FIGURE 9-6

This leads to our next important result:

V. *If the net resultant external torques about a given axis vanish, then the total angular momentum of the system about that axis remains constant in time.*

Note also that the term

$$\sum_{\beta} \mathbf{r}_\alpha \times \mathbf{f}_{\alpha\beta} \tag{9.32}$$

is the torque on the αth particle due to all the internal forces—that is, it is the *internal torque*. Because the sum of this quantity over all the particles α vanishes (see Equation 9.28),

$$\sum_{\alpha,\ \beta \neq \alpha} (\mathbf{r}_\alpha \times \mathbf{f}_{\alpha\beta}) = \sum_{\alpha < \beta} (\mathbf{r}_{\alpha\beta} \times \mathbf{f}_{\alpha\beta}) = 0 \tag{9.33}$$

the total internal torque must vanish, which we can state as

VI. *The total internal torque must vanish if the internal forces are central— that is, if $\mathbf{f}_{\alpha\beta} = -\mathbf{f}_{\beta\alpha}$, and the angular momentum of an isolated system cannot be altered without the application of external forces.*

EXAMPLE **9.3**

A light string of length a has bobs of mass m_1 and m_2 ($m_2 > m_1$) on its ends. The end with m_1 is held and whirled vigorously by hand above the head and then released. Describe the subsequent motion, and find the tension in the string after release.

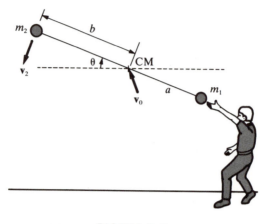

FIGURE 9-7

Solution: The system is shown in Figure 9-7. The center of mass is a distance b = $[m_1/(m_1 + m_2)]a$ from mass m_2. After being released, the only forces on the system are the gravitational forces on m_1 and m_2. Assume that v_0 is the initial velocity of the center of mass CM. The CM will continue in a parabolic path under the influence of gravity as if all the mass ($m_1 + m_2$) were concentrated at the CM. But when released, mass m_2 is rotating around m_1 rapidly. Because no external torque exists, the system will continue to rotate. But now both m_1 and m_2 rotate about the CM, and the angular momentum is conserved. If mass m_2 is traveling with the linear velocity v_2 when released, then we must have $v_2 = b\dot{\theta}$ [similarly, $v_1 = (a - b)\dot{\theta}$]. The tension in the string is, however, due to the centrifugal reaction of the masses rotating, which is, in this case,

$$\text{Centrifugal force} = \frac{m_2(b\dot{\theta})^2}{b} = \text{Tension}$$

$$\text{Tension} = m_2 b\dot{\theta}^2 = m_2\left(\frac{m_1 a}{m_1 + m_2}\right)\dot{\theta}^2 = \frac{m_1 m_2 a\dot{\theta}^2}{m_1 + m_2}$$

9.5 ENERGY OF THE SYSTEM

The final conservation theorem, that of energy, may be derived for a system of particles as follows. Consider the work done on the system in moving it from a Configuration 1, in which all the coordinates r_α are specified, to a Configuration 2, in which the coordinates r_α have some different specification. (Note that the individual particles may just be rearranged in such a process, and that, for example, the position of the center of mass could remain stationary.) In analogy with Equation 2.84, we write

$$W_{12} = \sum_\alpha \int_1^2 \mathbf{F}_\alpha \cdot d\mathbf{r}_\alpha \tag{9.34}$$

where $\mathbf{F}_\alpha$ is the net resultant force acting on particle α. Using a procedure similar to that used to obtain Equation 2.86, we have

$$W_{12} = \sum_\alpha \int_1^2 d(\tfrac{1}{2}m_\alpha v_\alpha^2) = T_2 - T_1 \tag{9.35}$$

where

$$T = \sum_\alpha T_\alpha = \sum_\alpha \tfrac{1}{2}m_\alpha v_\alpha^2 \tag{9.36}$$

Using the relation (see Equation 9.19)

$$\dot{\mathbf{r}}_\alpha = \dot{\mathbf{r}}'_\alpha + \dot{\mathbf{R}} \tag{9.37}$$

we have

$$\begin{aligned}
\dot{\mathbf{r}}_\alpha \cdot \dot{\mathbf{r}}_\alpha = v_\alpha^2 &= (\dot{\mathbf{r}}'_\alpha + \dot{\mathbf{R}}) \cdot (\dot{\mathbf{r}}'_\alpha + \dot{\mathbf{R}}) \\
&= (\dot{\mathbf{r}}'_\alpha \cdot \dot{\mathbf{r}}'_\alpha) + 2(\dot{\mathbf{r}}'_\alpha \cdot \dot{\mathbf{R}}) + (\dot{\mathbf{R}} \cdot \dot{\mathbf{R}}) \\
&= v_\alpha'^2 + 2(\dot{\mathbf{r}}'_\alpha \cdot \dot{\mathbf{R}}) + V^2
\end{aligned}$$

where $\mathbf{v}' \equiv \dot{\mathbf{r}}'$ and where V is the velocity of the center of mass. Then

$$\begin{aligned}
T &= \sum_\alpha \tfrac{1}{2}m_\alpha v_\alpha^2 \\
&= \sum_\alpha \tfrac{1}{2}m_\alpha v_\alpha'^2 + \sum_\alpha \tfrac{1}{2}m_\alpha V^2 + \dot{\mathbf{R}} \cdot \frac{d}{dt} \sum_\alpha m_\alpha \mathbf{r}'_\alpha
\end{aligned} \tag{9.38}$$

But, by a previous argument, $\sum_\alpha m_\alpha \mathbf{r}'_\alpha = 0$, and the last term vanishes. Thus,

$$\boxed{T = \sum_\alpha \tfrac{1}{2}m_\alpha v_\alpha'^2 + \tfrac{1}{2}MV^2} \tag{9.39}$$

which can be stated:

VII. *The total kinetic energy of the system is equal to the sum of the kinetic energy of a particle of mass M moving with the velocity of the center of mass and the kinetic energy of motion of the individual particles relative to the center of mass.*

Next, the total force in Equation 9.34 can be separated as in Equation 9.6:

$$W_{12} = \sum_\alpha \int_1^2 \mathbf{F}_\alpha^{(e)} \cdot d\mathbf{r}_\alpha + \sum_{\alpha,\,\beta \neq \alpha} \int_1^2 \mathbf{f}_{\alpha\beta} \cdot d\mathbf{r}_\alpha \tag{9.40}$$

If the forces $\mathbf{F}_\alpha^{(e)}$ and $\mathbf{f}_{\alpha\beta}$ are conservative, then they are derivable from potential functions, and we can write

$$\left.\begin{aligned}
\mathbf{F}_\alpha^{(e)} &= -\nabla_\alpha U_\alpha \\
\mathbf{f}_{\alpha\beta} &= -\nabla_\alpha \bar{U}_{\alpha\beta}
\end{aligned}\right\} \tag{9.41}$$

where U_α and $\bar{U}_{\alpha\beta}$ are the potential functions but which do not necessarily have the same form. The notation ∇_α means that the gradient operation is performed with respect to the coordinates of the αth particle.

The first term in Equation 9.40 becomes

$$\sum_\alpha \int_1^2 \mathbf{F}_\alpha^{(e)} \cdot d\mathbf{r}_\alpha = -\sum_\alpha \int_1^2 (\nabla_\alpha U_\alpha) \cdot d\mathbf{r}_\alpha$$

$$= -\sum_\alpha U_\alpha \Big|_1^2 \qquad (9.42)$$

The second term* in Equation 9.40 is

$$\sum_{\alpha,\,\beta\neq\alpha} \int_1^2 \mathbf{f}_{\alpha\beta} \cdot d\mathbf{r}_\alpha = \sum_{\alpha<\beta} \int_1^2 (\mathbf{f}_{\alpha\beta} \cdot d\mathbf{r}_\alpha + \mathbf{f}_{\beta\alpha} \cdot d\mathbf{r}_\beta)$$

$$= \sum_{\alpha<\beta} \int_1^2 \mathbf{f}_{\alpha\beta} \cdot (d\mathbf{r}_\alpha - d\mathbf{r}_\beta) = \sum_{\alpha<\beta} \int_1^2 \mathbf{f}_{\alpha\beta} \cdot d\mathbf{r}_{\alpha\beta} \qquad (9.43)$$

where, following the definition in Equation 9.27, $d\mathbf{r}_{\alpha\beta} = d\mathbf{r}_\alpha - d\mathbf{r}_\beta$.

Because $\bar{U}_{\alpha\beta}$ is a function only of the distance between m_α and m_β, it therefore depends on six quantities—that is, the three coordinates of m_α (the $x_{\alpha,i}$) and the three coordinates of m_β (the $x_{\beta,i}$). The total derivative of $\bar{U}_{\alpha\beta}$ is therefore the sum of six partial derivatives and is given by

$$d\bar{U}_{\alpha\beta} = \sum_i \left(\frac{\partial \bar{U}_{\alpha\beta}}{\partial x_{\alpha,i}} dx_{\alpha,i} + \frac{\partial \bar{U}_{\alpha\beta}}{\partial x_{\beta,i}} dx_{\beta,i} \right) \qquad (9.44)$$

where the $x_{\beta,i}$ are held constant in the first term and the $x_{\alpha,i}$ are held constant in the second. Thus,

$$d\bar{U}_{\alpha\beta} = (\nabla_\alpha \bar{U}_{\alpha\beta}) \cdot d\mathbf{r}_\alpha + (\nabla_\beta \bar{U}_{\alpha\beta}) \cdot d\mathbf{r}_\beta \qquad (9.45)$$

Now

$$\nabla_\alpha \bar{U}_{\alpha\beta} = -\mathbf{f}_{\alpha\beta} \qquad (9.46)$$

but $\bar{U}_{\alpha\beta} = \bar{U}_{\beta\alpha}$, so

$$\nabla_\beta \bar{U}_{\alpha\beta} = \nabla_\beta \bar{U}_{\beta\alpha} = -\mathbf{f}_{\beta\alpha} = \mathbf{f}_{\alpha\beta} \qquad (9.47)$$

Therefore,

$$d\bar{U}_{\alpha\beta} = -\mathbf{f}_{\alpha\beta} \cdot (d\mathbf{r}_\alpha - d\mathbf{r}_\beta)$$

$$= -\mathbf{f}_{\alpha\beta} \cdot d\mathbf{r}_{\alpha\beta} \qquad (9.48)$$

---

* Note that, unlike the term $\sum_{\alpha,\,\beta\neq\alpha} \mathbf{f}_{\alpha\beta}$ that appears in Equation 9.9, the term

$$\sum_{\alpha,\,\beta\neq\alpha} \int_1^2 \mathbf{f}_{\alpha\beta} \cdot d\mathbf{r}_\alpha$$

is *not* antisymmetric in α and β and therefore does not, in general, vanish.

Using this result in Equation 9.43, we have

$$\sum_{\alpha,\ \beta \neq \alpha} \int_1^2 \mathbf{f}_{\alpha\beta} \cdot d\mathbf{r}_\alpha = -\sum_{\alpha<\beta} \int_1^2 d\bar{U}_{\alpha\beta} = -\sum_{\alpha<\beta} \bar{U}_{\alpha\beta} \Big|_1^2 \qquad (9.49)$$

Combining Equations 9.42 and 9.49 to evaluate W_{12} in Equation 9.40, we find

$$W_{12} = -\sum_\alpha U_\alpha \Big|_1^2 - \sum_{\alpha<\beta} \bar{U}_{\alpha\beta} \Big|_1^2 \qquad (9.50)$$

We obtained this equation assuming that both the external and internal forces were derivable from potentials. In such a case, the *total potential energy* (both internal and external) for the system can be written as

$$U = \sum_\alpha U_\alpha + \sum_{\alpha<\beta} \bar{U}_{\alpha\beta} \qquad (9.51)$$

Then,

$$W_{12} = -U \Big|_1^2 = U_1 - U_2 \qquad (9.52)$$

Combining this result with Equation 9.35, we have

$$T_2 - T_1 = U_1 - U_2$$

or

$$T_1 + U_1 = T_2 + U_2$$

so that

$$\boxed{E_1 = E_2} \qquad (9.53)$$

which expresses the conservation of energy for the system. This result is valid for a system in which all the forces are derivable from potentials that do not depend explicitly on the time; we say that such a system is *conservative*.

VIII. *The total energy for a conservative system is constant.*

In Equation 9.51, the term

$$\sum_{\alpha<\beta} \bar{U}_{\alpha\beta}$$

represents the **internal potential energy** of the system. If the system is a rigid body with the constituent particles restrained to maintain their relative positions, then, in any process involving the body, the internal potential energy remains constant. In such a case, the internal potential energy can be ignored when computing the total potential energy of the system. This amounts simply to redefining the position of zero potential energy, but this position is arbitrarily chosen anyway; that is, it is only the *difference* in potential energy that is physically significant. The absolute value of the potential energy is an arbitrary quantity.

EXAMPLE 9.4 --

A projectile of mass M explodes while in flight into three fragments (Figure 9-8). One mass ($m_1 = M/2$) travels in the original direction of the projectile, mass m_2 ($= M/6$) travels in the opposite direction, and mass m_3 ($= M/3$) comes to rest. The energy E released in the explosion is equal to five times the projectile's kinetic energy at explosion. What are the velocities?

Solution: Let the velocity of the projectile of mass M be **v**. The three fragments have the following masses and velocities:

$$m_1 = \frac{M}{2}, \quad \mathbf{v}_1 = k_1\mathbf{v} \qquad \text{Forward direction, } k_1 > 0$$

$$m_2 = \frac{M}{6}, \quad \mathbf{v}_2 = -k_2\mathbf{v} \qquad \text{Opposite direction, } k_2 > 0$$

$$m_3 = \frac{M}{3}, \quad \mathbf{v}_3 = 0 \qquad \text{At rest}$$

The conservation of linear momentum and energy give

$$Mv = \frac{M}{2}k_1v - \frac{M}{6}k_2v \tag{9.54}$$

$$E + \frac{1}{2}Mv^2 = \frac{1}{2}\frac{M}{2}(k_1v)^2 + \frac{1}{2}\frac{M}{6}(k_2v)^2 \tag{9.55}$$

From Equation 9.54, $k_2 = 3k_1 - 6$, which we can insert into Equation 9.55:

$$5\left(\frac{1}{2}Mv^2\right) + \frac{1}{2}Mv^2 = \frac{Mv^2}{4}k_1^2 + \frac{Mv^2}{12}(3k_1 - 6)^2$$

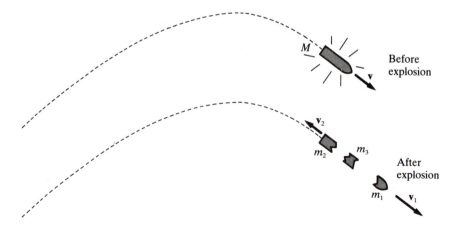

Before explosion

$\mathbf{v}_2$

m_2 m_3

After explosion

m_1 $\mathbf{v}_1$

FIGURE 9-8

which reduces to $k_1^2 - 3k_1 = 0$, giving the results $k_1 = 0$ and $k_1 = 3$. For $k_1 = 0$, the value of $k_2 = -6$, which is inconsistent with $k_2 > 0$. For $k_1 = 3$, the value of $k_2 = 3$. The velocities become

$$\mathbf{v}_1 = 3\mathbf{v}$$

$$\mathbf{v}_2 = -3\mathbf{v}$$

$$\mathbf{v}_3 = 0$$

●

EXAMPLE 9.5

A rope of uniform linear density ρ and mass m is wrapped one complete turn around a hollow cylinder of mass M and radius R. The cylinder rotates freely about its axis as the rope unwraps (Figure 9-9). The rope ends are at $x = 0$ (one fixed, one loose) when point P is at $\theta = 0$, and the system is slightly displaced from equilibrium at rest. Find the angular velocity of the cylinder.

Solution: Gravity has done work on the system to unwind the rope. Consider a section dx of the rope located a distance x from where it unwinds. The mass of this section is $\rho\, dx$. If we were to perform work by reaching up and wrapping this loose end of the rope against the cylinder, how far up would the section dx actually travel? The distance x would be on the circumference of the cylinder (see Figure 9-9), and dx would be $R \sin (x/R)$ below $x = 0$. The total distance the section dx would move up is

$$\text{Distance } dx \text{ moves } = x - R \sin\left(\frac{x}{R}\right)$$

$$\text{Work done} = (\rho\, dx)g\left[x - R \sin\left(\frac{x}{R}\right)\right] \qquad \textbf{(9.56)}$$

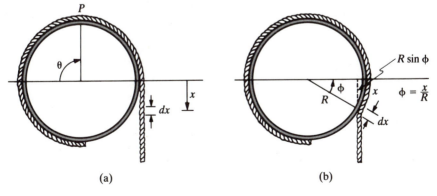

(a)   (b)

FIGURE 9-9

The total work done by gravity in unwrapping the rope is, therefore,

$$W = \int_0^{R\theta} \rho g \left[x - R \sin \left(\frac{x}{R} \right) \right] dx$$

$$W = \rho g R^2 \left(\frac{\theta^2}{2} + \cos \theta - 1 \right) \tag{9.57}$$

The work done by gravity must equal the kinetic energy gained by the rope and the cylinder.

$$T = \frac{1}{2} m (R\dot\theta)^2 + \frac{1}{2} M (R\dot\theta)^2 \tag{9.58}$$

Because $W = T$ and $\rho = m/(2\pi R)$,

$$\frac{mgR}{2\pi} \left[\frac{\theta^2}{2} + \cos \theta - 1 \right] = \frac{1}{2} (m + M) R^2 \dot\theta^2$$

and

$$\dot\theta^2 = \frac{mg(\theta^2 + 2 \cos \theta - 2)}{2\pi R(m + M)} \tag{9.59}$$

9.6 ELASTIC COLLISIONS OF TWO PARTICLES

In the rest of the chapter, we apply the conservation laws to the interaction of two particles. When two particles interact, the motion of one particle relative to the other is governed by the force law that describes the interaction. This interaction may result from actual contact, as in the collision of two billiard balls, or the interaction may take place through the intermediary of a force field. For example, a *free* object (i.e., one not bound in a solar orbit) may "scatter" from the sun by a gravitational interaction, or an α-particle may be scattered by the electric field of an atomic nucleus. We demonstrated in the previous chapter that once the force law is known, the two-body problem can be completely solved. But even if the force of interaction between two particles is not known, a great deal can still be learned about the relative motion by using only the results of the conservation of momentum and energy. Thus, if the initial state of the system is known (i.e., if the velocity vector of each of the particles is specified), the conservation laws allow us to obtain information regarding the velocity vectors in the final state.*

On the basis of the conservation theorems alone, it is not possible to predict, for example, the angle between the initial and final velocity vectors of one of the

---

* The "initial state" of the system is the condition of the particles when they are not yet sufficiently close to interact appreciably; the "final state" is the condition after the interaction has taken place. For a contact interaction, these conditions are obvious. But if the interaction takes place by a force field, then the rate of decrease of the force with distance must be taken into account in specifying the initial and final states.

particles; knowledge of the force law is required for such details. In this section and the next, we derive those relationships that require only the conservation of momentum and energy. Then, we examine the features of the collision process, which demand that the force law be specified. We limit our discussion primarily to elastic collisions, because the essential features of two-particle kinematics are adequately demonstrated by elastic collisions. The results obtained under the assumption only of momentum and energy conservation are valid (in the nonrelativistic velocity region) even for quantum mechanical systems, because these conservation theorems are applicable to quantum as well as to classical systems.

We demonstrated on several occasions that the description of many physical processes is considerably simplified if one chooses coordinate systems at rest with respect to the system's center of mass. In the problem we now discuss—the elastic collision of two particles—the usual situation (and the one to which we confine our attention) is one in which the collision is between a moving particle and a particle at rest.* Although it is indeed simpler to describe the effects of the collision in a coordinate system in which the *center of mass is at rest,* the actual measurements are made in the **laboratory coordinate system** in which the observer is at rest. In this system, one of the particles is normally moving, and the struck particle is normally at rest. We here refer to these two coordinate systems simply as the **CM** and the **LAB** systems.

We wish to take advantage of the simplifications that result by describing an elastic collision in the CM system. It is therefore necessary to derive the equations connecting the CM and LAB systems.

We use the following notation:

$$\begin{matrix} m_1 = \\ m_2 = \end{matrix} \text{Mass of the } \begin{Bmatrix} \text{moving} \\ \text{struck} \end{Bmatrix} \text{ particle}$$

In general, primed quantities refer to the CM system:

$$\begin{matrix} \mathbf{u}_1 = \text{Initial} \\ \mathbf{v}_1 = \text{Final} \end{matrix} \Bigg\} \text{ velocity of } m_1 \text{ in the LAB system}$$

$$\begin{matrix} \mathbf{u}_1' = \text{Initial} \\ \mathbf{v}_1' = \text{Final} \end{matrix} \Bigg\} \text{ velocity of } m_1 \text{ in the CM system}$$

and similarly for $\mathbf{u}_2$, $\mathbf{v}_2$, $\mathbf{u}_2'$, and $\mathbf{v}_2'$ (but $\mathbf{u}_2 = 0$):

$$\begin{matrix} T_0 = \\ T_0' = \end{matrix} \text{Total initial kinetic energy in } \begin{Bmatrix} \text{LAB} \\ \text{CM} \end{Bmatrix} \text{ system}$$

$$\begin{matrix} T_1 = \\ T_1' = \end{matrix} \text{Final kinetic energy of } m_1 \text{ in } \begin{Bmatrix} \text{LAB} \\ \text{CM} \end{Bmatrix} \text{ system}$$

---

* A collision is elastic if no change in the internal energy of the particles results; thus, the conservation of energy may be applied without regard to the internal energy. Notice that heat may be generated when two mechanical bodies collide inelastically. Heat is just a manifestation of the agitation of a body's constituent particles and may therefore be considered a part of the internal energy. The laws governing the elastic collision of two bodies were first investigated by John Wallis (1668), Wren (1668), and Huygens (1669).

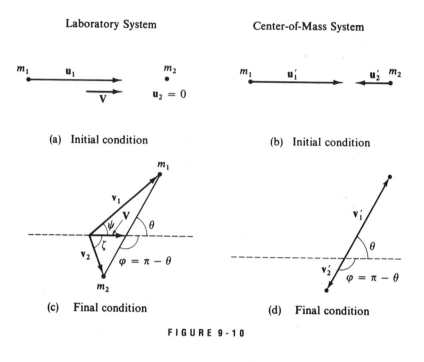

Laboratory System

Center-of-Mass System

(a) Initial condition

(b) Initial condition

(c) Final condition

(d) Final condition

FIGURE 9-10

and similarly for T_2 and T_2',

$\mathbf{V}$ = velocity of the center of mass in the LAB system

ψ = angle through which m_1 is deflected in the LAB system

ζ = angle through which m_2 is deflected in the LAB system

θ = angle through which m_1 and m_2 are deflected in the CM system

Figure 9-10 illustrates the geometry of an elastic collision* in both the LAB and CM systems. The final state in the LAB and CM systems for the scattered particle m_1 may be conveniently summarized by the diagrams in Figure 9-11. We can interpret these diagrams in the following manner. To the velocity $\mathbf{V}$ of the CM, we can add the final CM velocity $\mathbf{v}_1'$ of the scattered particle. Depending on the angle θ at which the scattering takes place, the possible vectors $\mathbf{v}_1'$ lie on the circle of radius v_1' whose center is at the terminus of the vector $\mathbf{V}$. The LAB velocity $\mathbf{v}_1$ and LAB scattering angle ψ are then obtained by connecting the point of origin of $\mathbf{V}$ with the terminus of $\mathbf{v}_1'$.

If $V < v_1'$, only one possible relationship exists between $\mathbf{V}$, $\mathbf{v}_1$, $\mathbf{v}_1'$ and θ (see Figure 9-11a). But if $V > v_1'$, then for every set $\mathbf{V}$, $\mathbf{v}_1'$, there exist two possible

---

* We assume throughout that the scattering is axially symmetric so that no azimuthal angle need be introduced. However, axial symmetry is not always found in scattering problems; this is particularly true in certain quantum mechanical systems.

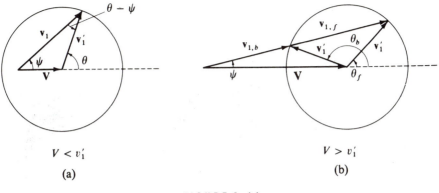

$$V < v_1'$$

(a)

$$V > v_1'$$

(b)

FIGURE 9-11

scattering angles and laboratory velocities: $\mathbf{v}_{1,b}$, θ_b and $\mathbf{v}_{1,f}$, θ_f (see Figure 9-11b), where the designations b and f stand for *backward* and *forward*. This situation results from the fact that if the final CM velocity $\mathbf{v}_1'$ is insufficient to overcome the velocity $\mathbf{V}$ of the center of mass, then, even if m_1 is scattered into the backward direction in the CM system ($\theta > \pi/2$), the particle will appear at a forward angle in the LAB system ($\psi < \pi/2$). Thus, for $V > v_1'$, the velocity $\mathbf{v}_1$ in the LAB system is a double-valued function of $\mathbf{v}_1'$. In an experiment, we usually measure ψ, not the velocity *vector* $\mathbf{v}_1$, so that a single value of ψ can correspond to two different values of θ. Note, however, that a specification of the vectors $\mathbf{V}$ and $\mathbf{v}_1'$ always leads to a unique combination $\mathbf{v}_1$, θ; but a specification of $\mathbf{V}$ and only the *direction* of $\mathbf{v}_1$ (i.e., ψ) allows the possibility of two final vectors, $\mathbf{v}_{1,b}$ and $\mathbf{v}_{1,f}$, if $V > v_1'$.

Having given a qualitative description of the scattering process, we now obtain some of the equations relating the various quantities.

According to the definition of the center of mass (Equation 9.3), we have

$$m_1\mathbf{r}_1 + m_2\mathbf{r}_2 = M\mathbf{R} \tag{9.60}$$

Differentiating with respect to the time, we find

$$m_1\mathbf{u}_1 + m_2\mathbf{u}_2 = M\mathbf{V} \tag{9.61}$$

But $\mathbf{u}_2 = 0$ and $M = m_1 + m_2$; the center of mass must therefore be moving (in the LAB system) toward m_2 with a velocity

$$\mathbf{V} = \frac{m_1\mathbf{u}_1}{m_1 + m_2} \tag{9.62}$$

By the same reasoning, because m_2 is initially at rest, the initial CM speed of m_2 must just equal V:

$$u_2' = V = \frac{m_1 u_1}{m_1 + m_2} \tag{9.63}$$

Note, however, that the motion and the velocities are opposite in direction and that vectorially $\mathbf{u}_2' = -\mathbf{V}$.

The great advantage of using the CM coordinate system is because the total linear momentum in such a system is zero, so that before the collision the particles move directly toward each other and after the collision they move in exactly opposite directions. If the collision is elastic, as we have specified, then the masses do not change, and the conservation of linear momentum and kinetic energy is sufficient to provide that the CM speeds before and after collision are equal:

$$u_1' = v_1', \qquad u_2' = v_2' \qquad (9.64)$$

Term u_1 is the *relative* speed of the two particles in either the CM or the LAB system, $u_1 = u_1' + u_2'$. We therefore have, for the final CM speeds,

$$v_2' = \frac{m_1 u_1}{m_1 + m_2} \qquad (9.65a)$$

$$v_1' = u_1 - u_2' = \frac{m_2 u_1}{m_1 + m_2} \qquad (9.65b)$$

We have (see Figure 9-11a)

$$v_1' \sin \theta = v_1 \sin \psi \qquad (9.66a)$$

and

$$v_1' \cos \theta + V = v_1 \cos \psi \qquad (9.66b)$$

Dividing Equation 9.66a by Equation 9.66b,

$$\tan \psi = \frac{v_1' \sin \theta}{v_1' \cos \theta + V} = \frac{\sin \theta}{\cos \theta + (V/v_1')} \qquad (9.67)$$

According to Equations 9.62 and 9.65b, V/v_1' is given by

$$\frac{V}{v_1'} = \frac{m_1 u_1/(m_1 + m_2)}{m_2 u_1/(m_1 + m_2)} = \frac{m_1}{m_2} \qquad (9.68)$$

Thus, the ratio m_1/m_2 governs whether Figure 9-11a or Figure 9-11b describes the scattering process:

Figure 9-11a: $V < v_1'$, $\quad m_1 < m_2$
Figure 9-11b: $V > v_1'$, $\quad m_1 > m_2$

If we combine Equations 9.67 and 9.68 and write

$$\tan \psi = \frac{\sin \theta}{\cos \theta + (m_1/m_2)} \qquad (9.69)$$

we see that if $m_1 \ll m_2$, the LAB and CM scattering angles are approximately equal; that is, the particle m_2 is but little affected by the collision with m_1 and acts essentially as a fixed scattering center. Thus

$$\psi \cong \theta, \qquad m_1 \ll m_2 \qquad (9.70)$$

However, if $m_1 = m_2$, then

$$\tan \psi = \frac{\sin \theta}{\cos \theta + 1} = \tan \frac{\theta}{2}$$

so that

$$\boxed{\psi = \frac{\theta}{2}, \quad m_1 = m_2} \tag{9.71}$$

and the LAB scattering angle is one half the CM scattering angle. Because the maximum value of θ is 180°, Equation 9.71 indicates that for $m_1 = m_2$ there can be no scattering in the LAB system at angles greater than 90°.

Let us now refer to Figure 9-10c and construct a diagram for the recoil particle m_2 similar to Figure 9-11a. The situation is illustrated in Figure 9-12, from which we find

$$v_2 \sin \zeta = v_2' \sin \theta \tag{9.72a}$$

$$v_2 \cos \zeta = V - v_2' \cos \theta \tag{9.72b}$$

Dividing Equation 9.72a by Equation 9.72b, we have

$$\tan \zeta = \frac{v_2' \sin \theta}{V - v_2' \cos \theta} = \frac{\sin \theta}{(V/v_2') - \cos \theta}$$

But, according to Equations 9.63 and 9.65a, V and v_2' are equal. Therefore,

$$\tan \zeta = \frac{\sin \theta}{1 - \cos \theta} = \cot \frac{\theta}{2} \tag{9.73}$$

which we may write as

$$\tan \zeta = \tan \left(\frac{\pi}{2} - \frac{\theta}{2} \right)$$

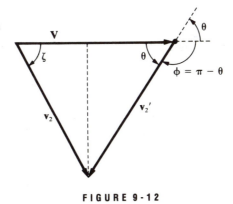

FIGURE 9-12

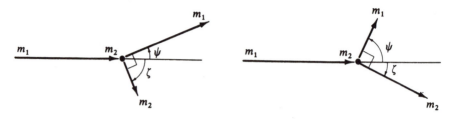

FIGURE 9-13

Thus,

$$2\zeta = \pi - \theta = \phi \tag{9.74}$$

For particles with equal mass, $m_1 = m_2$, we have $\theta = 2\psi$. Combining this result in Equation 9.74, we have

$$\boxed{\zeta + \psi = \frac{\pi}{2}, \qquad m_1 = m_2} \tag{9.75}$$

Hence, the scattering of particles of equal mass always produces a final state in which the velocity vectors of the particles are at right angles if one of the particles is initially at rest (see Figure 9-13).*

EXAMPLE 9.6 ---

What is the maximum angle that ψ can attain for the case $V > v_1'$? What is ψ_{max} for $m_1 \gg m_2$ and $m_1 = m_2$?

Solution: For the case of ψ_{max}, Figure 9-11b becomes as shown in Figure 9-14. The angle between v_1' and v_1 is 90° for ψ to be a maximum.

$$\sin \psi_{max} = \frac{v_1'}{V} \tag{9.76}$$

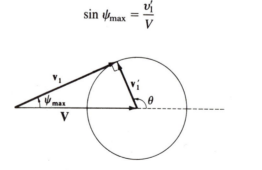

FIGURE 9-14

---

* This result is valid only in the nonrelativistic limit; see Equation 14.131 for the relativistic expression governing this case.

According to Equation 9.68, this is just

$$\sin \psi_{\max} = \frac{m_2}{m_1}$$

from which

$$\psi_{\max} = \sin^{-1}\left(\frac{m_2}{m_1}\right) \tag{9.77}$$

For $m_1 \gg m_2$, $\psi_{\max} = 0$ (no scattering), and for $m_1 = m_2$, $\psi_{\max} = 90°$. Generally, for $m_1 > m_2$, no scattering of m_1 backward of $90°$ can occur.

9.7 KINEMATICS OF ELASTIC COLLISIONS

Relationships involving the energies of the particles may be obtained as follows. First, we have simply

$$T_0 = \tfrac{1}{2} m_1 u_1^2 \tag{9.78}$$

and, in the CM system,

$$T_0' = \tfrac{1}{2}(m_1 u_1'^2 + m_2 u_2'^2)$$

which, on using Equations 9.65a and 9.65b, becomes

$$T_0' = \frac{1}{2} \frac{m_1 m_2}{m_1 + m_2} u_1^2 = \frac{m_2}{m_1 + m_2} T_0 \tag{9.79}$$

This result shows that the initial kinetic energy in the CM system T_0' is always a fraction $m_2/(m_1 + m_2) < 1$ of the initial LAB energy. For the final CM energies, we find

$$T_1' = \frac{1}{2} m_1 v_1'^2 = \frac{1}{2} m_1 \left(\frac{m_2}{m_1 + m_2}\right)^2 u_1^2 = \left(\frac{m_2}{m_1 + m_2}\right)^2 T_0 \tag{9.80}$$

and

$$T_2' = \frac{1}{2} m_2 v_2'^2 = \frac{1}{2} m_2 \left(\frac{m_1}{m_1 + m_2}\right)^2 u_1^2 = \frac{m_1 m_2}{(m_1 + m_2)^2} T_0 \tag{9.81}$$

To obtain T_1 in terms of T_0, we write

$$\frac{T_1}{T_0} = \frac{\tfrac{1}{2} m_1 v_1^2}{\tfrac{1}{2} m_1 u_1^2} = \frac{v_1^2}{u_1^2} \tag{9.82}$$

Referring to Figure 9-11a and using the cosine law, we can write

$$v_1'^2 = v_1^2 + V^2 - 2v_1 V \cos \psi$$

or

$$\frac{T_1}{T_0} = \frac{v_1^2}{u_1^2} = \frac{v_1'^2}{u_1^2} - \frac{V^2}{u_1^2} + 2\frac{v_1 V}{u_1^2}\cos\psi \qquad (9.83)$$

From the previous definitions, we have

$$\frac{v_1'}{u_1} = \frac{m_2}{m_1 + m_2} \qquad \text{and} \qquad \frac{V}{u_1} = \frac{m_1}{m_1 + m_2} \qquad (9.84)$$

The squares of these quantities give the desired expressions for the first two terms on the right-hand side of Equation 9.83. To evaluate the third term, we write, using Equation 9.66a.

$$2\frac{v_1 V}{u_1^2}\cos\psi = 2\left(v_1'\frac{\sin\theta}{\sin\psi}\right)\cdot\frac{V}{u_1^2}\cos\psi \qquad (9.85)$$

The quantity $v_1'V/u_1^2$ can be obtained from the product of the equations in Equation 9.84, and using Equation 9.69, we have

$$\frac{\sin\theta\cos\psi}{\sin\psi} = \frac{\sin\theta}{\tan\psi} = \cos\theta + \frac{m_1}{m_2}$$

so that

$$2\frac{v_1 V}{u_1^2}\cos\psi = \frac{2m_1 m_2}{(m_1 + m_2)^2}\left(\cos\theta + \frac{m_1}{m_2}\right) \qquad (9.86)$$

Substituting Equations 9.84 and 9.86 into Equation 9.83, we obtain

$$\frac{T_1}{T_0} = \left(\frac{m_2}{m_1 + m_2}\right)^2 - \left(\frac{m_1}{m_1 + m_2}\right)^2 + \frac{2m_1 m_2}{(m_1 + m_2)^2}\left(\cos\theta + \frac{m_1}{m_2}\right)$$

which simplifies to

$$\frac{T_1}{T_0} = 1 - \frac{2m_1 m_2}{(m_1 + m_2)^2}(1 - \cos\theta) \qquad (9.87a)$$

Similarly, we can also obtain the ratio T_1/T_0 in terms of the LAB scattering angle ψ:

$$\frac{T_1}{T_0} = \frac{m_1^2}{(m_1 + m_2)^2}\left[\cos\psi \pm \sqrt{\left(\frac{m_2}{m_1}\right)^2 - \sin^2\psi}\right]^2 \qquad (9.87b)$$

where the plus ($+$) sign for the radical is to be taken unless $m_1 > m_2$—in which case the result is double-valued, and Equation 9.77 specifies the maximum value allowed for ψ.

The LAB energy of the recoil particle m_2 can be calculated from

$$\frac{T_2}{T_0} = 1 - \frac{T_1}{T_0} = \frac{4m_1 m_2}{(m_1 + m_2)^2}\cos^2\zeta, \qquad \zeta \leq \pi/2 \qquad (9.88)$$

If $m_1 = m_2$, we have the simple relation

$$\boxed{\frac{T_1}{T_0} = \cos^2 \psi, \qquad m_1 = m_2}$$

(9.89a)

with the restriction noted in the discussion following Equation 9.71 that $\psi \leq 90°$. Also,

$$\boxed{\frac{T_2}{T_0} = \sin^2 \psi, \qquad m_1 = m_2}$$

(9.89b)

Several further relationships are

$$\sin \zeta = \sqrt{\frac{m_1 T_1}{m_2 T_2}} \sin \psi$$

(9.90)

$$\tan \psi = \frac{\sin 2\zeta}{(m_1/m_2) - \cos 2\zeta}$$

(9.91)

$$\tan \psi = \frac{\sin \phi}{(m_1/m_2) - \cos \phi}$$

(9.92)

As an example of applying the kinematic relations we have derived, consider the following situation. Suppose that we have a beam of projectiles, all with mass m_1 and energy T_0. We direct this beam toward a target consisting of a group of particles whose masses m_2 may not all be the same. Some of the incident particles interact with the target particles and are scattered. The incident particles all move in the same direction in a beam of small cross-sectional area, and we assume that the target particles are localized in space so that the scattered particles emerge from a small region. If we position a detector at, say, 90° to the incident beam and with this detector measure the energies of the scattered particles, we can display the results as in the lower portion of Figure 9-15. This graph is a **histogram** that plots the number of particles detected within a range of energy ΔT at the energy T. This particle histogram shows that three energy groups were observed in the particles detected at $\psi = 90°$. The upper portion of the figure shows a curve giving the scattered energy T_1 in terms of T_0 as a function of the mass ratio m_2/m_1 (Equation 9.87b). The curve can be used to determine the mass m_2 of the particle from which one of the incident particles was scattered to fall into one of the three energy groups. Thus, the energy group with $T_1 \cong 0.8T_0$ results from the scattering by target particles with mass $m_2 = 10m_1$, and the other two groups result from target masses $5m_1$ and $2m_1$.

The measurement of the energies of scattered particles is therefore a method of *qualitative analysis* of the target material. Indeed, this method is useful in practice when the incident beam consists of particles (protons, say) that have been given high velocities in an accelerator of some sort. If the detector is capable of precise

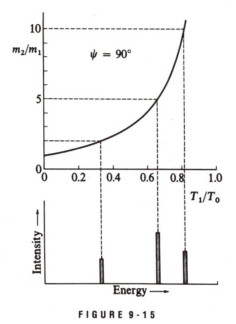

FIGURE 9-15

energy measurements, the method yields accurate information regarding the composition of the target. **Quantitative analysis** can also be made from the intensities of the groups if the cross sections are known (see the following section). Applying this technique has been useful in determining the composition of air pollution.

EXAMPLE 9.7 -

In a head-on elastic collision of two particles with masses m_1 and m_2, the initial velocities are u_1 and $u_2 = \alpha u_1$ ($\alpha > 0$). If the initial kinetic energies of the two particles are equal in the LAB system, find the conditions on u_1/u_2 and m_1/m_2 so that m_1 will be at rest in the LAB system after the collision.

Solution: Because the initial kinetic energies are equal, we have

$$\tfrac{1}{2}m_1u_1^2 = \tfrac{1}{2}m_2u_2^2 = \tfrac{1}{2}\alpha^2 m_2u_1^2$$

or

$$\frac{m_1}{m_2} = \alpha^2 \qquad\qquad (9.93)$$

If m_1 is at rest after the collision, the conservation of energy requires

$$\tfrac{1}{2}m_1u_1^2 + \tfrac{1}{2}m_2u_2^2 = \tfrac{1}{2}m_2v_2^2$$

or

$$m_1 u_1^2 = \tfrac{1}{2} m_2 v_2^2 \tag{9.94}$$

The conservation of linear momentum states that

$$m_1 \mathbf{u}_1 + m_2 \mathbf{u}_2 = (m_1 + \alpha m_2)\, \mathbf{u}_1 = m_2 \mathbf{v}_2 \tag{9.95}$$

Substituting v_2 from Equation 9.95 into Equation 9.94 gives

$$m_1 u_1^2 = \frac{1}{2} m_2 \left(\frac{m_1 + \alpha m_2}{m_2} \right)^2 u_1^2$$

or

$$m_1 = \frac{1}{2} m_2 \left(\frac{m_1}{m_2} + \alpha \right)^2 \tag{9.96}$$

Substituting $m_1/m_2 = \alpha^2$ from Equation 9.93 gives

$$2\alpha^2 = (\alpha^2 + \alpha)^2$$

with the result

$$\alpha = \sqrt{2} - 1 = 0.414$$

$$\alpha^2 = 0.172$$

so that

$$\frac{m_1}{m_2} = \alpha^2 = 0.172$$

and

$$\frac{u_2}{u_1} = \alpha = 0.414$$

Because $\alpha > 0$, both particles are traveling in the same direction; the collision is shown in Figure 9-16.

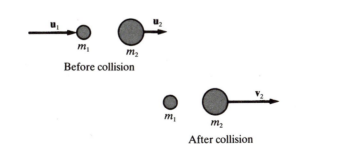

Before collision

After collision

FIGURE 9-16

EXAMPLE   9.8   -

Particles of mass m_1 elastically scatter from particles of mass m_2 at rest. (a) At what LAB angle should a magnetic spectrometer be set to detect particles that lose one third of their momentum? (b) Over what range m_1/m_2 is this possible? (c) Calculate the scattering angle for $m_1/m_2 = 1$.

Solution: We have

$$m_1 v_1 = \frac{2}{3} m_1 u_1 \quad \text{and} \quad v_1 = \frac{2}{3} u_1$$

Using Equations 9.82 and 9.87a, we have

$$\frac{T_1}{T_0} = \frac{v_1^2}{u_1^2} = \left(\frac{2}{3}\right)^2 = 1 - \frac{2m_1 m_2}{(m_1 + m_2)^2}(1 - \cos\theta) \tag{9.97}$$

This equation can be solved for $\cos\theta$, yielding

$$\cos\theta = 1 - \frac{5(m_1 + m_2)^2}{18 m_1 m_2} = 1 - y \tag{9.98}$$

where

$$y = \frac{5(m_1 + m_2)^2}{18 m_1 m_2} \tag{9.99}$$

But we need ψ, which can be obtained from Equation 9.69.

$$\tan\psi = \frac{\sin\theta}{\cos\theta + m_1/m_2} = \frac{\sqrt{2y - y^2}}{1 - y + m_1/m_2} \tag{9.100}$$

where we have used Equation 9.98 for $\cos\theta$ and found $\sin\theta = \sqrt{2y - y^2}$.

Because $\tan\psi$ must be a real number, only values for m_1/m_2 where $2 - y \geq 0$ are possible. Therefore,

$$2 - \frac{5(m_1 + m_2)^2}{18 m_1 m_2} \geq 0 \tag{9.101}$$

which can be reduced to

$$-5\left(\frac{m_1}{m_2}\right)^2 + 26\left(\frac{m_1}{m_2}\right) - 5 \geq 0$$

or

$$-5x^2 + 26x - 5 \geq 0 \tag{9.102}$$

where $x = m_1/m_2$. The solutions for x when Equation 9.102 is equal to zero are $x = 1/5, 5$. Substitution verifies that

$$\frac{1}{5} \le \frac{m_1}{m_2} \le 5$$

satisfies Equation 9.101, but values of m_1/m_2 outside this range do not.
Substituting $m_1/m_2 = 1$ into Equation 9.99 gives

$$y = \frac{5(m_1 + m_2)^2}{18m_1m_2} = \frac{5\left(\dfrac{m_1}{m_2} + 1\right)^2}{18m_1/m_2}$$

$$= \frac{5(1 + 1)^2}{18} = \frac{10}{9}$$

and substituting for y into Equation 9.100 gives $\psi = 48°$.

-- ●

9.8 INELASTIC COLLISIONS

When two particles interact, many results are possible, depending on the forces involved. In the previous two sections, we were restricted to elastic collisions. But, in general, multiparticles may be produced if large changes of energy are involved. For example, when a proton collides with some nuclei, energy may be released. In addition, the proton may be absorbed, and the collision may produce a neutron or alpha particle instead. All these possibilities are handled with the same methods: conservation of energy and linear momentum. We continue to restrict our considerations to the same particles in the final system as were considered in the initial system. In general, the conservation of energy is

$$Q + \tfrac{1}{2}m_1u_1^2 + \tfrac{1}{2}m_2u_2^2 = \tfrac{1}{2}m_1v_1^2 + \tfrac{1}{2}m_2v_2^2 \tag{9.103}$$

where Q is called the Q-value and represents the energy loss or gain in the collision.

$Q = 0$: Elastic collision, kinetic energy is conserved

$Q > 0$: Exoergic collision, kinetic energy is gained

$Q < 0$: Endoergic collision, kinetic energy is lost

An inelastic collision is an example of an endoergic collision. The kinetic energy may be converted to mass-energy, as, for example, in a nuclear collision. Or it may be lost as heat energy, as, for example, by frictional forces in a collision. The collisions of all macroscopic bodies are endoergic (inelastic) to some degree. Two silly putty balls with equal masses and speeds striking head-on may come to a complete stop, a totally inelastic collision. Even two billiard balls colliding do not completely conserve kinetic energy; some small fraction of the initial kinetic energy is converted to heat.

A measure of the inelasticity of two bodies colliding may be considered by referring to a direct head-on collision (see Figure 9-17) in which no rotations are involved (translational kinetic energy only). Newton found experimentally that the

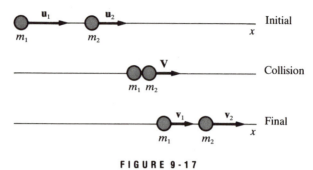

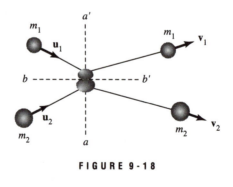

FIGURE 9-17

ratio of the relative initial velocities to the relative final velocities was nearly constant for any two bodies. This ratio, called the **coefficient of restitution (ε)**, is defined by

$$\varepsilon = \frac{|v_2 - v_1|}{|u_2 - u_1|} \tag{9.104}$$

This is sometimes called Newton's rule. For a perfectly elastic collision, $\varepsilon = 1$; and for a totally inelastic collision, $\varepsilon = 0$. Values of ε have the limits 0 and 1.

We must be careful when applying Equation 9.104 to oblique collisions, because Newton's rule applies only to the velocity components along the normal (aa') to the plane of contact (bb') between the two bodies as shown in Figure 9-18. For smooth surfaces, the velocity components along the plane of contact are hardly changed by the collision.

FIGURE 9-18

EXAMPLE 9.9

For an elastic head-on collision described in Sections 9.6 and 9.7, show that $\varepsilon = 1$. The mass m_2 is initially at rest.

Solution: Because the final velocities are along the same direction as $\mathbf{u}_1$, we state the conservation of linear momentum and energy as

$$m_1 u_1 = m_1 v_1 + m_2 v_2 \tag{9.105}$$

$$\tfrac{1}{2}m_1u_1^2 = \tfrac{1}{2}m_1v_1^2 + \tfrac{1}{2}m_2v_2^2 \tag{9.106}$$

We solve Equation 9.105 for v_2 and substitute into the equation for ε

$$\varepsilon = \frac{v_2 - v_1}{u_1} = \frac{\dfrac{m_1u_1 - m_1v_1}{m_2} - v_1}{u_1} = \frac{m_1}{m_2} - \frac{m_1}{m_2}\frac{v_1}{u_1} - \frac{v_1}{u_1} \tag{9.107}$$

We can find the ratio v_1/u_1 from Equation 9.106 after substituting for v_2 from Equation 9.105:

$$\frac{1}{2}m_1u_1^2 = \frac{1}{2}m_1v_1^2 + \frac{1}{2}m_2\left(\frac{m_1u_1 - m_1v_1}{m_2}\right)^2$$

$$m_1u_1^2 = m_1v_1^2 + \frac{m_1^2}{m_2}(u_1^2 + v_1^2 - 2u_1v_1)$$

Dividing by $m_1u_1^2$ and letting $x = v_1/u_1$ gives

$$1 = x^2 + \frac{m_1}{m_2}(1 + x^2 - 2x)$$

Collecting terms,

$$\left(1 + \frac{m_1}{m_2}\right)x^2 - \frac{2m_1}{m_2}x + \left(\frac{m_1}{m_2} - 1\right) = 0$$

Using the quadratic equation to solve for x, we find

$$x = 1$$

and

$$x = \frac{\dfrac{m_1}{m_2} - 1}{\dfrac{m_1}{m_2} + 1}$$

The solution $x = 1$ is trivial ($v_1 = u_1$, $v_2 = 0$), so we substitute the other solution for x into Equation 9.107:

$$\varepsilon = \frac{m_1}{m_2} - \frac{\dfrac{m_1}{m_2}\dfrac{m_1}{m_2} - 1}{\dfrac{m_1}{m_2} + 1} + \frac{1 + \dfrac{m_1}{m_2}}{\dfrac{m_1}{m_2} + 1}$$

$$\varepsilon = \frac{\dfrac{m_1^2}{m_2^2} + \dfrac{m_1}{m_2} - \dfrac{m_1^2}{m_2^2} + \dfrac{m_1}{m_2} + 1 - \dfrac{m_1}{m_2}}{\dfrac{m_1}{m_2} + 1} = 1$$

During a collision (elastic or inelastic), the forces involved may act over a very short period of time and are called **impulsive forces**. A hammer striking a nail and two billiard balls colliding are examples of impulsive forces. Newton's Second Law is still valid throughout the time period Δt of the collision:

$$\mathbf{F} = \frac{d}{dt}(m\mathbf{v}) \tag{9.108}$$

After multiplying by dt and integrating, we have

$$\int_{t_1}^{t_2} \mathbf{F}\, dt = m\mathbf{v} - m\mathbf{u} \equiv \mathbf{P} \tag{9.109}$$

where $\Delta t = t_2 - t_1$ and $\mathbf{u}$, $\mathbf{v}$ represent the velocities before and after the collision, respectively. Equation 9.109 defines the term **impulse P**. The impulse may be measured experimentally by the change of momentum. An ideal impulse represented by no displacement during the collision would be caused by an infinite force acting during an infinitesimal time.

E X A M P L E 9.10

Consider a rope of mass per unit length ρ and length a suspended just above a table as shown in Figure 9-19. If the rope is released from rest at the top, find the force on the table when a length x of the rope has dropped to the table.

Solution: We have a gravitational force of $mg = \rho x g$ because the rope lies on the table, but we need to consider the impulsive force as well.

$$F = \frac{dp}{dt} \tag{9.110}$$

During the time interval dt the mass of rope equal to $\rho(v\,dt)$ drops to the floor. The change in momentum imparted to the table is

$$dp = (\rho v\, dt)v = \rho v^2\, dt$$

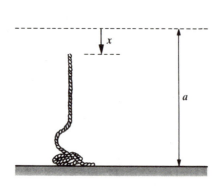

FIGURE 9-19

and

$$\frac{dp}{dt} = \rho v^2 = F_{\text{impulse}} \tag{9.111}$$

The velocity v is related to x at time t by $v^2 = 2gx$, because each part of the remaining rope is under constant acceleration g.

$$F_{\text{impulse}} = \rho v^2 = 2\rho x g \tag{9.112}$$

The total force is the sum of the gravitational and impulsive forces:

$$F = F_g + F_{\text{impulse}} = 3\rho x g \tag{9.113}$$

which is equivalent to the weight of a length $3x$ of the rope.

-- ●

9.9 CROSS SECTIONS

In the preceding sections, we derived various relationships connecting the initial state of a moving particle with the final states of the original particle and a struck particle. Only kinematic relationships were involved; that is, no attempt was made to *predict* a scattering angle or a final velocity—only equations *connecting* these quantities were obtained. We now look more closely at the collision process and investigate the scattering in the event that the particles interact with a specified force field. Consider the situation depicted in Figure 9-20, which illustrates such a collision in the LAB coordinate system when a repulsive force exists between m_1 and m_2. The particle m_1 approaches the vicinity of m_2 in such a way that if there were no force acting between the particles, m_1 would pass m_2 with a distance of closest approach b. The quantity b is called the **impact parameter**. If the velocity of m_1 is u_1, then the impact parameter b clearly specifies the angular momentum l of particle m_1 about m_2:

$$l = m_1 u_1 b \tag{9.114}$$

We may express u_1 in terms of the incident energy T_0 by using Equation 9.78:

$$l = b \sqrt{2m_1 T_0} \tag{9.115}$$

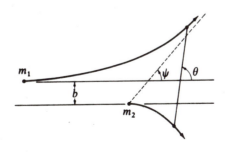

FIGURE 9-20

Evidently, for a given energy T_0, the angular momentum and hence the scattering angle θ (or ψ) is uniquely specified by the impact parameter b if the force law is known.*

We now consider the distribution of scattering angles that result from collisions with various impact parameters. To accomplish this, let us assume that we have a narrow beam of particles, each having mass m_1 and energy T_0. We direct this beam toward a small region of space containing a collection of particles, each of which has mass m_2 and is at rest (in the LAB system). We define the **intensity** (or **flux density**) I of the incident particles as the number of particles passing in unit time through a unit area normal to the direction of the beam. If we assume that the force law between m_1 and m_2 falls off with distance sufficiently rapidly, then after an encounter, the motion of a scattered particle asymptotically approaches a straight line with a well-defined angle θ between the initial and final directions of motion. We now define a **differential scattering cross section** $\sigma(\theta)$ in the CM system for the scattering into an element of solid angle $d\Omega'$ at a particular CM angle θ:

$$\sigma(\theta) = \frac{\left(\begin{array}{c}\text{Number of interactions per target particle that} \\ \text{lead to scattering into } d\Omega' \text{ at the angle } \theta\end{array}\right)}{\text{Number of incident particles per unit area}} \qquad \textbf{(9.116)}$$

If dN is the number of particles scattered into $d\Omega'$ per unit time, then

$$\sigma(\theta)\, d\Omega' = \frac{dN}{I} \qquad \textbf{(9.117a)}$$

We sometimes write, alternatively,

$$\sigma(\theta) = \frac{d\sigma}{d\Omega'} = \frac{1}{I}\frac{dN}{d\Omega'} \qquad \textbf{(9.117b)}$$

(The fact that $\sigma(\theta)$ has the dimensions of *area* gives rise to the term *cross section*.) If the scattering has axial symmetry (as for central forces), we can immediately perform the integration over the azimuthal angle to obtain 2π, and then the element of solid angle $d\Omega'$ is given by

$$d\Omega' = 2\pi \sin\theta\, d\theta \qquad \textbf{(9.118)}$$

If we return, for the moment, to the equivalent one-body problem discussed in the preceding chapter, we can consider the scattering of a particle of mass μ by a force center. For such a case, Figure 9-21 shows that the number of particles with impact parameters within a range db at a distance b must correspond to the number of particles scattered into the angular range $d\theta$ at an angle θ. Therefore,

$$I \cdot 2\pi b\, db = -I \cdot \sigma(\theta) \cdot 2\pi \sin\theta\, d\theta \qquad \textbf{(9.119)}$$

---

* In the scattering of atomic or nuclear particles, we can neither choose nor measure directly the impact parameter. We are therefore reduced, in such situations, to speaking in terms of the probability for scattering at various angles θ.

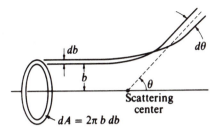

FIGURE 9-21

where $db/d\theta$ is negative, because we assume that the force law is such that the amount of angular deflection decreases (monotonically) with increasing impact parameter. Hence,

$$\sigma(\theta) = \frac{b}{\sin \theta} \left| \frac{db}{d\theta} \right| \tag{9.120}$$

We can obtain the relationship between the impact parameter b and the scattering angle θ by using Figure 9-22. In the preceding chapter, we found (in Equation 8.31) that the change in angle for a particle of mass μ moving in a central-force field was given by

$$\Delta\Theta = \int_{r_{\min}}^{r_{\max}} \frac{(l/r^2) \, dr}{\sqrt{2\mu[E - U - (l^2/2\mu r^2)]}} \tag{9.121}$$

The motion of a particle in a central-force field is symmetric about the point of closest approach to the force center (see point A in Figure 9-22). The angles α and β are therefore equal and, in fact, are equal to Θ. Thus,

$$\theta = \pi - 2\Theta \tag{9.122}$$

For the case that $r_{\max} = \infty$, the angle Θ is given by

$$\Theta = \int_{r_{\min}}^{\infty} \frac{(b/r^2) \, dr}{\sqrt{1 - (b^2/r^2) - (U/T_0')}} \tag{9.123}$$

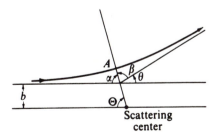

FIGURE 9-22

where use has been made of the one-body equivalent of Equation 9.115:

$$l = b \sqrt{2\mu T_0'}$$

where, as in Equation 9.79, $T_0' = \frac{1}{2}\mu u_1^2$. We have also used $E = T_0'$ because the total energy E must equal the kinetic energy T_0' at $r = \infty$ where $U = 0$. The value of r_{min} is a root of the radical in the denominator in Equations 9.121 or 9.123— that is, r_{min} is a turning point of the motion and corresponds to the distance of closest approach of the particle to the force center. Thus, Equations 9.122 and 9.123 give the dependence of the scattering angle θ on the impact parameter b. Once we know $b = b(\theta)$ for a given potential $U(r)$ and a given value of T_0', we can calculate the scattering cross section from Equation 9.120. This procedure leads to the scattering cross section in the CM system, because we have been considering m_2 as a fixed force center. If $m_2 \gg m_1$, the cross section so obtained is very close to the LAB system cross section; but if m_1 cannot be considered negligible compared with m_2, the proper transformation of solid angles must be made. We now obtain the general relations.

Because the total number of particles scattered into a unit solid angle must be the same in the LAB system as in the CM system, we have

$$\sigma(\theta) \, d\Omega' = \sigma(\psi) \, d\Omega$$

$$\sigma(\theta) \cdot 2\pi \sin\theta \, d\theta = \sigma(\psi) \cdot 2\pi \sin\psi \, d\psi \tag{9.124}$$

where θ and ψ represent the *same* scattering angle but measured in the CM or LAB system, respectively, and where $d\Omega'$ and $d\Omega$ represent the *same* element of solid angle but measured in the CM or LAB system, respectively. Therefore, $\sigma(\theta)$ and $\sigma(\psi)$ are the differential cross sections for the scattering in the CM and LAB systems, respectively. Thus,

$$\sigma(\psi) = \sigma(\theta) \cdot \frac{\sin\theta \, d\theta}{\sin\psi \, d\psi} \tag{9.125}$$

The derivative $d\theta/d\psi$ can be evaluated by first referring to Figure 9-11a and writing, from the sine law,

$$\frac{\sin(\theta - \psi)}{\sin\psi} = \frac{m_1}{m_2} \equiv x \tag{9.126}$$

Differentiating this equation, we find

$$\frac{d\theta}{d\psi} = \frac{\sin(\theta - \psi)\cos\psi}{\cos(\theta - \psi)\sin\psi} + 1$$

Expanding $\sin(\theta - \psi)$ and simplifying, we have

$$\frac{d\theta}{d\psi} = \frac{\sin\theta}{\cos(\theta - \psi)\sin\psi}$$

and so

$$\sigma(\psi) = \sigma(\theta) \cdot \frac{\sin^2\theta}{\cos(\theta - \psi)\sin^2\psi} \tag{9.127}$$

Multiplying both sides of Equation 9.126 by cos ψ and then adding $\cos(\theta - \psi)$ to both sides, we have

$$\frac{\sin(\theta - \psi)\cos \psi}{\sin \psi} + \cos(\theta - \psi) = x \cos \psi + \cos(\theta - \psi)$$

Expanding $\sin(\theta - \psi)$ and $\cos(\theta - \psi)$ on the left-hand side, we obtain

$$\frac{\sin \theta}{\sin \psi} = x \cos \psi + \cos(\theta - \psi)$$

Substituting this result into Equation 9.127,

$$\sigma(\psi) = \sigma(\theta) \cdot \frac{[x \cos \psi + \cos(\theta - \psi)]^2}{\cos(\theta - \psi)}, \qquad (x < 1) \qquad \textbf{(9.128)}$$

And from Equation 9.126, we have

$$\cos(\theta - \psi) = \sqrt{1 - x^2 \sin^2\psi}$$

Hence,

$$\sigma(\psi) = \sigma(\theta) \cdot \frac{\left[x \cos \psi + \sqrt{1 - x^2 \sin^2\psi}\right]^2}{\sqrt{1 - x^2 \sin^2\psi}} \qquad \textbf{(9.129)}$$

Equation 9.126 can be used to write

$$\theta = \sin^{-1}(x \sin \psi) + \psi \qquad \textbf{(9.130)}$$

Equations 9.129 and 9.130 therefore specify the cross section entirely in terms of the angle ψ.* For the general case (i.e., for an arbitrary value of x), the evaluation of $\sigma(\psi)$ is complicated. Tables exist, however, so the particular cases can be computed with relative ease.[†]

The transformation represented by Equations 9.129 and 9.130 assumes a simple form for two cases. For $x = m_1/m_2 = 1$, we have from Equation 9.71, $\theta = 2\psi$, and Equation 9.129 becomes

$$\sigma(\psi) = \sigma(\theta)|_{\theta = 2\psi} \cdot 4 \cos \psi, \qquad m_1 = m_2 \qquad \textbf{(9.131)}$$

and for $m_1 \ll m_2$, $x \cong 0$, and $\theta \cong \psi$, so that

$$\sigma(\psi) \cong \sigma(\theta)|_{\theta = \psi}, \qquad m_1 \ll m_2 \qquad \textbf{(9.132)}$$

---

* These equations apply not only for elastic collisions but also for inelastic collisions (in which the internal potential energy of one or both of the particles is altered as a result of the interaction) if the parameter x is written as V/v_1' instead of m_1/m_2 (see Equation 9.68). Note that the preceding equations refer only to the usual case $x < 1$.

[†] See, for example, the tables by Marion et al. (Ma59).

9.10   RUTHERFORD SCATTERING FORMULA*

One of the most important problems that makes use of the formulas developed in the preceding section is the scattering of charged particles in a Coulomb or electrostatic field. The potential for this case is

$$U(r) = \frac{k}{r} \tag{9.133}$$

where $k = q_1 q_2 / 4\pi \varepsilon_0$, with q_1 and q_2 the amounts of charge that the two particles carry (k may be either positive or negative, depending on whether the charges are of the same or opposite sign; $k > 0$ corresponds to a repulsive force and $k < 0$ to an attractive force). Equation 9.123 then becomes

$$\Theta = \int_{r_{min}}^{\infty} \frac{(b/r)\, dr}{\sqrt{r^2 - (k/T_0')r - b^2}} \tag{9.134}$$

which can be integrated to obtain (see the integration of Equation 8.38):

$$\cos \Theta = \frac{(\kappa/b)}{\sqrt{1 + (\kappa/b)^2}} \tag{9.135}$$

where

$$\kappa \equiv \frac{k}{2T_0'} \tag{9.136}$$

Equation 9.135 can be rewritten as

$$b^2 = \kappa^2 \tan^2 \Theta \tag{9.137}$$

But Equation 9.122 states that $\Theta = \pi/2 - \theta/2$, so

$$b = \kappa \cot(\theta/2) \tag{9.138}$$

Thus,

$$\frac{db}{d\theta} = -\frac{\kappa}{2} \frac{1}{\sin^2(\theta/2)} \tag{9.139}$$

Equation 9.120 thus becomes

$$\sigma(\theta) = \frac{\kappa^2}{2} \cdot \frac{\cot(\theta/2)}{\sin \theta \sin^2(\theta/2)}$$

Now,

$$\sin \theta = 2 \sin (\theta/2) \cos (\theta/2)$$

---

* E. Rutherford, *Phil. Mag.* **21**, 669 (1911).

Hence,

$$\sigma(\theta) = \frac{\kappa^2}{4} \cdot \frac{1}{\sin^4(\theta/2)}$$

or

$$\sigma(\theta) = \frac{k^2}{(4T_0')^2} \cdot \frac{1}{\sin^4(\theta/2)} \qquad (9.140)$$

which is the Rutherford scattering formula* and demonstrates the dependence of the CM scattering cross section on the inverse fourth power of $\sin(\theta/2)$. Note that $\sigma(\theta)$ is independent of the sign of k, so that the form of the scattering distribution is the same for an attractive force as for a repulsive one. It is also rather remarkable that the quantum-mechanical treatment of Coulomb scattering leads to exactly the same result as does the classical derivation.[†] This is indeed a fortunate circumstance because, if it were otherwise, the disagreement at this early stage between classical theory and experiment might have seriously delayed the progress of nuclear physics.

For the case $m_1 = m_2$, Equation 9.79 states that $T_0' = \frac{1}{2}T_0$, so that

$$\sigma(\theta) = \frac{k^2}{4T_0^2} \cdot \frac{1}{\sin^4(\theta/2)}, \qquad m_1 = m_2 \qquad (9.141)$$

Or, from Equation 9.131,

$$\sigma(\psi) = \frac{k^2}{T_0^2} \frac{\cos\psi}{\sin^4\psi}, \qquad m_1 = m_2 \qquad (9.142)$$

All the preceding discussion applies to the calculation of *differential* scattering cross sections. If it is desired to know the probability that *any* interaction *whatsoever* will take place, it is necessary to integrate $\sigma(\theta)$ [or $\sigma(\psi)$] over all possible scattering angles. The resulting quantity is called the **total scattering cross section** (σ_t) and is equal to the effective area of the target particle for producing a scattering event:

$$\sigma_t = \int_{4\pi} \sigma(\theta)\, d\Omega' = 2\pi \int_0^\pi \sigma(\theta) \sin\theta\, d\theta \qquad (9.143)$$

where the integration over θ runs from 0 to π. The *total* cross section is the same in the LAB as in the CM system. If we wish to express the total cross section in

---

* This form of the scattering law was verified for the interaction of α particles and heavy nuclei by the experiments of H. Geiger and E. Marsden, *Phil. Mag.* **25**, 605 (1913).

[†] N. Bohr showed that the identity of the results is a consequence of the $1/r^2$ nature of the force; it cannot be expected for any other type of force law.

terms of an integration over the LAB quantities,

$$\sigma_t = \int \sigma(\psi)\, d\Omega$$

then if $m_1 < m_2$, ψ also runs from 0 to π. If $m_1 \geq m_2$, ψ runs only up to ψ_{max} (given by Equation 9.77), and we have

$$\sigma_t = 2\pi \int_0^{\psi_{max}} \sigma(\psi) \sin \psi\, d\psi \qquad \qquad \text{(9.144)}$$

If we attempt to calculate σ_t for the case of Rutherford scattering, we find that the result is infinite. This occurs because the Coulomb potential, which varies as $1/r$, falls off so slowly that, as the impact parameter b becomes indefinitely large, the decrease in scattering angle is too slow to prevent the integral from diverging. We have, however, pointed out in Example 8.6 that the Coulomb field of a real atomic nucleus is screened by the surrounding electrons so that the potential is effectively cut off at large distances. The evaluation of the scattering cross section for a screened Coulomb potential according to the classical theory is quite complicated and is not discussed here; the quantum-mechanical treatment is actually easier for this case.

P R O B L E M S

9-1. Find the center of mass of a hemispherical shell of constant density and inner radius r_1 and outer radius r_2.

▷ **9-2.** Find the center of mass of a uniformly solid cone of base $2a$ and height h.

▷▷ **9-3.** Find the center of mass of a uniformly solid cone of base $2a$ and height h and a solid hemisphere of radius a where the two bases are touching.

9-4. Find the center of mass of a uniform wire that subtends an arc θ if the radius of the circular arc is a, as shown in Figure 9-A.

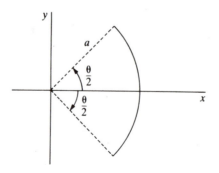

FIGURE 9-A

9-5. The center of gravity of a system of particles is the point about which external gravitational forces exert no net torque. For a uniform gravitational force, show that the center of gravity is identical to the center of mass for the system of particles.

9-6. Consider two particles of equal mass m. The forces on the particles are $\mathbf{F}_1 = 0$ and $\mathbf{F}_2 = F_0\mathbf{i}$. If the particles are initially at rest at the origin, what is the position, velocity, and acceleration of the center of mass?

▷ **9-7.** A model of the water molecule H_2O is shown in Figure 9-B. Where is the center of mass?

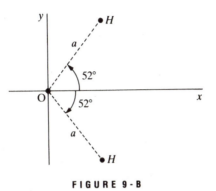

FIGURE 9-B

9-8. Where is the center of mass of the isosceles right triangle of uniform areal density shown in Figure 9-C?

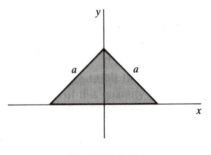

FIGURE 9-C

9-9. A projectile is fired at an angle of 45° with initial kinetic energy E_0. At the top of its trajectory, the projectile explodes with additional energy E_0 into two fragments. One fragment of mass m_1 travels straight down. What is the velocity (magnitude and direction) of the second fragment of mass m_2 and the velocity of the first? What is the ratio of m_1/m_2 when m_1 is a maximum?

9-10. A cannon in a fort overlooking the ocean fires a shell of mass M at an elevation angle θ and muzzle velocity v_0. At the highest point, the shell explodes into two fragments (masses $m_1 + m_2 = M$), with an additional energy E, traveling in the original horizontal direction. Find the distance separating the two fragments when they land in the ocean. For simplicity, assume the cannon is at sea level.

9-11. Verify that the second term on the right-hand side of Equation 9.9 indeed vanishes for the case $n = 3$.

▷ **9-12.** Astronaut Stumblebum wanders too far away from the space shuttle orbiter while repairing a broken communications satellite. Stumblebum realizes that the orbiter is moving away from him at 3 m/s. Stumblebum and his maneuvering unit have a mass of 100 kg, including a pressurized tank of mass 10 kg. The tank includes only 2 kg of gas which is used to propel him in space. The gas escapes with a constant velocity of 100 m/s.
(a) Will Stumblebum run out of gas before he reaches the orbiter?
(b) With what velocity will Stumblebum have to throw the empty tank away to reach the orbiter?

9-13. Even though the total force on a system of particles (Equation 9.9) is zero, the net torque may not be zero. Show that the net torque has the same value in any coordinate system.

9-14. Consider a system of particles interacting by magnetic forces. Are Equations 9.11 and 9.31 valid? Explain.

9-15. A smooth rope is placed above a hole in a table (Figure 9-D). One end of the rope falls through the hole at $t = 0$, pulling steadily on the remainder of the rope. Find the velocity of the rope as a function of the distance to the end of the rope x. Ignore friction of the rope as it unwinds. Then find the acceleration of the falling rope and the energy lost from the system as the end of the rope length L and mass m leaves the table.

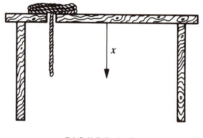

FIGURE 9-D

9-16. For the energy conserving case of the falling chain in Example 9.2 show that the tension on either side of the bottom bend is equal and has the value $\mu \dot{x}^2/4$.

9-17. Integrate Equation 9.17 in Example 9.2 numerically and make a plot of the speed versus the time using dimensionless parameters, $\dot{x}/\sqrt{2gb}$ vs. $t/\sqrt{2b/g}$ where $\sqrt{2b/g}$ is the free fall time, $t_{\text{free fall}}$. Find the time it takes for the free end to reach the bottom. Define natural units by $\tau \equiv \sqrt{g/2b}$, $\alpha \equiv x/2b$ and integrate $d\tau/d\alpha$ from $\alpha = \varepsilon$ (some small number greater than 0) to $\alpha = 1/2$. One can't integrate numerically from $\alpha = 0$ because of a singularity in $d\tau/d\alpha$. The expression $d\tau/d\alpha$ is

$$\frac{d\tau}{d\alpha} = \sqrt{\frac{1 - 2\alpha}{2\alpha(1 - \alpha)}}$$

9-18. Use a computer to make a plot of the tension versus time for the falling chain in Example 9.2. Use dimensionless parameters (T/Mg) versus $t/t_{\text{free fall}}$, where $t_{\text{free fall}} = \sqrt{2b/g}$. Stop the plot before T/Mg becomes greater than 50.

9-19. A chain such as the one in Example 9.2 (with the same parameters) of length b and mass ρb is suspended from one end at a point that is a height b above a table so that the free end barely touches the tabletop. At time $t = 0$, the fixed end of the chain is released. Find the force that the tabletop exerts on the chain after the original fixed end has fallen a distance x.

9-20. A uniform rope of total length $2a$ hangs in equilibrium over a smooth nail. A very small impulse causes the rope to slowly roll off the nail. Find the velocity of the rope as it just clears the nail. Assume the rope is prevented from lifting off the nail and is in free fall.

9-21. A particle of mass m_1 and velocity u_1 collides with a particle of mass m_2 at rest. The two particles stick together. What fraction of the original kinetic energy is lost in the collision?

9-22. A particle of mass m at the end of a light string wraps itself about a fixed vertical cylinder of radius a (Figure 9-E). All the motion is in the horizontal plane (disregard gravity). The angular velocity of the cord is ω_0 when the distance from the particle to the point of contact of the string and cylinder is b. Find the angular velocity and tension in the string after the cord has turned through an additional angle θ.

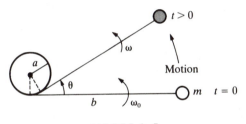

FIGURE 9-E

9-23. Slow-moving neutrons have a much larger absorption rate in ^{235}U than fast neutrons produced by $^{236}U^*$ fission in a nuclear reactor. For that reason, reactors consist of moderators to slow down neutrons by elastic collisions. What elements are best to be used as moderators? Explain.

9-24. The force of attraction between two particles is given by

$$\mathbf{f}_{12} = k\left[(\mathbf{r}_2 - \mathbf{r}_1) - \frac{r}{v_0}(\dot{\mathbf{r}}_2 - \dot{\mathbf{r}}_1) \right]$$

where k is a constant, v_0 is a constant velocity, and $r \equiv |\mathbf{r}_2 - \mathbf{r}_1|$. Calculate the internal torque for the system; why does this quantity not vanish? Is the system conservative?

9-25. Derive Equation 9.90.

▷ **9-26.** A particle of mass m_1 elastically collides with a particle of mass m_2 at rest. What is the maximum fraction of kinetic energy loss for m_1? Describe the reaction.

9-27. Derive Equation 9.91.

9-28. A tennis player strikes an incoming tennis ball of mass 60 g as shown in Figure 9-F (See page 378). The incoming tennis ball velocity is $v_i = 8$ m/s, and the outgoing velocity is $v_f = 16$ m/s.
(a) What impulse was given to the tennis ball?
(b) If the collision time was 0.01 s, what was the average force exerted by the tennis racket?

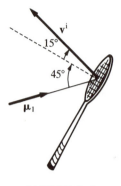

FIGURE 9-F

9-29. Derive Equation 9.92.

9-30. A particle of mass m and velocity u_1 makes a head-on collision with another particle of mass $2m$ at rest. If the coefficient of restitution is such to make the loss of total kinetic energy a maximum, what are the velocities v_1 and v_2 after the collision?

9-31. Show that T_1/T_0 can be expressed in terms of $m_2/m_1 \equiv \alpha$ and $\cos \psi \equiv y$ as

$$\frac{T_1}{T_0} = (1 + \alpha)^{-2} (2y^2 + \alpha^2 - 1 + 2y \sqrt{\alpha^2 + y^2 - 1})$$

Plot T_1/T_0 as a function of ψ for $\alpha = 1, 2, 4,$ and 12. These plots correspond to the energies of protons or neutrons after scattering from hydrogen ($\alpha = 1$), deuterium ($\alpha = 2$), helium ($\alpha = 4$), and carbon ($\alpha = 12$), or of alpha particles scattered from helium ($\alpha = 1$), oxygen ($\alpha = 4$), and so forth.

▷ **9-32.** A billiard ball of initial velocity u_1 collides with another billiard ball (same mass) initially at rest. The first ball moves off at $\psi = 45°$. For an elastic collision, what are the velocities of both balls after the collision? At what LAB angle does the second ball emerge?

9-33. A particle of mass m_1 with initial laboratory velocity u_1 collides with a particle of mass m_2 at rest in the LAB system. The particle m_1 is scattered through a LAB angle ψ and has a final velocity v_1, where $v_1 = v_1(\psi)$. Find the surface such that the time of travel of the scattered particle from the point of collision to the surface is independent of the scattering angle. Consider the cases (a) $m_2 = m_1$, (b) $m_2 = 2m_1$, (c) $m_2 = \infty$. Suggest an application of this result in terms of a detector for nuclear particles.

9-34. In an elastic collision of two particles with masses m_1 and m_2, the initial velocities are $\mathbf{u}_1$ and $\mathbf{u}_2 = \alpha\mathbf{u}_1$. If the initial kinetic energies of the two particles are equal, find the conditions on u_1/u_2 and m_1/m_2 such that m_1 is at rest after the collision. Examine both cases for the sign of α.

9-35. When a bullet fires in a gun, the explosion subsides quickly. Suppose the force on the bullet is $F = (360 - 10^7 t^2)$N until the force becomes zero (and remains zero). The mass of the bullet is 3 g.
(a) What impulse acts on the bullet?
(b) What is the muzzle velocity of the gun?

9-36. Show that

$$\frac{T_1}{T_0} = \frac{m_1^2}{(m_1 + m_2)^2} \cdot S^2$$

where

$$S \equiv \cos \psi + \frac{\cos(\theta - \psi)}{\left(\dfrac{m_1}{m_2}\right)}$$

9-37. A particle of mass m strikes a smooth wall at an angle θ from the normal. The coefficient of restitution is ε. Find the velocity and the rebound angle of the particle after leaving the wall.

9-38. A particle of mass m_1 and velocity u_1 strikes head-on a particle of mass m_2 at rest. The coefficient of restitution is ε. Particle m_2 is tied to a point a distance a away as shown in Figure 9-G. Find the velocity (magnitude and direction) of m_1 and m_2 after the collision.

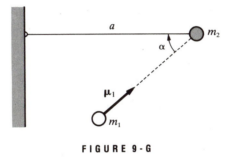

FIGURE 9-G

9-39. A rubber ball is dropped from rest onto a linoleum floor a distance h_1 away. The rubber ball bounces up to a height h_2. What is the coefficient of restitution? What fraction of the original kinetic energy is lost in terms of ε?

9-40. A steel ball of velocity 5 m/s strikes a smooth, heavy steel plate at an angle of 30° from the normal. If the coefficient of restitution is 0.8, at what angle and velocity does the steel ball bounce off the plate?

9-41. A proton (mass m) of kinetic energy T_0 collides with a helium nucleus (mass $4m$) at rest. Find the recoil angle of the helium if $\psi = 45°$ and the inelastic collision has $Q = -T_0/6$.

9-42. A uniform dense rope of length b and mass density μ is coiled on a smooth table. One end is lifted by hand with a constant velocity v_0. Find the force of the rope held by the hand when the rope is a distance a above the table ($b > a$).

9-43. Show that the equivalent of Equation 9.129 expressed in terms of θ rather ψ is

$$\sigma(\theta) = \sigma(\psi) \cdot \frac{1 + x \cos \theta}{(1 + 2x \cos \theta + x^2)^{3/2}}$$

9-44. Calculate the differential cross section $\sigma(\theta)$ and the total cross section σ_t for the elastic scattering of a particle from an impenetrable sphere; the potential is given by

$$U(r) = \begin{cases} 0, & r > a \\ \infty, & r < a \end{cases}$$

9-45. Show that the Rutherford scattering cross section (for the case $m_1 = m_2$) can be expressed in terms of the recoil angle as

$$\sigma_{LAB}(\zeta) = \frac{k^2}{T_0^2} \cdot \frac{1}{\cos^3 \zeta}$$

9-46. Consider the case of Rutherford scattering in the event that $m_1 \gg m_2$. Obtain an approximate expression for the differential cross section in the LAB coordinate system.

9-47. Consider the case of Rutherford scattering in the event that $m_2 \gg m_1$. Obtain an expression of the differential cross section in the CM system that is correct to first order in the quantity m_1/m_2. Compare this result with Equation 9.140.

9-48. A fixed force center scatters a particle of mass m according to the force law $F(r) = k/r^3$. If the initial velocity of the particle is u_0, show that the differential scattering cross section is

$$\sigma(\theta) = \frac{k\pi^2(\pi - \theta)}{mu_0^2\theta^2(2\pi - \theta)^2 \sin \theta}$$

9-49. It is found experimentally that in the elastic scattering of neutrons by protons $(m_n \cong m_p)$ at relatively low energies, the energy distribution of the recoiling protons in the LAB system is constant up to a maximum energy, which is the energy of the incident neutrons. What is the angular distribution of the scattering in the CM system?

9-50. Show that the energy distribution of particles recoiling from an elastic collision is always directly proportional to the differential scattering cross section in the CM system.

10

MOTION IN A NONINERTIAL REFERENCE FRAME

10.1 INTRODUCTION

The advantage of choosing an inertial reference frame to describe dynamic processes was made evident in the discussions in Chapters 2 and 7. It is always possible to express the equations of motion for a system in an inertial frame. But there are types of problems for which these equations would be extremely complex, and it becomes easier to treat the motion of the system in a noninertial frame of reference.

To describe, for example, the motion of a particle on or near the surface of the Earth, it is tempting to do so by choosing a coordinate system fixed with respect to the Earth. We know, however, that the Earth undergoes a complicated motion, compounded of many different rotations (and hence accelerations) with respect to an inertial reference frame identified with the "fixed" stars. The Earth coordinate system is, therefore, a *noninertial* frame of reference; and, although the solutions to many problems can be obtained to the desired degree of accuracy by ignoring this distinction, many important effects result from the noninertial nature of the Earth coordinate system.

In analyzing the motion of rigid bodies in the following chapter, we also find it convenient to use noninertial reference frames and therefore make use of much of the development presented here.

10.2 ROTATING COORDINATE SYSTEMS

Let us consider two sets of coordinate axes. Let one set be the "fixed" or inertial axes, and let the other be an arbitrary set that may be in motion with respect to the inertial system. We designate these axes as the "fixed" and "rotating" axes, respectively. We use x_i' as coordinates in the fixed system and x_i as coordinates in the rotating system. If we choose some point P, as in Figure 10-1, we have

$$\mathbf{r}' = \mathbf{R} + \mathbf{r} \tag{10.1}$$

where $\mathbf{r}'$ is the radius vector of P in the fixed system and $\mathbf{r}$ is the radius vector of P in the rotating system. The vector $\mathbf{R}$ locates the origin of the rotating system in the fixed system.

We may always represent an arbitrary infinitesimal displacement by a pure rotation about some axis called the **instantaneous axis of rotation**. For example, the instantaneous motion of a disk rolling down an inclined plane can be described as a rotation about the point of contact between the disk and the plane. Therefore, if the x_i system undergoes an infinitesimal rotation $\delta\boldsymbol{\theta}$, corresponding to some arbitrary infinitesimal displacement, the motion of P (which, for the moment, we consider to be at rest in the x_i system) can be described in terms of Equation 1.106 as

$$(d\mathbf{r})_{\text{fixed}} = d\boldsymbol{\theta} \times \mathbf{r} \tag{10.2}$$

where the designation "fixed" is explicitly included to indicate that the quantity $d\mathbf{r}$ is measured in the x_i', or *fixed*, coordinate system. Dividing this equation by dt, the time interval during which the infinitesimal rotation takes place, we obtain the time of rate change of $\mathbf{r}$ as measured in the fixed coordinate system:

$$\left(\frac{d\mathbf{r}}{dt}\right)_{\text{fixed}} = \frac{d\boldsymbol{\theta}}{dt} \times \mathbf{r} \tag{10.3}$$

or, because the angular velocity of the rotation is

$$\boldsymbol{\omega} \equiv \frac{d\boldsymbol{\theta}}{dt} \tag{10.4}$$

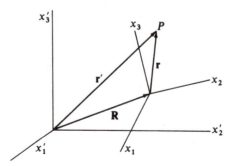

FIGURE 10-1

we have

$$\left(\frac{d\mathbf{r}}{dt}\right)_{\text{fixed}} = \boldsymbol{\omega} \times \mathbf{r} \qquad \text{(for } P \text{ fixed in } x_i \text{ system)} \qquad \textbf{(10.5)}$$

This same result was determined in Section 1.15.

If we allow the point P to have a velocity $(d\mathbf{r}/dt)_{\text{rotating}}$ with respect to the x_i system, this velocity must be added to $\boldsymbol{\omega} \times \mathbf{r}$ to obtain the time rate of change of $\mathbf{r}$ in the fixed system:

$$\left(\frac{d\mathbf{r}}{dt}\right)_{\text{fixed}} = \left(\frac{d\mathbf{r}}{dt}\right)_{\text{rotating}} + \boldsymbol{\omega} \times \mathbf{r} \qquad \textbf{(10.6)}$$

EXAMPLE **10.1** -

Consider a vector $\mathbf{r} = x_1\mathbf{e}_1 + x_2\mathbf{e}_2 + x_3\mathbf{e}_3$ in the rotating system. Let the fixed and rotating systems have the same origin. Find $\dot{\mathbf{r}}'$ in the fixed system by direct differentiation if the angular velocity of the rotating system is $\boldsymbol{\omega}$ in the fixed system.

Solution: We begin by taking the time derivative directly

$$\left(\frac{d\mathbf{r}}{dt}\right)_{\text{fixed}} = \frac{d}{dt}\left(\sum_i x_i\mathbf{e}_i\right)$$

$$= \sum_i (\dot{x}_i\mathbf{e}_i + x_i\dot{\mathbf{e}}_i) \qquad \textbf{(10.7)}$$

The first term is simply $\dot{\mathbf{r}}_r$ in the rotating system, but what are the $\dot{\mathbf{e}}_i$?

$$\dot{\mathbf{r}}_r = \left(\frac{d\mathbf{r}}{dt}\right)_{\text{rotating}}$$

$$\left(\frac{d\mathbf{r}}{dt}\right)_{\text{fixed}} = \dot{\mathbf{r}}_r + \sum_i x_i\dot{\mathbf{e}}_i \qquad \textbf{(10.8)}$$

Look at Figure 10-2 and examine which components of ω_i tend to rotate $\mathbf{e}_1$. We see that ω_2 tends to rotate $\mathbf{e}_1$ toward the $-\mathbf{e}_3$ direction and that ω_3 tends to rotate $\mathbf{e}_1$ toward the $+\mathbf{e}_2$ direction. We therefore have

$$\frac{d\mathbf{e}_1}{dt} = \omega_3\mathbf{e}_2 - \omega_2\mathbf{e}_3 \qquad \textbf{(10.9a)}$$

Similarly, we have

$$\frac{d\mathbf{e}_2}{dt} = -\omega_3\mathbf{e}_1 + \omega_1\mathbf{e}_3 \qquad \textbf{(10.9b)}$$

$$\frac{d\mathbf{e}_3}{dt} = \omega_2\mathbf{e}_1 - \omega_1\mathbf{e}_2 \qquad \textbf{(10.9c)}$$

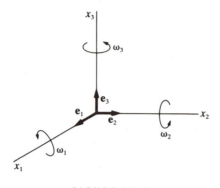

FIGURE 10-2

In each case, the direction of the time derivative of the unit vector must be perpendicular to the unit vector in order not to change its magnitude.

Equations 10.9a–c can be written

$$\dot{\mathbf{e}}_i = \boldsymbol{\omega} \times \mathbf{e}_i \tag{10.10}$$

and Equation 10.8 becomes

$$\left(\frac{d\mathbf{r}}{dt}\right)_{\text{fixed}} = \dot{\mathbf{r}}_r + \sum_i \boldsymbol{\omega} \times x_i \mathbf{e}_i$$

$$= \dot{\mathbf{r}}_r + \boldsymbol{\omega} \times \mathbf{r} \tag{10.11}$$

which is the same result as Equation 10.6.

-- ●

Although we choose the displacement vector $\mathbf{r}$ for the derivation of Equation 10.6, the validity of this expression is not limited to the vector $\mathbf{r}$. In fact, for an arbitrary vector $\mathbf{Q}$, we have

$$\boxed{\left(\frac{d\mathbf{Q}}{dt}\right)_{\text{fixed}} = \left(\frac{d\mathbf{Q}}{dt}\right)_{\text{rotating}} + \boldsymbol{\omega} \times \mathbf{Q}} \tag{10.12}$$

Equation 10.12 is an important result.

We note, for example, that the angular acceleration $\dot{\boldsymbol{\omega}}$ is the same in both the fixed and rotating systems:

$$\left(\frac{d\boldsymbol{\omega}}{dt}\right)_{\text{fixed}} = \left(\frac{d\boldsymbol{\omega}}{dt}\right)_{\text{rotating}} + \boldsymbol{\omega} \times \boldsymbol{\omega} \equiv \dot{\boldsymbol{\omega}} \tag{10.13}$$

because $\boldsymbol{\omega} \times \boldsymbol{\omega}$ vanishes and $\dot{\boldsymbol{\omega}}$ designates the common value in the two systems.

Equation 10.12 may now be used to obtain the expressions for the velocity of the point P as measured in the fixed coordinate system. From Equation 10.1, we have

$$\left(\frac{d\mathbf{r}'}{dt}\right)_{\text{fixed}} = \left(\frac{d\mathbf{R}}{dt}\right)_{\text{fixed}} + \left(\frac{d\mathbf{r}}{dt}\right)_{\text{fixed}} \tag{10.14}$$

so that

$$\left(\frac{d\mathbf{r}'}{dt}\right)_{\text{fixed}} = \left(\frac{d\mathbf{R}}{dt}\right)_{\text{fixed}} + \left(\frac{d\mathbf{r}}{dt}\right)_{\text{rotating}} + \boldsymbol{\omega} \times \mathbf{r} \tag{10.15}$$

If we define

$$\mathbf{v}_f \equiv \dot{\mathbf{r}}_f \equiv \left(\frac{d\mathbf{r}'}{dt}\right)_{\text{fixed}} \tag{10.16a}$$

$$\mathbf{V} \equiv \dot{\mathbf{R}}_f \equiv \left(\frac{d\mathbf{R}}{dt}\right)_{\text{fixed}} \tag{10.16b}$$

$$\mathbf{v}_r \equiv \dot{\mathbf{r}}_r \equiv \left(\frac{d\mathbf{r}}{dt}\right)_{\text{rotating}} \tag{10.16c}$$

we may write

$$\boxed{\mathbf{v}_f = \mathbf{V} + \mathbf{v}_r + \boldsymbol{\omega} \times \mathbf{r}} \tag{10.17}$$

where

$\mathbf{v}_f =$ Velocity relative to the fixed axes

$\mathbf{V} =$ Linear velocity of the moving origin

$\mathbf{v}_r =$ Velocity relative to the rotating axes

$\boldsymbol{\omega} =$ Angular velocity of the rotating axes

$\boldsymbol{\omega} \times \mathbf{r} =$ Velocity due to the rotation of the moving axes

10.3 CENTRIFUGAL AND CORIOLIS FORCES

We have seen that Newton's equation $\mathbf{F} = m\mathbf{a}$ is valid only in an inertial frame of reference. The expression for the force on a particle can therefore be obtained from

$$\mathbf{F} = m\mathbf{a}_f = m\left(\frac{d\mathbf{v}_f}{dt}\right)_{\text{fixed}} \tag{10.18}$$

where the differentiation must be carried out with respect to the fixed system. Differentiating Equation 10.17, we have

$$\left(\frac{d\mathbf{v}_f}{dt}\right)_{\text{fixed}} = \left(\frac{d\mathbf{V}}{dt}\right)_{\text{fixed}} + \left(\frac{d\mathbf{v}_r}{dt}\right)_{\text{fixed}} + \dot{\boldsymbol{\omega}} \times \mathbf{r} + \boldsymbol{\omega} \times \left(\frac{d\mathbf{r}}{dt}\right)_{\text{fixed}} \tag{10.19}$$

We denote the first term by $\ddot{\mathbf{R}}_f$:

$$\ddot{\mathbf{R}}_f \equiv \left(\frac{d\mathbf{V}}{dt}\right)_{\text{fixed}} \tag{10.20}$$

The second term can be evaluated by substituting $\mathbf{v}_r$ for $\mathbf{Q}$ in Equation 10.12:

$$\left(\frac{d\mathbf{v}_r}{dt}\right)_{\text{fixed}} = \left(\frac{d\mathbf{v}_r}{dt}\right)_{\text{rotating}} + \boldsymbol{\omega} \times \mathbf{v}_r$$

$$= \mathbf{a}_r + \boldsymbol{\omega} \times \mathbf{v}_r \tag{10.21}$$

where $\mathbf{a}_r$ is the acceleration in the rotating coordinate system. The last term in Equation 10.19 can be obtained directly from Equation 10.6:

$$\boldsymbol{\omega} \times \left(\frac{d\mathbf{r}}{dt}\right)_{\text{fixed}} = \boldsymbol{\omega} \times \left(\frac{d\mathbf{r}}{dt}\right)_{\text{rotating}} + \boldsymbol{\omega} \times (\boldsymbol{\omega} \times \mathbf{r})$$

$$= \boldsymbol{\omega} \times \mathbf{v}_r + \boldsymbol{\omega} \times (\boldsymbol{\omega} \times \mathbf{r}) \tag{10.22}$$

Combining Equations 10.18–10.22, we obtain

$$\mathbf{F} = m\mathbf{a}_f = m\ddot{\mathbf{R}}_f + m\mathbf{a}_r + m\dot{\boldsymbol{\omega}} \times \mathbf{r} + m\boldsymbol{\omega} \times (\boldsymbol{\omega} \times \mathbf{r}) + 2m\boldsymbol{\omega} \times \mathbf{v}_r \tag{10.23}$$

To an observer in the rotating coordinate system, however, the effective force on a particle is given by*

$$\mathbf{F}_{\text{eff}} \equiv m\mathbf{a}_r \tag{10.24}$$

$$= \mathbf{F} - m\ddot{\mathbf{R}}_f - m\dot{\boldsymbol{\omega}} \times \mathbf{r} - m\boldsymbol{\omega} \times (\boldsymbol{\omega} \times \mathbf{r}) - 2m\boldsymbol{\omega} \times \mathbf{v}_r \tag{10.25}$$

The first term, $\mathbf{F}$, is the sum of the forces acting on the particle as measured in the fixed inertial system. The second ($-m\ddot{\mathbf{R}}_f$) and third ($-m\dot{\boldsymbol{\omega}} \times \mathbf{r}$) terms result because of the translational and angular acceleration, respectively, of the moving coordinate system relative to the fixed system.

The quantity $-m\boldsymbol{\omega} \times (\boldsymbol{\omega} \times \mathbf{r})$ is the usual *centrifugal force* term and reduces to $-m\omega^2 r$ for the case in which $\boldsymbol{\omega}$ is normal to the radius vector. Note that the minus sign implies that the centrifugal force is directed *outward* from the center of rotation (Figure 10-3).

The last term in Equation 10.25 is a totally new quantity that arises from the motion of the particle in the rotating coordinate system. This term is called the **Coriolis force**. Note that the Coriolis force does indeed arise from the *motion* of the particle, because the force is proportional to v_r and hence vanishes if there is no motion.

Because we have used (on several occasions) the term *centrifugal force* and have now introduced the Coriolis force, we must now inquire about the physical meaning of these quantities. It is important to realize that the centrifugal and Coriolis forces are not forces in the usual sense of the word; they have been introduced

---

* This result was published by G. G. Coriolis in 1835. The theory of the composition of accelerations was an outgrowth of Coriolis's study of water wheels.

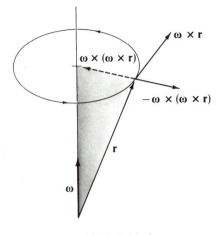

FIGURE 10-3

in an artificial manner as a result of our arbitrary requirement that we be able to write an equation resembling Newton's equation that is at the same time valid in a noninertial reference frame; that is, the equation

$$\mathbf{F} = m\mathbf{a}_f$$

is valid only in an inertial frame. If, in a rotating reference frame, we wish to write (let $\ddot{\mathbf{R}}_f$ and $\dot{\boldsymbol{\omega}}$ be zero for simplicity)

$$\mathbf{F}_{\text{eff}} = m\mathbf{a}_r$$

then we can express such an equation in terms of the real force $m\mathbf{a}_f$ as

$$\mathbf{F}_{\text{eff}} = m\mathbf{a}_f + (\text{noninertial terms})$$

where the "noninertial terms" are identified as the centrifugal and Coriolis "forces." Thus, for example, if a body rotates about a fixed force center, the only real force on the body is the force of attraction toward the force center (and gives rise to the *centripetal* acceleration). An observer moving with the rotating body, however, measures this central force and also notes that the body does not fall toward the force center. To reconcile this result with the requirement that the net force on the body vanish, the observer must postulate an additional force—the centrifugal force. But the "requirement" is artificial; it arises solely from an attempt to extend the form of Newton's equation to a noninertial system, and this can be done only by introducing a fictitious "correction force." The same comments apply for the Coriolis force; this "force" arises when an attempt is made to describe motion relative to the rotating body.

Despite their artificiality, the concepts of centrifugal and Coriolis forces are useful. To describe the motion of a particle relative to a body rotating with respect to an inertial reference frame is a complicated matter. But the problem can be made relatively easy by the simple expedient of introducing the "noninertial forces," which then allows the use of an equation of motion resembling Newton's equation.

EXAMPLE **10.2** ---

A student is performing measurements with a hockey puck on a large merry-go-round with a smooth (frictionless) horizontal, flat surface. The merry-go-round has a constant angular velocity ω and rotates counterclockwise as seen from above. (a) Find the effective force on the hockey puck after it is given a push. (b) Plot the path for various initial directions and velocities of the puck as observed by the person on the merry-go-round that pushes the puck.

Solution: The first three terms for $\mathbf{F}_{eff}$ in Equation 10.25 are zero, so the effective force as observed by the person on the merry-go-round is

$$\mathbf{F}_{eff} = -m\boldsymbol{\omega} \times (\boldsymbol{\omega} \times \mathbf{r}) - 2m\boldsymbol{\omega} \times \mathbf{v}_r \tag{10.26}$$

We have taken the frictional force to be zero. Remember that $\mathbf{v}_r$ is the velocity as measured by the observer on the rotating surface. The effective acceleration is

$$\mathbf{a}_{eff} = \frac{\mathbf{F}_{eff}}{m} = -\boldsymbol{\omega} \times (\boldsymbol{\omega} \times \mathbf{r}) - 2\boldsymbol{\omega} \times \mathbf{v}_r \tag{10.27}$$

The velocity and position are given by integration, in turn, of the acceleration.

$$\mathbf{v}_{eff} = \int \mathbf{a}_{eff}\, dt \tag{10.28a}$$

$$\mathbf{r}_{eff} = \int \mathbf{v}_{eff}\, dt \tag{10.28b}$$

We put the origin of our rotating coordinate system at the center of the merry-go-round. We will need the initial positions and velocities of the puck to plot the motion. For this example, we let the radius of the merry-go-round be R and the velocities be in units of ωR. The initial position of the puck will always be at an (x, y) position of $(-0.5R, 0)$.

We perform a numerical calculation to determine the motion and show the results for several directions and values of the initial velocity in Figure 10-4. For purposes of calculation, we let $\omega = 1$ rad/s and $R = 1$ m, so the units of v_0 (initial speed) and T (time for puck to slide off the surface) shown in Figure 10-4 are in m/s and s, respectively. For parts (a)–(d), the initial velocity is in the $+y$-direction, and the initial speed decreases in each succeeding view. In (a), the puck slides off quickly. For (b) and (d), the puck slides off at similar positions, but note the differences in initial speeds as well as the time it takes the puck to reach the edge. For a speed intermediate between these two speeds, as seen in (c), the puck may make several paths around the merry-go-round; at some speed, the puck must stay on. The last two views show the initial velocity at an angle of $45°$ to the x-axis. In (e), the puck loops around its path along the way to exiting the merry-go-round, and in (f), it changes direction rather abruptly.

The real challenge is to perform such experiments to compare the actual paths in the fixed and rotating coordinate systems with the computer calculations. In each of the cases above, the puck will move in a straight line in the fixed system, because there is no friction or external force in the plane.

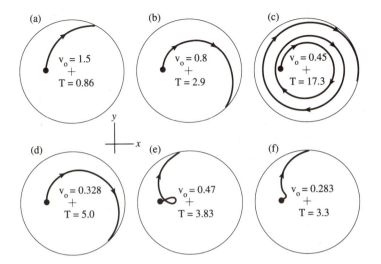

FIGURE 10-4   The motion of the hockey puck of Example 10.2 as observed in the rotating system for various initial velocities and directions. The angular velocity ω (1 rad/s) is out of the page.

10.4   MOTION RELATIVE TO THE EARTH

The motion of the Earth with respect to an inertial reference frame is dominated by the Earth's rotation about its own axis. The effects of the other motions (e.g., the revolution about the sun and the motion of the solar system with respect to the local galaxy) are small by comparison. If we place the fixed inertial frame $x'y'z'$ at the center of the Earth and the moving reference frame xyz on the surface of the Earth, we can describe the motion of a moving object close to the surface of the Earth as shown in Figure 10-5. We then apply Equation 10.25 to the dynamical

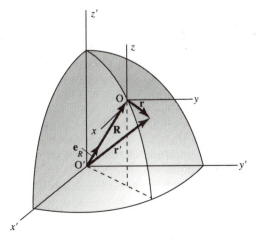

FIGURE 10-5

motion. We denote the forces as measured in the fixed inertial system as $\mathbf{F} = \mathbf{S} + m\mathbf{g}_0$, where $\mathbf{S}$ represents the sum of the external forces (e.g., impulse, electromagnetic, friction) other than gravitation and $m\mathbf{g}_0$ represents the gravitational attraction to the Earth. In this case, $\mathbf{g}_0$ represents the Earth's gravitational field vector (Equation 5.3),

$$\mathbf{g}_0 = -G\frac{M_E}{R^2}\mathbf{e}_R \tag{10.29}$$

where M_E is the mass of the Earth, R is the radius of the Earth, and the unit vector $\mathbf{e}_R$ is a unit vector along the direction of $\mathbf{R}$ in Figure 10-5. We are assuming the Earth is spherical and isotropic and that $\mathbf{R}$ originates from the center of the Earth. The acceleration of gravity varies over the surface of the Earth due to the Earth's oblateness, density nonuniformities, and altitude. We choose, at the present time, not to add this complexity to motion relative to the Earth, but we have previously pointed out that effects such as these can be considered in due course by performing computer calculations.

The effective force $\mathbf{F}_{\text{eff}}$ as measured in the moving system placed on the surface of the Earth becomes, from Equation 10.25,

$$\mathbf{F}_{\text{eff}} = \mathbf{S} + m\mathbf{g}_0 - m\ddot{\mathbf{R}}_f - m\dot{\boldsymbol{\omega}} \times \mathbf{r} - m\boldsymbol{\omega} \times (\boldsymbol{\omega} \times \mathbf{r}) - 2m\boldsymbol{\omega} \times \mathbf{v}_r \tag{10.30}$$

We let the Earth's angular velocity $\boldsymbol{\omega}$ be along the inertial system's z' direction ($\mathbf{e}'_z$). The value of ω is 7.3×10^{-5} rad/s, which is a relatively slow rotation, but it is 365 times greater than the rotation frequency of the Earth about the sun. The value of $\boldsymbol{\omega}$ is practically constant in time, and the term $\dot{\boldsymbol{\omega}} \times \mathbf{r}$ will be neglected.

According to Equation 10.12, we have for the third term above,

$$\ddot{\mathbf{R}}_f = \boldsymbol{\omega} \times \dot{\mathbf{R}}_f$$

$$\ddot{\mathbf{R}}_f = \boldsymbol{\omega} \times (\boldsymbol{\omega} \times \mathbf{R}) \tag{10.31}$$

Equation 10.30 now becomes

$$\mathbf{F}_{\text{eff}} = \mathbf{S} + m\mathbf{g}_0 - m\boldsymbol{\omega} \times [\boldsymbol{\omega} \times (\mathbf{r} + \mathbf{R})] - 2m\boldsymbol{\omega} \times \mathbf{v}_r \tag{10.32}$$

The second and third terms (divided by m) are what we experience (and measure) on the surface of the Earth as the effective $\mathbf{g}$, and we will henceforth denote it as $\mathbf{g}$. Its value is

$$\mathbf{g} = \mathbf{g}_0 - \boldsymbol{\omega} \times [\boldsymbol{\omega} \times (\mathbf{r} + \mathbf{R})] \tag{10.33}$$

The second term of Equation 10.33 is the centrifugal force. Because we are limiting our present consideration to motion near the surface of the Earth, we have $r \ll R$, and the $\boldsymbol{\omega} \times (\boldsymbol{\omega} \times \mathbf{R})$ term totally dominates the centrifugal force. For situations far away from the surface of the Earth, we would have to consider both the variation of g with altitude as well as the $\boldsymbol{\omega} \times (\boldsymbol{\omega} \times \mathbf{r})$ term. The centrifugal force is responsible for the oblateness of the Earth. The Earth is not really a solid spheroid; it is more like a strongly viscous liquid with a solid crust. Because of the Earth's rotation, the Earth has deformed so that its equatorial radius is 21.4 km greater than its polar radius, and the acceleration of gravity is 0.052 m/s^2 greater at the poles than at the equator. The surface of calm ocean water is perpendicular to **g**, not **g₀**, and on the average, the plane of the Earth's surface is also perpendicular to **g**.

We rewrite Equation 10.32 in simpler terms as

$$\boxed{\mathbf{F}_{\text{eff}} = \mathbf{S} + m\mathbf{g} - 2m\boldsymbol{\omega} \times \mathbf{v}_r} \tag{10.34}$$

It is this equation that we will use to discuss the motion of objects close to the surface of the Earth.

But first, let's return to the effective **g** of Equation 10.33. The period of a pendulum determines the magnitude of **g**, and the direction of a plumb bob in equilibrium determines the direction of **g**. The value of $\omega^2 R$ is 0.034 m/s², and this is a significant enough amount (3.5%) of the magnitude of **g** to be considered. We determined the direction of the centrifugal term $\boldsymbol{\omega} \times [\boldsymbol{\omega} \times (\mathbf{r} + \mathbf{R})]$ in Figure 10-3 (where the **r** is our **r′** of Figure 10-5). The direction of the centrifugal term $(-\boldsymbol{\omega} \times [\boldsymbol{\omega} \times (\mathbf{r} + \mathbf{R})]$ is outward from the axis of the rotating Earth. The direction of a plumb bob will include the centrifugal term. Because of this fact, the direction of **g** at a given point is in general slightly different from the true vertical (defined as the direction of the line connecting the point with center of the Earth; see Problem 10-12). The situation is represented schematically (with considerable exaggeration) in Figure 10-6.

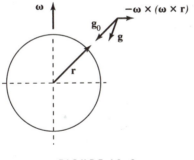

FIGURE 10-6

Coriolis Force Effects

The angular velocity vector **ω**, which represents the Earth's rotation about its axis, is directed in a northerly direction. Therefore, in the Northern Hemisphere, **ω** has a component ω_z directed *outward* along the local vertical. If a particle is projected in a horizontal plane (in the local coordinate system at the surface of the Earth) with a velocity **v**ᵣ, then the Coriolis force $-2m\boldsymbol{\omega} \times \mathbf{v}_r$ has a component in the plane of magnitude $2m\omega_z v_r$ directed toward the *right* of the particle's motion (see Figure 10-7), and a deflection from the original direction of motion results.*

---

* Poisson discussed the deviation of projectile motion in 1837.

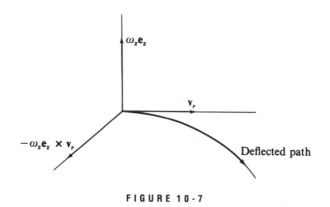

FIGURE 10-7

Because the magnitude of the horizontal component of the Coriolis force is proportional to the vertical component of **ω**, the portion of the Coriolis force producing deflections depends on the latitude, being a maximum at the North Pole and zero at the equator. In the Southern Hemisphere, the component ω_z is directed *inward* along the local vertical, and hence all deflections are in the opposite sense from those in the Northern Hemisphere.*

Perhaps the most noticeable effect of the Coriolis force is that on the air masses. As air flows from high-pressure regions to low pressure, the Coriolis force deflects the air toward the right in the Northern Hemisphere, producing cyclonic motion (Figure 10-8). The air rotates with high pressure on the right and low pressure on the left. The high pressure prevents the Coriolis force from deflecting the air masses farther to the right, resulting in a counterclockwise flow of air. In the temperate regions, the airflow does not tend to be along the pressure gradients, but rather along the pressure isobars due to the Coriolis force and the associated centrifugal force of the rotation.

Near the equatorial regions, the sun heating the Earth's surface causes hot surface air to rise. In the Northern Hemisphere, this results in cooler air moving in a southerly direction toward the equator. The Coriolis force deflects this moving air to the right, resulting in the *trade winds,* which provide a breeze toward the southwest in the Northern Hemisphere and toward the northwest in the Southern Hemisphere. Note that this particular effect does not occur *at the equator* because of the directions of **ω** and the air's surface **v**.

The actual motion of air masses is much more complicated than the simple picture described here, but the qualitative features of cyclonic motion and the trade winds are correctly given by considering the effects of the Coriolis force. The motion of water in whirlpools is (at least in principle) a similar situation, but in

---

* During the naval engagement near the Falkland Islands early in World War I, the British gunners were surprised to see their accurately aimed salvos falling 100 yards to the left of the German ships. The designers of the sighting mechanisms were well aware of the Coriolis deflection and had carefully taken this into account, but they apparently were under the impression that all sea battles took place near 50° N latitude and never near 50° S latitude. The British shots, therefore, fell at a distance from the targets equal to *twice* the Coriolis deflection.

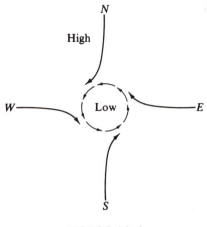

FIGURE 10-8

actuality, other factors (various perturbations and residual angular momentum) dominate the Coriolis force, and whirlpools are found with both directions of flow. Even under laboratory conditions, it is extremely difficult to isolate the Coriolis effect. (Reports of water in flush toilets and bathtubs circulating in opposite directions as cruise ships cross the equator are most likely highly exaggerated.)

EXAMPLE **10.3**

Find the horizontal deflection from the plumb line caused by the Coriolis force acting on a particle falling freely in the Earth's gravitational field from a height h above the Earth's surface.

Solution: We use Equation 10.34 with the applied forces $\mathbf{S} = 0$. If we set $\mathbf{F}_{\text{eff}} = m\mathbf{a}_r$, we can solve for the acceleration of the particle in the rotating coordinate system fixed on the Earth.

$$\mathbf{a}_r = \mathbf{g} - 2\boldsymbol{\omega} \times \mathbf{v}_r$$

The acceleration due to gravity $\mathbf{g}$ is the effective one and is along the plumb line. We choose a z-axis directed vertically outward (along $-\mathbf{g}$) from the surface of the Earth. With this definition of $\mathbf{e}_z$, we complete the construction of a right-hand coordinate system by specifying that $\mathbf{e}_x$ be in a southerly and $\mathbf{e}_y$ in an easterly direction, as in Figure 10-9. We make the approximation that the distance of fall is sufficiently small that g remains constant during the process.

Because we have chosen the origin O of the rotating coordinate system to lie in the Northern Hemisphere, we have

$$\omega_x = -\omega \cos \lambda$$

$$\omega_y = 0$$

$$\omega_z = \omega \sin \lambda$$

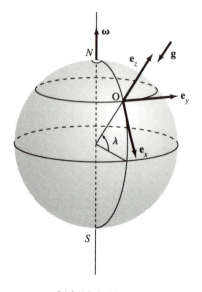

FIGURE 10-9

Although the Coriolis force produces small velocity components in the $\mathbf{e}_y$ and $\mathbf{e}_x$ directions, we can certainly neglect $\dot{x}$ and $\dot{y}$ compared with $\dot{z}$, the vertical velocity. Then, approximately,

$$\dot{x} \cong 0$$

$$\dot{y} \cong 0$$

$$\dot{z} \cong -gt$$

where we obtain $\dot{z}$ by considering a fall from rest. Therefore, we have

$$\boldsymbol{\omega} \times \mathbf{v}_r \cong \begin{vmatrix} \mathbf{e}_x & \mathbf{e}_y & \mathbf{e}_z \\ -\omega \cos \lambda & 0 & \omega \sin \lambda \\ 0 & 0 & -gt \end{vmatrix}$$

$$\cong -(\omega gt \cos \lambda)\mathbf{e}_y$$

The components of **g** are

$$g_x = 0$$

$$g_y = 0$$

$$g_z = -g$$

so the equations for the components of $\mathbf{a}_r$ (neglecting terms in ω^2; see Problem 10-13) become

$$(\mathbf{a}_r)_x = \ddot{x} \cong 0$$

$$(\mathbf{a}_r)_y = \ddot{y} \cong 2\omega gt \cos \lambda$$

$$(\mathbf{a}_r)_z = \ddot{z} \cong -g$$

Thus, the effect of the Coriolis force is to produce an acceleration in the $\mathbf{e}_y$, or easterly, direction. Integrating $\ddot{y}$ twice, we have

$$y(t) \cong \tfrac{1}{3}\omega g t^3 \cos \lambda$$

where $y = 0$ and $\dot{y} = 0$ at $t = 0$. The integration of $\dot{z}$ yields the familiar result for the distance of fall,

$$z(t) \cong z(0) - \tfrac{1}{2}g t^2$$

and the time of fall from a height $h = z(0)$ is given by

$$t \cong \sqrt{2h/g}$$

Hence the result for the eastward deflection d of a particle dropped from rest at a height h and at a northern latitude λ is*

$$d \cong \tfrac{1}{3}\,\omega \cos \lambda \sqrt{\frac{8h^3}{g}} \qquad (10.35)$$

An object dropped from a height of 100 m at latitude 45° is deflected approximately 1.55 cm (neglecting the effects of air resistance).

-- ●

E X A M P L E   10.4 --

To demonstrate the power of the Coriolis method for obtaining the equations of motion in a noninertial reference frame, rework the last example but use only the formalism previously developed—the theory of central-force motion.

Solution: If we release a particle of small mass from a tower of height h above the Earth's surface, the path the particle describes is a conic section—an ellipse with $\varepsilon \cong 1$ and with one focus very close to the Earth's center. If R is the Earth's radius and λ the (northern) latitude, then at the moment of release, the particle has a horizontal velocity in the eastward direction:

$$v_{\text{hor}} = r\omega \cos \lambda = (R + h)\omega \cos \lambda$$

and the angular momentum about the polar axis is

$$l = mr v_{\text{hor}} = m(R + h)^2 \omega \cos \lambda \qquad (10.36)$$

----

* The eastward deflection was predicted by Newton (1679), and several experiments (notably those of Robert Hooke) appeared to confirm the results. The most careful measurements were probably those of F. Reich (1831; published 1833), who dropped pellets down a mine shaft 188 m deep and observed a mean deflection of 28 mm. This is smaller than the value calculated from Equation 10.35, the decrease being due to air resistance effects. In all the experiments, a small southerly component of the deflection was observed—and remained unaccounted for until Coriolis's theorem was appreciated (see Problems 10-13 and 10-14).

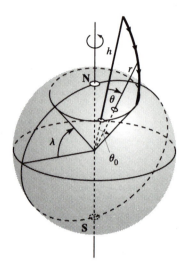

FIGURE 10-10

The equation of the path is*

$$\frac{\alpha}{r} = 1 - \varepsilon \cos \theta \qquad (10.37)$$

if we measure θ from the initial position of the particle (see Figure 10-10). At $t = 0$, we have

$$\frac{\alpha}{R + h} = 1 - \varepsilon$$

so Equation 10.37 can be written as

$$r = \frac{(1 - \varepsilon)(R + h)}{1 - \varepsilon \cos \theta} \qquad (10.38)$$

From Equation 8.12 for the areal velocity, we can write

$$\frac{1}{2} r^2 \frac{d\theta}{dt} = \frac{l}{2m}$$

Thus, the time t required to describe an angle θ is

$$t = \frac{m}{l} \int_0^\theta r^2 \, d\theta$$

---

* Notice that there is a change of sign between Equation 10.37 and Equation 8.41 due to the different origins for θ in the two cases.

Substituting into this expression the value of l from Equation 10.36 and r from Equation 10.38, we find

$$t = \frac{1}{\omega \cos \lambda} \int_0^\theta \left(\frac{1 - \varepsilon}{1 - \varepsilon \cos \theta} \right)^2 d\theta \qquad (10.39)$$

If we let $\theta = \theta_0$ when the particle has reached the Earth's surface ($r = R$), then Equation 10.38 becomes

$$\frac{R}{R + h} = \frac{1 - \varepsilon}{1 - \varepsilon \cos \theta_0}$$

or, inverting,

$$1 + \frac{h}{R} = \frac{1 - \varepsilon \cos \theta_0}{1 - \varepsilon}$$

$$= \frac{1 - \varepsilon[1 - 2 \sin^2(\theta_0/2)]}{1 - \varepsilon}$$

$$= 1 + \frac{2\varepsilon}{1 - \varepsilon} \sin^2 \frac{\theta_0}{2} \qquad (10.40)$$

from which we have

$$\frac{h}{R} = \frac{2\varepsilon}{1 - \varepsilon} \sin^2 \frac{\theta_0}{2}$$

Because the path described by the particle is almost vertical, little change occurs in the angle θ between the position of release and the point at which the particle reaches the surface of the Earth; θ_0 is therefore small and $\sin(\theta_0/2)$ can be approximated by its argument:

$$\frac{h}{R} \cong \frac{\varepsilon \theta_0^2}{2(1 - \varepsilon)} \qquad (10.41)$$

If we expand the integrand in Equation 10.39 by the same method used to obtain Equation 10.40, we find

$$t = \frac{1}{\omega \cos \lambda} \int_0^\theta \frac{d\theta}{\{1 + [2\varepsilon/(1 - \varepsilon)] \sin^2(\theta/2)\}^2}$$

and because θ is small, we have

$$t \cong \frac{1}{\omega \cos \lambda} \int_0^\theta \frac{d\theta}{[1 + \varepsilon \theta^2/2(1 - \varepsilon)]^2}$$

Substituting for $\varepsilon/2(1 - \varepsilon)$ from Equation 10.41 and writing $t(\theta = \theta_0) = T$ for the total time of fall, we obtain

$$T \cong \frac{1}{\omega \cos \lambda} \int_0^{\theta_0} \frac{d\theta}{[1 + (h\theta^2/R\theta_0^2)]^2}$$

$$\cong \frac{1}{\omega \cos \lambda} \int_0^{\theta_0} \left(1 - \frac{2h}{R\theta_0^2} \theta^2 \right) d\theta$$

$$= \frac{1}{\omega \cos \lambda} \left(1 - \frac{2h}{3R} \right) \theta_0$$

Solving for θ_0, we find

$$\theta_0 \cong \frac{\omega T \cos \lambda}{1 - 2h/3R} \cong \omega T \cos \lambda \left(1 + \frac{2h}{3R} \right)$$

During the time of fall T, the Earth turns through an angle ωT, so the point on the Earth directly beneath the initial position of the particle moves toward the east by an amount $R\omega T \cos \lambda$. During the same time, the particle is deflected toward the east by an amount $R\theta_0$. Thus, the net easterly deviation d is

$$d = R\theta_0 - R\omega T \cos \lambda$$

$$= \tfrac{2}{3} h\omega T \cos \lambda$$

and using $T \cong \sqrt{2h/g}$ as in the preceding example, we have, finally,

$$d \cong \frac{1}{3} \omega \cos \lambda \sqrt{\frac{8h^3}{g}}$$

which is identical with the result obtained previously (Equation 10.35).

---●

EXAMPLE **10.5** --

The effect of the Coriolis force on the motion of a pendulum produces a *precession*, or rotation with time of the plane of oscillation. Describe the motion of this system, called a *Foucault pendulum*.*

Solution: To describe this effect, let us select a set of coordinate axes with origin at the equilibrium point of the pendulum and z-axis along the local vertical. We are interested only in the rotation of the plane of oscillation—that is, we wish to con-

---

* Devised in 1851 by the French physicist Jean-Bernard-Léon Foucault (1819–1868).

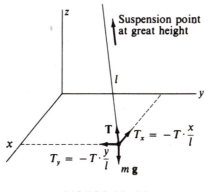

Suspension point
at great height

$T_x = -T \cdot \dfrac{x}{l}$

$T_y = -T \cdot \dfrac{y}{l}$

$m\,\mathbf{g}$

FIGURE 10-11

sider the motion of the pendulum bob in the x–y plane (the horizontal plane). We therefore limit the motion to oscillations of small amplitude, with the horizontal excursions small compared with the length of the pendulum. Under this condition, $\dot{z}$ is small compared with $\dot{x}$ and $\dot{y}$ and can be neglected.

The equation of motion is

$$\mathbf{a}_r = \mathbf{g} + \frac{\mathbf{T}}{m} - 2\boldsymbol{\omega} \times \mathbf{v}_r \tag{10.42}$$

where $\mathbf{T}/m$ is the acceleration produced by the force of tension $\mathbf{T}$ in the pendulum suspension (Figure 10-11). We therefore have, approximately,

$$\left.\begin{aligned} T_x &= -T \cdot \frac{x}{l} \\[2mm] T_y &= -T \cdot \frac{y}{l} \\[2mm] T_z &\cong T \end{aligned}\right\} \tag{10.43}$$

As before,

$$g_x = 0$$

$$g_y = 0$$

$$g_z = -g$$

and

$$\omega_x = -\omega \cos \lambda$$

$$\omega_y = 0$$

$$\omega_z = \omega \sin \lambda$$

with

$$(\mathbf{v}_r)_x = \dot{x}$$

$$(\mathbf{v}_r)_y = \dot{y}$$

$$(\mathbf{v}_r)_z = \dot{z} \cong 0$$

Therefore,

$$\boldsymbol{\omega} \times \mathbf{v}_r \cong \begin{vmatrix} \mathbf{e}_x & \mathbf{e}_y & \mathbf{e}_z \\ -\omega\cos\lambda & 0 & \omega\sin\lambda \\ \dot{x} & \dot{y} & 0 \end{vmatrix}$$

so that

$$\left.\begin{aligned} (\boldsymbol{\omega} \times \mathbf{v}_r)_x &\cong -\dot{y}\omega\sin\lambda \\ (\boldsymbol{\omega} \times \mathbf{v}_r)_y &\cong \dot{x}\omega\sin\lambda \\ (\boldsymbol{\omega} \times \mathbf{v}_r)_z &\cong -\dot{y}\omega\cos\lambda \end{aligned}\right\} \qquad (10.44)$$

Thus, the equations of interest are

$$\left.\begin{aligned} (\mathbf{a}_r)_x &= \ddot{x} \cong -\frac{T}{m}\cdot\frac{x}{l} + 2\dot{y}\omega\sin\lambda \\ (\mathbf{a}_r)_y &\cong \ddot{y} \cong -\frac{T}{m}\cdot\frac{y}{l} - 2\dot{x}\omega\sin\lambda \end{aligned}\right\} \qquad (10.45)$$

For small displacements, $T \cong mg$. Defining $\alpha^2 \equiv T/ml \cong g/l$, and writing $\omega_z = \omega\sin\lambda$, we have

$$\left.\begin{aligned} \ddot{x} + \alpha^2 x &\cong 2\omega_z\dot{y} \\ \ddot{y} + \alpha^2 y &\cong -2\omega_z\dot{x} \end{aligned}\right\} \qquad (10.46)$$

We note that the equation for $\ddot{x}$ contains a term in $\dot{y}$ and that the equation for $\ddot{y}$ contains a term in $\dot{x}$. Such equations are called **coupled equations**. A solution for this pair of coupled equations can be effected by adding the first of the above equations to i times the second:

$$(\ddot{x} + i\ddot{y}) + \alpha^2(x + iy) \cong -2\omega_z(i\dot{x} - \dot{y}) = -2i\omega_z(\dot{x} + i\dot{y})$$

If we write

$$q \equiv x + iy$$

we then have

$$\ddot{q} + 2i\omega_z\dot{q} + \alpha^2 q \cong 0$$

This equation is identical with the equation that describes damped oscillations (Equation 3.35), except that here the term corresponding to the damping factor is purely imaginary. The solution (see Equation 3.37) is

$$q(t) \cong \exp[-i\omega_z t][A\exp(\sqrt{-\omega_z^2 - \alpha^2}\,t) + B\exp(-\sqrt{-\omega_z^2 - \alpha^2}\,t)] \qquad (10.47)$$

If the Earth were not rotating, so that $\omega_z = 0$, then the equation for q would become

$$\ddot{q}' + \alpha^2 q' \cong 0, \qquad \omega_z = 0$$

from which it is seen that α corresponds to the oscillation frequency of the pendulum. This frequency is clearly much greater than the angular frequency of the Earth's rotation. Therefore, $\alpha \gg \omega_z$, and the equation for $q(t)$ becomes

$$q(t) \cong e^{-i\omega_z t}(Ae^{i\alpha t} + Be^{-i\alpha t}) \qquad \text{(10.48)}$$

We can interpret this equation more easily if we note that the equation for q' has the solution

$$q'(t) = x'(t) + iy'(t) = Ae^{i\alpha t} + Be^{-i\alpha t}$$

Thus,

$$q(t) = q'(t) \cdot e^{-i\omega_z t}$$

or

$$x(t) + iy(t) = \left[(x'(t) + iy'(t)) \right] \cdot e^{-i\omega_z t}$$

$$= (x' + iy')(\cos \omega_z t - i \sin \omega_z t)$$

$$= (x' \cos \omega_z t + y' \sin \omega_z t)$$

$$+ i(-x' \sin \omega_z t + y' \cos \omega_z t)$$

Equating real and imaginary parts,

$$\left. \begin{array}{l} x(t) = x' \cos \omega_z t + y' \sin \omega_z t \\ y(t) = -x' \sin \omega_z t + y' \cos \omega_z t \end{array} \right\}$$

We can write these equations in matrix form as

$$\begin{pmatrix} x(t) \\ y(t) \end{pmatrix} = \begin{pmatrix} \cos \omega_z t & \sin \omega_z t \\ -\sin \omega_z t & \cos \omega_z t \end{pmatrix} \begin{pmatrix} x'(t) \\ y'(t) \end{pmatrix} \qquad \text{(10.49)}$$

from which (x, y) may be obtained from (x', y') by the application of a rotation matrix of the familiar form

$$\boldsymbol{\lambda} = \begin{pmatrix} \cos \theta & \sin \theta \\ -\sin \theta & \cos \theta \end{pmatrix} \qquad \text{(10.50)}$$

Thus, the angle of rotation is $\theta = \omega_z t$, and the plane of oscillation of the pendulum therefore rotates with a frequency $\omega_z = \omega \sin \lambda$. The observation of this rotation gives a clear demonstration of the rotation of the Earth.*

---●

---

* Vincenzo Viviani (1622–1703), a pupil of Galileo, had noticed as early as about 1650 that a pendulum undergoes a slow rotation, but there is no evidence that he correctly interpreted the phenomenon. Foucault's invention of the gyroscope in the year following the demonstration of his pendulum provided an even more striking visual proof of the Earth's rotation.

P R O B L E M S

▷ **10-1.** Calculate the centrifugal acceleration, due to the Earth's rotation, on a particle on the surface of the Earth at the equator. Compare this result with the gravitational acceleration. Compute also the centrifugal acceleration due to the motion of the Earth about the sun and justify the remark made in the text that this acceleration may be neglected compared with the acceleration caused by axial rotation.

▷ **10-2.** An automobile drag racer drives a car with acceleration a and instantaneous velocity v. The tires (of radius r_0) are not slipping. Find which point on the tire has the greatest acceleration relative to the ground. What is this acceleration?

▷ **10-3.** In Example 10.2, assume that the coefficient of static friction between the hockey puck and a horizontal rough surface (on the merry-go-round) is μ_s. How far away from the center of the merry-go-round can the hockey puck be placed without sliding?

10-4. In Example 10.2, for what initial velocity and direction in the rotating system will the hockey puck appear to be subsequently motionless in the fixed system? What will be the motion in the rotating system? Let the initial position be the same as in Example 10.2. You may choose to do a numerical calculation.

10-5. Perform a numerical calculation using the parameters in Example 10.2 and Figure 10-4e, but find the initial velocity for which the path of motion passes back over the initial position in the rotating system. At what time does the puck exit the merry-go-round?

▷ **10-6.** A bucket of water is set spinning about its symmetry axis. Determine the shape of the water in the bucket.

10-7. Determine how much greater the gravitational field strength g is at the pole than at the equator. Assume a spherical earth. If the actual measured difference is $\Delta g = 52$ mm/s^2, explain the difference. How might you calculate this difference between the measured result and your calculation?

10-8. If a particle is projected vertically upward to a height h above a point on the Earth's surface at a northern latitude λ, show that it strikes the ground at a point $\frac{4}{3}\omega \cos \lambda \sqrt{8h^3/g}$ to the west. (Neglect air resistance, and consider only small vertical heights.)

10-9. If a projectile is fired due east from a point on the surface of the Earth at a northern latitude λ with a velocity of magnitude V_0 and at an angle of inclination to the horizontal of α, show that the lateral deflection when the projectile strikes the Earth is

$$d = \frac{4V_0^3}{g^2} \cdot \omega \sin \lambda \cdot \sin^2 \alpha \cos \alpha$$

where ω is the rotation frequency of the Earth.

10-10. In the preceding problem, if the range of the projectile is R_0' for the case $\omega = 0$, show that the change of range due to the rotation of the Earth is

$$\Delta R' = \sqrt{\frac{2R_0'^3}{g}} \cdot \omega \cos \lambda \left(\cot^{1/2}\alpha - \frac{1}{3}\tan^{3/2}\alpha \right)$$

10-11. Obtain an expression for the angular deviation of a particle projected from the North Pole in a path that lies close to the Earth. Is the deviation significant for a missile that makes a 4,800-km flight in 10 minutes? What is the "miss distance" if the missile is aimed directly at the target? Is the miss distance greater for a 19,300-km flight at the same velocity?

10-12. Show that the small angular deviation ε of a plumb line from the true vertical (i.e., toward the center of the Earth) at a point on the Earth's surface at a latitude λ is

$$\varepsilon = \frac{R\omega^2 \sin \lambda \cos \lambda}{g_0 - R\omega^2 \cos^2 \lambda}$$

where R is the radius of the Earth. What is the value (in seconds of arc) of the maximum deviation? Note that the entire denominator in the answer is actually the effective g, and g_0 denotes the pure gravitational component.

▷ **10-13.** Refer to Example 10.3 concerning the deflection from the plumb line of a particle falling in the Earth's gravitational field. Take g to be defined at ground level and use the zeroth order result for the time-of-fall, $T = \sqrt{2h/g}$. Perform a calculation in second approximation (i.e., retain terms in ω^2) and calculate the *southerly* deflection. There are three components to consider: (a) Coriolis force to second order (C_1), (b) variation of centrifugal force with height (C_2), and (c) variation of gravitational force with height (C_3). Show that each of these components gives a result equal to

$$C_i \frac{h^2}{g} \omega^2 \sin \lambda \cos \lambda$$

with $C_1 = 2/3$, $C_2 = 5/6$, and $C_3 = 5/2$. The total southerly deflection is therefore $(4h^2\omega^2 \sin \lambda \cos \lambda)/g$.

10-14. Refer to Example 10.3 and the previous problem, but drop the particle at the Earth's surface down a mineshaft to a depth h. Show that in this case there is no southerly deflection due to the variation of gravity and that the total southerly deflection is only

$$\frac{3}{2} \frac{h^2 \omega^2}{g} \sin \lambda \cos \lambda$$

10-15. Consider a particle moving in a potential $U(r)$. Rewrite the Lagrangian in terms of a coordinate system in uniform rotation with respect to an inertial frame. Calculate the Hamiltonian and determine whether $H = E$. Is H a constant of the motion? If E is not a constant of motion, why isn't it? The expression for the Hamiltonian thus obtained is the standard formula $1/2 \, mv^2 + U$ plus an additional term. Show that the extra term is the *centrifugal potential energy*. Use the Lagrangian you obtained to reproduce the equations of motion given in Equation 10.25 (without the second and third terms).

10-16. Consider Problem 2-57 but include the effects of the Coriolis force on the probe. The probe is launched at a latitude of 45° straight up. Determine the horizontal deflection in the probe at its maximum height for each part of Problem 2-57.

11

DYNAMICS OF RIGID BODIES

11.1 INTRODUCTION

We define a rigid body as a collection of particles whose relative distances are constrained to remain absolutely fixed. Such bodies do not exist in nature, because the ultimate component particles composing every body (the atoms) are always undergoing some relative motion. This motion, however, is microscopic, and it therefore usually may be ignored when describing the macroscopic motion of the body. However, macroscopic displacement within the body (such as elastic deformations) can take place. For many bodies of interest, we can safely neglect the changes in size and shape caused by such deformations and obtain equations of motion valid to a high degree of accuracy.

It should also be clear that there is a relativistic limitation to the concept of an absolutely rigid body. Consider, for example, a long bar of some material. If we strike a blow at one end of the bar and if the bar were absolutely rigid, the effect would be felt instantaneously at the opposite end. But this corresponds to the transmission of a signal with an infinite velocity—a situation that, from relativity theory, we know is impossible. (Actually, the velocity of transmission of such a signal in a metal bar is rather low compared with the velocity of light—$\sim 10^7$ m/s—and depends on the elastic properties of the material.)

We here use the idealized concept of a rigid body as a collection of discrete particles or as a continuous distribution of matter interchangeably. The only change is the replacement of summations over particles by integrations over mass density distributions. The equations of motion are equally valid for either viewpoint.

To describe the motion of a rigid body, we use two coordinate systems—an inertial frame and a coordinate system fixed with respect to the body. Six quantities must be specified to denote the position of the body. These can be taken to be the coordinates of the center of mass (which can often conveniently be made to coincide with the origin of the body coordinate system) and three independent angles that give the orientation of the body coordinate system with respect to the fixed (or inertial) system.* The three independent angles can conveniently be taken to be the **Eulerian angles**, described in Section 11.7.

It should be intuitively obvious that any arbitrary finite motion of a rigid body can be considered to be the sum of two independent motions—a linear translation of some point of the body plus a rotation about that point.[†] If the point is chosen to be the center of mass of the body, then such a separation of the motion into two parts allows the use of the development in Chapter 9, which indicates that the angular momentum (see Equation 9.23) and the kinetic energy (see Equation 9.39) can be separated into portions relating to the motion *of* the center of mass and to the motion *around* the center of mass.

If the potential energy can also be separated (as is always the case, for example, for the potential energy in a uniform force field), then the Lagrangian separates, and the entire problem conveniently divides into two parts, one involving only translation and the other only rotation. Each portion of the problem can then be solved independently of the other.[‡] This type of separation is essential for a relatively uncomplicated description of rigid-body motion.

11.2 INERTIA TENSOR

We now direct our attention to a rigid body composed of n particles of masses m_α, $\alpha = 1, 2, 3, \ldots, n$. If the body rotates with an instantaneous angular velocity $\boldsymbol{\omega}$ about some point fixed with respect to the body coordinate system and if this point moves with an instantaneous linear velocity $\mathbf{V}$ with respect to the fixed coordinate system, then the instantaneous velocity of the αth particle in the fixed system can be obtained by using Equation 10.17. But we are now considering a rigid body, so

$$\mathbf{v}_r = \left(\frac{d\mathbf{r}}{dt}\right)_{\text{rotating}} \equiv 0$$

Therefore,

$$\boxed{\mathbf{v}_\alpha = \mathbf{V} + \boldsymbol{\omega} \times \mathbf{r}_\alpha} \qquad (11.1)$$

---

* In this chapter, we use the designation *body system* in place of the term *rotating system* used in the preceding chapter. The term *fixed system* will be retained.

[†] *Chasles' theorem*, which is even more general than this statement (it says that the line of translation and the axis of rotation can be made to coincide), was proven by the French mathematician Michel Chasles (1793–1880) in 1830. The proof is given, e.g., by E. T. Whittaker (Wh37, p. 4).

[‡] This important point was first realized by Euler in 1749.

where the subscript f, denoting the fixed coordinate system, has been deleted from the velocity $\mathbf{v}_\alpha$, it now being understood that all velocities are measured in the fixed system. All velocities with respect to the rotating or body system now vanish because the body is *rigid*.

Because the kinetic energy of the αth particle is given by

$$T_\alpha = \tfrac{1}{2} m_\alpha v_\alpha^2 \tag{11.2}$$

we have, for the total kinetic energy,

$$T = \tfrac{1}{2} \sum_\alpha m_\alpha (\mathbf{V} + \boldsymbol{\omega} \times \mathbf{r}_\alpha)^2 \tag{11.3}$$

Expanding the squared term, we find

$$T = \tfrac{1}{2} \sum_\alpha m_\alpha V^2 + \sum_\alpha m_\alpha \mathbf{V} \cdot \boldsymbol{\omega} \times \mathbf{r}_\alpha + \tfrac{1}{2} \sum_\alpha m_\alpha (\boldsymbol{\omega} \times \mathbf{r}_\alpha)^2 \tag{11.4}$$

This is a general expression for the kinetic energy and is valid for any choice of the origin from which the vectors $\mathbf{r}_\alpha$ are measured. But if we make the origin of the body coordinate system coincide with the center of mass of the object, a considerable simplification results. First, we note that in the second term on the right-hand side of this equation neither $\mathbf{V}$ nor $\boldsymbol{\omega}$ is characteristic of the αth particle, and therefore, these quantities may be taken outside the summation:

$$\sum_\alpha m_\alpha \mathbf{V} \cdot \boldsymbol{\omega} \times \mathbf{r}_\alpha = \mathbf{V} \cdot \boldsymbol{\omega} \times \left(\sum_\alpha m_\alpha \mathbf{r}_\alpha \right) \tag{11.5}$$

But now the term

$$\sum_\alpha m_\alpha \mathbf{r}_\alpha = M\mathbf{R}$$

is the center-of-mass vector (see Equation 9.3), which vanishes in the body system because the vectors $\mathbf{r}_\alpha$ are measured from the center of mass. The kinetic energy can then be written as

$$T = T_{\text{trans}} + T_{\text{rot}}$$

where

$$T_{\text{trans}} = \tfrac{1}{2} \sum_\alpha m_\alpha V^2 = \tfrac{1}{2} MV^2 \tag{11.6a}$$

$$T_{\text{rot}} = \tfrac{1}{2} \sum_\alpha m_\alpha (\boldsymbol{\omega} \times \mathbf{r}_\alpha)^2 \tag{11.6b}$$

T_{trans} and T_{rot} designate the translational and rotational kinetic energies, respectively. Thus, the kinetic energy separates into two independent parts.

The rotational kinetic energy term can be evaluated by noting that

$$(\mathbf{A} \times \mathbf{B})^2 = (\mathbf{A} \times \mathbf{B}) \cdot (\mathbf{A} \times \mathbf{B})$$
$$= A^2B^2 - (\mathbf{A} \cdot \mathbf{B})^2$$

Therefore,

$$T_{\text{rot}} = \frac{1}{2} \sum_\alpha m_\alpha [\omega^2 r_\alpha^2 - (\boldsymbol{\omega} \cdot \mathbf{r}_\alpha)^2] \qquad (11.7)$$

We now express T_{rot} by using the components ω_i and $r_{\alpha,i}$ of the vectors $\boldsymbol{\omega}$ and $\mathbf{r}_\alpha$. We also note that $\mathbf{r}_\alpha = (x_{\alpha,1}, x_{\alpha,2}, x_{\alpha,3})$ in the body system, so we can write $r_{\alpha,i} = x_{\alpha,i}$. Thus,

$$T_{\text{rot}} = \frac{1}{2} \sum_\alpha m_\alpha \left[\left(\sum_i \omega_i^2 \right) \left(\sum_k x_{\alpha,k}^2 \right) - \left(\sum_i \omega_i x_{\alpha,i} \right) \left(\sum_j \omega_j x_{\alpha,j} \right) \right] \qquad (11.8)$$

Now, we can write $\omega_i = \sum_j \omega_j \delta_{ij}$, so that

$$T_{\text{rot}} = \frac{1}{2} \sum_\alpha \sum_{i,j} m_\alpha \left[\omega_i \omega_j \delta_{ij} \left(\sum_k x_{\alpha,k}^2 \right) - \omega_i \omega_j x_{\alpha,i} x_{\alpha,j} \right]$$
$$= \frac{1}{2} \sum_{i,j} \omega_i \omega_j \sum_\alpha m_\alpha \left(\delta_{ij} \sum_k x_{\alpha,k}^2 - x_{\alpha,i} x_{\alpha,j} \right) \qquad (11.9)$$

If we define the ijth element of the sum over α to be I_{ij},

$$\boxed{I_{ij} \equiv \sum_\alpha m_\alpha \left(\delta_{ij} \sum_k x_{\alpha,k}^2 - x_{\alpha,i} x_{\alpha,j} \right)} \qquad (11.10)$$

then we have

$$\boxed{T_{\text{rot}} = \frac{1}{2} \sum_{i,j} I_{ij} \omega_i \omega_j} \qquad (11.11)$$

This equation in its most restricted form becomes

$$T_{\text{rot}} = \frac{1}{2} I \omega^2 \qquad (11.12)$$

where I is the (scalar) moment of inertia about the axis of rotation. This equation will be recognized as the familiar expression for the rotational kinetic energy given in elementary treatments.

The nine terms I_{ij} constitute the elements of a quantity we designated by $\{\mathbf{I}\}$. In form, $\{\mathbf{I}\}$ is similar to a 3×3 matrix. It is the proportionality factor between the rotational kinetic energy and the angular velocity and has the dimensions (mass) $\times$ (length)^2. Because $\{\mathbf{I}\}$ relates two quite different physical quantities, we expect that it is a member of a somewhat higher class of functions than has heretofore been encountered. Indeed, $\{\mathbf{I}\}$ is a tensor and is known as the **inertia tensor**.* Note, however, that T_{rot} can be calculated without regard to any of the special properties of tensors, by using Equation 11.9, which completely specifies the necessary operations.

The elements of $\{\mathbf{I}\}$ can be obtained directly from Equation 11.10. We write the elements in a 3×3 array for clarity:

$$\{\mathbf{I}\} = \begin{Bmatrix} \sum_\alpha m_\alpha(x_{\alpha,2}^2 + x_{\alpha,3}^2) & -\sum_\alpha m_\alpha x_{\alpha,1}\, x_{\alpha,2} & -\sum_\alpha m_\alpha x_{\alpha,1} x_{\alpha,3} \\ -\sum_\alpha m_\alpha x_{\alpha,2}\, x_{\alpha,1} & \sum_\alpha m_\alpha(x_{\alpha,1}^2 + x_{\alpha,3}^2) & -\sum_\alpha m_\alpha x_{\alpha,2} x_{\alpha,3} \\ -\sum_\alpha m_\alpha x_{\alpha,3} x_{\alpha,1} & -\sum_\alpha m_\alpha x_{\alpha,3}\, x_{\alpha,2} & \sum_\alpha m_\alpha(x_{\alpha,1}^2 + x_{\alpha,2}^2) \end{Bmatrix}$$

(11.13a)

Equation 11.10 is a compact way to write the inertia tensor components, but Equation 11.13a is an imposing equation. By using components $(x_\alpha, y_\alpha, z_\alpha)$ instead of $(x_{\alpha,1}, x_{\alpha,2}, x_{\alpha,3})$ and letting $r_\alpha^2 = x_\alpha^2 + y_\alpha^2 + z_\alpha^2$, Equation 11.13a can be written as

$$\{\mathbf{I}\} = \begin{Bmatrix} \sum_\alpha m_\alpha(r_\alpha^2 - x_\alpha^2) & -\sum_\alpha m_\alpha x_\alpha y_\alpha & -\sum_\alpha m_\alpha x_\alpha z_\alpha \\ -\sum_\alpha m_\alpha y_\alpha x_\alpha & \sum_\alpha m_\alpha(r_\alpha^2 - y_\alpha^2) & -\sum_\alpha m_\alpha y_\alpha z_\alpha \\ -\sum_\alpha m_\alpha z_\alpha x_\alpha & -\sum_\alpha m_\alpha z_\alpha y_\alpha & \sum_\alpha m_\alpha(r_\alpha^2 - z_\alpha^2) \end{Bmatrix}$$

(11.13b)

which is less imposing and more recognizable. We continue, however, with the $x_{\alpha,i}$ notation because of its utility.

The diagonal elements, I_{11}, I_{22}, and I_{33}, are called the **moments of inertia** about the x_1-, x_2-, and x_3-axes, respectively, and the negatives of the off-diagonal elements I_{12}, I_{13}, and so forth, are termed the **products of inertia**.[†] It should be clear that the inertia tensor is symmetric; that is,

$$I_{ij} = I_{ji} \tag{11.14}$$

and, therefore, that there are only six independent elements in $\{\mathbf{I}\}$. Furthermore, the inertia tensor is composed of additive elements; the inertia tensor for a body can be considered to be the sum of the tensors for the various portions of the body. Therefore, if we consider a body as a continuous distribution of matter with mass density $\rho = \rho(\mathbf{r})$, then

$$\boxed{I_{ij} = \int_V \rho(\mathbf{r}) \left(\delta_{ij} \sum_k x_k^2 - x_i x_j \right) dv} \tag{11.15}$$

---

* The true test of a tensor lies in its behavior under a coordinate transformation (see Section 11.6).
[†] Introduced by Huygens in 1673; Euler coined the name.

where $dv = dx_1\, dx_2\, dx_3$ is the element of volume at the position defined by the vector $\mathbf{r}$, and where V is the volume of the body.

E X A M P L E **11.1** -

Calculate the inertia tensor of a homogeneous cube of density ρ, mass M, and side of length b. Let one corner be at the origin, and let three adjacent edges lie along the coordinate axes (Figure 11-1). (For this choice of the coordinate axes, it should be obvious that the origin does not lie at the center of mass; we return to this point later.)

Solution: According to Equation 11.15, we have

$$I_{11} = \rho \int_0^b dx_3 \int_0^b dx_2 (x_2^2 + x_3^2) \int_0^b dx_1$$

$$= \tfrac{2}{3}\rho b^5 = \tfrac{2}{3} Mb^2$$

$$I_{12} = -\rho \int_0^b x_1\, dx_1 \int_0^b x_2\, dx_2 \int_0^b dx_3$$

$$= -\tfrac{1}{4}\rho b^5 = -\tfrac{1}{4}Mb^2$$

It should be easy to see that all the diagonal elements are equal and, furthermore, that all the off-diagonal elements are equal. If we define $\beta \equiv Mb^2$, we have

$$\left. \begin{array}{c} I_{11} = I_{22} = I_{33} = \tfrac{2}{3}\beta \\[2mm] I_{12} = I_{13} = I_{23} = -\tfrac{1}{4}\beta \end{array} \right\}$$

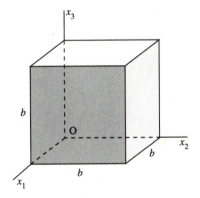

FIGURE 11-1

The moment-of-inertia tensor then becomes

$$\{I\} = \begin{Bmatrix} \frac{2}{3}\beta & -\frac{1}{4}\beta & -\frac{1}{4}\beta \\ -\frac{1}{4}\beta & \frac{2}{3}\beta & -\frac{1}{4}\beta \\ -\frac{1}{4}\beta & -\frac{1}{4}\beta & \frac{2}{3}\beta \end{Bmatrix}$$

We shall continue the investigation of the moment-of-inertia tensor for the cube in later sections.

●

11.3 ANGULAR MOMENTUM

With respect to some point O fixed in the body coordinate system, the angular momentum of the body is

$$L = \sum_\alpha r_\alpha \times p_\alpha \tag{11.16}$$

The most convenient choice for the position of the point O depends on the particular problem. Only two choices are important: (a) if one or more points of the body are fixed (in the fixed coordinate system), O is chosen to coincide with one such point (as in the case of the rotating top, Section 11.10); (b) if no point of the body is fixed, O is chosen to be the center of mass.

Relative to the body coordinate system, the linear momentum p_α is

$$p_\alpha = m_\alpha v_\alpha = m_\alpha \omega \times r_\alpha$$

Hence, the angular momentum of the body is

$$L = \sum_\alpha m_\alpha r_\alpha \times (\omega \times r_\alpha) \tag{11.17}$$

The vector identity

$$A \times (B \times A) = A^2 B - A(A \cdot B)$$

can be used to express L:

$$\boxed{L = \sum_\alpha m_\alpha [r_\alpha^2 \omega - r_\alpha (r_\alpha \cdot \omega)]} \tag{11.18}$$

The same technique we used to write T_{rot} in tensor form can now be applied here. But the angular momentum is a vector, so for the ith component, we write

$$L_i = \sum_\alpha m_\alpha \left(\omega_i \sum_k x_{\alpha,k}^2 - x_{\alpha,i} \sum_j x_{\alpha,j} \omega_j \right)$$

$$= \sum_\alpha m_\alpha \sum_j \left(\omega_j \delta_{ij} \sum_k x_{\alpha,k}^2 - \omega_j x_{\alpha,i} x_{\alpha,j} \right)$$

$$= \sum_j \omega_j \sum_\alpha m_\alpha \left(\delta_{ij} \sum_k x_{\alpha,k}^2 - x_{\alpha,i} x_{\alpha,j} \right) \tag{11.19}$$

The summation over α can be recognized (see Equation 11.10) as the ijth element of the inertia tensor. Therefore,

$$\boxed{L_i = \sum_j I_{ij}\omega_j} \tag{11.20a}$$

or, in tensor notation,

$$\mathbf{L} = \{\mathbf{I}\} \cdot \boldsymbol{\omega} \tag{11.20b}$$

Thus, the inertia tensor relates a *sum* over the components of the angular velocity vector to the ith component of the angular momentum vector. This may at first seem a somewhat unexpected result; for, if we consider a rigid body for which the inertia tensor has nonvanishing off-diagonal elements, then even if $\boldsymbol{\omega}$ is directed along, say, the x_1-direction, $\boldsymbol{\omega} = (\omega_1, 0, 0)$, the angular momentum vector in general has nonvanishing components in all three directions: $\mathbf{L} = (L_1, L_2, L_3)$; that is, the angular momentum vector does not in general have the same direction as the angular velocity vector. (It should be emphasized that this statement depends on $I_{ij} \neq 0$ for $i \neq j$; we return to this point in the next section.)

As an example of $\boldsymbol{\omega}$ and $\mathbf{L}$ not being colinear, consider the rotating dumbbell in Figure 11-2. (We consider the shaft connecting m_1 and m_2 to be weightless and extensionless.) The relation connecting $\mathbf{r}_\alpha$, $\mathbf{v}_\alpha$, and $\boldsymbol{\omega}$ is

$$\mathbf{v}_\alpha = \boldsymbol{\omega} \times \mathbf{r}_\alpha$$

and the relation connecting $\mathbf{r}_\alpha$, $\mathbf{v}_\alpha$, and $\mathbf{L}$ is

$$\mathbf{L} = \sum_\alpha m_\alpha \mathbf{r}_\alpha \times \mathbf{v}_\alpha$$

It should be clear that $\boldsymbol{\omega}$ is directed along the axis of rotation and that $\mathbf{L}$ is perpendicular to the line connecting m_1 and m_2.

We note, for this example, that the angular-momentum vector $\mathbf{L}$ does not remain constant in time but rotates with an angular velocity ω in such a way that it traces out a cone whose axis is the axis of rotation. Therefore $\dot{\mathbf{L}} \neq 0$. But Equation 9.31 states that

$$\dot{\mathbf{L}} = \mathbf{N} \tag{11.21}$$

where $\mathbf{N}$ is the external torque applied to the body. Thus, to keep the dumbbell rotating as in Figure 11-2, we must constantly apply a torque.

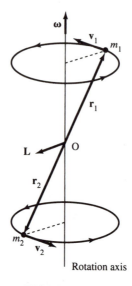

FIGURE 11-2

We can obtain another result from Equation 11.20a by multiplying L_i by $\frac{1}{2}\omega_i$ and summing over i:

$$\frac{1}{2}\sum_i \omega_i L_i = \frac{1}{2}\sum_{i,j} I_{ij}\omega_i\omega_j = T_{rot} \qquad (11.22a)$$

where the second equality is just Equation 11.11. Thus,

$$\boxed{T_{rot} = \frac{1}{2}\boldsymbol{\omega} \cdot \mathbf{L}} \qquad (11.22b)$$

Equations 11.20b and 11.22b illustrate two important properties of tensors. The product of a tensor and a vector yields a vector, as in

$$\mathbf{L} = \{\mathbf{I}\} \cdot \boldsymbol{\omega}$$

and the product of a tensor and two vectors yields a scalar, as in

$$T_{rot} = \frac{1}{2}\boldsymbol{\omega} \cdot \mathbf{L} = \frac{1}{2}\boldsymbol{\omega} \cdot \{\mathbf{I}\} \cdot \boldsymbol{\omega}$$

We shall not, however, have occasion to use tensor equations in this form. We use only the summation (or integral) expressions as in Equations 11.11, 11.15, and 11.20a.

EXAMPLE 11.2 --

Consider the pendulum shown in Figure 11-3 composed of a rigid rod of length b with a mass m_1 at its end. Another mass (m_2) is placed halfway down the rod. Find the frequency of small oscillations if the pendulum swings in a plane.

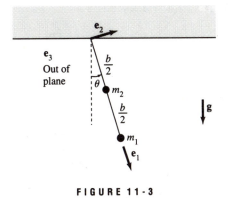

FIGURE 11-3

Solution: We use the methods of this chapter to analyze the system. Let the fixed and body systems have their origin at the pendulum pivot point. Let e_1 be along the rod, e_2 be in the plane, and e_3 be out of the plane (Figure 11-3). The angular velocity is

$$\boldsymbol{\omega} = \omega_3 e_3 = \dot{\theta} e_3 \tag{11.23}$$

We use Equation 11.10 to find the inertia tensor. All the mass is along e_1, with $x_{1,1} = b$ and $x_{2,1} = b/2$. All other components of $x_{\alpha,k}$ equal zero.

$$I_{ij} = m_1(\delta_{ij} x_{1,1}^2 - x_{1,i} x_{1,j}) + m_2(\delta_{ij} x_{2,1}^2 - x_{2,i} x_{2,j}) \tag{11.24}$$

The inertia tensor, Equation 11.13a, becomes

$$\{\mathbf{I}\} = \left\{ \begin{matrix} 0 & 0 & 0 \\ 0 & m_1 b^2 + m_2 \dfrac{b^2}{4} & 0 \\ 0 & 0 & m_1 b^2 + m_2 \dfrac{b^2}{4} \end{matrix} \right\} \tag{11.25}$$

We determine the angular momentum from Equation 11.20a:

$$\left. \begin{aligned} L_1 &= 0 \\ L_2 &= 0 \\ L_3 &= I_{33}\omega_3 = \left(m_1 b^2 + m_2 \dfrac{b^2}{4} \right) \dot{\theta} \end{aligned} \right\} \tag{11.26}$$

The only external force is gravity, which causes a torque $\mathbf{N}$ on the system. Because $\dot{\mathbf{L}} = \mathbf{N}$, we have

$$\left(m_1 b^2 + m_2 \dfrac{b^2}{4} \right) \ddot{\theta} e_3 = \sum_\alpha \mathbf{r}_\alpha \times \mathbf{F}_\alpha \tag{11.27}$$

Because the gravitational force is down,

$$\mathbf{g} = g \cos \theta \mathbf{e}_1 - g \sin \theta \mathbf{e}_2$$

Thus,

$$\mathbf{r}_1 \times \mathbf{F}_1 = b\mathbf{e}_1 \times (\cos \theta \mathbf{e}_1 - \sin \theta \mathbf{e}_2)m_1 g = -m_1 g b \sin \theta \mathbf{e}_3$$

$$\mathbf{r}_2 \times \mathbf{F}_2 = \frac{b}{2}\mathbf{e}_1 \times (\cos \theta \mathbf{e}_1 - \sin \theta \mathbf{e}_2)m_2 g = -m_2 g \frac{b}{2} \sin \theta \mathbf{e}_3$$

Equation 11.27 becomes

$$b^2 \left(m_1 + \frac{m_2}{4} \right) \ddot{\theta} = -bg \sin \theta \left(m_1 + \frac{m_2}{2} \right) \tag{11.28}$$

and the frequency of small oscillations is

$$\omega_0^2 = \frac{m_1 + \dfrac{m_2}{2}}{m_1 + \dfrac{m_2}{4}} \frac{g}{b} \tag{11.29}$$

We can check Equation 11.29 by noting that $\omega_0^2 \approx g/b$ for $m_1 \gg m_2$ and $\omega_0^2 \approx 2g/b$ for $m_2 \gg m_1$ as it should.

 This example could have just as easily been solved by finding the kinetic energy from Equation 11.22a and using Lagrange's equations of motion. We would then have

$$T_{\text{rot}} = \tfrac{1}{2}\omega_3 L_3 = \tfrac{1}{2}\omega_3^2 I_{33}$$

$$= \frac{1}{2}\left(m_1 b^2 + m_2 \frac{b^2}{4} \right) \dot{\theta}^2 \tag{11.30}$$

$$U = -m_1 g b \cos \theta - m_2 g \frac{b}{2} \cos \theta \tag{11.31}$$

Where $U = 0$ at the origin. The equation of motion (Equation 11.28) follows directly from a straightforward application of the Lagrangian technique.

 --- ●

11.4   PRINCIPAL AXES OF INERTIA*

It should be clear that a considerable simplification in the expressions for T and $\mathbf{L}$ would result if the inertia tensor consisted only of diagonal elements. If we could

---

* Discovered by Euler in 1750.

write

$$I_{ij} = I_i \delta_{ij} \tag{11.32}$$

then the inertia tensor would be

$$\{I\} = \begin{Bmatrix} I_1 & 0 & 0 \\ 0 & I_2 & 0 \\ 0 & 0 & I_3 \end{Bmatrix} \tag{11.33}$$

We would then have

$$L_i = \sum_j I_i \delta_{ij} \omega_j = I_i \omega_i \tag{11.34}$$

and

$$T_{\text{rot}} = \frac{1}{2} \sum_{i,j} I_i \delta_{ij} \omega_i \omega_j = \frac{1}{2} \sum_i I_i \omega_i^2 \tag{11.35}$$

Thus, the condition that $\{I\}$ have only diagonal elements provides quite simple expressions for the angular momentum and the rotational kinetic energy. We now determine the conditions under which Equation 11.32 becomes the description of the inertia tensor. This involves finding a set of body axes for which the products of inertia (i.e., the off-diagonal elements of $\{I\}$) vanish. We call such axes the **principal axes of inertia**.

If a body rotates around a principal axis, both the angular velocity and the angular momentum are, according to Equation 11.34, directed along this axis. Then, if I is the moment of inertia about this axis, we can write

$$\mathbf{L} = I\boldsymbol{\omega} \tag{11.36}$$

Equating the components of $\mathbf{L}$ in Equations 11.20a and 11.36, we have

$$\left. \begin{aligned} L_1 &= I\omega_1 = I_{11}\omega_1 + I_{12}\omega_2 + I_{13}\omega_3 \\ L_2 &= I\omega_2 = I_{21}\omega_1 + I_{22}\omega_2 + I_{23}\omega_3 \\ L_3 &= I\omega_3 = I_{31}\omega_1 + I_{32}\omega_2 + I_{33}\omega_3 \end{aligned} \right\} \tag{11.37}$$

Or, collecting terms, we obtain

$$\left. \begin{aligned} (I_{11} - I)\omega_1 + I_{12}\omega_2 + I_{13}\omega_3 &= 0 \\ I_{21}\omega_1 + (I_{22} - I)\omega_2 + I_{23}\omega_3 &= 0 \\ I_{31}\omega_1 + I_{32}\omega_2 + (I_{33} - I)\omega_3 &= 0 \end{aligned} \right\} \tag{11.38}$$

The condition that these equations have a nontrivial solution is that the determinant of the coefficients vanish:

$$\begin{vmatrix} (I_{11} - I) & I_{12} & I_{13} \\ I_{21} & (I_{22} - I) & I_{23} \\ I_{31} & I_{32} & (I_{33} - I) \end{vmatrix} = 0 \tag{11.39}$$

The expansion of this determinant leads to the **secular equation*** for I, which is a cubic. Each of the three roots corresponds to a moment of inertia about one of the principal axes. These values, I_1, I_2, and I_3, are called the **principal moments of inertia**. If the body rotates about the axis corresponding to the principal moment I_1, then Equation 11.36 becomes $\mathbf{L} = I_1\boldsymbol{\omega}$—that is, both $\boldsymbol{\omega}$ and $\mathbf{L}$ are directed along this axis. The direction of $\boldsymbol{\omega}$ with respect to the body coordinate system is then the same as the direction of the principal axis corresponding to I_1. Therefore, we can determine the direction of this principal axis by substituting I_1 for I in Equation 11.38 and determining the ratios of the components of the angular-velocity vector: $\omega_1:\omega_2:\omega_3$. We thereby determine the direction cosines of the axis about which the moment of inertia is I_1. The directions corresponding to I_2 and I_3 can be found in a similar fashion. That the principal axes determined in this manner are indeed *real* and *orthogonal* is proved in Section 11.6; these results also follow from the more general considerations given in Section 12.4.

The fact that the diagonalization procedure just described yields only the *ratios* of the components of $\boldsymbol{\omega}$ is no handicap, because the ratios completely determine the direction of each of the principal axes, and it is only the directions of these axes that is required. Indeed, we would not expect the *magnitudes* of the ω_i to be determined, because the actual rate of the body's angular motion cannot be specified by the geometry alone. We are free to impress on the body any magnitude of the angular velocity we wish.

For most of the problems encountered in rigid-body dynamics, the bodies are of some regular shape, so we can determine the principal axes merely by examining the symmetry of the body. For example, any body that is a solid of revolution (e.g., a cylindrical rod) has one principal axis that lies along the symmetry axis (e.g., the center line of the cylindrical rod), and the other two axes are in a plane perpendicular to the symmetry axis. It should be obvious that because the body is symmetrical, the choice of the angular placement of these other two axes is arbitrary. If the moment of inertia along the symmetry axis is I_1, then $I_2 = I_3$ for a solid of revolution—that is, the secular equation has a double root.

If a body has $I_1 = I_2 = I_3$, it is termed a **spherical top**; if $I_1 = I_2 \neq I_3$, it is termed a **symmetric top**; if the principal moments of inertia are all distinct, it is termed an **asymmetric top**. If a body has $I_1 = 0$, $I_2 = I_3$, as, for example, two point masses connected by a weightless shaft, or a diatomic molecule, it is called a **rotor**.

E X A M P L E   11.3   ---

Find the principal moments of inertia and the principal axes for the cube in Example 11.1.

---

* So called because a similar equation describes secular perturbations in celestial mechanics. The mathematical terminology is the *characteristic polynomial*.

Solution: In Example 11.1, we found that the moment-of-inertia tensor for a cube (with origin at one corner) had nonzero off-diagonal elements. Evidently, the coordinate axes chosen for that calculation were not principal axes. If, for example, the cube rotates about the x_3-axis, then $\boldsymbol{\omega} = \omega_3 \mathbf{e}_3$ and the angular momentum vector **L** (see Equation 11.37) has the components

$$L_1 = -\tfrac{1}{4}\beta\omega_3$$

$$L_2 = -\tfrac{1}{4}\beta\omega_3$$

$$L_3 = \tfrac{2}{3}\beta\omega_3$$

Thus,

$$\mathbf{L} = Mb^2\omega_3(-\tfrac{1}{4}\mathbf{e}_1 - \tfrac{1}{4}\mathbf{e}_2 + \tfrac{2}{3}\mathbf{e}_3)$$

which is not in the same direction as $\boldsymbol{\omega}$.

To find the principal moments of inertia, we must solve the secular equation

$$\begin{vmatrix} \tfrac{2}{3}\beta - I & -\tfrac{1}{4}\beta & -\tfrac{1}{4}\beta \\ -\tfrac{1}{4}\beta & \tfrac{2}{3}\beta - I & -\tfrac{1}{4}\beta \\ -\tfrac{1}{4}\beta & -\tfrac{1}{4}\beta & \tfrac{2}{3}\beta - I \end{vmatrix} = 0 \qquad \textbf{(11.40)}$$

The value of a determinant is not affected by adding (or subtracting) any row (or column) from any other row (or column). Equation 11.40 can be solved more easily if we subtract the first row from the second:

$$\begin{vmatrix} \tfrac{2}{3}\beta - I & -\tfrac{1}{4}\beta & -\tfrac{1}{4}\beta \\ -\tfrac{11}{12}\beta + I & \tfrac{11}{12}\beta - I & 0 \\ -\tfrac{1}{4}\beta & -\tfrac{1}{4}\beta & \tfrac{2}{3}\beta - I \end{vmatrix} = 0$$

We can factor $(\tfrac{11}{12}\beta - I)$ from the second row:

$$(\tfrac{11}{12}\beta - I)\begin{vmatrix} \tfrac{2}{3}\beta - I & -\tfrac{1}{4}\beta & -\tfrac{1}{4}\beta \\ -1 & 1 & 0 \\ -\tfrac{1}{4}\beta & -\tfrac{1}{4}\beta & \tfrac{2}{3}\beta - I \end{vmatrix} = 0$$

Expanding, we have

$$(\tfrac{11}{12}\beta - I)[(\tfrac{2}{3}\beta - I)^2 - \tfrac{1}{8}\beta^2 - \tfrac{1}{4}\beta(\tfrac{2}{3}\beta - I)] = 0$$

which can be factored to obtain

$$(\tfrac{1}{6}\beta - I)(\tfrac{11}{12}\beta - I)(\tfrac{11}{12}\beta - I) = 0$$

Thus, we have the following roots, which give the principal moments of inertia:

$$I_1 = \tfrac{1}{6}\beta, \qquad I_2 = \tfrac{11}{12}\beta, \qquad I_3 = \tfrac{11}{12}\beta$$

The diagonalized moment-of-inertia tensor becomes

$$\{\mathbf{I}\} = \begin{Bmatrix} \tfrac{1}{6}\beta & 0 & 0 \\ 0 & \tfrac{11}{12}\beta & 0 \\ 0 & 0 & \tfrac{11}{12}\beta \end{Bmatrix} \tag{11.41}$$

Because two of the roots are identical, $I_2 = I_3$, the principal axis associated with I_1 must be an axis of symmetry.

To find the direction of the principal axis associated with I_1, we substitute for I in Equation 11.38 the value $I = I_1 = \tfrac{1}{6}\beta$:

$$\left.\begin{array}{r} (\tfrac{2}{3}\beta - \tfrac{1}{6}\beta)\omega_{11} - \tfrac{1}{4}\beta\omega_{21} - \tfrac{1}{4}\beta\omega_{31} = 0 \\[4pt] -\tfrac{1}{4}\beta\omega_{11} + (\tfrac{2}{3}\beta - \tfrac{1}{6}\beta)\omega_{21} - \tfrac{1}{4}\beta\omega_{31} = 0 \\[4pt] -\tfrac{1}{4}\beta\omega_{11} - \tfrac{1}{4}\beta\omega_{21} + (\tfrac{2}{3}\beta - \tfrac{1}{6}\beta)\omega_{31} = 0 \end{array}\right\}$$

where the second subscript 1 on the ω_i signifies that we are considering the principal axis associated with I_1. Dividing the first two of these equations by $\beta/4$, we have

$$\left.\begin{array}{r} 2\omega_{11} - \omega_{21} - \omega_{31} = 0 \\[4pt] -\omega_{11} + 2\omega_{21} - \omega_{31} = 0 \end{array}\right\} \tag{11.42}$$

Subtracting the second of these equations from the first, we find $\omega_{11} = \omega_{21}$. Using this result in either of the Equations 11.42, we obtain $\omega_{11} = \omega_{21} = \omega_{31}$, and the desired ratios are

$$\omega_{11}:\omega_{21}:\omega_{31} = 1:1:1$$

Therefore, when the cube rotates about an axis that has associated with it the moment of inertia $I_1 = \tfrac{1}{6}\beta = \tfrac{1}{6}Mb^2$, the projections of $\boldsymbol{\omega}$ on the three coordinate axes are all equal. Hence, this principal axis corresponds to the diagonal of the cube.

Because the moments I_2 and I_3 are equal, the orientation of the principal axes associated with these moments is arbitrary; they need only lie in a plane normal to the diagonal of the cube.

11.5 MOMENTS OF INERTIA FOR DIFFERENT BODY COORDINATE SYSTEMS

For the kinetic energy to be separable into translational and rotational portions (see Equation 11.6), it is, in general, necessary to choose a body coordinate system whose origin is the center of mass of the body. For certain geometrical shapes, it may not always be convenient to compute the elements of the inertia tensor using such a coordinate system. We therefore consider some other set of coordinate axes X_i, also fixed with respect to the body and having the same orientation as the x_i-axes but with an origin Q that does not correspond with the origin O (located at the center of mass of the body coordinate system). Origin Q may be located either within or outside the body under consideration.

The elements of the inertia tensor relative to the X_i-axes can be written as

$$J_{ij} = \sum_\alpha m_\alpha \left(\delta_{ij} \sum_k X_{\alpha,k}^2 - X_{\alpha,i} X_{\alpha,j} \right) \tag{11.43}$$

If the vector connecting Q with O is $\mathbf{a}$, then the general vector $\mathbf{R}$ (Figure 11-4) can be written as

$$\mathbf{R} = \mathbf{a} + \mathbf{r} \tag{11.44}$$

with components

$$X_i = a_i + x_i \tag{11.45}$$

Using Equation 11.45, the tensor element J_{ij} becomes

$$J_{ij} = \sum_\alpha m_\alpha \left(\delta_{ij} \sum_k (x_{\alpha,k} + a_k)^2 - (x_{\alpha,i} + a_i)(x_{\alpha,j} + a_j) \right)$$

$$= \sum_\alpha m_\alpha \left(\delta_{ij} \sum_k x_{\alpha,k}^2 - x_{\alpha,i} x_{\alpha,j} \right)$$

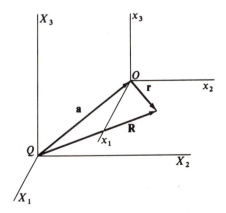

FIGURE 11-4

$$+ \sum_\alpha m_\alpha \left(\delta_{ij} \sum_k (2x_{\alpha,k}a_k + a_k^2) - (a_i x_{\alpha,j} + a_j x_{\alpha,i} + a_i a_j) \right) \quad \textbf{(11.46)}$$

Identifying the first summation as I_{ij}, we have, on regrouping,

$$J_{ij} = I_{ij} + \sum_\alpha m_\alpha \left(\delta_{ij} \sum_k a_k^2 - a_i a_j \right)$$

$$+ \sum_\alpha m_\alpha \left(2\delta_{ij} \sum_k x_{\alpha,k}a_k - a_i x_{\alpha,i} - a_j x_{\alpha,i} \right) \quad \textbf{(11.47)}$$

But each term in the last summation involves a sum of the form

$$\sum_\alpha m_\alpha x_{\alpha,k}$$

We know, however, that because O is located at the center of mass,

$$\sum_\alpha m_\alpha \mathbf{r}_\alpha = 0$$

or, for the kth component,

$$\sum_\alpha m_\alpha x_{\alpha,k} = 0$$

Therefore, all such terms in Equation 11.47 vanish and we have

$$J_{ij} = I_{ij} + \sum_\alpha m_\alpha \left(\delta_{ij} \sum_k a_k^2 - a_i a_j \right) \quad \textbf{(11.48)}$$

But

$$\sum_\alpha m_\alpha = M \quad \text{and} \quad \sum_k a_k^2 \equiv a^2$$

Solving for I_{ij}, we have the result

$$\boxed{I_{ij} = J_{ij} - M(a^2 \delta_{ij} - a_i a_j)} \quad \textbf{(11.49)}$$

which allows the calculation of the elements I_{ij} of the desired inertia tensor (with origin at the center of mass) once those with respect to the X_i-axes are known. The second term on the right-hand side of Equation 11.49 is the inertia tensor referred to the origin Q for a point mass M.

Equation 11.49 is the general form of **Steiner's parallel-axis theorem,*** the simplified form of which is given in elementary treatments. Consider, for example,

---

* Jacob Steiner (1796–1863).

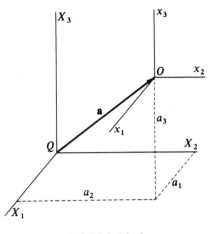

FIGURE 11-5

Figure 11-5. Element I_{11} is

$$I_{11} = J_{11} - M[(a_1^2 + a_2^2 + a_3^2)\,\delta_{11} - a_1^2]$$

$$= J_{11} - M(a_2^2 + a_3^2)$$

which states that the difference between the elements is equal to the mass of the body multiplied by the square of the distance between the parallel axes (in this case, between the x_1- and X_1-axes).

E X A M P L E   11.4   ---

Find the inertia tensor of the cube of Example 11.1 in a coordinate system with origin at the center of mass.

Solution: In Example 11.1, with the origin at the corner of the cube, we found the inertia tensor to be

$$\{\mathbf{J}\} = \begin{Bmatrix} \frac{2}{3}Mb^2 & -\frac{1}{4}Mb^2 & -\frac{1}{4}Mb^2 \\ -\frac{1}{4}Mb^2 & \frac{2}{3}Mb^2 & -\frac{1}{4}Mb^2 \\ -\frac{1}{4}Mb^2 & -\frac{1}{4}Mb^2 & \frac{2}{3}Mb^2 \end{Bmatrix} \tag{11.50}$$

We may now use Equation 11.49 to obtain the inertia tensor $\{\mathbf{I}\}$ referred to a coordinate system with origin at the center of mass. In keeping with the notation of this section, we call the new axes x_i with origin O and call the previous axes X_i with origin Q at one corner of the cube (Figure 11-6).

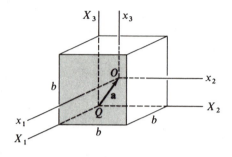

FIGURE 11-6

The center of mass of the cube is at the point $(b/2, b/2, b/2)$ in the X_i coordinate system, and the components of the vector **a** therefore are

$$a_1 = a_2 = a_3 = b/2$$

From Equation 11.50, we have

$$\left.\begin{aligned} J_{11} = J_{22} = J_{33} &= \tfrac{2}{3} Mb^2 \\[6pt] J_{12} = J_{13} = J_{23} &= -\tfrac{1}{4} Mb^2 \end{aligned}\right\}$$

And applying Equation 11.49, we find

$$\begin{aligned} I_{11} &= J_{11} - M(a^2 - a_1^2) \\ &= J_{11} - M(a_2^2 + a_3^2) \\ &= \tfrac{2}{3} Mb^2 - \tfrac{1}{2} Mb^2 = \tfrac{1}{6} Mb^2 \end{aligned}$$

and

$$\begin{aligned} I_{12} &= J_{12} - M(-a_1 a_2) \\ &= -\tfrac{1}{4} Mb^2 + \tfrac{1}{4} Mb^2 = 0 \end{aligned}$$

Altogether, we have

$$I_{12} = I_{22} = I_{33} = \tfrac{1}{6} Mb^2$$

$$I_{12} = I_{13} = I_{23} = 0$$

The inertia tensor is therefore diagonal:

$$\{\mathbf{I}\} = \left\{ \begin{matrix} \tfrac{1}{6} Mb^2 & 0 & 0 \\[6pt] 0 & \tfrac{1}{6} Mb^2 & 0 \\[6pt] 0 & 0 & \tfrac{1}{6} Mb^2 \end{matrix} \right\} \qquad\qquad \textbf{(11.51)}$$

If we factor out the common term $\frac{1}{6}Mb^2$ from this expression, we can write

$$\{I\} = \tfrac{1}{6} Mb^2 \{1\} \tag{11.52}$$

where $\{1\}$ is the **unit tensor**:

$$\{1\} \equiv \begin{Bmatrix} 1 & 0 & 0 \\ 0 & 1 & 0 \\ 0 & 0 & 1 \end{Bmatrix} \tag{11.53}$$

Thus, we find that, for the choice of the origin at the center of mass of the cube, the principal axes are perpendicular to the faces of the cube. Because, from a physical standpoint, nothing distinguishes any one of these axes from another, the principal moments of inertia are all equal for this case. We note further that, as long as we maintain the origin at the center of mass, then the inertia tensor is the same for *any* orientation of the coordinate axes and these axes are equally valid principal axes.*

11.6 FURTHER PROPERTIES OF THE INERTIA TENSOR

Before attacking the problems of rigid-body dynamics by obtaining the general equations of motion, we should consider the fundamental importance of some of the operations we have been discussing. Let us begin by examining the properties of the inertia tensor under coordinate transformations.[†]

We have already obtained the fundamental relation connecting the inertia tensor and the angular momentum and angular velocity vectors (Equation 11.20), which we can write as

$$L_k = \sum_l I_{kl}\omega_l \tag{11.54a}$$

Because this is a vector equation, in a coordinate system rotated with respect to the system for which Equation 11.54a applies, we must have an entirely analogous relation,

$$L_i' = \sum_j I_{ij}'\omega_j' \tag{11.54b}$$

where the primed quantities all refer to the rotated system. Both $\mathbf{L}$ and $\boldsymbol{\omega}$ obey the standard transformation equation for vectors (Equation 1.8):

$$x_i = \sum_j \lambda_{ij}^t x_j' = \sum_j \lambda_{ji} x_j'$$

---

* In this regard, the cube is similar to a sphere as far as the inertia tensor is concerned (i.e., for an origin at the center of mass, the structure of the inertia tensor elements is not sufficiently detailed to discriminate between a cube and a sphere).

[†] We confine our attention to rectangular coordinate systems so that we may ignore some of the more complicated properties of tensors that manifest themselves in general curvilinear coordinates.

We can therefore write

$$L_k = \sum_m \lambda_{mk} L_m' \tag{11.55a}$$

and

$$\omega_l = \sum_j \lambda_{jl} \omega_j' \tag{11.55b}$$

If we substitute Equations 11.55a and b into Equation 11.54a, we obtain

$$\sum_m \lambda_{mk} L_m' = \sum_l I_{kl} \sum_j \lambda_{jl} \omega_j' \tag{11.56}$$

Next, we multiply both sides of this equation by λ_{ik} and sum over k:

$$\sum_m \left(\sum_k \lambda_{ik} \lambda_{mk} \right) L_m' = \sum_j \left(\sum_{k,l} \lambda_{ik} \lambda_{jl} I_{kl} \right) \omega_j' \tag{11.57}$$

The term in parentheses on the left-hand side is just δ_{im}, so performing the summation over m we obtain

$$L_i' = \sum_j \left(\sum_{k,l} \lambda_{ik} \lambda_{jl} I_{kl} \right) \omega_j' \tag{11.58}$$

For this equation to be identical with Equation 11.54b, we must have

$$I_{ij}' = \sum_{k,l} \lambda_{ik} \lambda_{jl} I_{kl} \tag{11.59}$$

This is therefore the rule that the inertia tensor must obey under a coordinate transformation. Equation 11.59 is, in fact, the *general* rule specifying the manner in which any second-rank tensor must transform. For a tensor {**T**} of arbitrary rank, the statement is*

$$T_{abcd\ldots}' = \sum_{i,j,k,l,\ldots} \lambda_{ai} \lambda_{bj} \lambda_{ck} \lambda_{dl} \ldots T_{ijkl\ldots} \tag{11.60}$$

Note that we can write Equation 11.59 as

$$\boxed{I_{ij}' = \sum_{k,l} \lambda_{ik} I_{kl} \lambda_{lj}^t} \tag{11.61}$$

Although matrices and tensors are distinct types of mathematical objects, the

---

* Note that a tensor of the *first rank* transforms as

$$T_a' = \sum_i \lambda_{ai} T_i$$

Such a tensor is in fact a *vector*. A tensor of zero rank implies that $T' = T$, or that such a tensor is a scalar. The properties of quantities that transform in this manner were first discussed by C. Niven in 1874. The application of the term *tensor* to such quantities can be traced to J. Willard Gibbs.

manipulation of tensors is in many respects the same as for matrices. Thus, Equation 11.61 can be expressed as a matrix equation:

$$\mathbf{I'} = \boldsymbol{\lambda} \mathbf{I} \boldsymbol{\lambda}^t \qquad (11.62)$$

where we understand **I** to be the matrix consisting of the elements of the tensor {**I**}. Because we are considering only orthogonal transformation matrices, the transpose of $\boldsymbol{\lambda}$ is equal to its inverse, so we can express Equation 11.62 as

$$\boxed{\mathbf{I'} = \boldsymbol{\lambda} \mathbf{I} \boldsymbol{\lambda}^{-1}} \qquad (11.63)$$

A transformation of this general type is called a **similarity transformation** (**I'** is similar to **I**).

E X A M P L E 11.5

Prove the assertion stated in Example 11.4 that the inertia tensor for a cube (with origin at the center of mass) is independent of the orientation of the axes.

Solution: The change in the inertia tensor under a rotation of the coordinate axes can be computed by making a similarity transformation. Thus, if the rotation is described by the matrix $\boldsymbol{\lambda}$, we have

$$\mathbf{I'} = \boldsymbol{\lambda} \mathbf{I} \boldsymbol{\lambda}^{-1} \qquad (11.64)$$

But the matrix **I**, which is derived from the elements of the tensor {**I**} (Equation 11.52 of Example 11.4), is just the identity matrix **1** multiplied by a constant:

$$\mathbf{I} = \tfrac{1}{6} M b^2 \begin{pmatrix} 1 & 0 & 0 \\ 0 & 1 & 0 \\ 0 & 0 & 1 \end{pmatrix} = \tfrac{1}{6} M b^2 \mathbf{1} \qquad (11.65)$$

Therefore, the operations specified in Equation 11.64 are trivial:

$$\mathbf{I'} = \tfrac{1}{6} M b^2 \boldsymbol{\lambda} \mathbf{1} \boldsymbol{\lambda}^{-1} = \tfrac{1}{6} M b^2 \boldsymbol{\lambda} \boldsymbol{\lambda}^{-1} = \tfrac{1}{6} M b^2 \mathbf{1} = \mathbf{I} \qquad (11.66)$$

Thus, the transformed inertia tensor is identical to the original tensor, independent of the details of the rotation.

Let us next determine what condition must be satisfied if we take an arbitrary inertia tensor and perform a coordinate rotation in such a way that the transformed inertia tensor is diagonal. Such an operation implies that the quantity I'_{ij} in Equation 11.59 must satisfy (see Equation 11.32) the relation

$$I'_{ij} = I_i \delta_{ij} \qquad (11.67)$$

Thus,

$$I_i \delta_{ij} = \sum_{k,l} \lambda_{ik} \lambda_{jl} I_{kl} \tag{11.68}$$

If we multiply both sides of this equation by λ_{im} and sum over i, we obtain

$$\sum_i I_i \lambda_{im} \delta_{ij} = \sum_{k,l} \left(\sum_i \lambda_{im} \lambda_{ik} \right) \lambda_{jl} I_{kl} \tag{11.69}$$

The term in parentheses is just δ_{mk}, so the summation over i on the left-hand side of the equation and the summation over k on the right-hand side yield

$$I_j \lambda_{jm} = \sum_l \lambda_{jl} I_{ml} \tag{11.70}$$

Now the left-hand side of this equation can be written as

$$I_j \lambda_{jm} = \sum_l I_j \lambda_{jl} \delta_{ml} \tag{11.71}$$

so Equation 11.70 becomes

$$\sum_l I_j \lambda_{jl} \delta_{ml} = \sum_l \lambda_{jl} I_{ml} \tag{11.72a}$$

or

$$\sum_l (I_{ml} - I_j \delta_{ml}) \lambda_{jl} = 0 \tag{11.72b}$$

This is a set of simultaneous linear algebraic equations; for each value of j there are three such equations, one for each of the three possible values of m. For a nontrivial solution to exist, the determinant of the coefficients must vanish, so the principal moments of inertia, I_1, I_2, and I_3, are obtained as roots of the secular determinant for I:

$$\boxed{|I_{ml} - I\delta_{ml}| = 0} \tag{11.73}$$

This equation is just Equation 11.39; it is a cubic equation that yields the principal moments of inertia.

Thus, for any inertia tensor, the elements of which are computed for a given origin, it is possible to perform a rotation of the coordinate axes about that origin in such a way that the inertia tensor becomes diagonal. The new coordinate axes are then the principal axes of the body, and the new moments are the principal moments of inertia. Thus, for any body and for any choice of origin, there always exists a set of principal axes.

E X A M P L E 11.6 ---

For the cube of Example 11.1, diagonalize the inertia tensor by rotating the coordinate axes.

Solution: We choose the origin to lie at one corner and perform the rotation in such a manner that the x_1-axis coincides with the diagonal of the cube. Such a rotation can conveniently be made in two steps: first, we rotate through an angle of 45° about the x_3-axis; second, we rotate through an angle of $\cos^{-1}\left(\sqrt{\frac{2}{3}}\right)$ about the x_2'-axis. The first rotation matrix is

$$\lambda_1 = \begin{pmatrix} \dfrac{1}{\sqrt{2}} & \dfrac{1}{\sqrt{2}} & 0 \\[2mm] -\dfrac{1}{\sqrt{2}} & \dfrac{1}{\sqrt{2}} & 0 \\[2mm] 0 & 0 & 1 \end{pmatrix} \tag{11.74}$$

and the second rotation matrix is

$$\lambda_2 = \begin{pmatrix} \sqrt{\dfrac{2}{3}} & 0 & \dfrac{1}{\sqrt{3}} \\[2mm] 0 & 1 & 0 \\[2mm] -\dfrac{1}{\sqrt{3}} & 0 & \sqrt{\dfrac{2}{3}} \end{pmatrix} \tag{11.75}$$

The complete rotation matrix is

$$\lambda = \lambda_2\lambda_1 = \begin{pmatrix} \dfrac{1}{\sqrt{3}} & \dfrac{1}{\sqrt{3}} & \dfrac{1}{\sqrt{3}} \\[2mm] -\dfrac{1}{\sqrt{2}} & \dfrac{1}{\sqrt{2}} & 0 \\[2mm] -\dfrac{1}{\sqrt{6}} & -\dfrac{1}{\sqrt{6}} & \sqrt{\dfrac{2}{3}} \end{pmatrix} = \frac{1}{\sqrt{3}}\begin{pmatrix} 1 & 1 & 1 \\[2mm] -\sqrt{\dfrac{3}{2}} & \sqrt{\dfrac{3}{2}} & 0 \\[2mm] -\dfrac{1}{\sqrt{2}} & -\dfrac{1}{\sqrt{2}} & \sqrt{2} \end{pmatrix} \tag{11.76}$$

The matrix form of the transformed inertia tensor (see Equation 11.62) is

$$\mathbf{I}' = \lambda\mathbf{I}\lambda' \tag{11.77}$$

or, factoring β out of $\mathbf{I}$,

$$\mathbf{I}' = \frac{\beta}{3}\begin{pmatrix} 1 & 1 & 1 \\[2mm] -\sqrt{\dfrac{3}{2}} & \sqrt{\dfrac{3}{2}} & 0 \\[2mm] -\dfrac{1}{\sqrt{2}} & -\dfrac{1}{\sqrt{2}} & \sqrt{2} \end{pmatrix}\begin{pmatrix} \dfrac{2}{3} & -\dfrac{1}{4} & -\dfrac{1}{4} \\[2mm] -\dfrac{1}{4} & \dfrac{2}{3} & -\dfrac{1}{4} \\[2mm] -\dfrac{1}{4} & -\dfrac{1}{4} & \dfrac{2}{3} \end{pmatrix}\begin{pmatrix} 1 & -\sqrt{\dfrac{3}{2}} & -\dfrac{1}{\sqrt{2}} \\[2mm] 1 & \sqrt{\dfrac{3}{2}} & -\dfrac{1}{\sqrt{2}} \\[2mm] 1 & 0 & \sqrt{2} \end{pmatrix}$$

$$
\mathbf{I'} = \frac{\beta}{3}
\begin{pmatrix}
1 & 1 & 1 \\
-\sqrt{\dfrac{3}{2}} & \sqrt{\dfrac{3}{2}} & 0 \\
-\dfrac{1}{\sqrt{2}} & -\dfrac{1}{\sqrt{2}} & \sqrt{2}
\end{pmatrix}
\begin{pmatrix}
\dfrac{1}{6} & -\dfrac{11}{12}\sqrt{\dfrac{3}{2}} & -\dfrac{11}{12}\dfrac{\sqrt{2}}{2} \\
\dfrac{1}{6} & \dfrac{11}{12}\sqrt{\dfrac{3}{2}} & -\dfrac{11}{12}\dfrac{\sqrt{2}}{2} \\
\dfrac{1}{6} & 0 & \dfrac{11}{12}\sqrt{2}
\end{pmatrix}
$$

$$
=
\begin{pmatrix}
\dfrac{1}{6}\beta & 0 & 0 \\
0 & \dfrac{11}{12}\beta & 0 \\
0 & 0 & \dfrac{11}{12}\beta
\end{pmatrix}
\tag{11.78}
$$

Equation 11.78 is just the matrix form of the inertia tensor found by the diagonalization procedure using the secular determinant (Equation 11.41 of Example 11.3).

-- ●

We have demonstrated two general procedures to diagonalize the inertia tensor. We previously pointed out that these methods are not limited to the inertia tensor but are generally valid. Either procedure can be very complicated. For example, if we wish to use the rotation procedure in the most general case, we must first construct a matrix that describes an arbitrary rotation. This entails three separate rotations, one about each of the coordinate axes. This rotation matrix must then be applied to the tensor in a similarity transformation. The off-diagonal elements of the resulting matrix* must then be examined and values of the rotation angles determined so that these off-diagonal elements vanish. The actual use of such a procedure can tax the limits of human patience, but in some simple situations, this method of diagonalization can be used with profit. This is particularly true if the geometry of the problem indicates that only a simple rotation about one of the coordinate axes is necessary; the rotation angle can then be evaluated without difficulty (see, for example, Problems 11-16, 11-18, and 11-19).

In practice, there are systematic procedures for finding principal moments and principal axes of any inertia tensor. Standard computer programs and hand-calculator methods are available to find the n roots of an nth-order polynomial and to diagonalize a matrix. When the principal moments are known, the principal axes are easily found.

---

* A *large* sheet of paper should be used!

The example of the cube illustrates the important point that the elements of the inertia tensor, the values of the principal moments of inertia, and the orientation of the principal axes for a rigid body all depend on the choice of origin for the system. Recall, however, that for the kinetic energy to be separable into translational and rotational portions, the origin of the body coordinate system must, in general, be taken to coincide with the center of mass of the body. However, for *any* choice of the origin for *any* body, there always exists an orientation of the axes that diagonalizes the inertia tensor. Hence, these axes become principal axes for that particular origin.

Next, we seek to prove that the principal axes actually form an orthogonal set. Let us assume that we have solved the secular equation and have determined the principal moments of inertia, all of which are distinct. We know that for each principal moment there exists a corresponding principal axis with the property that, if the angular velocity vector $\boldsymbol{\omega}$ lies along this axis, then the angular momentum vector $\mathbf{L}$ is similarly oriented; that is, to each I_j there corresponds an angular velocity $\boldsymbol{\omega}_j$ with components ω_{1j}, ω_{2j}, ω_{3j}. (We use the subscript on the vector $\boldsymbol{\omega}$ and the second subscript on the components of $\boldsymbol{\omega}$ to designate the principal moment with which we are concerned.) For the mth principal moment, we have

$$L_{im} = I_m \omega_{im} \tag{11.79}$$

In terms of the elements of the moment-of-inertia tensor, we also have

$$L_{im} = \sum_k I_{ik} \omega_{km} \tag{11.80}$$

Combining these two relations, we have

$$\sum_k I_{ik} \omega_{km} = I_m \omega_{im} \tag{11.81a}$$

Similarly, we can write for the nth principal moment:

$$\sum_i I_{ki} \omega_{in} = I_n \omega_{kn} \tag{11.81b}$$

If we multiply Equation 11.81a by ω_{in} and sum over i and then multiply Equation 11.81b by ω_{km} and sum over k, we have

$$\left.\begin{array}{l} \sum_{i,k} I_{ik} \omega_{km} \omega_{in} = \sum_i I_m \omega_{im} \omega_{in} \\[2mm] \sum_{i,k} I_{ki} \omega_{in} \omega_{km} = \sum_k I_n \omega_{kn} \omega_{km} \end{array}\right\} \tag{11.82}$$

The left-hand sides of these equations are identical, because the inertia tensor is symmetrical ($I_{ik} = I_{ki}$). Therefore, on subtracting the second equation from the first, we have

$$I_m \sum_i \omega_{im} \omega_{in} - I_n \sum_k \omega_{km} \omega_{kn} = 0 \tag{11.83}$$

Because i and k are both dummy indices, we can replace them by l, say, and obtain

$$(I_m - I_n) \sum_l \omega_{lm} \omega_{ln} = 0 \tag{11.84}$$

By hypothesis, the principal moments are distinct, so that $I_m \neq I_n$. Therefore, Equation 11.84 can be satisfied only if

$$\sum_l \omega_{lm}\omega_{ln} = 0 \tag{11.85}$$

But this summation is just the definition of the scalar product of the vectors $\boldsymbol{\omega}_m$ and $\boldsymbol{\omega}_n$. Hence,

$$\boldsymbol{\omega}_m \cdot \boldsymbol{\omega}_n = 0 \tag{11.86}$$

Because the principal moments I_m and I_n were picked arbitrarily from the set of three moments, we conclude that each pair of principal axes is perpendicular; the three principal axes therefore constitute an orthogonal set.

If a double root of the secular equation exists, so that the principal moments are $I_1, I_2 = I_3$, then the preceding analysis shows that the angular velocity vectors satisfy the relations

$$\boldsymbol{\omega}_1 \perp \boldsymbol{\omega}_2, \qquad \boldsymbol{\omega}_1 \perp \boldsymbol{\omega}_3$$

but that nothing may be said regarding the angle between $\boldsymbol{\omega}_2$ and $\boldsymbol{\omega}_3$. But the fact that $I_2 = I_3$ implies that the body possesses an axis of symmetry. Therefore, $\boldsymbol{\omega}_1$ lies along the symmetry axis, and $\boldsymbol{\omega}_2$ and $\boldsymbol{\omega}_3$ are required only to lie in the plane perpendicular to $\boldsymbol{\omega}_1$. Consequently, there is no loss of generality if we also choose $\boldsymbol{\omega}_2 \perp \boldsymbol{\omega}_3$. Thus, the principal axes for a rigid body with an axis of symmetry can also be chosen to be an orthogonal set.

We have previously shown that the principal moments of inertia are obtained as the roots of the secular equation—a cubic equation. Mathematically, at least one of the roots of a cubic equation must be real, but there may be two imaginary roots. If the diagonalization procedures for the inertia tensor are to be physically meaningful, we must always obtain only real values for the principal moments. We can show in the following way that this is a general result. First, we assume the roots to be complex and use a procedure similar to that used in the preceding proof. But now we must also allow the quantities ω_{km} to become complex. There is no mathematical reason why we cannot do this, and we are not concerned with any physical interpretation of these quantities. We therefore write Equation 11.81a as before, but we take the complex conjugate of Equation 11.81b:

$$\left.\begin{array}{c} \sum_k I_{ik}\omega_{km} = I_m\omega_{im} \\[2mm] \sum_i I_{ki}^*\omega_{in}^* = I_n^*\omega_{kn}^* \end{array}\right\} \tag{11.87}$$

Next, we multiply the first of these equations by ω_{in}^* and sum over i and multiply the second by ω_{km} and sum over k. The inertia tensor is symmetrical, and its elements are all real, so that $I_{ik} = I_{ki}^*$. Therefore, subtracting the second of these equations from the first, we find

$$(I_m - I_n^*)\sum_l \omega_{lm}\omega_{ln}^* = 0 \tag{11.88}$$

For the case $m = n$, we have

$$(I_m - I_m^*) \sum_l \omega_{lm} \omega_{lm}^* = 0 \qquad (11.89)$$

The sum is just the definition of the scalar product of $\boldsymbol{\omega}_m$ and $\boldsymbol{\omega}_m^*$:

$$\boldsymbol{\omega}_m \cdot \boldsymbol{\omega}_m^* = |\boldsymbol{\omega}_m|^2 \geq 0 \qquad (11.90)$$

Therefore, because the squared magnitude of $\boldsymbol{\omega}_m$ is in general positive, it must be true that $I_m = I_m^*$ for Equation 11.89 to be satisfied. If a quantity and its complex conjugate are equal, then the imaginary parts must vanish identically. Thus, the principal moments of inertia are all real. Because $\{I\}$ is real, the vectors $\boldsymbol{\omega}_m$ must also be real.

If $m \neq n$ in Equation 11.88 and if $I_m \neq I_n$, then the equation can be satisfied only if $\boldsymbol{\omega}_m \cdot \boldsymbol{\omega}_n = 0$; that is, these vectors are orthogonal, as before.

In all the proofs carried out in this section, we have referred to the inertia tensor. But examining these proofs reveals that the only properties of the inertia tensor that have actually been used are the facts that the tensor is symmetrical and that the elements are real. We may therefore conclude that *any* real, symmetric tensor* has the following properties:

1. Diagonalization may be accomplished by an appropriate rotation of axes, that is, a similarity transformation.

2. The eigenvalues[†] are obtained as roots of the secular determinant and are real.

3. The eigenvectors[†] are real and orthogonal.

11.7 EULERIAN ANGLES

The transformation from one coordinate system to another can be represented by a matrix equation of the form

$$\mathbf{x} = \boldsymbol{\lambda} \mathbf{x}'$$

If we identify the fixed system with $\mathbf{x}'$ and the body system with $\mathbf{x}$, then the rotation matrix $\boldsymbol{\lambda}$ completely describes the relative orientation of the two systems. The rotation matrix $\boldsymbol{\lambda}$ contains three independent angles. There are many possible choices for these angles; we find it convenient to use the **Eulerian angles**[‡] ϕ, θ, and ψ.

---

* To be more precise, we require only that the elements of the tensor obey the relation $I_{ik} = I_{ki}^*$; thus we allow the possibility of complex quantities. Tensors (and matrices) with this property are said to be **Hermitean**.

[†] The terms *eigenvalues* and *eigenvectors* are the generic names of the quantities, which, in the case of the inertia tensor, are the principal moments and the principal axes, respectively. We shall encounter these terms again in the discussion of small oscillations in Chapter 12.

[‡] The rotation scheme of Euler was first published in 1776.

The Eulerian angles are generated in the following series of rotations, which takes the x_i' system into the x_i system*:

1. The first rotation is counterclockwise through an angle ϕ about the x_3'-axis (Figure 11-7a) to transform the x_i' into the x_i''. Because the rotation takes place in the x_1'-x_2' plane, the transformation matrix is

$$\boldsymbol{\lambda}_\phi = \begin{pmatrix} \cos\phi & \sin\phi & 0 \\ -\sin\phi & \cos\phi & 0 \\ 0 & 0 & 1 \end{pmatrix} \qquad (11.91)$$

and

$$\mathbf{x}'' = \boldsymbol{\lambda}_\phi \mathbf{x}' \qquad (11.92)$$

2. The second rotation is counterclockwise through an angle θ about the x_1''-axis (Figure 11-7b) to transform the x_i'' into the x_i'''. Because the rotation is now

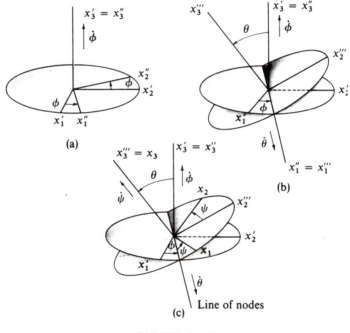

(a)

(b)

(c)   Line of nodes

FIGURE 11-7

---

* The designations of the Euler angles and even the manner in which they are generated are not universally agreed upon. Therefore, some care must be taken in comparing any results from different sources. The notation used here is that most commonly found in modern texts.

in the x_2''-x_3'' plane, the transformation matrix is

$$\boldsymbol{\lambda}_\theta = \begin{pmatrix} 1 & 0 & 0 \\ 0 & \cos\theta & \sin\theta \\ 0 & -\sin\theta & \cos\theta \end{pmatrix} \tag{11.93}$$

and

$$\mathbf{x}''' = \boldsymbol{\lambda}_\theta \mathbf{x}'' \tag{11.94}$$

3. The third rotation is counterclockwise through an angle ψ about the x_3'''-axis (Figure 11-7c) to transform the x_i''' into the x_i. The transformation matrix is

$$\boldsymbol{\lambda}_\psi = \begin{pmatrix} \cos\psi & \sin\psi & 0 \\ -\sin\psi & \cos\psi & 0 \\ 0 & 0 & 1 \end{pmatrix} \tag{11.95}$$

and

$$\mathbf{x} = \boldsymbol{\lambda}_\psi \mathbf{x}''' \tag{11.96}$$

The line common to the planes containing the x_1- and x_2-axes and the x_1'- and x_2'-axes is called the **line of nodes**. The complete transformation from the x_i' system to the x_i system is given by

$$\mathbf{x} = \boldsymbol{\lambda}_\psi \mathbf{x}''' = \boldsymbol{\lambda}_\psi \boldsymbol{\lambda}_\theta \mathbf{x}'' \tag{11.97}$$
$$= \boldsymbol{\lambda}_\psi \boldsymbol{\lambda}_\theta \boldsymbol{\lambda}_\phi \mathbf{x}'$$

and the rotation matrix $\boldsymbol{\lambda}$ is

$$\boldsymbol{\lambda} = \boldsymbol{\lambda}_\psi \boldsymbol{\lambda}_\theta \boldsymbol{\lambda}_\phi \tag{11.98}$$

The components of this matrix are

$$\left. \begin{aligned}
\lambda_{11} &= \cos\psi\cos\phi - \cos\theta\sin\phi\sin\psi \\
\lambda_{21} &= -\sin\psi\cos\phi - \cos\theta\sin\phi\cos\psi \\
\lambda_{31} &= \sin\theta\sin\phi \\
\lambda_{12} &= \cos\psi\sin\phi + \cos\theta\cos\phi\sin\psi \\
\lambda_{22} &= -\sin\psi\sin\phi + \cos\theta\cos\phi\cos\psi \\
\lambda_{32} &= -\sin\theta\cos\phi \\
\lambda_{13} &= \sin\psi\sin\theta \\
\lambda_{23} &= \cos\psi\sin\theta \\
\lambda_{33} &= \cos\theta
\end{aligned} \right\} \tag{11.99}$$

(The components λ_{ij} are off-set in the preceding equation to assist in the visualization of the complete $\boldsymbol{\lambda}$ matrix.)

Because we can associate a vector with an infinitesimal rotation, we can associate the time derivatives of these rotation angles with the components of the angular velocity vector **ω**. Thus,

$$\left.\begin{array}{c} \omega_\phi = \dot{\phi} \\[2mm] \omega_\theta = \dot{\theta} \\[2mm] \omega_\psi = \dot{\psi} \end{array}\right\} \tag{11.100}$$

The rigid-body equations of motion are most conveniently expressed in the body coordinate system (i.e., the x_i system), and therefore we must express the components of **ω** in this system. We note that in Figure 11-7 the angular velocities $\dot{\phi}$, $\dot{\theta}$, and $\dot{\psi}$ are directed along the following axes:

$\dot{\phi}$ along the x_3'- (fixed) axis

$\dot{\theta}$ along the line of nodes

$\dot{\psi}$ along the x_3- (body) axis

The components of these angular velocities along the body coordinate axes are

$$\left.\begin{array}{c} \dot{\phi}_1 = \dot{\phi}\sin\theta\sin\psi \\[2mm] \dot{\phi}_2 = \dot{\phi}\sin\theta\cos\psi \\[2mm] \dot{\phi}_3 = \dot{\phi}\cos\theta \end{array}\right\} \tag{11.101a}$$

$$\left.\begin{array}{c} \dot{\theta}_1 = \dot{\theta}\cos\psi \\[2mm] \dot{\theta}_2 = -\dot{\theta}\sin\psi \\[2mm] \dot{\theta}_3 = 0 \end{array}\right\} \tag{11.101b}$$

$$\left.\begin{array}{c} \dot{\psi}_1 = 0 \\[2mm] \dot{\psi}_2 = 0 \\[2mm] \dot{\psi}_3 = \dot{\psi} \end{array}\right\} \tag{11.101c}$$

Collecting the individual components of **ω,** we have, finally,

$$\boxed{\begin{array}{l} \omega_1 = \dot{\phi}_1 + \dot{\theta}_1 + \dot{\psi}_1 = \dot{\phi}\sin\theta\sin\psi + \dot{\theta}\cos\psi \\[2mm] \omega_2 = \dot{\phi}_2 + \dot{\theta}_2 + \dot{\psi}_2 = \dot{\phi}\sin\theta\cos\psi - \dot{\theta}\sin\psi \\[2mm] \omega_3 = \dot{\phi}_3 + \dot{\theta}_3 + \dot{\psi}_3 = \dot{\phi}\cos\theta + \dot{\psi} \end{array}} \tag{11.102}$$

These relations will be of use later in expressing the components of the angular momentum in the body coordinate system.

E X A M P L E **11.7**

Using the Eulerian angles, find the transformation that moves the original x_1'-axis to the x_2'-x_3' plane halfway between x_2' and x_3' and moves x_2' perpendicular to the x_2'-x_3' plane (Figure 11-8).

Solution: The key to transformations using Eulerian angles is the second rotation about the line of nodes, because this single rotation must move x_3' to x_3. From the statement of the problem, x_3 must be in the x_2'-x_3' plane, rotated 45° from x_3'. The first rotation must move x_1' to x_1'' to have the correct position to rotate $x_3' = x_3''$ to $x_3''' = x_3$.

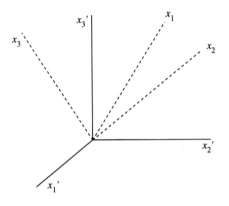

FIGURE 11-8

In this case, $x_3' = x_3''$ is rotated $\theta = 45°$ about the original $x_1' = x_1''$-axis so that $\phi = 0$ and

$$\lambda_\phi = 1 \tag{11.103}$$

$$\lambda_\theta = \begin{pmatrix} 1 & 0 & 0 \\ 0 & 1/\sqrt{2} & 1/\sqrt{2} \\ 0 & -1/\sqrt{2} & 1/\sqrt{2} \end{pmatrix} \tag{11.104}$$

The last rotation, $\psi = 90°$, moves $x_1' = x_1'' = x_1'''$ to x_1 to the position desired in the original x_2'-x_3' plane.

$$\lambda_\psi = \begin{pmatrix} 0 & 1 & 0 \\ -1 & 0 & 0 \\ 0 & 0 & 1 \end{pmatrix} \tag{11.105}$$

The transformation matrix λ is $\lambda = \lambda_\psi \lambda_\theta \lambda_\phi = \lambda_\psi \lambda_\theta$:

$$\lambda = \begin{pmatrix} 0 & 1 & 0 \\ -1 & 0 & 0 \\ 0 & 0 & 1 \end{pmatrix} \begin{pmatrix} 1 & 0 & 0 \\ 0 & 1/\sqrt{2} & 1/\sqrt{2} \\ 0 & -1/\sqrt{2} & 1/\sqrt{2} \end{pmatrix}$$

$$\boldsymbol{\lambda} = \begin{pmatrix} 0 & 1/\sqrt{2} & 1/\sqrt{2} \\ -1 & 0 & 0 \\ 0 & -1/\sqrt{2} & 1/\sqrt{2} \end{pmatrix} \qquad \textbf{(11.106)}$$

Direction comparison between the x_i- and x_i'-axes shows that $\boldsymbol{\lambda}$ represents a single rotation describing the transformation.

--- ●

11.8 EULER'S EQUATIONS FOR A RIGID BODY

Let us first consider the force-free motion of a rigid body. In such a case, the potential energy U vanishes and the Lagrangian L becomes identical with the rotational kinetic energy T.* If we choose the x_i-axes to correspond to the principal axes of the body, then from Equation 11.35 we have

$$T = \tfrac{1}{2} \sum_i I_i \omega_i^2 \qquad \textbf{(11.107)}$$

If we choose the Eulerian angles as the generalized coordinates, then Lagrange's equation for the coordinate ψ is

$$\frac{\partial T}{\partial \psi} - \frac{d}{dt} \frac{\partial T}{\partial \dot{\psi}} = 0 \qquad \textbf{(11.108)}$$

which can be expressed as

$$\sum_i \frac{\partial T}{\partial \omega_i} \frac{\partial \omega_i}{\partial \psi} - \frac{d}{dt} \sum_i \frac{\partial T}{\partial \omega_i} \frac{\partial \omega_i}{\partial \dot{\psi}} = 0 \qquad \textbf{(11.109)}$$

If we differentiate the components of $\boldsymbol{\omega}$ (Equation 11.102) with respect to ψ and $\dot{\psi}$, we have

$$\left. \begin{aligned} \frac{\partial \omega_1}{\partial \psi} &= \dot{\phi} \sin\theta \cos\psi - \dot{\theta} \sin\psi = \omega_2 \\[2mm] \frac{\partial \omega_2}{\partial \psi} &= -\dot{\phi} \sin\theta \sin\psi - \dot{\theta} \cos\psi = -\omega_1 \\[2mm] \frac{\partial \omega_3}{\partial \psi} &= 0 \end{aligned} \right\} \qquad \textbf{(11.110)}$$

and

$$\left. \begin{aligned} \frac{\partial \omega_1}{\partial \dot{\psi}} &= \frac{\partial \omega_2}{\partial \dot{\psi}} = 0 \\[2mm] \frac{\partial \omega_3}{\partial \dot{\psi}} &= 1 \end{aligned} \right\} \qquad \textbf{(11.111)}$$

---

* Because the motion is force free, the translational kinetic energy is unimportant for our purposes here. (We can always transform to a coordinate system in which the center of mass of the body is at rest.)

From Equation 11.107, we also have

$$\frac{\partial T}{\partial \omega_i} = I_i \omega_i \qquad \textbf{(11.112)}$$

Equation 11.109 therefore becomes

$$I_1 \omega_1 \omega_2 + I_2 \omega_2 (-\omega_1) - \frac{d}{dt} I_3 \omega_3 = 0$$

or

$$(I_1 - I_2)\omega_1 \omega_2 - I_3 \dot{\omega}_3 = 0 \qquad \textbf{(11.113)}$$

Because the designation of any particular principal axis as the x_3-axis is entirely arbitrary, Equation 11.113 can be permuted to obtain relations for $\dot{\omega}_1$ and $\dot{\omega}_2$:

$$\left.\begin{array}{l} (I_2 - I_3)\omega_2 \omega_3 - I_1 \dot{\omega}_1 = 0 \\[6pt] (I_3 - I_1)\omega_3 \omega_1 - I_2 \dot{\omega}_2 = 0 \\[6pt] (I_1 - I_2)\omega_1 \omega_2 - I_3 \dot{\omega}_3 = 0 \end{array}\right\} \qquad \textbf{(11.114)}$$

Equations 11.114 are called **Euler's equations** for force-free motion.* It must be noted that, although Equation 11.113 for $\dot{\omega}_3$ is indeed the Lagrange equation for the coordinate ψ, the Euler equations for $\dot{\omega}_1$ and $\dot{\omega}_2$ are *not* the Lagrange equations for θ and ϕ.

To obtain Euler's equations for motion in a force field, we may start with the fundamental relation (see Equation 2.83) for the torque **N**:

$$\left(\frac{d\mathbf{L}}{dt}\right)_{\text{fixed}} = \mathbf{N} \qquad \textbf{(11.115)}$$

where the designation "fixed" has been explicitly appended to $\dot{\mathbf{L}}$ because this relation is derived from Newton's equation and is therefore valid only in an inertial frame of reference. From Equation 10.12, we have

$$\left(\frac{d\mathbf{L}}{dt}\right)_{\text{fixed}} = \left(\frac{d\mathbf{L}}{dt}\right)_{\text{body}} + \boldsymbol{\omega} \times \mathbf{L} \qquad \textbf{(11.116)}$$

or

$$\left(\frac{d\mathbf{L}}{dt}\right)_{\text{body}} + \boldsymbol{\omega} \times \mathbf{L} = \mathbf{N} \qquad \textbf{(11.117)}$$

The component of this equation along the x_3-axis (note that this is a *body* axis) is

$$\dot{L}_3 + \omega_1 L_2 - \omega_2 L_1 = N_3 \qquad \textbf{(11.118)}$$

---

* Leonard Euler, 1758.

But because we have chosen the x_i-axes to coincide with the principal axes of the body, we have, from Equation 11.34,

$$L_i = I_i \omega_i$$

so that

$$I_3 \dot{\omega}_3 - (I_1 - I_2)\omega_1 \omega_2 = N_3 \qquad \textbf{(11.119)}$$

By permuting the subscripts, we can write all three components of **N**:

$$\left. \begin{aligned} I_1 \dot{\omega}_1 - (I_2 - I_3)\omega_2 \omega_3 &= N_1 \\ I_2 \dot{\omega}_2 - (I_3 - I_1)\omega_3 \omega_1 &= N_2 \\ I_3 \dot{\omega}_3 - (I_1 - I_2)\omega_1 \omega_2 &= N_3 \end{aligned} \right\} \qquad \textbf{(11.120)}$$

Using the permutation symbol, we can write, in general

$$\boxed{(I_i - I_j)\omega_i \omega_j - \sum_k (I_k \dot{\omega}_k - N_k)\varepsilon_{ijk} = 0} \qquad \textbf{(11.121)}$$

Equations 11.120 and 11.121 are the desired Euler equations for the motion of a rigid body in a force field.

The motion of a rigid body depends on the structure of the body only through the three numbers I_1, I_2, and I_3—that is, the principal moments of inertia. Thus, any two bodies with the same principal moments move in exactly the same manner, regardless of the fact that they may have quite different shapes. (However, effects such as frictional retardation may depend on the shape of a body.) The simplest geometrical shape that a body having three given principal moments may possess is a homogeneous ellipsoid. The motion of any rigid body can therefore be represented by the motion of the **equivalent ellipsoid**.* The treatment of rigid-body dynamics from this point of view was originated by Poinsot in 1834. The **Poinsot construction** is sometimes useful for depicting the motion of a rigid body geometrically.[†]

E X A M P L E 〔 **11.8** 〕 ---

Consider the dumbbell of Section 11.3. Find the angular momentum of the system and the torque required to maintain the motion shown in Figures 11-2 and 11-9.

Solution: Let $|\mathbf{r}_1| = |\mathbf{r}_2| = b$. Let the body fixed coordinate system have its origin at O and the symmetry axis x_3 be along the weightless shaft toward m_1.

---

* The momental ellipsoid was introduced by the French mathematician Baron Augustin Louis Cauchy (1789–1857) in 1827.

[†] See, for example, Goldstein (Go80, p. 205).

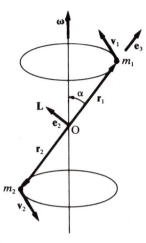

FIGURE 11-9

$$L = \sum_\alpha m_\alpha \mathbf{r}_\alpha \times \mathbf{v}_\alpha \qquad (11.122)$$

Because L is perpendicular to the shaft and L rotates around $\boldsymbol{\omega}$ as the shaft rotates, let $\mathbf{e}_2$ be along L:

$$L = L_2 \mathbf{e}_2 \qquad (11.123)$$

If α is the angle between $\boldsymbol{\omega}$ and the shaft, the components of $\boldsymbol{\omega}$ are

$$\left. \begin{aligned} \omega_1 &= 0 \\ \omega_2 &= \omega \sin \alpha \\ \omega_3 &= \omega \cos \alpha \end{aligned} \right\} \qquad (11.124)$$

The principal axes are x_1, x_2, and x_3, and the principal moments of inertia are, from Equation 11.13a,

$$\left. \begin{aligned} I_1 &= (m_1 + m_2)b^2 \\ I_2 &= (m_1 + m_2)b^2 \\ I_3 &= 0 \end{aligned} \right\} \qquad (11.125)$$

Combining Equations 11.124 and 11.125

$$\left. \begin{aligned} L_1 &= I_1 \omega_1 = 0 \\ L_2 &= I_2 \omega_2 = (m_1 + m_2)b^2 \omega \sin \alpha \\ L_3 &= I_3 \omega_3 = 0 \end{aligned} \right\} \qquad (11.126)$$

which agrees with Equation 11.123.

Using Euler's equations (Equation 11.120) and $\dot{\omega} = 0$, the torque components are

$$
\left.
\begin{aligned}
N_1 &= -(m_1 + m_2)b^2\omega^2 \sin\alpha \cos\alpha \\
N_2 &= 0 \\
N_3 &= 0
\end{aligned}
\right\}
\qquad (11.127)
$$

The torque required to maintain the motion if $\dot{\omega} = 0$ is directed along the x_1-axis.

- ●

11.9 FORCE-FREE MOTION OF A SYMMETRIC TOP

If we consider a symmetric top, that is, a rigid body with $I_1 = I_2 \neq I_3$, then the force-free Euler equations (Equation 11.114) become

$$
\left.
\begin{aligned}
(I_1 - I_3)\omega_2\omega_3 - I_1\dot{\omega}_1 &= 0 \\
(I_3 - I_1)\omega_3\omega_1 - I_1\dot{\omega}_2 &= 0 \\
I_3\dot{\omega}_3 &= 0
\end{aligned}
\right\}
\qquad (11.128)
$$

where I_1 has been substituted for I_2. Because for force-free motion the center of mass of the body is either at rest or in uniform motion with respect to the fixed or inertial frame of reference, we can, without loss of generality, specify that the body's center of mass is at rest and located at the origin of the fixed coordinate system. We consider the case in which the angular velocity vector $\boldsymbol{\omega}$ does not lie along a principal axis of the body, otherwise, the motion is trivial.

The first result for the motion follows from the third part of Equations 11.128, $\dot{\omega}_3 = 0$, or

$$
\omega_3(t) = \text{const.}
\qquad (11.129)
$$

The first two parts of Equation 11.128 can be written as

$$
\left.
\begin{aligned}
\dot{\omega}_1 &= -\left(\frac{I_3 - I_1}{I_1}\,\omega_3\right)\omega_2 \\
\dot{\omega}_2 &= \left(\frac{I_3 - I_1}{I_1}\,\omega_3\right)\omega_1
\end{aligned}
\right\}
\qquad (11.130)
$$

Because the terms in the parentheses are identical and composed of constants, we may define

$$
\Omega \equiv \frac{I_3 - I_1}{I_1}\,\omega_3
\qquad (11.131)
$$

so that

$$
\left.
\begin{aligned}
\dot{\omega}_1 + \Omega\omega_2 &= 0 \\
\dot{\omega}_2 - \Omega\omega_1 &= 0
\end{aligned}
\right\}
\qquad (11.132)
$$

These are coupled equations of familiar form, and we can effect a solution by multiplying the second equation by i and adding to the first:

$$(\dot{\omega}_1 + i\dot{\omega}_2) - i\Omega(\omega_1 + i\omega_2) = 0 \qquad \textbf{(11.133)}$$

If we define

$$\eta \equiv \omega_1 + i\omega_2 \qquad \textbf{(11.134)}$$

then

$$\dot{\eta} - i\Omega\eta = 0 \qquad \textbf{(11.135)}$$

with solution*

$$\eta(t) = Ae^{i\Omega t} \qquad \textbf{(11.136)}$$

Thus,

$$\omega_1 + i\omega_2 = A\cos\Omega t + iA\sin\Omega t \qquad \textbf{(11.137)}$$

and therefore

$$\left.\begin{array}{l}\omega_1(t) = A\cos\Omega t \\ \omega_2(t) = A\sin\Omega t\end{array}\right\} \qquad \textbf{(11.138)}$$

Because $\omega_3 = $ constant, we note that the magnitude of $\boldsymbol{\omega}$ is also constant:

$$|\boldsymbol{\omega}| = \omega = \sqrt{\omega_1^2 + \omega_2^2 + \omega_3^2} = \sqrt{A^2 + \omega_3^2} = \text{constant} \qquad \textbf{(11.139)}$$

Equations 11.138 are the parametric equations of a circle, so the projection of the vector $\boldsymbol{\omega}$ (which is of constant magnitude) onto the x_1-x_2 plane describes a circle with time (Figure 11-10).

The x_3-axis is the symmetry axis of the body, so we find that the angular velocity vector $\boldsymbol{\omega}$ revolves or *precesses* about the body x_3-axis with a constant angular frequency Ω. Thus, to an observer in the body coordinate system, $\boldsymbol{\omega}$ traces out a cone around the body symmetry axis, called the **body cone**.

Because we are considering force-free motion, the angular-momentum vector **L** is stationary in the fixed coordinate system and is constant in time. An additional constant of the motion for the force-free case is the kinetic energy, or in particular, because the body's center of mass is fixed, the *rotational* kinetic energy is constant:

$$T_{\text{rot}} = \tfrac{1}{2}\boldsymbol{\omega} \cdot \mathbf{L} = \text{constant} \qquad \textbf{(11.140)}$$

But we have $\mathbf{L} = $ constant, so $\boldsymbol{\omega}$ must move such that its projection on the stationary angular-momentum vector is constant. Thus, $\boldsymbol{\omega}$ precesses around and makes a constant angle with the vector **L**. In such a case, **L**, $\boldsymbol{\omega}$, and the x_3- (body) axis (i.e., the unit vector $\mathbf{e}_3$) all lie in a *plane*. We can show this by proving that

---

* In general, the constant coefficient is complex, so we should properly write $A\exp(i\delta)$. For simplicity, however, we set the phase δ equal to zero; this can always be done by choosing an appropriate instant to call $t = 0$.

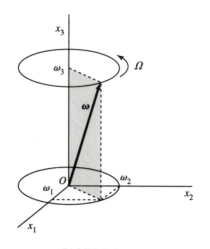

FIGURE 11-10

$\mathbf{L} \cdot (\boldsymbol{\omega} \times \mathbf{e}_3) = 0$. First, $\boldsymbol{\omega} \times \mathbf{e}_3 = \omega_2 \mathbf{e}_1 - \omega_1 \mathbf{e}_2$. If we take the scalar product of this result with $\mathbf{L}$, we have $\mathbf{L} \cdot (\boldsymbol{\omega} \times \mathbf{e}_3) = I_1 \omega_1 \omega_2 - I_2 \omega_1 \omega_2 = 0$, because $I_1 = I_2$ for the symmetric top. Therefore, if we designate the x_3'-axis in the fixed coordinate system to coincide with $\mathbf{L}$, then to an observer in the fixed system, $\boldsymbol{\omega}$ traces out a cone around the fixed x_3'-axis, called the **space cone**. The situation is then described (Figure 11-11) by one cone rolling on another, such that $\boldsymbol{\omega}$ precesses around the x_3-axis in the body system and around the x_3'-axis (or $\mathbf{L}$) in the space-fixed system.

The rate at which $\boldsymbol{\omega}$ precesses around the body symmetry axis is given by Equation 11.131:

$$\Omega = \frac{I_3 - I_1}{I_1} \omega_3$$

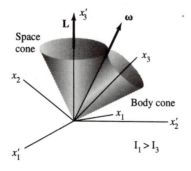

FIGURE 11-11

If $I_1 \cong I_3$, then Ω becomes very small compared with ω_3. The Earth is slightly flattened near the poles,* so its shape can be approximated by an oblate spheroid with $I_1 \cong I_3$, but with $I_3 > I_1$. If the Earth is considered to be a rigid body, then the moments I_1 and I_3 are such that $\Omega \cong \omega_3/300$. Because the period of the Earth's rotation is $2\pi/\omega = 1$ day, and because $\omega_3 \cong \omega$, the period predicted for the precession of the axis of rotation is $1/\Omega \cong 300$ days. The observed precession has an irregular period about 50% greater than that predicted on the basis of this simple theory; the deviation is ascribed to the facts that (1) the Earth is not a *rigid* body and (2) the shape is not exactly that of an oblate spheroid, but rather has a higher-order deformation and actually resembles a flattened pear.

The Earth's equatorial "bulge" together with the fact that the Earth's rotational axis is inclined at an angle of approximately 23.5° to the plane of the Earth's orbit around the sun (the **plane of the ecliptic**) produces a gravitational torque (caused by both the sun and the moon), which produces a slow precession of the Earth's axis. The period of this precessional motion is approximately 26,000 years. Thus, in different epochs, different stars become the "pole star."*

EXAMPLE 11.9

Show that the motion depicted in Figure 11-11 actually refers to the motion of a prolate object such as an elongated rod ($I_1 > I_3$), whereas for a flat disk ($I_3 > I_1$) the space cone would be inside the body cone rather than outside.

Solution: If **L** is along x_3', then the Euler angle θ (between the x_3- and x_3'-axes) is the angle between **L** and the x_3-axis. At a given instant, we align e_2 to be in the plane defined by **L**, **ω**, and e_3. Then, at this same instant,

$$\left. \begin{array}{l} L_1 = 0 \\ L_2 = L\sin\theta \\ L_3 = L\cos\theta \end{array} \right\} \tag{11.141}$$

Let α be the angle between **ω** and the x_3-axis. Then, at this same instant, we have

$$\left. \begin{array}{l} \omega_1 = 0 \\ \omega_2 = \omega\sin\alpha \\ \omega_3 = \omega\cos\alpha \end{array} \right\} \tag{11.142}$$

---

* The flattening at the poles was shown by Newton to be caused by the Earth's rotation; the resulting precessional motion was first calculated by Euler.

* This precession of the equinoxes was apparently discovered by the Babylonian astronomer Cidenas in about 343 B.C.

We can also determine the components of $\mathbf{L}$ from Equation 11.34:

$$\left.\begin{aligned} L_1 &= I_1\omega_1 = 0 \\ L_2 &= I_1\omega_2 = I_1\omega \sin \alpha \\ L_3 &= I_3\omega_3 = I_3\omega \cos \alpha \end{aligned}\right\} \qquad \textbf{(11.143)}$$

We can obtain the ratio L_2/L_3 from Equations 11.141 and 11.143,

$$\frac{L_2}{L_3} = \tan \theta = \frac{I_1}{I_3} \tan \alpha \qquad \textbf{(11.144)}$$

so we have

Prolate spheroid

$$I_1 > I_3, \qquad \theta > \alpha \qquad \textbf{(11.145a)}$$

Oblate spheroid

$$I_3 > I_1, \qquad \alpha > \theta \qquad \textbf{(11.145b)}$$

The two cases are shown in Figure 11-12. From Equation 11.131, we determine that Ω and ω_3 have the same sign if $I_3 > I_1$ but have opposite signs if $I_1 > I_3$. Thus, the sense of precession is opposite for the two cases. This fact and Equation 11.145 can be reconciled only if the space cone is outside the body cone for the prolate case but inside the body cone for the oblate case. The angular velocity $\boldsymbol{\omega}$ defines both cones as it rotates about $\mathbf{L}$ (space cone) and the symmetry axis x_3 (body cone). The line of contact between the space and body cones is the instantaneous axis of rotation (along $\boldsymbol{\omega}$). At any instant, this axis is at rest, so that the body cone rolls around the space cone without slipping. In both cases, the space cone is fixed, because $\mathbf{L}$ is constant.

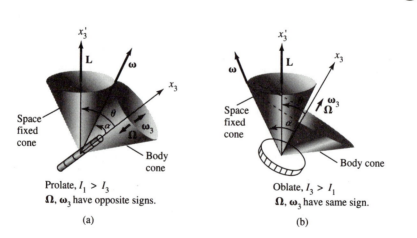

Prolate, $I_1 > I_3$
Ω, ω_3 have opposite signs.

(a)

Oblate, $I_3 > I_1$
Ω, ω_3 have same sign.

(b)

FIGURE 11-12

E X A M P L E **11.10** --

With what angular velocity does the symmetry axis (x_3) and $\boldsymbol{\omega}$ rotate about the fixed angular momentum L?

Solution: Because $\mathbf{e}_3$, $\boldsymbol{\omega}$, and $\mathbf{L}$ are in the same plane, $\mathbf{e}_3$ and $\boldsymbol{\omega}$ precess about $\mathbf{L}$ with the same angular velocity. In Section 11.7 we learned that $\dot{\phi}$ is the angular velocity along the x_3'-axis. If we use the same instant of time considered in the previous example (when $\mathbf{e}_2$ was in the plane of $\mathbf{e}_3$, $\boldsymbol{\omega}$, and $\mathbf{L}$), then the Euler angle $\psi = 0$, and from Equation 11.102

$$\omega_2 = \dot{\phi}\sin\theta$$

and

$$\dot{\phi} = \frac{\omega_2}{\sin\theta} \tag{11.146}$$

Substituting for ω_2 from Equation 11.142, we have

$$\dot{\phi} = \frac{\omega\sin\alpha}{\sin\theta} \tag{11.147}$$

We can rewrite $\dot{\phi}$ by substituting $\sin\alpha$ from Equation 11.143 and $\sin\theta$ from Equation 11.141:

$$\dot{\phi} = \omega\frac{L_2}{I_1\omega}\frac{L}{L_2} = \frac{L}{I_1} \tag{11.148}$$

-- ●

11.10 MOTION OF A SYMMETRIC TOP WITH ONE POINT FIXED

Consider a symmetric top with tip held fixed* rotating in a gravitational field. In our previous development, we have been able to separate the kinetic energy into translational and rotational parts by taking the body's center of mass to be the origin of the rotating or body coordinate system. Alternatively, if we can choose the origins of the fixed and the body coordinate systems to coincide, then the translational kinetic energy vanishes, because $\mathbf{V} = \dot{\mathbf{R}} = 0$. Such a choice is quite convenient for discussing the top, because the stationary tip may then be taken as the origin for both coordinate systems. Figure 11-13 shows the Euler angles for this situation. The x_3'- (fixed) axis corresponds to the vertical, and we choose the x_3- (body) axis to be the symmetry axis of the top. The distance from the fixed tip to the center of mass is h, and the mass of the top is M.

---

* This problem was first solved in detail by Lagrange in *Mécanique analytique*.

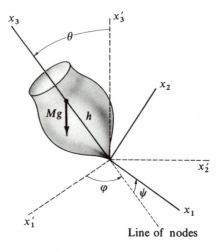

Line of nodes

FIGURE 11-13

Because we have a symmetric top, the principal moments of inertia about the x_1- and x_2-axes are equal: $I_1 = I_2$. We assume $I_3 \neq I_1$. The kinetic energy is then given by

$$T = \tfrac{1}{2} \sum_i I_i \omega_i^2 = \tfrac{1}{2}I_1(\omega_1^2 + \omega_2^2) + \tfrac{1}{2}I_3\omega_3^2 \tag{11.149}$$

According to Equation 11.102, we have

$$\omega_1^2 = (\dot{\phi}\sin\theta\sin\psi + \dot{\theta}\cos\psi)^2$$
$$= \dot{\phi}^2\sin^2\theta\sin^2\psi + 2\dot{\phi}\dot{\theta}\sin\theta\sin\psi\cos\psi + \dot{\theta}^2\cos^2\psi$$
$$\omega_2^2 = (\dot{\phi}\sin\theta\cos\psi - \dot{\theta}\sin\psi)^2$$
$$= \dot{\phi}^2\sin^2\theta\cos^2\psi - 2\dot{\phi}\dot{\theta}\sin\theta\sin\psi\cos\psi + \dot{\theta}^2\sin^2\psi$$

so that

$$\omega_1^2 + \omega_2^2 = \dot{\phi}^2\sin^2\theta + \dot{\theta}^2 \tag{11.150a}$$

and

$$\omega_3^2 = (\dot{\phi}\cos\theta + \dot{\psi})^2 \tag{11.150b}$$

Therefore,

$$T = \tfrac{1}{2}I_1(\dot{\phi}^2\sin^2\theta + \dot{\theta}^2) + \tfrac{1}{2}I_3(\dot{\phi}\cos\theta + \dot{\psi})^2 \tag{11.151}$$

Because the potential energy is $Mgh\cos\theta$, the Lagrangian becomes

$$L = \tfrac{1}{2}I_1(\dot{\phi}^2\sin^2\theta + \dot{\theta}^2) + \tfrac{1}{2}I_3(\dot{\phi}\cos\theta + \dot{\psi})^2 - Mgh\cos\theta \tag{11.152}$$

The Lagrangian is cyclic in both the ϕ- and ψ-coordinates. The momenta conjugate to these coordinates are therefore constants of the motion:

$$p_\phi = \frac{\partial L}{\partial \dot{\phi}} = (I_1 \sin^2\theta + I_3 \cos^2\theta)\dot{\phi} + I_3 \dot{\psi}\cos\theta = \text{constant} \quad \textbf{(11.153)}$$

$$p_\psi = \frac{\partial L}{\partial \dot{\psi}} = I_3(\dot{\psi} + \dot{\phi}\cos\theta) = \text{constant} \quad \textbf{(11.154)}$$

Because the cyclic coordinates are *angles*, the conjugate momenta are *angular momenta*—the angular momenta along the axes for which ϕ and ψ are the rotation angles, that is, the x_3'- (or vertical) axis and the x_3- (or body symmetry) axis, respectively. We note that this result is ensured by the construction shown in Figure 11-13, because the gravitational torque is directed along the line of nodes. Hence, the torque can have no component along either the x_3'- or the x_3-axis, both of which are perpendicular to the line of nodes. Thus, the angular momenta along these axes are constants of the motion.

Equations 11.153 and 11.154 can be solved for $\dot{\phi}$ and $\dot{\psi}$ in terms of θ. From Equation 11.154, we can write

$$\dot{\psi} = \frac{p_\psi - I_3 \dot{\phi}\cos\theta}{I_3} \quad \textbf{(11.155)}$$

and substituting this result into Equation 11.153, we find

$$(I_1 \sin^2\theta + I_3 \cos^2\theta)\dot{\phi} + (p_\psi - I_3\dot{\phi}\cos\theta)\cos\theta = p_\phi$$

or

$$(I_1 \sin^2\theta)\dot{\phi} + p_\psi \cos\theta = p_\phi$$

so that

$$\dot{\phi} = \frac{p_\phi - p_\psi \cos\theta}{I_1 \sin^2\theta} \quad \textbf{(11.156)}$$

Using this expression for $\dot{\phi}$ in Equation 11.155, we have

$$\dot{\psi} = \frac{p_\psi}{I_3} - \frac{(p_\phi - p_\psi \cos\theta)\cos\theta}{I_1 \sin^2\theta} \quad \textbf{(11.157)}$$

By hypothesis, the system we are considering is conservative; we therefore have the further property that the total energy is a constant of the motion:

$$E = \tfrac{1}{2}I_1 (\dot{\phi}^2 \sin^2\theta + \dot{\theta}^2) + \tfrac{1}{2}I_3\omega_3^2 + Mgh \cos\theta = \text{constant} \quad \textbf{(11.158)}$$

Using the expression for ω_3 (e.g., see Equation 11.102), we note that Equation 11.154 can be written as

$$p_\psi = I_3\omega_3 = \text{constant} \quad \textbf{(11.159a)}$$

or

$$I_3 \omega_3^2 = \frac{p_\psi^2}{I_3} = \text{constant} \qquad (11.159b)$$

Therefore, not only is E a constant of the motion, but so is $E - \frac{1}{2}I_3\omega_3^2$; we let this quantity be E':

$$E' \equiv E - \frac{1}{2}I_3\omega_3^2 = \frac{1}{2}I_1(\dot{\phi}^2 \sin^2\theta + \dot{\theta}^2) + Mgh \cos\theta = \text{constant} \qquad (11.160)$$

Substituting into this equation the expression for $\dot{\phi}$ (Equation 11.156), we have

$$E' = \frac{1}{2}I_1\dot{\theta}^2 + \frac{(p_\phi - p_\psi \cos\theta)^2}{2I_1 \sin^2\theta} + Mgh \cos\theta \qquad (11.161)$$

which we can write as

$$E' = \frac{1}{2}I_1 \dot{\theta}^2 + V(\theta) \qquad (11.162)$$

where $V(\theta)$ is an "effective potential" given by

$$V(\theta) \equiv \frac{(p_\phi - p_\psi \cos\theta)^2}{2I_1 \sin^2\theta} + Mgh \cos\theta \qquad (11.163)$$

Equation 11.162 can be solved to yield $t(\theta)$:

$$t(\theta) = \int \frac{d\theta}{\sqrt{(2/I_1)[E' - V(\theta)]}} \qquad (11.164)$$

This integral can (formally, at least) be inverted to obtain $\theta(t)$, which, in turn, can be substituted into Equations 11.156 and 11.157 to yield $\phi(t)$ and $\psi(t)$. Because the Euler angles θ, ϕ, ψ completely specify the orientation of the top, the results for $\theta(t)$, $\phi(t)$, and $\psi(t)$ constitute a complete solution for the problem. It should be clear that such a procedure is complicated and not very illuminating. But we can obtain some qualitative features of the motion by examining the preceding equations in a manner analogous to that used for treating the motion of a particle in a central-force field (see Section 8.6).

Figure 11-14 shows the form of the effective potential $V(\theta)$ in the range $0 \le \theta \le \pi$, which clearly is the physically limited region for θ. This energy diagram indicates that for any general values of E' (e.g., the value represented by E_1') the motion is limited by two extreme values of θ—that is, θ_1 and θ_2, which correspond to the turning points of the central-force problem and are roots of the denominator in Equation 11.164. Thus we find that the inclination of the rotating top is, in general, confined to the region $\theta_1 \le \theta \le \theta_2$. For the case that $E' = E_2' = V_{\min}$, θ is limited to the single value θ_0, and the motion is a steady precession at a fixed angle of inclination. Such motion is similar to the occurrence of circular orbits in the central-force problem.

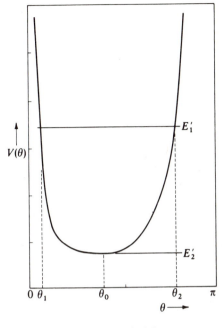

FIGURE 11-14

The value of θ_0 can be obtained by setting the derivative of $V(\theta)$ equal to zero. Thus,

$$\left.\frac{\partial V}{\partial \theta}\right|_{\theta=\theta_0} = \frac{-\cos\theta_0(p_\phi - p_\psi\cos\theta_0)^2 + p_\psi\sin^2\theta_0(p_\phi - p_\psi\cos\theta_0)}{I_1\sin^3\theta_0}$$
$$- Mgh\sin\theta_0 = 0$$

$$(11.165)$$

If we define

$$\beta \equiv p_\phi - p_\psi\cos\theta_0 \qquad (11.166)$$

then Equation 11.165 becomes

$$(\cos\theta_0)\,\beta^2 - (p_\psi\sin^2\theta_0)\beta + (MghI_1\sin^4\theta_0) = 0 \qquad (11.167)$$

This is a quadratic in β and can be solved with the result

$$\beta = \frac{p_\psi\sin^2\theta_0}{2\cos\theta_0}\left(1 \pm \sqrt{1 - \frac{4MghI_1\cos\theta_0}{p_\psi^2}}\right) \qquad (11.168)$$

Because β must be a real quantity, the radicand in Equation 11.168 must be positive. If $\theta_0 < \pi/2$, we have

$$p_\psi^2 \geq 4MghI_1\cos\theta_0 \qquad (11.169)$$

But from Equation 11.159a, $p_\psi = I_3\omega_3$; thus,

$$\omega_3 \geq \frac{2}{I_3} \sqrt{MghI_1 \cos\theta_0} \tag{11.170}$$

We therefore conclude that a steady precession can occur at the fixed angle of inclination θ_0 only if the angular velocity of spin is larger than the limiting value given by Equation 11.170.

From Equation 11.156, we note that we can write (for $\theta = \theta_0$)

$$\dot\phi_0 = \frac{\beta}{I_1 \sin^2\theta_0} \tag{11.171}$$

We therefore have two possible values of the precessional angular velocity $\dot\phi_0$, one for each of the values of β given by Equation 11.168:

$$\dot\phi_{0(+)} \rightarrow \text{Fast precession}$$

and

$$\dot\phi_{0(-)} \rightarrow \text{Slow precession}$$

If ω_3 (or p_ψ) is large (a fast top), then the second term in the radicand of Equation 11.168 is small, and we may expand the radical. Retaining only the first nonvanishing term in each case, we find

$$\left.\begin{aligned} \dot\phi_{0(+)} &\cong \frac{I_3\omega_3}{I_1 \cos\theta_0} \\[2mm] \dot\phi_{0(-)} &\cong \frac{Mgh}{I_3\omega_3} \end{aligned}\right\} \tag{11.172}$$

It is the slower of the two possible precessional angular velocities, $\dot\phi_{0(-)}$, that is usually observed.

The preceding results apply if $\theta_0 < \pi/2$; but if* $\theta_0 > \pi/2$, the radicand in Equation 11.168 is always positive and there is no limiting condition on ω_3. Because the radical is greater than unity in such a case, the values of $\dot\phi_0$ for fast and slow precession have opposite signs; that is, for $\theta_0 > \pi/2$, the fast precession is in the same direction as that for $\theta_0 < \pi/2$, but the slow precession takes place in the opposite sense.

For the general case, in which $\theta_1 < \theta < \theta_2$, Equation 11.156 indicates that $\dot\phi$ may or may not change sign as θ varies between its limits—depending on the values of p_ϕ and p_ψ. If $\dot\phi$ does not change sign, the top precesses monotonically around the x_3'-axis (see Figure 11-13), and the x_3- (or symmetry) axis oscillates between $\theta = \theta_1$ and $\theta = \theta_2$. This phenomenon is called **nutation**; the path

---

* If $\theta_0 > \pi/2$, the fixed tip of the top is at a position *above* the center of mass. Such motion is possible, for example, with a gyroscopic top whose tip is actually a ball and rests in a cup that is fixed atop a pedestal.

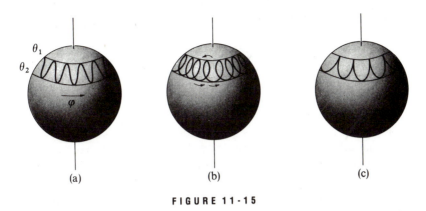

FIGURE 11-15

described by the projection of the body symmetry axis on a unit sphere in the fixed system is shown in Figure 11-15a.

If $\dot\phi$ does change sign between the limiting values of θ, the precessional angular velocity must have opposite signs at $\theta = \theta_1$ and $\theta = \theta_2$. Thus, the nutational-precessional motion produces the looping motion of the symmetry axis depicted in Figure 11-15b.

Finally, if the values of p_ϕ and p_ψ are such that

$$(p_\phi - p_\psi \cos \theta)|_{\theta=\theta_1} = 0 \tag{11.173}$$

then

$$\dot\phi|_{\theta=\theta_1} = 0, \qquad \dot\theta|_{\theta=\theta_1} = 0 \tag{11.174}$$

Figure 11-15c shows the resulting cusplike motion. It is just this case that corresponds to the usual method of starting a top. First, the top is spun around its axis, then it is given a certain initial tilt and released. Thus, initial conditions are $\theta = \theta_1$ and $\dot\theta = 0 = \dot\phi$. Because the first motion of the top is to begin to fall in the gravitational field, the conditions are exactly those of Figure 11-15c, and the cusplike motion ensues. Figures 11-15a and 11-15b correspond to the motion in the event that there is an initial angular velocity $\dot\phi$ either in the direction of or opposite to the direction of precession.

11.11 STABILITY OF RIGID-BODY ROTATIONS

We now consider a rigid body undergoing force-free rotation around one of its principal axes and inquire whether such motion is stable. "Stability" here means, as before (see Section 8.10), that if a small perturbation is applied to the system, the motion will either return to its former mode or will perform small oscillations about it.

We choose for our discussion a general rigid body for which all the principal moments of inertia are distinct, and we label them such that $I_3 > I_2 > I_1$. We let the body axes coincide with the principal axes, and we start with the body rotating around the x_1-axis—that is, around the principal axis associated with the moment of inertia I_1. Then,

$$\boldsymbol{\omega} = \omega_1 \mathbf{e}_1 \tag{11.175}$$

If we apply a small perturbation, the angular velocity vector assumes the form

$$\boldsymbol{\omega} = \omega_1 \mathbf{e}_1 + \lambda \mathbf{e}_2 + \mu \mathbf{e}_3 \tag{11.176}$$

where λ and μ are small quantities and correspond to the parameters used previously in other perturbation expansions. (λ and μ are sufficiently small so that their product can be neglected compared with all other quantities of interest to the discussion.)

The Euler equations (see Equation 11.114) become

$$\left. \begin{array}{l} (I_2 - I_3)\lambda\mu - I_1\dot{\omega}_1 = 0 \\[2mm] (I_3 - I_1)\mu\omega_1 - I_2\dot{\lambda} = 0 \\[2mm] (I_1 - I_2)\lambda\omega_1 - I_3\dot{\mu} = 0 \end{array} \right\} \tag{11.177}$$

Because $\lambda\mu \approx 0$, the first of these equations requires $\dot{\omega}_1 = 0$, or $\omega_1 = $ constant. Solving the other two equations for $\dot{\lambda}$ and $\dot{\mu}$, we find

$$\dot{\lambda} = \left(\frac{I_3 - I_1}{I_2} \omega_1 \right) \mu \tag{11.178}$$

$$\dot{\mu} = \left(\frac{I_1 - I_2}{I_3} \omega_1 \right) \lambda \tag{11.179}$$

where the terms in parentheses are both constants. These are coupled equations, but they cannot be solved by the method used in Section 11.9, because the constants in the two equations are different. The solution can be obtained by first differentiating the equation for $\dot{\lambda}$:

$$\ddot{\lambda} = \left(\frac{I_3 - I_1}{I_2} \omega_1 \right) \dot{\mu} \tag{11.180}$$

The expression for $\dot{\mu}$ can now be substituted in this equation:

$$\ddot{\lambda} + \left(\frac{(I_1 - I_3)(I_1 - I_2)}{I_2 I_3} \omega_1^2 \right) \lambda = 0 \tag{11.181}$$

The solution to this equation is

$$\lambda(t) = A e^{i\Omega_{1\lambda}t} + B e^{-i\Omega_{1\lambda}t} \tag{11.182}$$

where

$$\Omega_{1\lambda} \equiv \omega_1 \sqrt{\frac{(I_1 - I_3)(I_1 - I_2)}{I_2 I_3}} \tag{11.183}$$

and where the subscripts 1 and λ indicate that we are considering the solution for λ when the rotation is around the x_1-axis.

By hypothesis, $I_1 < I_3$ and $I_1 < I_2$, so $\Omega_{1\lambda}$ is real. The solution for $\lambda(t)$ therefore represents oscillatory motion with a frequency $\Omega_{1\lambda}$. We can similarly investigate $\mu(t)$, with the result that $\Omega_{1\mu} = \Omega_{1\lambda} \equiv \Omega_1$. Thus, the small perturbations introduced by forcing small x_2- and x_3-components on $\boldsymbol{\omega}$ do not increase with time but oscillate around the equilibrium values $\lambda = 0$ and $\mu = 0$. Consequently, the rotation around the x_1-axis is stable.

If we consider rotations around the x_2- and x_3-axes, we can obtain expressions for Ω_2 and Ω_3 from Equation 11.183 by permutation:

$$\Omega_1 = \omega_1 \sqrt{\frac{(I_1 - I_3)(I_1 - I_2)}{I_2 I_3}} \qquad \textbf{(11.184a)}$$

$$\Omega_2 = \omega_2 \sqrt{\frac{(I_2 - I_1)(I_2 - I_3)}{I_1 I_3}} \qquad \textbf{(11.184b)}$$

$$\Omega_3 = \omega_3 \sqrt{\frac{(I_3 - I_2)(I_3 - I_1)}{I_1 I_2}} \qquad \textbf{(11.184c)}$$

But because $I_1 < I_2 < I_3$, we have

$$\Omega_1, \Omega_3 \text{ real}, \qquad \Omega_2 \text{ imaginary}$$

Thus, when the rotation takes place around either the x_1- or x_3-axes, the perturbation produces oscillatory motion and the rotation is stable. When the rotation takes place around x_2, however, the fact that Ω_2 is imaginary results in the perturbation increasing with time without limit; such motion is unstable.

Because we have assumed a completely arbitrary rigid body for this discussion, we conclude that rotation around the principal axis corresponding to either the greatest or smallest moment of inertia is stable and that rotation around the principal axis corresponding to the intermediate moment is unstable. We can demonstrate this effect with, say, a book (kept closed by tape or a rubber band). If we toss the book into the air with an angular velocity around one of the principal axes, the motion is unstable for rotation around the intermediate axis and stable for the other two axes.

If two of the moments of inertia are equal ($I_1 = I_2$, say), then the coefficient of λ in Equation 11.179 vanishes, and we have $\dot{\mu} = 0$ or $\mu(t) = $ constant. Equation 11.178 for λ can therefore be integrated to yield

$$\lambda(t) = C + Dt \qquad \textbf{(11.185)}$$

and the perturbation increases linearly with the time; the motion around the x_1-axis is therefore unstable. We find a similar result for motion around the x_2-axis. Stability exists only for the x_3-axis, independent of whether I_3 is greater or less than $I_1 = I_2$.

A good example of the stability of rotating objects is seen by the satellites put into space by the space shuttle orbiter. When the satellites are ejected from the payload bay, they are normally spinning in a stable configuration. In May 1992, when the astronauts attempted to grab in space the Intelsat satellite to attach a rocket that would insert it into geosynchronous orbit, the spinning satellite was slowed down and stopped before the astronaut attempted to attach a grappling fixture to bring it into the payload bay. After each futile attempt, when the grappling fixture failed, the satellite tumbled even more. After each of two unsuccessful days of trying to attach the grappling fixture, the astronauts had to abort their attempts because of the increased tumbling. Ground controllers would then require a few hours to restabilize the satellite using jet thrusters. The satellite would be left in a stable configuration of spinning slowly about its cyclindrical symmetry axis (a principal axis). Finally, on the third day, three astronauts went outside the orbiter, grabbed the slightly rotating satellite, stopped it, and put it into the payload bay where the rocket skirt was attached. The Intelsat satellite was finally successfully placed into orbit in time to broadcast the 1992 Barcelona Olympic summer games.

P R O B L E M S

11-1. Calculate the moments of inertia I_1, I_2, and I_3 for a homogeneous sphere of radius R and mass M. (Choose the origin at the center of the sphere.)

11-2. Calculate the moments of inertia I_1, I_2, and I_3 for a homogeneous cone of mass M whose height is h and whose base has a radius R. Choose the x_3-axis along the axis of symmetry of the cone. Choose the origin at the apex of the cone, and calculate the elements of the inertia tensor. Then make a transformation such that the center of mass of the cone becomes the origin, and find the principal moments of inertia.

11-3. Calculate the moments of inertia I_1, I_2, and I_3 for a homogeneous ellipsoid of mass M with axes' lengths $2a > 2b > 2c$.

11-4. Consider a thin rod of length l and mass m pivoted about one end. Calculate the moment of inertia. Find the point at which, if all the mass were concentrated, the moment of inertia about the pivot axis would be the same as the real moment of inertia. The distance from this point to the pivot is called the **radius of gyration.**

11-5. (a) Find the height at which a billiard ball should be struck so that it will roll with no initial slipping. (b) Calculate the optimum height of the rail of a billiard table. On what basis is the calculation predicated?

11-6. Two spheres are of the same diameter and same mass, but one is solid and the other is a hollow shell. Describe in detail a nondestructive experiment to determine which is solid and which is hollow.

11-7. A homogeneous disk of radius R and mass M rolls without slipping on a horizontal surface and is attracted to a point a distance d below the plane. If the force of attraction is proportional to the distance from the disk's center of mass to the force center, find the frequency of oscillations around the position of equilibrium.

11-8. A door is constructed of a thin homogeneous slab of material; it has a width of 1 m. If the door is opened through 90°, it is found that on release it closes itself in 2 s. Assume that the hinges are frictionless, and show that the line of hinges must make an angle of approximately 3° with the vertical.

11-9. A homogeneous slab of thickness a is placed atop a fixed cylinder of radius R whose axis is horizontal. Show that the condition for stable equilibrium of the slab, assuming no slipping, is $R > a/2$. What is the frequency of small oscillations? Sketch the potential energy U as a function of the angular displacement θ. Show that there is a minimum at $\theta = 0$ for $R > a/2$ but not for $R < a/2$.

11-10. A solid sphere of mass M and radius R rotates freely in space with an angular velocity ω about a fixed diameter. A particle of mass m, initially at one pole, moves with a constant velocity v along a great circle of the sphere. Show that, when the particle has reached the other pole, the rotation of the sphere will have been retarded by an angle

$$\alpha = \omega T \left(1 - \sqrt{\frac{2M}{2M + 5m}} \right)$$

where T is the total time required for the particle to move from one pole to the other.

11-11. A homogeneous cube, each edge of which has a length l, is initially in a position of unstable equilibrium with one edge in contact with a horizontal plane. The cube is then given a small displacement and allowed to fall. Show that the angular velocity of the cube when one face strikes the plane is given by

$$\omega^2 = A \frac{g}{l} (\sqrt{2} - 1)$$

where $A = 3/2$ if the edge cannot slide on the plane and where $A = 12/5$ if sliding can occur without friction.

11-12. Show that none of the principal moments of inertia can exceed the sum of the other two.

11-13. A three-particle system consists of masses m_i and coordinates (x_1, x_2, x_3) as follows:

$$m_1 = 3m, \quad (b, 0, b)$$

$$m_2 = 4m, \quad (b, b, -b)$$

$$m_3 = 2m, \quad (-b, b, 0)$$

Find the inertia tensor, principal axes, and principal moments of inertia.

11-14. Determine the principal axes and principal moments of inertia of a uniformly solid hemisphere of radius b and mass m about its center of mass.

11-15. If a physical pendulum has the same period of oscillation when pivoted about either of two points of unequal distances from the center of mass, show that the length of the simple pendulum with the same period is equal to the separation of the pivot points. Such a physical pendulum, called **Kater's reversible pendulum**, at one time provided the most accurate way (to about 1 part in 10^5) to measure the acceleration of gravity.* Discuss the advantages of Kater's pendulum over a simple pendulum for such a purpose.

---

* First used in 1818 by Captain Henry Kater (1777–1835), but the method was apparently suggested somewhat earlier by Bohnenberger. The theory of Kater's pendulum was treated in detail by Friedrich Wilhelm Bessel (1784–1846) in 1826.

▷ **11-16.** Consider the following inertia tensor:

$$\{I\} = \begin{Bmatrix} \frac{1}{2}(A+B) & \frac{1}{2}(A-B) & 0 \\ \frac{1}{2}(A-B) & \frac{1}{2}(A+B) & 0 \\ 0 & 0 & C \end{Bmatrix}$$

Perform a rotation of the coordinate system by an angle θ about the x_3-axis. Evaluate the transformed tensor elements, and show that the choice $\theta = \pi/4$ renders the inertia tensor diagonal with elements A, B, and C.

▷ **11-17.** Consider a thin homogeneous plate that lies in the x_1–x_2 plane. Show that the inertia tensor takes the form

$$\{I\} = \begin{Bmatrix} A & -C & 0 \\ -C & B & 0 \\ 0 & 0 & A+B \end{Bmatrix}$$

▷ **11-18.** If, in the previous problem, the coordinate axes are rotated through an angle θ about the x_3-axis, show that the new inertia tensor is

$$\{I\} = \begin{Bmatrix} A' & -C' & 0 \\ -C' & B' & 0 \\ 0 & 0 & A'+B' \end{Bmatrix}$$

where

$$A' = A\cos^2\theta - C\sin 2\theta + B\sin^2\theta$$

$$B' = A\sin^2\theta + C\sin 2\theta + B\cos^2\theta$$

$$C' = C\cos 2\theta - \frac{1}{2}(B-A)\sin 2\theta$$

and hence show that the x_1- and x_2-axes become principal axes if the angle of rotation is

$$\theta = \frac{1}{2}\tan^{-1}\left(\frac{2C}{B-A}\right)$$

11-19. Consider a plane homogeneous plate of density ρ bounded by the logarithmic spiral $r = ke^{\alpha\theta}$ and the radii $\theta = 0$ and $\theta = \pi$. Obtain the inertia tensor for the origin at $r = 0$ if the plate lies in the x_1–x_2 plane. Perform a rotation of the coordinate axes to obtain the principal moments of inertia, and use the results of the previous problem to show that they are

$$I_1' = \rho k^4 P(Q-R), \qquad I_2' = \rho k^4 P(Q+R), \qquad I_3' = I_1' + I_2'$$

where

$$P = \frac{e^{4\pi\alpha} - 1}{16(1+4\alpha^2)}, \qquad Q = \frac{1+4\alpha^2}{2\alpha}, \qquad R = \sqrt{1+4\alpha^2}$$

11-20. A uniform rod of strength b stands vertically upright on a rough floor and then tips over. What is the rod's angular velocity when it hits the floor?

11-21. The proof represented by Equations 11.54–11.61 is expressed entirely in the summation convention. Rewrite this proof in matrix notation.

▷ **11-22.** The **trace** of a tensor is defined as the sum of the diagonal elements:

$$\text{tr}\{\mathbf{I}\} \equiv \sum_k I_{kk}$$

Show, by performing a similarity tranformation, that the trace is an invariant quantity. In other words, show that

$$\text{tr}\{\mathbf{I}\} = \text{tr}\{\mathbf{I}'\}$$

where $\{\mathbf{I}\}$ is the tensor in one coordinate system and $\{\mathbf{I}\}'$ is the tensor in a coordinate system rotated with respect to the first system. Verify this result for the different forms of the inertia tensor for a cube given in several examples in the text.

11-23. Show by the method used in the previous problem that the *determinant* of the elements of a tensor is an invariant quantity under a similarity transformation. Verify this result also for the case of the cube.

▷ **11-24.** Find the frequency of small oscillations for a thin homogeneous plate if the motion takes place in the plane of the plate and if the plate has the shape of an equilateral triangle and is suspended (a) from the midpoint of one side and (b) from one apex.

11-25. Consider a thin disk composed of two homogeneous halves connected along a diameter of the disk. If one half has density ρ and the other has density 2ρ, find the expression for the Lagrangian when the disk rolls without slipping along a horizontal surface. (The rotation takes place in the plane of the disk.)

11-26. Obtain the components of the angular velocity vector $\boldsymbol{\omega}$ (see Equation 11.102) directly from the transformation matrix $\boldsymbol{\lambda}$ (Equation 11.99).

11-27. A symmetric body moves without the influence of forces or torques. Let x_3 be the symmetry axis of the body and $\mathbf{L}$ be along x_3'. The angle between $\boldsymbol{\omega}$ and x_3 is α. Let $\boldsymbol{\omega}$ and $\mathbf{L}$ initially be in the x_2–x_3 plane. What is the angular velocity of the symmetry axis about $\mathbf{L}$ in terms of I_1, I_3, ω, and α?

11-28. Show from Figure 11-7c that the components of $\boldsymbol{\omega}$ along the fixed (x_i') axes are

$$\omega_1' = \dot{\theta} \cos \phi + \dot{\psi} \sin \theta \sin \phi$$

$$\omega_2' = \dot{\theta} \sin \phi - \dot{\psi} \sin \theta \cos \phi$$

$$\omega_3' = \dot{\psi} \cos \theta + \dot{\phi}$$

11-29. Investigate the motion of the symmetric top discussed in Section 11.10 for the case in which the axis of rotation is vertical (i.e., the x_3'- and x_3-axes coincide). Show that the motion is either stable or unstable depending on whether the quantity $4I_1 \, Mgh/I_3^2\omega_3^2$ is less than or greater than unity. Sketch the effective potential $V(\theta)$ for the two cases, and point out the features of these curves that determine whether the motion is stable. If the top is set spinning in the stable configuration, what is the effect as friction gradually reduces the value of ω_3? (This is the case of the "sleeping top.")

11-30. Refer to the discussion of the symmetric top in Section 11.10. Investigate the equation for the turning points of the nutational motion by setting $\dot{\theta} = 0$ in Equation 11.162. Show that the resulting equation is a cubic in $\cos \theta$ and has two real roots and one imaginary root for θ.

11-31. Consider a thin homogeneous plate with principal momenta of inertia

$$I_1 \quad \text{along the principal axis } x_1$$

$$I_2 > I_1 \quad \text{along the principal axis } x_2$$

$$I_3 = I_1 + I_2 \quad \text{along the principal axis } x_3$$

Let the origins of the x_i and x_i' systems coincide and be located at the center of mass 0 of the plate. At time $t = 0$, the plate is set rotating in a force-free manner with an angular velocity Ω about an axis inclined at an angle α from the plane of the plate and perpendicular to the x_2-axis. If $I_1/I_2 \equiv \cos 2\alpha$, show that at time t the angular velocity about the x_2-axis is

$$\omega_2(t) = \Omega \cos \alpha \tanh(\Omega t \sin \alpha)$$

12

COUPLED
OSCILLATIONS

12.1 INTRODUCTION

In Chapter 3, we examined the motion of an oscillator subjected to an external driving force. The discussion was limited to the case in which the driving force is periodic; that is, the driver is itself a harmonic oscillator. We considered the action of the driver on the oscillator, but we did not include the feedback effect of the oscillator on the driver. In many instances, ignoring this effect is unimportant, but if two (or many) oscillators are connected in such a way that energy can be transferred back and forth between (or among) them, the situation becomes the more complicated case of **coupled oscillations**.* Motion of this type can be quite complex (the motion may not even be periodic), but we can always describe the motion of any oscillatory system in terms of **normal coordinates**, which have the property that each oscillates with a single, well-defined frequency; that is, the normal coordinates are constructed in such a way that no coupling occurs among them, even though there is coupling among the ordinary (rectangular) coordinates describing the positions of particles. Initial conditions can always be prescribed for the system so that in the subsequent motion only one normal coordinate varies with time. In this circumstance, we say that one of the **normal modes** of the system has been

---

*The general theory of the oscillatory motion of a system of particles with a finite number of degrees of freedom was formulated by Lagrange during the period 1762–1765, but the pioneering work had been done in 1753 by Daniel Bernoulli (1700–1782).

excited. If the system has n degrees of freedom (e.g., n-coupled one-dimensional oscillators or $n/3$-coupled three-dimensional oscillators), there are in general n normal modes, some of which may be identical. The general motion of the system is a complicated superposition of all the normal modes of oscillation, but we can always find initial conditions such that any given one of the normal modes is independently excited. Identifying each of a system's normal modes allows us to construct a revealing picture of the motion, even though the system's *general* motion is a complicated combination of all the normal modes.

In the following chapter, we shall continue the development begun here and discuss the motion of vibrating strings. This example by no means exhausts the usefulness of the normal-mode approach to the description of oscillatory systems; indeed, applications can be found in many areas of mathematical physics, such as the microscopic motions in crystalline solids and the oscillations of the electromagnetic field.

12.2   TWO COUPLED HARMONIC OSCILLATORS

A physical example of a coupled system is a solid in which the atoms interact by elastic forces between each other and oscillate about their equilibrium positions. Springs between the atoms represent the elastic forces. A molecule composed of a few such interacting atoms would be an even simpler model. We begin by considering a similar system of coupled motion in one dimension: two masses connected by a spring to each other and by springs to fixed positions (Figure 12-1). We return to this example throughout the chapter as we describe various instances of coupled motion.

We let each of the oscillator springs have a force constant* κ; the force constant of the coupling spring is κ_{12}. We restrict the motion to the line connecting the masses, so the system has only two degrees of freedom, represented by the coordinates x_1 and x_2. Each coordinate is measured from the position of equilibrium.

If m_1 and m_2 are displaced from their equilibrium positions by amounts x_1 and x_2, respectively, the force on m_1 is $-\kappa x_1 - \kappa_{12}(x_1 - x_2)$, and the force on m_2 is

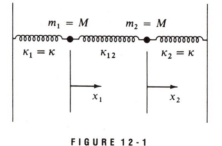

FIGURE 12-1

---

*Henceforth, we denote force constants by κ rather than by k as heretofore. The symbol k is reserved for (beginning in Chapter 13) an entirely different context.

$-\kappa x_2 - \kappa_{12}(x_2 - x_1)$. Therefore the equations of motion are

$$\left. \begin{array}{l} M\ddot{x}_1 + (\kappa + \kappa_{12})x_1 - \kappa_{12}x_2 = 0 \\ M\ddot{x}_2 + (\kappa + \kappa_{12})x_2 - \kappa_{12}x_1 = 0 \end{array} \right\} \tag{12.1}$$

Because we expect the motion to be oscillatory, we attempt a solution of the form

$$\left. \begin{array}{l} x_1(t) = B_1 e^{i\omega t} \\ x_2(t) = B_2 e^{i\omega t} \end{array} \right\} \tag{12.2}$$

where the frequency ω is to be determined and where the amplitudes B_1 and B_2 may be complex.* These trial solutions are complex functions. Thus, in the final step of the solution, the real parts of $x_1(t)$ and $x_2(t)$ will be taken, because the real part is all that is physically significant. We use this method of solution because of its great efficiency, and we use it again later, leaving out most of the details. Substituting these expressions for the displacements into the equations of motion, we find

$$\left. \begin{array}{l} -M\omega^2 B_1 e^{i\omega t} + (\kappa + \kappa_{12})B_1 e^{i\omega t} - \kappa_{12}B_2 e^{i\omega t} = 0 \\ -M\omega^2 B_2 e^{i\omega t} + (\kappa + \kappa_{12})B_2 e^{i\omega t} - \kappa_{12}B_1 e^{i\omega t} = 0 \end{array} \right\} \tag{12.3}$$

Collecting terms and canceling the common exponential factor, we obtain

$$\left. \begin{array}{l} (\kappa + \kappa_{12} - M\omega^2) B_1 - \kappa_{12}B_2 = 0 \\ -\kappa_{12}B_1 + (\kappa + \kappa_{12} - M\omega^2) B_2 = 0 \end{array} \right\} \tag{12.4}$$

For a nontrivial solution to exist for this pair of simultaneous equations, the determinant of the coefficients of B_1 and B_2 must vanish:

$$\begin{vmatrix} \kappa + \kappa_{12} - M\omega^2 & -\kappa_{12} \\ -\kappa_{12} & \kappa + \kappa_{12} - M\omega^2 \end{vmatrix} = 0 \tag{12.5}$$

The expansion of this secular determinant yields

$$(\kappa + \kappa_{12} - M\omega^2)^2 - \kappa_{12}^2 = 0 \tag{12.6}$$

Hence,

$$\kappa + \kappa_{12} - M\omega^2 = \pm\kappa_{12}$$

Solving for ω, we obtain

$$\omega = \sqrt{\frac{\kappa + \kappa_{12} \pm \kappa_{12}}{M}} \tag{12.7}$$

---

*Because a complex amplitude has a *magnitude* and a *phase*, we have the two arbitrary constants necessary in the solution of a second-order differential equation; that is, we could equally well write $x(t) = |B| \exp [i(\omega t - \delta)]$ or $x(t) = |B| \cos (\omega t - \delta)$, as in Equation 3.6b. Later (see Equation 12.9), we shall find it more convenient to use two distinct *real* amplitudes and the time-varying factors $\exp(i\omega t)$ and $\exp(-i\omega t)$. These various forms of solution are all entirely equivalent.

We therefore have two **characteristic frequencies** (or **eigenfrequencies**) for the system:

$$\omega_1 = \sqrt{\frac{\kappa + 2\kappa_{12}}{M}}, \qquad \omega_2 = \sqrt{\frac{\kappa}{M}} \qquad \text{(12.8)}$$

Thus, the general solution to the problem is

$$\left.\begin{aligned}
x_1(t) &= B_{11}^+ e^{i\omega_1 t} + B_{11}^- e^{-i\omega_1 t} + B_{12}^+ e^{i\omega_2 t} + B_{12}^- e^{-i\omega_2 t} \\
x_2(t) &= B_{21}^+ e^{i\omega_1 t} + B_{21}^- e^{-i\omega_1 t} + B_{22}^+ e^{i\omega_2 t} + B_{22}^- e^{-i\omega_2 t}
\end{aligned}\right\} \qquad \text{(12.9)}$$

where we have explicitly written both positive and negative frequencies, because the radicals in Equations 12.7 and 12.8 can carry either sign.

In Equations 12.9, the amplitudes are not all independent, as we may verify by substituting ω_1 and ω_2 into Equations 12.4. We find

$$\text{for } \omega = \omega_1: \quad B_{11} = -B_{21}$$

$$\text{for } \omega = \omega_2: \quad B_{12} = B_{22}$$

The only subscripts on the Bs now necessary are those indicating the particular eigenfrequency (i.e., the *second* subscripts). We can therefore write the general solution as

$$\left.\begin{aligned}
x_1(t) &= B_1^+ e^{i\omega_1 t} + B_1^- e^{-i\omega_1 t} + B_2^+ e^{i\omega_2 t} + B_2^- e^{-i\omega_2 t} \\
x_2(t) &= -B_1^+ e^{i\omega_1 t} - B_1^- e^{-i\omega_1 t} + B_2^+ e^{i\omega_2 t} + B_2^- e^{-i\omega_2 t}
\end{aligned}\right\} \qquad \text{(12.10)}$$

Thus, we have *four* arbitrary constants in the general solution—just as we expect—because we have *two* equations of motion that are of *second* order.

We mentioned earlier that we could always define a set of coordinates that have a simple time dependence and that correspond to the excitation of the various oscillation modes of the system. Let us examine the pair of coordinates defined by

$$\left.\begin{aligned}
\eta_1 &\equiv x_1 - x_2 \\
\eta_2 &\equiv x_1 + x_2
\end{aligned}\right\} \qquad \text{(12.11)}$$

or

$$\left.\begin{aligned}
x_1 &= \tfrac{1}{2}(\eta_2 + \eta_1) \\
x_2 &= \tfrac{1}{2}(\eta_2 - \eta_1)
\end{aligned}\right\} \qquad \text{(12.12)}$$

Substituting these expressions for x_1 and x_2 into Equation 12.1, we find

$$\left.\begin{aligned}
M(\ddot{\eta}_1 + \ddot{\eta}_2) + (\kappa + 2\kappa_{12})\eta_1 + \kappa\eta_2 &= 0 \\
M(\ddot{\eta}_1 - \ddot{\eta}_2) + (\kappa + 2\kappa_{12})\eta_1 - \kappa\eta_2 &= 0
\end{aligned}\right\} \qquad \text{(12.13)}$$

which can be solved (by adding and subtracting) to yield

$$\left.\begin{aligned}
M\ddot{\eta}_1 + (\kappa + 2\kappa_{12})\eta_1 &= 0 \\
M\ddot{\eta}_2 + \kappa\eta_2 &= 0
\end{aligned}\right\} \qquad \text{(12.14)}$$

The coordinates η_1 and η_2 are now *uncoupled* and are therefore *independent*. The solutions are

$$\left.\begin{aligned}\eta_1(t) &= C_1^+ e^{i\omega_1 t} + C_1^- e^{-i\omega_1 t} \\ \eta_2(t) &= C_2^+ e^{i\omega_2 t} + C_2^- e^{-i\omega_2 t}\end{aligned}\right\} \qquad (12.15)$$

where the frequencies ω_1 and ω_2 are given by Equations 12.8. Thus, η_1 and η_2 are the *normal coordinates* of the problem. In a later section, we establish a general method for obtaining the normal coordinates.

If we impose the special initial conditions $x_1(0) = -x_2(0)$ and $\dot{x}_1(0) = -\dot{x}_2(0)$, we find $\eta_2(0) = 0$ and $\dot{\eta}_2(0) = 0$, which leads to $C_2^+ = C_2^- = 0$; that is, $\eta_2(t) \equiv 0$ for all values of t. Thus, the particles oscillate always *out of phase* and with frequency ω_1; this is the **antisymmetrical** mode of oscillation. However, if we begin with $x_1(0) = x_2(0)$ and $\dot{x}_1(0) = \dot{x}_2(0)$, we find $\eta_1(t) \equiv 0$, and the particles oscillate *in phase* and with frequency ω_2; this is the **symmetrical** mode of oscillation. These results are illustrated schematically in Figure 12-2. The general motion of the system is a linear combination of the symmetrical and antisymmetrical modes.

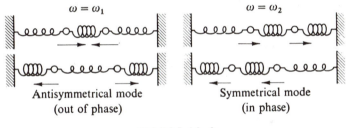

$\omega = \omega_1$ $\omega = \omega_2$

Antisymmetrical mode   Symmetrical mode
(out of phase)   (in phase)

FIGURE 12-2

The fact that the antisymmetrical mode has the higher frequency and the symmetrical mode has the lower frequency is actually a general result. In a complex system of linearly coupled oscillators, the mode possessing the highest degree of symmetry has the lowest frequency. If the symmetry is destroyed, then the springs must "work harder" in the antisymmetrical modes, and the frequency is raised.

Notice that if we were to hold m_2 fixed and allow m_1 to oscillate, the frequency would be $\sqrt{(\kappa + \kappa_{12})/M}$. We would obtain the same result for the frequency of oscillation of m_2 if m_1 were held fixed. The oscillators are identical and in the absence of coupling have the same oscillation frequency. The effect of coupling is to separate the common frequency, with one characteristic frequency becoming larger and one becoming smaller than the frequency for uncoupled motion. If we denote by ω_0 the frequency for uncoupled motion, then $\omega_1 > \omega_0 > \omega_2$, and we may schematically indicate the effect of the coupling as in Figure 12-3a. The solution for the characteristic frequencies in the problem of three coupled identical masses is illustrated in Figure 12-3b. Again, we have a splitting of the characteristic frequencies, with one greater and one smaller than ω_0. This is a general result: For an even number n of identical nearest neighbor coupled oscillators, $n/2$ characteristic frequencies are greater than ω_0, and $n/2$ characteristic frequencies are smaller

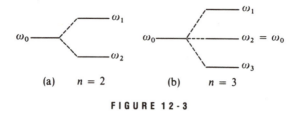

FIGURE 12-3

than ω_0. If n is odd, one characteristic frequency is equal to ω_0, and the remaining $n - 1$ characteristic frequencies are symmetrically distributed above and below ω_0. The reader familiar with the phenomenon of the Zeeman effect in atomic spectra will appreciate the similarity with this result: In each case, there is a symmetrical splitting of the frequency caused by the introduction of an interaction (in one case by the application of a magnetic field and in the other by the coupling of particles through the intermediary of the springs).

12.3 WEAK COUPLING

Some of the more interesting cases of coupled oscillations occur when the coupling is *weak*—that is, when the force constant of the coupling spring is small compared with that of the oscillator springs: $\kappa_{12} \ll \kappa$. According to Equation 12.8, the frequencies ω_1 and ω_2 are

$$\omega_1 = \sqrt{\frac{\kappa + 2\kappa_{12}}{M}}, \qquad \omega_2 = \sqrt{\frac{\kappa}{M}} \tag{12.16}$$

If the coupling is weak, we may expand the expression for ω_1:

$$\omega_1 = \sqrt{\frac{\kappa}{M}} \ \sqrt{1 + \frac{2\kappa_{12}}{\kappa}} = \sqrt{\frac{\kappa}{M}} \sqrt{1 + 4\varepsilon}$$

where

$$\varepsilon \equiv \frac{\kappa_{12}}{2\kappa} \ll 1 \tag{12.17}$$

The frequency ω_1 now reduces to

$$\omega_1 \cong \sqrt{\frac{\kappa}{M}}(1 + 2\varepsilon) \tag{12.18}$$

The natural frequency of either oscillator, when the other is held fixed, is

$$\omega_0 = \sqrt{\frac{\kappa + \kappa_{12}}{M}} \cong \sqrt{\frac{\kappa}{M}}(1 + \varepsilon) \tag{12.19}$$

or

$$\sqrt{\frac{\kappa}{M}} \cong \omega_0(1 - \varepsilon) \qquad (12.20)$$

Therefore, the two characteristic frequencies are given approximately by

$$\left.\begin{array}{ll} \omega_1 \cong \sqrt{\dfrac{\kappa}{M}}\,(1 + 2\varepsilon), & \omega_2 = \sqrt{\dfrac{\kappa}{M}} \\[2mm] \quad \cong \omega_0(1 - \varepsilon)(1 + 2\varepsilon) & \quad \cong \omega_0(1 - \varepsilon) \\[2mm] \quad \cong \omega_0(1 + \varepsilon) & \end{array}\right\} \qquad (12.21)$$

We can now examine the way a weakly coupled system behaves. If we displace Oscillator 1 a distance D and release it from rest, the initial conditions for the system are

$$x_1(0) = D, \qquad x_2(0) = 0, \qquad \dot{x}_1(0) = 0, \qquad \dot{x}_2(0) = 0 \qquad (12.22)$$

If we substitute these initial conditions into Equation 12.10 for $x_1(t)$ and $x_2(t)$, we find the amplitudes to be

$$B_1^+ = B_1^- = B_2^+ = B_2^- = \frac{D}{4} \qquad (12.23)$$

Then, $x_1(t)$ becomes

$$\begin{aligned} x_1(t) &= \frac{D}{4}\left[(e^{i\omega_1 t} + e^{-i\omega_1 t}) + (e^{i\omega_2 t} + e^{-i\omega_2 t})\right] \\[2mm] &= \frac{D}{2}\left(\cos \omega_1 t + \cos \omega_2 t\right) \\[2mm] &= D \cos\left(\frac{\omega_1 + \omega_2}{2}t\right)\cos\left(\frac{\omega_1 - \omega_2}{2}t\right) \qquad (12.24) \end{aligned}$$

But, according to Equation 12.21,

$$\frac{\omega_1 + \omega_2}{2} = \omega_0; \qquad \frac{\omega_1 - \omega_2}{2} = \varepsilon\omega_0 \qquad (12.25)$$

Therefore,*

$$x_1(t) = (D \cos \varepsilon\omega_0 t)\cos \omega_0 t \qquad (12.26a)$$

Similarly,

$$x_2(t) = (D \sin \varepsilon\omega_0 t)\sin \omega_0 t \qquad (12.26b)$$

---

*Note that in this fortuitous case, x_1 and x_2 were always real, so the real part did not have to be expressly taken in the final step as outlined after Equation 12.2.

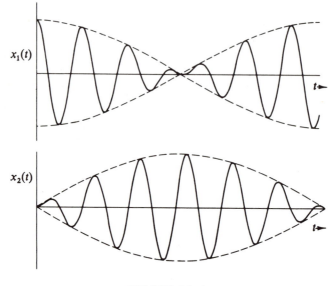

FIGURE 12-4

Because ε is small, the quantities $D \cos \varepsilon\omega_0 t$ and $D \sin \varepsilon\omega_0 t$ vary slowly with time. Therefore, $x_1(t)$ and $x_2(t)$ are essentially sinusoidal functions with slowly varying amplitudes. Although only x_1 is initially different from zero, as time increases the amplitude of x_1 decreases slowly with time, and the amplitude of x_2 increases slowly from zero. Hence, energy is transferred from the first oscillator to the second. When $t = \pi/2\varepsilon\omega_0$, then $D \cos \varepsilon\omega_0 t = 0$, and all the energy has been transferred. As time increases further, energy is transferred back to the first oscillator. This is the familiar phenomenon of *beats* and is illustrated in Figure 12-4. (In the case illustrated, $\varepsilon = 0.08$.)

12.4 GENERAL PROBLEM OF COUPLED OSCILLATIONS

In the preceding sections, we found that the effect of coupling in a simple system with two degrees of freedom produced two characteristic frequencies and two modes of oscillation. We now turn our attention to the general problem of coupled oscillations. Let us consider a conservative system described in terms of a set of generalized coordinates q_k and the time t. If the system has n degrees of freedom, then $k = 1, 2, \ldots, n$. We specify that a configuration of stable equilibrium exists for the system and that at equilibrium the generalized coordinates have values q_{k0}. In such a configuration, Lagrange's equations are satisfied by

$$q_k = q_{k0}, \qquad \dot{q}_k = 0, \qquad \ddot{q}_k = 0, \qquad k = 1, 2, \ldots, n$$

Every nonzero term of the form $(d/dt)(\partial L/\partial \dot{q}_k)$ must contain at least either $\dot{q}_k$ or $\ddot{q}_k$, so all such terms vanish at equilibrium. From Lagrange's equation, we therefore have

$$\left.\frac{\partial L}{\partial q_k}\right|_0 = \left.\frac{\partial T}{\partial q_k}\right|_0 - \left.\frac{\partial U}{\partial q_k}\right|_0 = 0 \tag{12.27}$$

where the subscript 0 designates that the quantity is evaluated at equilibrium.

We assume that the equations connecting the generalized coordinates and the rectangular coordinates do not explicitly contain the time; that is, we have

$$x_{\alpha,i} = x_{\alpha,i}(q_j) \qquad \text{or} \qquad q_j = q_j(x_{\alpha,i})$$

The kinetic energy is thus a homogeneous quadratic function of the generalized velocities (see Equation 7.121):

$$T = \frac{1}{2} \sum_{j,k} m_{jk} \dot{q}_j \dot{q}_k \tag{12.28}$$

Therefore, in general,

$$\left.\frac{\partial T}{\partial q_k}\right|_0 = 0, \qquad k = 1, 2, \ldots, n \tag{12.29}$$

and hence, from Equation 12.27, we have

$$\left.\frac{\partial U}{\partial q_k}\right|_0 = 0, \qquad k = 1, 2, \ldots, n \tag{12.30}$$

We may further specify that the generalized coordinates q_k be measured from the equilibrium positions; that is, we choose $q_{k0} = 0$. (If we originally had chosen a set of coordinates q_k' such that $q_{k0}' \neq 0$, we could always effect a simple linear transformation of the form $q_k = q_k' + \alpha_k$ such that $q_{k0} = 0$.)

The expansion of the potential energy in a Taylor series about the equilibrium configuration yields

$$U(q_1, q_2, \ldots, q_n) = U_0 + \sum_k \left.\frac{\partial U}{\partial q_k}\right|_0 q_k + \frac{1}{2} \sum_{j,k} \left.\frac{\partial^2 U}{\partial q_j \partial q_k}\right|_0 q_j q_k + \cdots \tag{12.31}$$

The second term in the expansion vanishes in view of Equation 12.30, and—without loss of generality—we may choose to measure U in such a way that $U_0 \equiv 0$. Then, if we restrict the motion of the generalized coordinates to be small, we may neglect all terms in the expansion containing products of the q_k of degree higher than second. This is equivalent to restricting our attention to simple harmonic oscillations, in which case only terms quadratic in the coordinates appear. Thus,

$$U = \frac{1}{2} \sum_{j,k} A_{jk} q_j q_k \tag{12.32}$$

where we define

$$A_{jk} \equiv \left.\frac{\partial^2 U}{\partial q_j \partial q_k}\right|_0 \tag{12.33}$$

Because the order of differentiation is immaterial (if U has continuous second partial derivatives), the quantity A_{jk} is symmetrical; that is, $A_{jk} = A_{kj}$.

We have specified that the motion of the system is to take place in the vicinity of the equilibrium configuration, and we have shown (Equation 12.30) that U must have a minimum value when the system is in this configuration. Because we have chosen $U = 0$ at equilibrium, we must have, in general, $U \geq 0$. It should be clear that we must also have $T \geq 0$.*

Equations 12.28 and 12.32 are of a similar form:

$$
\begin{array}{|c|}
\hline
T = \frac{1}{2} \sum_{j,k} m_{jk} \dot{q}_j \dot{q}_k \\[2ex]
U = \frac{1}{2} \sum_{j,k} A_{jk} q_j q_k \\
\hline
\end{array}
\qquad \text{(12.34)}
$$

The quantities A_{jk} are just numbers (see Equation 12.33); but the m_{jk} may be functions of the coordinates (see Equation 7.119):

$$
m_{jk} = \sum_{\alpha} m_{\alpha} \sum_i \frac{\partial x_{\alpha,i}}{\partial q_j} \frac{\partial x_{\alpha,i}}{\partial q_k}
$$

We can expand the m_{jk} about the equilibrium position with the result

$$
m_{jk}(q_1, q_2, \ldots, q_n) = m_{jk}(q_{l0}) + \sum_l \left. \frac{\partial m_{jk}}{\partial q_l} \right|_0 q_l + \cdots
\qquad \text{(12.35)}
$$

We wish to retain only the first nonvanishing term in this expansion; but, unlike the expansion of the potential energy (Equation 12.31), we cannot choose the constant term $m_{jk}(q_{l0})$ to be zero, so this leading term becomes the constant value of m_{jk} in this approximation. This is the same order of approximation as that used for U, because the next higher order term in T would involve the cubic quantity $q_j \dot{q}_k \dot{q}_l$ and the next higher order term in U would contain $q_j q_k q_l$. In the small oscillation approximation, T should be treated similarly to U, and just like U is normally expanded to order q^2, one needs to expand T to order $\dot{q}^2$, and m_{jk} is evaluated at equilibrium. Thus, in Equation 12.34, the m_{jk} and the A_{jk} are $n \times n$ arrays of *numbers* specifying the way the motions of the various coordinates are coupled. For example, if $m_{rs} \neq 0$ for $r \neq s$, then the kinetic energy contains a term proportional to $\dot{q}_r \dot{q}_s$, and a coupling exists between the rth and sth coordinate. If, however, m_{jk} is *diagonal*, so that† $m_{jk} \neq 0$ for $j = k$ but vanishes otherwise, then the kinetic energy is of the form

$$
T = \frac{1}{2} \sum_r m_r \dot{q}_r^2
$$

---

* That is, both U and T are *positive definite* quantities, in that they are always positive unless the coordinates (in the case of U) or the velocities (in the case of T) are zero, in which case they vanish.

† If a diagonal element of m_{jk} (say, m_{rr}) vanishes, then the problem can be reduced to one of $n - 1$ degrees of freedom.

where m_{rr} has been abbreviated to m_r. Thus, the kinetic energy is a simple sum of the kinetic energies associated with the various coordinates. As we see below, if, in addition, A_{jk} is diagonal so that U is also a simple sum of individual potential energies, then each coordinate behaves in an uncomplicated manner, undergoing oscillations with a single, well-defined frequency. The problem is therefore to find a coordinate transformation that simultaneously diagonalizes both m_{jk} and A_{jk} and thereby renders the system describable in the simplest possible terms. Such coordinates are the *normal coordinates*.

The equations of motion of the system with kinetic and potential energies given by Equation 12.34 are obtained from Lagrange's equation

$$\frac{\partial L}{\partial q_k} - \frac{d}{dt}\frac{\partial L}{\partial \dot{q}_k} = 0$$

But because T is a function only of the generalized velocities and U is a function only of the generalized coordinates, Lagrange's equation for the kth coordinate becomes

$$\frac{\partial U}{\partial q_k} + \frac{d}{dt}\frac{\partial T}{\partial \dot{q}_k} = 0 \tag{12.36}$$

From Equations 12.34, we evaluate the derivatives:

$$\left.\begin{aligned}\frac{\partial U}{\partial q_k} &= \sum_j A_{jk} q_j \\[2mm] \frac{\partial T}{\partial \dot{q}_k} &= \sum_j m_{jk} \dot{q}_j\end{aligned}\right\} \tag{12.37}$$

The equations of motion then become

$$\boxed{\sum_j (A_{jk} q_j + m_{jk}\ddot{q}_j) = 0} \tag{12.38}$$

This is a set of n second-order linear homogeneous differential equations with constant coefficients. Because we are dealing with an oscillatory system, we expect a solution of the form

$$q_j(t) = a_j e^{i(\omega t - \delta)} \tag{12.39}$$

where the a_j are real amplitudes and where the phase δ has been included to give the two arbitrary constants (a_j and δ) required by the second-order nature of each of the differential equations.* (Only the real part of the right-hand side is to be considered.) The frequency ω, and the phase δ are to be determined by the equations of motion. If ω is a real quantity, then the solution (Equation 12.39) represents

---

* This is entirely equivalent to our previous procedure of writing $x(t) = B \exp(i\omega t)$ (see Equations 12.2) with B allowed to be complex. In Equations 12.9, we exhibited the requisite arbitrary constants as real amplitudes by using $\exp(i\omega t)$ and $\exp(-i\omega t)$ rather than by incorporating a phase factor as in Equation 12.39.

oscillatory motion. That ω is indeed real may be seen by the following physical argument. Suppose that ω contains an imaginary part $i\omega_i$ (in which ω_i is real). This produces terms of the form $e^{\omega_i t}$ and $e^{-\omega_i t}$ in the expression of q_j. Thus, when the total energy of the system is computed, $T + U$ contains factors that increase or decrease monotonically with the time. But this violates the assumption that we are dealing with a conservative system; therefore, the frequency ω must be a real quantity.

With a solution of the form given by Equation 12.39, the equations of motion become

$$\sum_j (A_{jk} - \omega^2 m_{jk})\, a_j = 0 \qquad (12.40)$$

where the common factor $\exp[i(\omega t - \delta)]$ has been canceled. This is a set of n linear, homogeneous, *algebraic* equations that the a_j must satisfy. For a nontrivial solution to exist, the determinant of the coefficients must vanish:

$$|A_{jk} - \omega^2 m_{jk}| = 0 \qquad (12.41)$$

To be more explicit, this is an $n \times n$ determinant of the form

$$\begin{vmatrix} A_{11} - \omega^2 m_{11} & A_{12} - \omega^2 m_{12} & A_{13} - \omega^2 m_{13} \cdots \\ A_{12} - \omega^2 m_{12} & A_{22} - \omega^2 m_{22} & A_{23} - \omega^2 m_{23} \cdots \\ A_{13} - \omega^2 m_{13} & A_{23} - \omega^2 m_{23} & A_{33} - \omega^2 m_{33} \cdots \\ \vdots & \vdots & \vdots \end{vmatrix} = 0 \qquad (12.42)$$

where the symmetry of the A_{jk} and m_{jk} has been explicitly included.

The equation represented by this determinant is called the **characteristic equation** or **secular equation** of the system and is an equation of degree n in ω^2. There are, in general, n roots we may label ω_r^2. The ω_r are called the **characteristic frequencies** or **eigenfrequencies** of the system. (In some situations, two or more of the ω_r can be equal; this is the phenomenon of **degeneracy** and is discussed later.) Just as in the procedure for determining the directions of the principal axes for a rigid body, each of the roots of the characteristic equation may be substituted into Equation 12.40 to determine the ratios $a_1:a_2:a_3: \cdots : a_n$ for each value of ω_r. Because there are n values of ω_r, we can construct n sets of ratios of the a_j. Each of the sets defines the components of the n-dimensional vector $\mathbf{a}_r$, called an **eigenvector** of the system. Thus $\mathbf{a}_r$ is the eigenvector associated with the eigenfrequency ω_r. We designate by a_{jr} the jth component of the rth eigenvector.

Because the principle of superposition applies for the differential equation (Equation 12.38), we must write the general solution for q_j as a linear combination of the solutions for each of the n values of r:

$$q_j(t) = \sum_r a_{jr} e^{i(\omega_r t - \delta_r)} \qquad (12.43)$$

Because it is only the *real* part of $q_j(t)$ that is physically meaningful, we actually have*

$$q_j(t) = \text{Re} \sum_r a_{jr} e^{i(\omega_r t - \delta_r)} = \sum_r a_{jr} \cos(\omega_r t - \delta_r) \qquad \textbf{(12.44)}$$

The motion of the coordinate q_j is therefore compounded of motions with each of the n values of the frequencies ω_r. The q_j evidently are not the normal coordinates that simplify the problem. We continue the search for normal coordinates in Section 12.6.

E X A M P L E 12.1 -

Find the characteristic frequencies for the case of the two masses connected by springs of Section 12.2 by means of the general formalism just developed.

Solution: The situation is that shown in Figure 12-1. The potential energy of the system is

$$U = \tfrac{1}{2}\kappa x_1^2 + \tfrac{1}{2}\kappa_{12}(x_2 - x_1)^2 + \tfrac{1}{2}\kappa x_2^2$$

$$= \tfrac{1}{2}(\kappa + \kappa_{12})x_1^2 + \tfrac{1}{2}(\kappa + \kappa_{12})x_2^2 - \kappa_{12}x_1 x_2 \qquad \textbf{(12.45)}$$

The term proportional to $x_1 x_2$ is the factor that expresses the coupling in the system. Calculating the A_{jk}, we find

$$\left.\begin{array}{l}
A_{11} = \dfrac{\partial^2 U}{\partial x_1^2}\bigg|_0 = \kappa + \kappa_{12} \\[4mm]
A_{12} = \dfrac{\partial^2 U}{\partial x_1 \, \partial x_2}\bigg|_0 = -\kappa_{12} = A_{21} \\[4mm]
A_{22} = \dfrac{\partial^2 U}{\partial x_2^2}\bigg|_0 = \kappa + \kappa_{12}
\end{array}\right\} \qquad \textbf{(12.46)}$$

The kinetic energy of the system is

$$T = \tfrac{1}{2}M\dot{x}_1^2 + \tfrac{1}{2}M\dot{x}_2^2 \qquad \textbf{(12.47)}$$

According to Equation 12.28,

$$T = \tfrac{1}{2}\sum_{j,k} m_{jk}\dot{x}_j \dot{x}_k \qquad \textbf{(12.48)}$$

---

* Notice here, unlike the example of weak coupling described in Section 12.3 (Equation 12.26), the real part of $q_j(t)$ has to be explicitly taken so that the $q_j(t)$ in Equation 12.44 is not the same as the $q_j(t)$ in Equation 12.43. But here and elsewhere, because of their close relationship, we use the same symbol (e.g., $q_j(t)$) for convenience.

Identifying terms between these two expressions for T, we find

$$\left.\begin{array}{l} m_{11} = m_{22} = M \\ m_{12} = m_{21} = 0 \end{array}\right\} \qquad (12.49)$$

Thus, the secular determinant (Equation 12.42) becomes

$$\begin{vmatrix} \kappa + \kappa_{12} - M\omega^2 & -\kappa_{12} \\ -\kappa_{12} & \kappa + \kappa_{12} - M\omega^2 \end{vmatrix} = 0 \qquad (12.50)$$

This is exactly Equation 12.5, so the solutions are the same (see Equations 12.7 and 12.8) as before:

$$\omega = \sqrt{\frac{\kappa + \kappa_{12} \pm \kappa_{12}}{M}}$$

The eigenfrequencies are

$$\omega_1 = \sqrt{\frac{\kappa + 2\kappa_{12}}{M}}, \qquad \omega_2 = \sqrt{\frac{\kappa}{M}}$$

The results of the two procedures are identical.

---●

12.5 ORTHOGONALITY OF THE EIGENVECTORS (OPTIONAL)*

We now wish to show that the eigenvectors $\mathbf{a}_r$ form an orthonormal set. Rewriting Equation 12.40 for the sth root of the secular equation, we have

$$\omega_s^2 \sum_k m_{jk} a_{ks} = \sum_k A_{jk} a_{ks} \qquad (12.51)$$

Next, we write a comparable equation for the rth root by substituting r for s and interchanging j and k:

$$\omega_r^2 \sum_j m_{jk} a_{jr} = \sum_j A_{jk} a_{jr} \qquad (12.52)$$

where we have used the symmetry of the m_{jk} and A_{jk}. We now multiply Equation 12.51 by q_{jr} and sum over j and also multiply Equation 12.52 by a_{ks} and sum over k:

$$\left.\begin{array}{l} \omega_s^2 \sum_{j,k} m_{jk} a_{jk} a_{ks} = \sum_{j,k} A_{jk} a_{jr} a_{ks} \\ \omega_r^2 \sum_{j,k} m_{jk} a_{jr} a_{ks} = \sum_{j,k} A_{jk} a_{jr} a_{ks} \end{array}\right\} \qquad (12.53)$$

---

* Section 12.5 may be omitted without losing physical understanding. This highly mathematical section is included for completeness. The method used here is a generalization of the steps used in Section 11.6 for the inertia tensor.

The right-hand sides of Equations 12.53 are now equal, so subtracting the first of these equations from the second, we have

$$(\omega_r^2 - \omega_s^2) \sum_{j,k} m_{jk} a_{jr} a_{ks} = 0 \tag{12.54}$$

We now examine the two possibilities $r = s$ and $r \neq s$. For $r \neq s$, the term $(\omega_r^2 - \omega_s^2)$ is, in general, different from zero. (The case of degeneracy, or multiple roots, is discussed later.) Therefore the sum must vanish identically:

$$\sum_{j,k} m_{jk} a_{jr} a_{ks} = 0, \qquad r \neq s \tag{12.55}$$

For the case $r = s$, the term $(\omega_r^2 - \omega_s^2)$ vanishes and the sum is indeterminate. The sum, however, cannot vanish identically. To show this, we write the kinetic energy for the system and substitute the expressions for $\dot{q}_j$ and $\dot{q}_k$ from Equation 12.44:

$$T = \tfrac{1}{2} \sum_{j,k} m_{jk} \dot{q}_j \dot{q}_k$$
$$= \tfrac{1}{2} \sum_{j,k} m_{jk} \left[\sum_r \omega_r a_{jr} \sin(\omega_r t - \delta_r) \right] \left[\sum_s \omega_s a_{ks} \sin(\omega_s t - \delta_s) \right]$$
$$= \tfrac{1}{2} \sum_{r,s} \omega_r \omega_s \sin(\omega_r t - \delta_r) \sin(\omega_s t - \delta_s) \sum_{j,k} m_{jk} a_{jr} a_{ks}$$

Thus, for $r = s$, the kinetic energy becomes

$$T = \tfrac{1}{2} \sum_r \omega_r^2 \sin^2(\omega_r t - \delta_r) \sum_{j,k} m_{jk} a_{jr} a_{kr} \tag{12.56}$$

We note first that

$$\omega_r^2 \sin^2(\omega_r t - \delta_r) \geq 0$$

We also know that T is positive and can become zero only if all the velocities vanish identically. Therefore,

$$\sum_{j,k} m_{jk} a_{jr} a_{kr} \geq 0$$

Thus, the sum is, in general, positive and can vanish only in the trivial instance that the system is not in motion—that is, that the velocities vanish identically and $T \equiv 0$.

We previously remarked that only the ratios of the a_{jr} are determined when the ω_r are substituted into Equation 12.40. We now remove this indeterminacy by imposing an additional condition on the a_{jr}. We require that

$$\sum_{j,k} m_{jk} a_{jr} a_{kr} = 1 \tag{12.57}$$

The a_{jr} are then said to be *normalized*. Combining Equations 12.55 and 12.57, we may write

$$\boxed{\sum_{j,k} m_{kj} a_{jr} a_{ks} = \delta_{rs}} \tag{12.58}$$

Because a_{jr} is the jth component of the rth eigenvector, we represent $\mathbf{a}_r$ by

$$\mathbf{a}_r = \sum_j a_{jr}\mathbf{e}_j \tag{12.59}$$

The vectors $\mathbf{a}_r$ defined in this way constitute an **orthonormal** set; that is, they are *orthogonal* according to the result given by Equation 12.55, and they have been *normalized* by setting the sum in Equation 12.57 equal to unity.

All the preceding discussion bears a striking resemblance to the procedure given in Chapter 11 for determining the principal moments of inertia and the principal axes for a rigid body. Indeed, the problems are mathematically identical, except that we are now dealing with a system with n degrees of freedom. The quantities m_{jk} and A_{jk} are actually tensor elements, because m and A are two-dimensional arrays that relate different physical quantities,* and as such, we write them as $\{\mathbf{m}\}$ and $\{\mathbf{A}\}$. The secular equation for determining the eigenfrequencies is the same as that for obtaining the principal moments of inertia, and the eigenvectors $\mathbf{a}_r$ correspond to the principal axes. Indeed, the proof of the orthogonality of the eigenvectors is merely a generalization of the proof given in Section 11.6 of the orthogonality of the principal axes. Although we have made a physical argument regarding the reality of the eigenfrequencies, we could carry out a mathematical proof using the same procedure used to show that the principal moments of inertia are real.

12.6 NORMAL COORDINATES

As we have seen (Equation 12.43), the general solution for the motion of the coordinate q_j must be a sum over terms, each of which depends on an individual eigenfrequency. In the previous section, we showed that the vectors $\mathbf{a}_r$ are orthogonal (Equation 12.55) and, as a matter of convenience, we even normalized their components a_{jr} (Equation 12.57) to arrive at Equation 12.58; that is, we have removed all ambiguity in the solution for the q_j, so it is no longer possible to specify an arbitrary displacement for a particle. Because such a restriction is not physically meaningful, we must introduce a constant scale factor α (which depends on the initial conditions of the problem) to account for the loss of generality introduced by the arbitrary normalization. Thus,

$$q_j(t) = \sum_r \alpha a_{jr} e^{i(\omega_r t - \delta_r)} \tag{12.60}$$

To simplify the notation, we write

$$q_j(t) = \sum_r \beta_r a_{jr} e^{i\omega_r t} \tag{12.61}$$

where the quantities β_r are new scale factors† (now complex) that incorporate the phases of δ_r.

---

* See the discussion in Section 11.6 concerning the mathematical definition of a tensor.

† There is a certain advantage in normalizing the a_{jr} to unity and introducing the scale factors α and β_r rather than leaving the normalization unspecified. The a_{jr} are then independent of the initial conditions, and a simple orthonormality equation results.

We now define a quantity η_r,

$$n_r(t) \equiv \beta_r e^{i\omega_r t} \tag{12.62}$$

so that

$$q_j(t) = \sum_r a_{jr} \eta_r(t) \tag{12.63}$$

The η_r, by definition, are quantities that undergo oscillation at only one frequency. They may be considered as new coordinates, called *normal coordinates*, for the system. The η_r satisfy equations of the form

$$\ddot{\eta}_r + \omega_r^2 \eta_r = 0 \tag{12.64}$$

There are n independent such equations, so the equations of motion expressed in normal coordinates become completely separable.

E X A M P L E 12.2

Derive Equation 12.64 directly by using Lagrange's equations of motion.

Solution: We note that from Equations 12.61 and 12.62

$$\dot{q}_j = \sum_j a_{jr} \dot{\eta}_r$$

and from Equation 12.34 we have, for the kinetic energy,

$$T = \frac{1}{2} \sum_{j,k} m_{jk} \dot{q}_j \dot{q}_k$$

$$= \frac{1}{2} \sum_{j,k} m_{jk} \left(\sum_r a_{jr} \dot{\eta}_r \right) \left(\sum_s a_{ks} \dot{\eta}_s \right)$$

$$= \frac{1}{2} \sum_{r,s} \left(\sum_{j,k} m_{jk} a_{jr} a_{ks} \right) \dot{\eta}_r \dot{\eta}_s$$

The sum in the parentheses is just δ_{rs} according to the orthonormality condition (Equation 12.58). Therefore,

$$T = \frac{1}{2} \sum_{r,s} \dot{\eta}_r \dot{\eta}_s \delta_{rs} = \frac{1}{2} \sum_r \dot{\eta}_r^2 \tag{12.65}$$

Similarly, from Equations 12.34 we have for the potential energy,

$$U = \frac{1}{2} \sum_{j,k} A_{jk} q_j q_k$$

$$= \frac{1}{2} \sum_{r,s} \left(\sum_{j,k} A_{jk} a_{jr} a_{ks} \right) \eta_r \eta_s$$

The first equation in Equation 12.53 is

$$\sum_{j,k} A_{jk} a_{jr} a_{ks} = \omega_s^2 \sum_{j,k} m_{jk} a_{jr} a_{ks}$$
$$= \omega_s^2 \delta_{rs}$$

so the potential energy becomes

$$U = \frac{1}{2} \sum_{r,s} \omega_s^2 \eta_r \eta_s \delta_{rs} = \frac{1}{2} \sum_r \omega_r^2 \eta_r^2 \qquad \textbf{(12.66)}$$

Using Equations 12.65 and 12.66, the Lagrangian is

$$L = \frac{1}{2} \sum_r (\dot{\eta}_r^2 - \omega_r^2 \eta_r^2) \qquad \textbf{(12.67)}$$

and Lagrange's equations are

$$\frac{\partial L}{\partial \eta_r} - \frac{d}{dt} \frac{\partial L}{\partial \dot{\eta}_r} = 0$$

or

$$\ddot{\eta}_r + \omega_r^2 \eta_r = 0$$

as found in Equation 12.64.

Thus, when the configuration of a system is expressed in normal coordinates, both the potential and kinetic energies become simultaneously diagonal. Because it is the off-diagonal elements of $\{\mathbf{m}\}$ and $\{\mathbf{A}\}$ that give rise to the coupling of the particles' motions, it should be evident that a choice of coordinates that renders these tensors diagonal uncouples the coordinates and makes the problem completely separable into the independent motions of the normal coordinates, each with its particular normal frequency.*

The foregoing has been a mathematical description of the methods used to determine the characteristic frequencies ω_r and to describe the coordinates η_r of the normal mode motion. The actual application of the method can be summarized by several statements:

1. Choose generalized coordinates and find T and U in the normal Lagrangian method. This corresponds to using Equations 12.34.

2. Represent A_{jk} and m_{jk} as tensors in $n \times n$ arrays, and use Equation 12.42 to determine the n values of eigenfrequencies ω_r.

3. For each value of ω_r, determine the ratios $a_{1r} : a_{2r} : a_{3r} : \cdots : a_{nr}$ by substituting into Equation 12.40:

-----

* The German mathematician Karl Weierstrass (1815–1897) showed in 1858 that the motion of a dynamical system can always be expressed in terms of normal coordinates.

$$\sum_j (A_{jk} - \omega_r^2 m_{jk})a_{jr} = 0 \qquad \text{(12.68)}$$

4. If needed, determine the scale factors β_r (Equation 12.60) from the initial conditions.

5. Determine the normal coordinates η_r by appropriate linear combinations of the q_j coordinates that display oscillations at the single eigenfrequency ω_r. The description of motion for this single normal coordinate η_r is called a normal mode. The general motion (Equation 12.63) of the system is a complicated superposition of the normal modes.

We now apply these steps in several examples.

E X A M P L E   12.3 ---

Determine the eigenfrequencies, eigenvectors, and normal coordinates of the mass–spring example in Section 12.2 by using the procedure just described. Assume $\kappa_{12} \approx \kappa$.

Solution: The eigenfrequencies were determined in Example 12.1, where we found T and U (step 1). We can find the components for A_{jk} directly from Equation 12.46 or by inspection from Equation 12.45, making sure A_{jk} is symmetrical.

$$\{ \mathbf{A} \} = \begin{Bmatrix} \kappa + \kappa_{12} & -\kappa_{12} \\ -\kappa_{12} & \kappa + \kappa_{12} \end{Bmatrix} \qquad \text{(12.69)}$$

The array m_{jk} can easily be determined from Equation 12.47:

$$\{ \mathbf{m} \} = \begin{Bmatrix} M & 0 \\ 0 & M \end{Bmatrix} \qquad \text{(12.70)}$$

We use Equation 12.42 to determine the eigenfrequencies ω_r.

$$\begin{vmatrix} \kappa + \kappa_{12} - M\omega^2 & -\kappa_{12} \\ -\kappa_{12} & \kappa + \kappa_{12} - M\omega^2 \end{vmatrix} = 0$$

which is identical to Equation 12.50 with the results of Equation 12.8 for ω_1 and ω_2.

We use Equation 12.68 to determine the eigenvector components a_{jr}. We have two equations for each value of r, but because we can determine only the ratios a_{1r}/a_{2r}, one equation for each r is sufficient. For $r = 1$, $k = 1$, we have

$$(A_{11} - \omega_1^2 m_{11})a_{11} + (A_{21} - \omega_1^2 m_{21})a_{21} = 0 \qquad \text{(12.71)}$$

or, inserting the values for $A_{11}, A_{21}, \omega_1^2$, and m_{11}, and using the simplification that $\kappa_{12} \approx \kappa$,

$$\left(2\kappa - \frac{3\kappa}{M} \cdot M \right) a_{11} - \kappa a_{21} = 0$$

with the result

$$a_{11} = -a_{21} \qquad \text{(12.72)}$$

For $r = 2$, $k = 1$, we have

$$\left(2\kappa - \frac{\kappa}{M} \cdot M\right)a_{12} - \kappa a_{22} = 0$$

with the result

$$a_{12} = a_{22} \tag{12.73}$$

The general motion (Equation 12.63) becomes

$$\left.\begin{array}{l} x_1 = a_{11}\eta_1 + a_{12}\eta_2 \\ x_2 = a_{21}\eta_1 + a_{22}\eta_2 \end{array}\right\} \tag{12.74}$$

Using Equations 12.72 and 12.73, this becomes

$$\left.\begin{array}{l} x_1 = a_{11}\eta_1 + a_{22}\eta_2 \\ x_2 = -a_{11}\eta_1 + a_{22}\eta_2 \end{array}\right\} \tag{12.75}$$

Adding x_1 and x_2 gives

$$\eta_2 = \frac{1}{2a_{22}}(x_1 + x_2) \tag{12.76}$$

Subtracting x_2 from x_1 gives

$$\eta_1 = \frac{1}{2a_{11}}(x_1 - x_2) \tag{12.77}$$

The normal coordinate η_2 can be determined by finding the conditions when the other normal coordinate η_1 remains equal to zero. From Equation 12.77, $\eta_1 = 0$ when $x_1 = x_2$. Thus, for normal mode 2 (η_2), the two masses oscillate *in phase* (the symmetrical mode). The distance between the particles is always the same, and they oscillate as if the spring connecting them were a rigid, weightless rod.

Similarly, we can find the conditions for the normal coordinate η_1 by determining when $\eta_2 = 0$ ($x_2 = -x_1$). In normal mode 1 (η_1), the particles oscillate out of phase (the antisymmetrical mode).

This analysis (summarized in Table 12-1) confirms our previous results (Section 12.2), and the particle motion is as shown in Figure 12-2. Such motions for atoms in molecules are common. Remember that we let $\kappa = \kappa_{12}$ during this example.

We may determine the components of the eigenvectors (Equation 12.59),

$$\left.\begin{array}{ll} \omega_1: & \mathbf{a}_1 = a_{11}\mathbf{e}_1 + a_{21}\mathbf{e}_2 \\ \omega_2: & \mathbf{a}_2 = a_{12}\mathbf{e}_1 + a_{22}\mathbf{e}_2 \end{array}\right\} \tag{12.78}$$

by using Equations 12.72 and 12.73.

$$\left.\begin{array}{l} \mathbf{a}_1 = a_{11}(\mathbf{e}_1 - \mathbf{e}_2) \\ \mathbf{a}_2 = a_{22}(\mathbf{e}_1 + \mathbf{e}_2) \end{array}\right\} \tag{12.79}$$

TABLE 12-1

NORMAL MODE MOTIONS

Normal mode	Eigenfrequency	Particle oscillation	Particle velocities
1	$\omega_1 = \sqrt{\dfrac{3\kappa}{M}}$	Out of phase	Equal but opposite
2	$\omega_2 = \sqrt{\dfrac{\kappa}{M}}$	In phase	Equal

Although normally not required, we may determine the values of a_{11} and a_{22} from the orthonormality condition of Equation 12.58 with the result

$$\left.\begin{array}{c} a_{11} = -\, a_{21} = \dfrac{1}{\sqrt{2M}} \\[2ex] a_{12} = a_{22} = \dfrac{1}{\sqrt{2M}} \end{array}\right\} \qquad \textbf{(12.80)}$$

In this example, it was not necessary to determine the scale factors β_r nor to write down the complete solution, because the initial conditions were not given.

●

E X A M P L E (12.4) ---

Determine the eigenfrequencies and describe the normal mode motion for two pendula of equal lengths b and equal masses m connected by a spring of force constant κ as shown in Figure 12-5. The spring is unstretched in the equilibrium position.

Solution: We choose θ_1 and θ_2 (Figure 12-5) as the generalized coordinates. The potential energy is chosen to be zero in the equilibrium position. The kinetic and potential energies of the system are, for small angles,

$$T = \tfrac{1}{2}m(b\dot\theta_1)^2 + \tfrac{1}{2}m(b\dot\theta_2)^2 \qquad \textbf{(12.81)}$$

$$U = mgb(1 - \cos\theta_1) + mgb(1 - \cos\theta_2)$$

$$+\tfrac{1}{2}\kappa(b\sin\theta_1 - b\sin\theta_2)^2 \qquad \textbf{(12.82)}$$

Using the small oscillation assumption $\sin\theta \approx \theta$ and $\cos\theta \approx 1 - \theta^2/2$, we can write

$$U = \frac{mgb}{2}(\theta_1^2 + \theta_2^2) + \frac{\kappa b^2}{2}(\theta_1 - \theta_2)^2 \qquad \textbf{(12.83)}$$

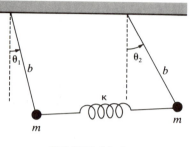

FIGURE 12-5

The components of $\{\mathbf{A}\}$ and $\{\mathbf{m}\}$ are

$$\{\mathbf{m}\} = \begin{Bmatrix} mb^2 & 0 \\ 0 & mb^2 \end{Bmatrix} \tag{12.84}$$

$$\{\mathbf{A}\} = \begin{Bmatrix} mgb^2 + \kappa b^2 & -\kappa b^2 \\ -\kappa b^2 & mgb + \kappa b^2 \end{Bmatrix} \tag{12.85}$$

The determinant needed to find the eigenfrequencies ω is

$$\begin{vmatrix} mgb + \kappa b^2 - \omega^2 mb^2 & -\kappa b^2 \\ -\kappa b^2 & mgb + \kappa b^2 - \omega^2 mb^2 \end{vmatrix} = 0 \tag{12.86}$$

which gives the characteristic equation

$$b^2(mg + \kappa b - \omega^2 mb)^2 - (\kappa b^2)^2 = 0$$

$$(mg + \kappa b - \omega^2 mb)^2 = (\kappa b)^2$$

or

$$mg + \kappa b - \omega^2 mb = \pm \kappa b \tag{12.87}$$

Taking the plus sign, $\omega = \omega_1$,

$$mg + \kappa b - \omega_1^2 mb = \kappa b$$

$$\omega_1^2 = \frac{g}{b} \tag{12.88}$$

Taking the minus sign in Equation 12.87, $\omega = \omega_2$,

$$mg + \kappa b - \omega_2^2 mb = -\kappa b$$

$$\omega_2^2 = \frac{g}{b} + \frac{2\kappa}{m} \tag{12.89}$$

Putting the values of ω_1 and ω_2 into Equation 12.40 gives, for $k = 1$,

$$(mgb + \kappa b^2 - \omega_r^2 mb^2)a_{1r} - \kappa b^2 a_{2r} = 0 \tag{12.90}$$

If $r = 1$, then

$$\left(mgb + \kappa b^2 - \frac{g}{b} mb^2 \right) a_{11} - \kappa b^2 a_{21} = 0$$

and

$$a_{11} = a_{21} \qquad \qquad \textbf{(12.91)}$$

If $r = 2$, then

$$\left(mgb + \kappa b^2 - \frac{g}{b} mb^2 - \frac{2\kappa}{m} mb^2 \right) a_{12} - \kappa b^2 a_{22} = 0$$

and

$$a_{12} = - a_{22} \qquad \qquad \textbf{(12.92)}$$

We write the coordinates θ_1 and θ_2 in terms of the normal coordinates by

$$\left. \begin{aligned} \theta_1 &= a_{11}\eta_1 + a_{12}\eta_2 \\ \theta_2 &= a_{21}\eta_1 + a_{22}\eta_2 \end{aligned} \right\} \qquad \textbf{(12.93)}$$

Using Equations 12.91 and 12.92, Equations 12.93 become

$$\left. \begin{aligned} \theta_1 &= a_{11}\eta_1 - a_{22}\eta_2 \\ \theta_2 &= a_{11}\eta_1 + a_{22}\eta_2 \end{aligned} \right\} \qquad \textbf{(12.94)}$$

The normal modes are easily determined, by adding and subtracting θ_1 and θ_2, to be

$$\left. \begin{aligned} \eta_1 &= \frac{1}{2a_{11}} (\theta_1 + \theta_2) \\ \eta_2 &= \frac{1}{2a_{22}} (\theta_2 - \theta_1) \end{aligned} \right\} \qquad \textbf{(12.95)}$$

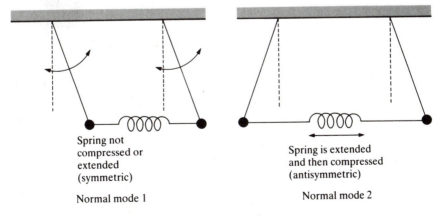

Spring not
compressed or
extended
(symmetric)

Normal mode 1

Spring is extended
and then compressed
(antisymmetric)

Normal mode 2

FIGURE 12-6

Because normal coordinate η_1 occurs when $\eta_2 = 0$, then $\theta_2 = \theta_1$ for normal mode 1 (symmetrical). Similarly, normal coordinate η_2 occurs when $\eta_1 = 0$ ($\theta_1 = -\theta_2$), and normal mode 2 is antisymmetrical. The normal mode motions are shown in Figure 12-6. Notice that for mode 1, the spring is neither compressed nor extended. The two pendula merely oscillate in unison with their natural frequencies ($\omega_1 = \omega_0 = \sqrt{g/b}$). These motions can be easily demonstrated in the laboratory or classroom. The higher frequency of normal mode 2 is easily displayed for a stiff spring.

12.7 MOLECULAR VIBRATIONS

We mentioned previously that molecular vibrations are good examples of the applications of the small oscillations discussed in this chapter. A molecule containing n atoms generally has $3n$ degrees of freedom. Three of these degrees of freedom are needed to describe the translational motion, and, generally, three are needed to describe rotations. Thus, there are $3n - 6$ vibrational degrees of freedom. For molecules with collinear atoms, only two possible rotational degrees of freedom exist, because rotation about the axis through the atoms is insignificant. In this case, there are $3n - 5$ degrees of freedom for vibrations.

We want to consider here only the vibrations occurring in a plane. We eliminate the translational and rotational degrees of freedom by appropriate transformations and choice of coordinate systems. For motion in a plane, there are $2n$ degrees of freedom. Because two are translational and one is rotational, generally $2n - 3$ normal vibrations occur in the plane [leaving $(3n - 6) - (2n - 3) = n - 3$ degrees of freedom for vibrations of the atoms out of the plane].

Linear molecules may have both longitudinal and transverse vibrations. The longitudinal vibrations occur along the line of the atoms. For n atoms, there are n degrees of freedom along the line, but one of them corresponds to translation. Thus, there are $n - 1$ possible vibrations in the longitudinal direction for n atoms in a linear molecule. If a total of $3n - 5$ vibrational degrees of freedom exist for a linear molecule, there must be $(3n - 5) - (n - 1) = 2n - 4$ transverse vibrations causing the atoms to vibrate perpendicular to the line of atoms. But from symmetry, any two mutually perpendicular directions suffice—so there are really only half the number of transverse frequencies, or $n - 2$.

EXAMPLE 12.5

Determine the eigenfrequencies and describe the normal mode motion of a symmetrical linear triatomic molecule (Figure 12-7). The central atom has mass M, and the symmetrical atoms have masses m. Both longitudinal and transverse vibrations are possible.

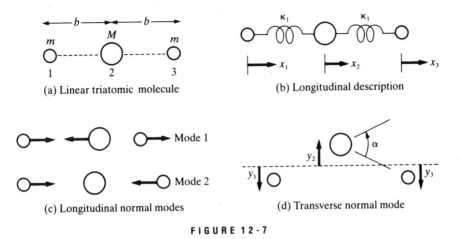

FIGURE 12-7

Solution: For three atoms, the preceding analysis indicates that we have two longitudinal and one transverse vibrational degrees of freedom if we eliminate the translational and rotational degrees of freedom.

We can solve the longitudinal and transverse motions separately, because they are independent. In Figure 12-7b, we represent the atomic displacements from equilibrium by x_1, x_2, x_3. The elastic forces between atoms are represented by springs of force constant κ_1. We have three longitudinal variables but only two degrees of freedom. We must eliminate the translational possibility by requiring the center of mass to be constant during the vibrations. This is satisfied if

$$m(x_1 + x_3) + M(x_2) = 0 \tag{12.96}$$

Therefore, we can eliminate the variable x_2:

$$x_2 = \frac{-m}{M}(x_1 + x_3) \tag{12.97}$$

The kinetic energy becomes

$$T = \tfrac{1}{2}m\dot{x}_1^2 + \tfrac{1}{2}m\dot{x}_3^2 + \tfrac{1}{2}M\dot{x}_2^2$$

$$= \tfrac{1}{2}m\dot{x}_1^2 + \tfrac{1}{2}m\dot{x}_3^2 + \tfrac{1}{2}\frac{m^2}{M}(\dot{x}_3^2 + \dot{x}_1^2 + 2\dot{x}_3\dot{x}_1) \tag{12.98}$$

Having the $\dot{x}_3\dot{x}_1$ coupling term in the kinetic energy (called "dynamic coupling") can be inconvenient when solving Equation 12.42 for the eigenfrequencies. We use a transformation to eliminate the dynamic coupling. Let

$$\left.\begin{aligned} q_1 &= x_3 + x_1 \\ q_2 &= x_3 - x_1 \end{aligned}\right\} \tag{12.99a}$$

Then

$$x_3 = \tfrac{1}{2}(q_1 + q_2)$$

$$x_1 = \tfrac{1}{2}(q_1 - q_2)$$

and

$$x_2 = \frac{-m}{M}q_1$$

(12.99b)

and the kinetic energy (Equation 12.98) becomes

$$T = \frac{m}{4}\dot{q}_2^2 + \frac{(mM + 2m^2)}{4M}\dot{q}_1^2$$

(12.100)

The potential energy is

$$U = \tfrac{1}{2}\kappa_1(x_2 - x_1)^2 + \tfrac{1}{2}\kappa_1(x_3 - x_2)^2$$

(12.101)

and with the transformations, Equations 12.99, the potential energy becomes (after considerable reduction)

$$U = \left(\frac{2m + M}{2M}\right)^2 \kappa_1 q_1^2 + \tfrac{1}{4}\kappa_1 q_2^2$$

(12.102)

The eigenfrequencies are determined by inspection, using Equation 12.42

$$\begin{vmatrix} \frac{1}{2}\left(\frac{2m + M}{M}\right)^2 \kappa_1 - \omega^2\left(\frac{mM + 2m^2}{2M}\right) & 0 \\ 0 & \frac{\kappa_1}{2} - \omega^2\frac{m}{2} \end{vmatrix} = 0 \quad (12.103)$$

to be

$$\omega_1^2 = \frac{(2m + M)}{mM}\kappa_1$$

$$\omega_2^2 = \frac{\kappa_1}{m}$$

(12.104)

Because the tensor formed by the coefficients of Equation 12.40 is already diagonal, the variables q_1 and q_2 represent the normal coordinates (unnormalized).

$$q_1 = a_{11}\eta_1 + a_{12}\eta_2$$

$$q_2 = a_{21}\eta_1 + a_{22}\eta_2$$

(12.105)

But

$$a_{12} = 0 \quad \text{and} \quad a_{21} = 0$$

$$q_1 = a_{11}\eta_1$$

$$q_2 = a_{22}\eta_2$$

TABLE 12-2

LONGITUDINAL NORMAL MODE MOTIONS

Mode	Eigenfrequencies	Variable	Motion	
1	$\dfrac{(2m + M)}{mM}\kappa_1$	$q_1 = x_3 + x_1$	$x_3 = x_1$	$(q_2 = 0)$
			$x_2 = \dfrac{-2m}{M}x_1$	
2	$\dfrac{\kappa_1}{m}$	$q_2 = x_3 - x_1$	$x_3 = -x_1$	$(q_1 = 0)$
			$x_2 = 0$	

As usual, we determine the motion of one normal mode when the other is zero. The descriptions of the longitudinal normal mode motion are given in Table 12-2. Normal mode 1 has the end atoms in symmetrical motion, but the central atom (from Equation 12.97) moves opposite to x_1 and x_3. Normal mode 2 has the end atoms vibrating antisymmetrically, but the central atom is at rest. This motion is displayed in Figure 12-7c.

Because we have eliminated rotations in our system, the transverse vibrations must be as shown in Figure 12-7d, with the end atoms moving in phase ($y_1 = y_3$) opposite to that of y_2. An equation similar to Equation 12.97 relates y_2 to y_1 and y_3 to keep the center of mass constant.

$$m(y_1 + y_3) + M(y_2) = 0 \qquad (12.106)$$

$$y_2 = \frac{-m}{M}(y_1 + y_3) \qquad (12.107)$$

We represent the single degree of freedom for the transverse vibration by the angle α representing the bending of the line of atoms.

$$\alpha = \frac{(y_1 - y_2) + (y_3 - y_2)}{b}$$

The kinetic energy for the transverse mode is

$$T = \tfrac{1}{2}m(\dot{y}_1^2 + \dot{y}_3^2) + \tfrac{1}{2}M\dot{y}_2^2$$

Because $y_1 = y_3$ and using Equation 12.107, α and T become

$$\alpha = \frac{2y_1}{bM}(2m + M) \qquad (12.108)$$

$$T = \frac{m}{M}(M + 2m)\dot{y}_1^2$$

$$T = \frac{mMb^2}{4(2m + M)}\dot{\alpha}^2 \qquad (12.109)$$

The potential energy represents the binding of the line of atoms. We assume the restoring force to be proportional to the total deviation from a straight line $(b\alpha)$, so the potential energy is

$$U = \tfrac{1}{2}\kappa_2(b\alpha)^2 \tag{12.110}$$

Equations 12.109 and 12.110 are similar to those for the mass–spring, with the vibrational frequency determined to be

$$\omega_3^2 = \frac{2(M + 2m)}{mM}\kappa_2 \tag{12.111}$$

The transverse normal mode is represented by

$$y_1 = y_3 \tag{12.112}$$

$$y_2 = \frac{-m}{M}(y_1 + y_3) \tag{12.113}$$

as already discussed and shown in Figure 12-7d.

The CO_2 molecule is an example of the symmetrical linear molecule just discussed. Radiation resulting from the first and third normal modes is observed, because the electrical center of the molecule deviates from the center of mass (m: O^-; M: C^{++}). But no radiation emanates from normal mode 2, because the electrical center is coincident with the center of mass and thus the system has no dipole moment*.

-- ●

12.8   THREE LINEARLY COUPLED PLANE PENDULA— AN EXAMPLE OF DEGENERACY

E X A M P L E   **12.6**   --

Consider three identical pendula suspended from a slightly yielding support. Because the support is not rigid, a coupling occurs between the pendula, and energy can be transferred from one pendulum to the other. Find the eigenfrequencies and eigenvectors and describe the normal mode motion. Figure 12-8 shows the geometry of the problem.

Solution: To simplify the notation, we adopt a system of units (sometimes called *natural units*) in which all lengths are measured in units of the length of the pendula l, all masses in units of the pendula masses M, and accelerations in units of g. Therefore, in our equations the values of the quantities M, l, and g are numerically

---

* For an interesting discussion of polyatomic molecules, see D. M. Dennison, *Rev. Mod. Phys.* **3**, 280 (1931).

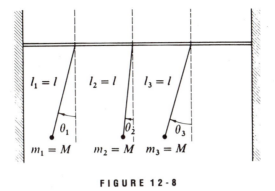

FIGURE 12-8

equal to unity. If the coupling between each pair of the pendula is the same, we have

$$T = \tfrac{1}{2}(\dot\theta_1^2 + \dot\theta_2^2 + \dot\theta_3^2) \left.\vphantom{\begin{matrix}a\\b\end{matrix}}\right\}$$
$$U = \tfrac{1}{2}(\theta_1^2 + \theta_2^2 + \theta_3^2 - 2\varepsilon\theta_1\theta_2 - 2\varepsilon\theta_1\theta_3 - 2\varepsilon\theta_2\theta_3)$$

(12.114)

Thus, the tensor $\{\mathbf{m}\}$ is diagonal,

$$\{\mathbf{m}\} = \begin{Bmatrix} 1 & 0 & 0 \\ 0 & 1 & 0 \\ 0 & 0 & 1 \end{Bmatrix}$$

(12.115)

but $\{\mathbf{A}\}$ has the form

$$\{\mathbf{A}\} = \begin{Bmatrix} 1 & -\varepsilon & -\varepsilon \\ -\varepsilon & 1 & -\varepsilon \\ -\varepsilon & -\varepsilon & 1 \end{Bmatrix}$$

(12.116)

The secular determinant is

$$\begin{vmatrix} 1 - \omega^2 & -\varepsilon & -\varepsilon \\ -\varepsilon & 1 - \omega^2 & -\varepsilon \\ -\varepsilon & -\varepsilon & 1 - \omega^2 \end{vmatrix} = 0$$

(12.117)

Expanding, we have

$$(1 - \omega^2)^3 - 2\varepsilon^3 - 3\varepsilon^2(1 - \omega^2) = 0$$

which can be factored to

$$(\omega^2 - 1 - \varepsilon)^2(\omega^2 - 1 + 2\varepsilon) = 0$$

and hence the roots are

$$\begin{aligned} \omega_1 &= \sqrt{1 + \varepsilon} \\ \omega_2 &= \sqrt{1 + \varepsilon} \\ \omega_3 &= \sqrt{1 - 2\varepsilon} \end{aligned} \left.\vphantom{\begin{matrix}a\\b\\c\end{matrix}}\right\}$$

(12.118)

Notice that we have a *double root*: $\omega_1 = \omega_2 = \sqrt{1 + \varepsilon}$. The normal modes corresponding to these frequencies are therefore *degenerate*—that is, these two modes are indistinguishable.

We now evaluate the quantities a_{jr}, beginning with a_{j3}. Again we note that, because the equations of motion determine only the ratios, we need consider only two of the three available equations; the third equation is automatically satisfied. Using the equation

$$\sum_j (A_{jk} - \omega_3^2 m_{jk})a_{j3} = 0$$

we find

$$\left.\begin{aligned} 2\varepsilon a_{13} - \varepsilon a_{23} - \varepsilon a_{33} = 0 \\ -\varepsilon a_{13} + 2\varepsilon a_{23} - \varepsilon a_{33} = 0 \end{aligned}\right\} \qquad \textbf{(12.119)}$$

Equations 12.119 yield

$$a_{13} = a_{23} = a_{33} \qquad \textbf{(12.120)}$$

and from the normalization condition we have

$$a_{13}^2 + a_{23}^2 + a_{33}^2 = 1$$

or

$$a_{13} = a_{23} = a_{33} = \frac{1}{\sqrt{3}} \qquad \textbf{(12.121)}$$

Thus, we find that for $r = 3$ there is no problem in evaluating the components of the eigenvector $\mathbf{a}_3$. (This is a general rule: There is no indefiniteness in evaluating the eigenvector components for a nondegenerate mode.) Because all the components of $\mathbf{a}_3$ are equal, this corresponds to the mode in which all three pendula oscillate in phase.

Let us now attempt to evaluate the a_{j1} and a_{j2}. From the six possible equations of motion (three values of j and two values of r), we obtain only two different relations:

$$\varepsilon(a_{11} + a_{21} + a_{31}) = 0 \qquad \textbf{*(12.122)}$$

$$\varepsilon(a_{12} + a_{22} + a_{32}) = 0 \qquad \textbf{*(12.123)}$$

The orthogonality equation is

$$\sum_{j,k} m_{jk}a_{jr}a_{ks} = 0, \qquad r \neq s$$

but, because $m_{jk} = \delta_{jk}$, this becomes

$$\sum_j a_{jr}a_{js} = 0, \qquad r \neq s \qquad \textbf{(12.124)}$$

which leads to only one new equation:

$$a_{11}a_{12} + a_{21}a_{22} + a_{31}a_{32} = 0 \qquad \textbf{*(12.125)}$$

(The other two possible equations are identical with Equations 12.122 and 12.123 above.) Finally, the normalization conditions yield

$$a_{11}^2 + a_{21}^2 + a_{31}^2 = 1 \qquad \text{*(12.126)}$$

$$a_{12}^2 + a_{22}^2 + a_{32}^2 = 1 \qquad \text{*(12.127)}$$

Thus, we have a total of only *five* (starred,*) equations for the six unknowns a_{j1} and a_{j2}. This indeterminacy in the eigenvectors corresponding to a double root is exactly the same as that encountered in constructing the principal axes for a rigid body with an axis of symmetry; the two equivalent principal axes may be placed in any direction as long as the set of three axes is orthogonal. Therefore, we are at liberty to arbitrarily specify the eigenvectors $\mathbf{a}_1$ and $\mathbf{a}_2$, as long as the orthogonality and normalizing relations are satisfied. For a simple system such as we are discussing, it should not be difficult to construct these vectors, so we do not give any general rules here.

If we arbitrarily choose $a_{31} = 0$, the indeterminacy is removed. We then find

$$\mathbf{a}_1 = \frac{1}{\sqrt{2}}(1, -1, 0), \qquad \mathbf{a}_2 = \frac{1}{\sqrt{6}}(1, 1, -2) \qquad \text{(12.128)}$$

from which we can verify that the starred relations are all satisfied.

Recall that the nondegenerate mode corresponds to the in-phase oscillation of all three pendula:

$$\mathbf{a}_3 = \frac{1}{\sqrt{3}}(1, 1, 1) \qquad \text{(12.129)}$$

We now see that the degenerate modes each correspond to out-of-phase oscillation. For example, $\mathbf{a}_2$ in Equation 12.128 represents two pendula oscillating together with a certain amplitude, whereas the third is out of phase and has twice the amplitude. Similarly, $\mathbf{a}_1$ in Equation 12.128 represents one pendulum stationary and the other two in out-of-phase oscillation. The eigenvectors $\mathbf{a}_1$ and $\mathbf{a}_2$ already given are only one set of an infinity of sets satisfying the conditions of the problem. But all such eigenvectors represent some sort of out-of-phase oscillation. (Further details of this example are examined in Problems 12-19 and 12-20.)

12.9 THE LOADED STRING*

We now consider a more complex system consisting of an elastic string (or a spring) on which a number of identical particles are placed at regular intervals. The ends of the string are constrained to remain stationary. Let the mass of each of the n

---

* The first attack on the problem of the loaded string (or one-dimensional lattice) was by Newton (in the *Principia*, 1687). The work was continued by Johann Bernoulli and his son Daniel, starting in 1727 and culminating in the latter's formulation of the principle of superposition in 1753. It is from this point that the theoretical treatment of the physics of *systems* (as distinct from *particles*) begins.

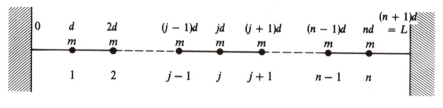

FIGURE 12-9

particles be m, and let the spacing between particles at equilibrium be d. Thus, the length of the string is $L = (n + 1)d$. The equilibrium situation is shown in Figure 12-9.

We wish to treat the case of small transverse oscillations of the particles about their equilibrium positions. First, we consider the vertical displacements of the masses numbered $j - 1$, j, and $j + 1$ (Figure 12-10). If the vertical displacements q_{j-1}, q_j, and q_{j+1} are small, then the tension τ in the string is approximately constant and equal to its value at equilibrium. For small displacements, the string section between any pair of particles makes only small angles with the equilibrium line. Approximating the sines of these angles by the tangents, the expression for the force that tends to restore the jth particle to its equilibrium position is

$$F_j = -\frac{\tau}{d}(q_j - q_{j-1}) - \frac{\tau}{d}(q_j - q_{j+1}) \tag{12.130}$$

The force F_j is, according to Newton's law, equal to $m\ddot{q}_j$; Equation 12.130 can therefore be written as

$$\ddot{q}_j = \frac{\tau}{md}(q_{j-1} - 2q_j + q_{j+1}) \tag{12.131}$$

which is the equation of motion for the jth particle. The system is coupled, because the force on the jth particle depends on the positions of the $(j - 1)$th and $(j + 1)$th particles; this is therefore an example of *nearest neighbor* interaction, in which the coupling is only to the adjacent particles. It is not necessary that the interaction be confined to nearest neighbors. If the force between pairs of particles were electrostatic, for example, then each particle would be coupled to *all* the other particles. The problem can then become quite complicated. But even if the force is electro-

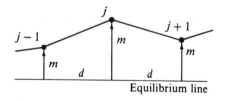

Equilibrium line

FIGURE 12-10

static, the $1/r^2$ dependence on distance frequently permits us to neglect interactions at distances greater than one interparticle spacing, so that the simple expression for the force given in Equation 12.130 is approximately correct.

We have considered only the motion perpendicular to the line of the string: transverse oscillations. It is easy to show that exactly the same form for the equations of motion results if we consider longitudinal vibrations—that is, motions along the line of the string. In this case, the factor τ/d is replaced by κ, the force constant of the string (see Problem 12-24).

Although we used Newton's equation to obtain the equations of motion (Equation 12.131), we may equally well use the Lagrangian method. The potential energy arises from the work done to stretch the $n + 1$ string segments*:

$$U = \frac{\tau}{2d} \sum_{j=1}^{n+1} (q_{j-1} - q_j)^2 \tag{12.132}$$

where q_0 and q_{n+1} are identically zero, because these positions correspond to the fixed ends of the string. We note that Equation 12.132 yields an expression for the force on the jth particle that is the same as the previous result (Equation 12.130):

$$F_j = -\frac{\partial U}{\partial q_j} = -\frac{\tau}{2d} \frac{\partial}{\partial q_j} [(q_{j-1} - q_j)^2 + (q_j - q_{j+1})^2]$$

$$= \frac{\tau}{d} (q_{j-1} - 2q_j + q_{j+1}) \tag{12.133}$$

The kinetic energy is given by the sum of the kinetic energies of the n individual particles:

$$T = \tfrac{1}{2} m \sum_{j=1}^{n} \dot{q}_j^2 \tag{12.134}$$

Because $\dot{q}_{n+1} \equiv 0$, we may extend the sum in Equation 12.134 to $j = n + 1$ so that the range of j is the same as that in the expression for the potential energy. Then, the Lagrangian becomes

$$L = \frac{1}{2} \sum_{j=1}^{n+1} \left[m\dot{q}_j^2 - \frac{\tau}{d}(q_{j-1} - q_j)^2 \right] \tag{12.135}$$

It should be obvious that the equation of motion for the jth particle must arise from only those terms in the Lagrangian containing q_j or $\dot{q}_j$. If we expand the sum in L, we find

$$L = \cdots + \frac{1}{2} m\dot{q}_j^2 - \frac{1}{2}\frac{\tau}{d}(q_{j-1} - q_j)^2 - \frac{1}{2}\frac{\tau}{d}(q_j - q_{j+1})^2 - \cdots \tag{12.136}$$

---

* We consider the potential energy to be only the elastic energy in the string; that is, we do not consider the individual masses to have any gravitational (or any other) potential energy.

where we have written only those terms that contain either q_j or $\dot{q}_j$. Applying Lagrange's equation for the coordinate q_j, we have

$$m\ddot{q}_j - \frac{\tau}{d}(q_{j-1} - 2q_j + q_{j+1}) = 0 \tag{12.137}$$

Thus, the result is the same as that obtained by using the Newtonian method.

To solve the equations of motion, we substitute, as usual,

$$q_j(t) = a_j e^{i\omega t} \tag{12.138}$$

where a_j can be *complex*. Substituting this expression for $q_j(t)$ into Equation 12.137, we find

$$-\frac{\tau}{d}a_{j-1} + \left(2\frac{\tau}{d} - m\omega^2\right)a_j - \frac{\tau}{d}a_{j+1} = 0 \tag{12.139}$$

where $j = 1, 2, \ldots, n$, but because the ends of the string are fixed, we must have $a_0 = a_{n+1} = 0$.

Equation 12.139 represents a **linear difference equation** that can be solved for the eigenfrequencies ω_r by setting the determinant of the coefficients equal to zero. We therefore have the following secular determinant:

$$\begin{vmatrix} \lambda & -\dfrac{\tau}{d} & 0 & 0 & 0 & \cdots \\ -\dfrac{\tau}{d} & \lambda & -\dfrac{\tau}{d} & 0 & 0 & \cdots \\ 0 & -\dfrac{\tau}{d} & \lambda & -\dfrac{\tau}{d} & 0 & \cdots \\ 0 & 0 & -\dfrac{\tau}{d} & \lambda & -\dfrac{\tau}{d} & \cdots \\ 0 & 0 & 0 & \cdot & \cdot & \\ \vdots & \vdots & \vdots & \vdots & \vdots & \end{vmatrix} = 0 \tag{12.140}$$

where we have used

$$\lambda \equiv 2\frac{\tau}{d} - m\omega^2 \tag{12.141}$$

This secular determinant is a special case of the general determinant (Equation 12.42) that results if the tensor **m** is diagonal and the tensor **A** involves a coupling only between adjacent particles. Thus, Equation 12.140 consists only of diagonal elements plus elements once-removed from the diagonal.

For the case $n = 1$ (i.e., a single mass suspended between two identical springs), we have $\lambda = 0$, or

$$\omega = \sqrt{\frac{2\tau}{md}}$$

We may adapt this result to the case of longitudinal motion by replacing τ/d by

κ; we then obtain the familiar expression,

$$\omega = \sqrt{\frac{2\kappa}{m}}$$

For the case $n = 2$, and with τ/d replaced by κ, we have $\lambda^2 = \kappa^2$, or

$$\omega = \sqrt{\frac{2\kappa \pm \kappa}{m}}$$

which are the same frequencies as those found in Section 12.2 for two coupled masses (Equation 12.8).

The secular equation should be relatively easy to solve directly for small values of n, but the solution becomes quite complicated for large n. In such cases, it is simpler to use the following method. We try a solution of the form

$$a_j = a e^{i(j\gamma - \delta)} \tag{12.142}$$

where a is *real*. The use of this device is justified if we can find a quantity γ and a phase δ such that the conditions of the problem are all satisfied. Substituting a_j in this form into Equation 12.139 and canceling the phase factor, we find

$$-\frac{\tau}{d}e^{-i\gamma} + \left(2\frac{\tau}{d} - m\omega^2\right) - \frac{\tau}{d}e^{i\gamma} = 0$$

Solving for ω^2, we obtain

$$\left.\begin{aligned} \omega^2 &= \frac{2\tau}{md} - \frac{\tau}{md}(e^{i\gamma} + e^{-i\gamma}) \\ &= \frac{2\tau}{md}(1 - \cos\gamma) \\ &= \frac{4\tau}{md}\sin^2\frac{\gamma}{2} \end{aligned}\right\} \tag{12.143}$$

Because we know that the secular determinant is of order n and therefore yields exactly n values for ω^2, we can write

$$\omega_r = 2\sqrt{\frac{\tau}{md}}\sin\frac{\gamma_r}{2}, \qquad r = 1, 2, \ldots, n \tag{12.144}$$

We now evaluate the quantity γ_r and the phase δ_r by applying the boundary condition that the ends of the string remain fixed. Thus, we have

$$a_{jr} = a_r e^{i(j\gamma_r - \delta_r)} \tag{12.145}$$

or, because it is only the real part that is physically meaningful,

$$a_{jr} = a_r \cos(j\gamma_r - \delta_r) \tag{12.146}$$

The boundary condition is

$$a_{0r} = a_{(n+1)r} \equiv 0 \tag{12.147}$$

For Equation 12.146 to yield $a_{jr} = 0$ for $j = 0$, it should be clear that δ_r must be $\pi/2$ (or some odd integer multiple thereof). Hence,

$$a_{jr} = a_r \cos\left(j\gamma_r - \frac{\pi}{2}\right)$$

$$= a_r \sin j\gamma_r \tag{12.148}$$

For $j = n + 1$, we have

$$a_{(n+1)r} = 0 = a_r \sin(n + 1)\gamma_r$$

Therefore,

$$(n + 1)\gamma_r = s\pi, \qquad s = 1, 2, \ldots$$

or

$$\gamma_r = \frac{s\pi}{n + 1}, \qquad s = 1, 2, \ldots$$

But there are just n distinct values of γ_r because Equation 12.144 requires n distinct values of ω_r. Therefore, the index s runs from 1 to n. Because there is a one-to-one correspondence between the values of s and the values of r, we can simply replace s in this last expression by the index r:

$$\gamma_r = \frac{r\pi}{n + 1}, \qquad r = 1, 2, \ldots, n \tag{12.149}$$

The a_{jr} (Equation 12.148) then become

$$\boxed{a_{jr} = a_r \sin\left(j\frac{r\pi}{n + 1}\right)} \tag{12.150}$$

The general solution for q_j (see Equation 12.61) is

$$q_j = \sum_r \beta'_r a_{jr} e^{i\omega_r t}$$

$$= \sum_r \beta'_r a_r \sin\left(j\frac{r\pi}{n + 1}\right) e^{i\omega_r t}$$

$$= \sum_r \beta_r \sin\left(j\frac{r\pi}{n + 1}\right) e^{i\omega_r t} \tag{12.151}$$

where we have written $\beta_r \equiv \beta'_r a_r$. Furthermore, for the frequency we have

$$\boxed{\omega_r = 2\sqrt{\frac{\tau}{md}} \sin\left(\frac{r\pi}{2(n + 1)}\right)} \tag{12.152}$$

We note that this expression yields the same results found for the case of two coupled oscillators (Equations 12.8) when we insert $n = 2$, $r = 1, 2$ and replace τ/d by κ ($= \kappa_{12}$).

Notice also that if either $r = 0$ or $r = n + 1$ is substituted into Equation 12.150, then all the amplitude factors a_{jr} vanish identically. These values of r therefore refer to *null modes*. Moreover, if r takes on the values $n + 2$, $n + 3$, . . . , $2n + 1$, then the a_{jr} are the same (except for a trivial sign change and in reverse order) as for $r = 1, 2, \ldots, n$; also, $r = 2n + 2$ yields the next null mode. We conclude, therefore, that there are indeed only n distinct modes and that increasing r beyond n merely duplicates the modes for smaller n. (A similar argument applies for $r < 0$.) These conclusions are illustrated in Figure 12-11 for the case $n = 3$. The distinct modes are specified by $r = 1, 2, 3$; $r = 4$ is a null mode. The displacement patterns are duplicated for $r = 7, 6, 5, 8$, but with a change in sign. In Figure 12-11, the dashed curves merely represent the sinusoidal behavior of the amplitude factors a_{rj} for various values of r; the only physically meaningful features of these curves are the values at the positions occupied by the particles ($j = 1, 2, 3$). The "high frequency" of the sine curves for $r = 5, 6, 7, 8$ is thus not at all related to the frequency of the particles' motions; these latter frequencies are the same as for $r = 1, 2, 3, 4$.

The normal coordinates of the system (Equation 12.62) are

$$\eta_r(t) \equiv \beta_r e^{i\omega_r t} \tag{12.153}$$

so that

$$q_j(t) = \sum_r \eta_r \sin\left(j\frac{r\pi}{n+1}\right) \tag{12.154}$$

This equation for q_j is similar to the previous expression (Equation 12.63) except that the quantities a_{jr} are now replaced by $\sin[j(r\pi)/(n+1)]$.

Because β_r may be complex, we write for the real part of q_j,

$$real: \quad q_j(t) = \sum_r \sin\left(j\frac{r\pi}{n+1}\right)(\mu_r \cos\omega_r t - \nu_r \sin\omega_r t) \tag{12.155}$$

where

$$\beta_r = \mu_r + i\nu_r \tag{12.156}$$

The initial value of $q_j(t)$ can be obtained from Equation 12.155:

$$q_j(0) = \sum_r \mu_r \sin\left(j\frac{r\pi}{n+1}\right) \tag{12.157}$$

$$\dot{q}_j(0) = -\sum_r \omega_r \nu_r \sin\left(j\frac{r\pi}{n+1}\right) \tag{12.158}$$

If we multiply Equation 12.157 by $\sin[j(s\pi)/(n+1)]$ and sum over j, we find

$$\sum_j q_j(0)\sin\left(j\frac{s\pi}{n+1}\right) = \sum_{j,r} \mu_r \sin\left(j\frac{r\pi}{n+1}\right)\sin\left(j\frac{s\pi}{n+1}\right) \tag{12.159}$$

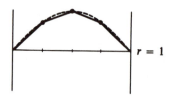

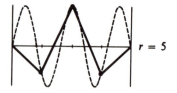

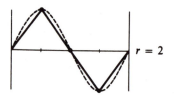

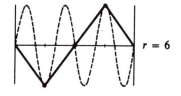

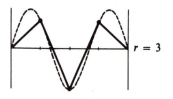

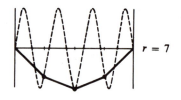

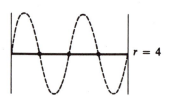

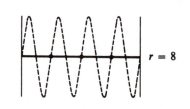

FIGURE 12-11

A relationship in the form of a trigonometric identity is available for the sine terms:

$$\sum_{j=1}^{n} \sin\left(j\frac{r\pi}{n+1}\right) \sin\left(j\frac{s\pi}{n+1}\right) = \frac{n+1}{2}\delta_{rs}, \qquad r, s = 1, 2, \ldots, n \quad \textbf{(12.160)}$$

so that Equation 12.159 becomes

$$\sum_{j} q_j(0) \sin\left(j\frac{s\pi}{n+1}\right) = \sum_{r} \mu_r \frac{n+1}{2}\delta_{rs}$$

$$= \frac{n+1}{2}\mu_s$$

or

$$\mu_s = \frac{2}{n+1}\sum_{j} q_j(0)\sin\left(j\frac{s\pi}{n+1}\right) \qquad \textbf{(12.161a)}$$

A similar procedure for ν_s yields

$$\nu_s = -\frac{2}{\omega_s(n+1)}\sum_{j} \dot{q}_j(0)\sin\left(j\frac{s\pi}{n+1}\right) \qquad \textbf{(12.161b)}$$

Thus, we have evaluated all the necessary quantities, and the description of the vibrations of a loaded string is therefore complete.

We should note the following point regarding the normalization procedures used here. First, in Equation 12.57 we arbitrarily normalized the a_{jr} to unity. Thus, the a_{jr} are *required* to be independent of the initial conditions imposed on the system. The scale factors α and β_r then allowed the magnitude of the oscillations to be varied by the selection of the initial conditions. Next, in the problem of the loaded string, we found that instead of the quantities a_{jr}, there arose the sine functions sin $[j(r\pi)/(n + 1)]$, and these functions possess a normalization property (Equation 12.160) that is specified by trigonometric identities. Therefore, in this case it is not possible arbitrarily to impose a normalization condition; we are automatically presented with the condition. But this is no restriction; it means only that the scale factors β_r for this case have a slightly different form. Thus, there are certain constants that occur in the two problems that, for convenience, are separated in different ways in the two cases.

EXAMPLE 12.7

Consider a loaded string consisting of three particles regularly spaced on the string. At $t = 0$ the center particle (only) is displaced a distance a and released from rest. Describe the subsequent motion.

Solution: The initial conditions are

$$\left.\begin{array}{c} q_2(0) = a, \quad q_1(0) = q_3(0) = 0 \\ \dot{q}_1(0) = \dot{q}_2(0) = \dot{q}_3(0) = 0 \end{array}\right\} \qquad \textbf{(12.162)}$$

Because the initial velocities are zero, the ν_r vanish. The μ_r are given by (Equation 12.161a):

$$\mu_r = \frac{2}{n+1}\sum_j q_j(0)\sin\left(j\frac{r\pi}{n+1}\right)$$

$$= \frac{1}{2}a\sin\left(\frac{r\pi}{2}\right) \tag{12.163}$$

because only the term $j = 2$ contributes to the sum. Thus,

$$\mu_1 = \tfrac{1}{2}a, \qquad \mu_2 = 0, \qquad \mu_3 = -\tfrac{1}{2}a \tag{12.164}$$

The quantities $\sin[j(r\pi)/(n+1)]$ that appear in the expression for $q_j(t)$ (Equation 12.155) are

	$r = 1$	2	3
$j = 1$	$\dfrac{\sqrt{2}}{2}$	1	$\dfrac{\sqrt{2}}{2}$
2	1	0	-1
3	$\dfrac{\sqrt{2}}{2}$	-1	$\dfrac{\sqrt{2}}{2}$

$$\tag{12.165}$$

The displacements of the three particles therefore are

$$q_1(t) = \frac{\sqrt{2}}{4}a\,(\cos\omega_1 t - \cos\omega_3 t)$$

$$q_2(t) = \frac{1}{2}a\,(\cos\omega_1 t - \cos\omega_3 t) \tag{12.166}$$

$$q_3(t) = \frac{\sqrt{2}}{4}a\,(\cos\omega_1 t + \cos\omega_3 t) = q_1(t)$$

where the characteristic frequencies are given by Equation 12.152:

$$\omega_r = 2\sqrt{\frac{\tau}{md}}\sin\left(\frac{r\pi}{8}\right), \qquad r = 1, 2, 3 \tag{12.167}$$

Notice that because the *middle* particle was initially displaced, no vibration mode occurs in which this particle is at rest; that is, mode 2 with frequency ω_2 (see Figure 12-11) is absent.

P R O B L E M S

12-1. Reconsider the problem of two coupled oscillators discussed in Section 12.2 in the event that the three springs all have different force constants. Find the two characteristic frequencies, and compare the magnitudes with the natural frequencies of the two oscillators in the absence of coupling.

12-2. Continue Problem 12-1, and investigate the case of weak coupling: $\kappa_{12} \ll \kappa_1, \kappa_2$. Show that the phenomenon of beats occurs but that the energy-transfer process is incomplete.

12-3. Two identical harmonic oscillators (with masses M and natural frequencies ω_0) are coupled such that by adding to the system a mass m common to both oscillators the equations of motion become

$$\ddot{x}_1 + (m/M)\ddot{x}_2 + \omega_0^2 x_1 = 0$$
$$\ddot{x}_2 + (m/M)\ddot{x}_1 + \omega_0^2 x_2 = 0$$

Solve this pair of coupled equations, and obtain the frequencies of the normal modes of the system.

12-4. Refer to the problem of the two coupled oscillators discussed in Section 12.2. Show that the total energy of the system is constant. (Calculate the kinetic energy of each of the particles and the potential energy stored in each of the three springs, and sum the results.) Notice that the kinetic and potential energy terms that have κ_{12} as a coefficient depend on C_1 and ω_1 but not on C_2 or ω_2. Why is such a result to be expected?

12-5. Find the normal coordinates for the problem discussed in Section 12.2 and in Example 12.2 if the two masses are different, $m_1 \neq m_2$. You may again assume the κ are equal.

12-6. Two identical harmonic oscillators are placed such that the two masses slide against one another, as in Figure 12-A. The frictional force provides a coupling of the motions proportional to the instantaneous relative velocity. Discuss the coupled oscillations of the system.

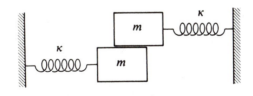

FIGURE 12-A

12-7. A particle of mass m is attached to a rigid support by a spring with force constant κ. At equilibrium, the spring hangs vertically downward. To this mass–spring combination is attached an identical oscillator, the spring of the latter being connected to the mass of the former. Calculate the characteristic frequencies for one-dimensional vertical oscillations, and compare with the frequencies when one or the other of the particles is held fixed while the other oscillates. Describe the normal modes of motion for the system.

12-8. A simple pendulum consists of a bob of mass m suspended by an inextensible (and massless) string of length l. From the bob of this pendulum is suspended a second, identical pendulum. Consider the case of small oscillations (so that $\sin \theta \cong \theta$), and calculate the characteristic frequencies. Describe also the normal modes of the system (refer to Problem 7-7).

12-9. The motion of a pair of coupled oscillators may be described by using a method similar to that used in constructing a phase diagram for a single oscillator (Section 3.3). For coupled oscillators, the two positions $x_1(t)$ and $x_2(t)$ may be represented by a point (the *system point*) in the two-dimensional *configuration space* x_1-x_2. As t increases, the locus of all such points defines a certain curve. The loci of the projection of the system points onto the x_1- and x_2-axes represent the motions of m_1 and m_2, respectively. In the general case, $x_1(t)$ and $x_2(t)$ are complicated functions, and so the curve is also complicated. But it is always possible to rotate the x_1-x_2 axes to a new set x_1'-x_2' in such a way that the projection of the system point onto each of the new axes is *simple harmonic*. The projected motions along the new axes take place with the characteristic frequencies and correspond to the normal modes of the system. The new axes are called *normal axes*. Find the normal axes for the problem discussed in Section 12.2 and verify the preceding statements regarding the motion relative to this coordinate system.

12-10. Consider two identical, coupled oscillators (as in Figure 12-1). Let each of the oscillators be damped, and let each have the same damping parameter β. A force $F_0 \cos \omega t$ is applied to m_1. Write down the pair of coupled differential equations that describe the motion. Obtain the solution by expressing the differential equations in terms of the normal coordinates given by Equation 12.11 and by comparing these equations with Equation 3.53. Show that the normal coordinates η_1 and η_2 exhibit resonance peaks at the characteristic frequencies ω_1 and ω_2, respectively.

12-11. Consider the electrical circuit in Figure 12-B. Use the developments in Section 12.2 to find the characteristic frequencies in terms of the capacitance C, inductance L, and mutual inductance M. The Kirchhoff circuit equations are

$$L\dot{I}_1 + \frac{q_1}{C} + M\dot{I}_2 = 0$$

$$L\dot{I}_2 + \frac{q_2}{C} + M\dot{I}_1 = 0$$

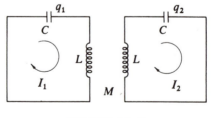

FIGURE 12-B

12-12. Show that the equations in Problem 12-11 can be put into the same form as Equation 12.1 by solving the second equation above for $\ddot{I}_2$ and substituting the result into the first equation. Similarly, substitute for $\ddot{I}_1$ in the second equation. The characteristic frequencies may then be written down immediately in analogy with Equation 12.8.

12-13. Find the characteristic frequencies of the coupled circuits of Figure 12-C.

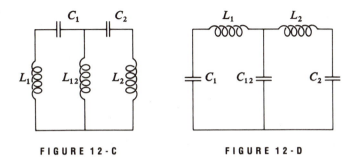

FIGURE 12-C FIGURE 12-D

12-14. Discuss the normal modes of the system shown in Figure 12-D.

12-15. In Figure 12-C, replace L_{12} by a resistor and analyze the oscillations.

12-16. A thin hoop of radius R and mass M oscillates in its own plane with one point of the hoop fixed. Attached to the hoop is a small mass M constrained to move (in a frictionless manner) along the hoop. Consider only small oscillations, and show that the eigenfrequencies are

$$\omega_1 = \sqrt{2}\sqrt{\frac{g}{R}}, \qquad \omega_2 = \frac{\sqrt{2}}{2}\sqrt{\frac{g}{R}}$$

Find the two sets of initial conditions that allow the system to oscillate in its normal modes. Describe the physical situation for each mode.

12-17. Find the eigenfrequencies and describe the normal modes for a system such as the one discussed in Section 12.2 but with three equal masses m and four springs (all with equal force constants) with the system fixed at the ends.

12-18. A mass M moves horizontally along a smooth rail. A pendulum is hung from M with a weightless rod and mass m at its end. Find the eigenfrequencies and describe the normal modes.

12-19. In the problem of the three coupled pendula, consider the three coupling constants as distinct, so that the potential energy may be written as

$$U = \tfrac{1}{2}(\theta_1^2 + \theta_2^2 + \theta_3^2 - 2\varepsilon_{12}\theta_1\theta_2 - 2\varepsilon_{13}\theta_1\theta_3 - 2\varepsilon_{23}\theta_2\theta_3)$$

with $\varepsilon_{12}, \varepsilon_{13}, \varepsilon_{23}$ all different. Show that no degeneracy occurs in such a system. Show also that degeneracy can occur *only* if $\varepsilon_{12} = \varepsilon_{13} = \varepsilon_{23}$.

12-20. Construct the possible eigenvectors for the degenerate modes in the case of the three coupled pendula by requiring $a_{11} = 2a_{21}$. Interpret this situation physically.

12-21. Three oscillators of equal mass m are coupled such that the potential energy of the system is given by

$$U = \tfrac{1}{2}[\kappa_1(x_1^2 + x_3^2) + \kappa_2 x_2^2 + \kappa_3(x_1 x_2 + x_2 x_3)]$$

where $\kappa_3 = \sqrt{2\kappa_1\kappa_2}$. Find the eigenfrequencies by solving the secular equation. What is the physical interpretation of the zero-frequency mode?

12-22. Consider a thin homogeneous plate of mass M that lies in the x_1–x_2 plane with its center at the origin. Let the length of the plate be $2A$ (in the x_2-direction) and let the width be $2B$ (in the x_1-direction). The plate is suspended from a fixed support by four springs of equal force constant κ at the four corners of the plate. The plate is free to oscillate but with the constraint that its center must remain on the x_3-axis. Thus, we have three degrees of freedom: (1) vertical motion, with the center of the plate moving along the x_3-axis; (2) a tipping motion lengthwise, with the x_1-axis serving as an axis of rotation (choose an angle θ to describe this motion); and (3) a tipping motion sidewise, with the x_2-axis serving as an axis of rotation (choose an angle ϕ to describe this motion). Assume only small oscillations and show that the secular equation has a double root, and hence that the system is degenerate. Discuss the normal modes of the system. (In evaluating the a_{jk} for the degenerate modes, arbitrarily set one of the a_{jk} equal to zero to remove the indeterminacy.) Show that the degeneracy can be removed by adding to the plate a thin bar of mass m and length $2A$ situated (at equilibrium) along the x_2-axis. Find the new eigenfrequencies for the system.

12-23. Evaluate the total energy associated with a normal mode, and show that it is constant in time. Show this explicitly for the case of Example 12.3.

12-24. Show that the equations of motion for *longitudinal* vibrations of a loaded string are of exactly the same form as the equations for transverse motion (Equation 12.131), except that the factor τ/d must be replaced by κ, the force constant of the string.

12-25. Rework the problem in Example 12.7, assuming that all three particles are displaced a distance a and released from rest.

13

CONTINUOUS SYSTEMS; WAVES

13.1 INTRODUCTION

We have so far been considering particles, systems of particles, or rigid bodies. Now, we want to consider bodies (gases, liquids, or solids) that are not rigid, that is, bodies whose particles move (however slightly) with respect to one another. The general study of such bodies is quite complex. However, one aspect of these *continuous* bodies is very important throughout physics—the ability to transmit wave motion. A disturbance on one part of the body can be transmitted by wave propagation throughout the body.

The simplest example of such phenomena is a vibrating string stretched under uniform tension between two fixed supports. As usual, the simple example represents many of the important results needed to understand other physical examples, such as stretched membranes and waves in solids. Waves may be either transverse or longitudinal. An example of a longitudinal wave is the vibration of molecules along the direction of propagation of a wave moving in a solid rod. Longitudinal waves occur in fluids and solids and are of great importance in acoustics.

Whereas both transverse and longitudinal waves may occur in solids, only longitudinal waves occur inside fluids, in which shearing forces are not possible. We have already considered (Chapter 12) both kinds of vibrations for a system of particles. A detailed study of the transverse vibrating string is important for several reasons. A study of a one-dimensional model of such string vibrations allows a mathematical solution with results that are applicable to more complex two- and three-dimensional problems. The modes of oscillation are similar. In particular, the application of boundary conditions (fixed ends), which are of extreme importance in many areas of physics, is easiest in one-dimensional problems. Boundary conditions play a role in the use of partial differential equations similar to the role initial conditions play in ordinary differential equations using Newtonian or Lagrangian techniques.

In this chapter, we extend the discussion of the vibrations of a loaded string presented in Chapter 12 by examining the consequences of allowing the number of particles on the string to become infinite (while maintaining a constant linear mass density). In this way, we pass to the case of a **continuous string**. All the results of interest for such a string can be obtained by this limiting process—including the derivation of the important **wave equation**, one of the truly fundamental equations of mathematical physics.

The solutions of the wave equation are in general subject to limitations imposed by certain physical restrictions peculiar to a given problem. These limitations frequently take the form of conditions on the solution that must be met at the extremes of the intervals of space and time that are of physical interest. We must therefore deal with a **boundary-value problem** involving a partial differential equation. Indeed, such a description characterizes essentially the whole of what we call *mathematical physics*.

We confine ourselves here to the solution to a one-dimensional wave equation. Such waves can describe a two-dimensional wave in two dimensions and can describe, for example, the motion of a vibrating string. The compression (or sound) waves that may be transmitted through an elastic medium, such as a gas, can also be approximated as one-dimensional waves if the medium is large enough that the edge effects are unimportant. In such a case, the condition of the medium is approximately the same at every point on a plane, and the properties of the wave motion are then functions only of the distance along a line normal to the plane. Such a wave in an extended medium, called a *plane wave*, is mathematically identical to the one-dimensional waves treated here.

13.2 CONTINUOUS STRING AS A LIMITING CASE OF THE LOADED STRING

In the preceding chapter, we considered a set of equally spaced point masses suspended by a string. We now wish to allow the number of masses to become infinite so that we have a continuous string. To do this, we must require that as $n \to \infty$ we simultaneously let the mass of each particle and the distance between each particle approach zero ($m \to 0$, $d \to 0$) in such a manner that the ratio m/d remains constant. We note that $m/d \equiv \rho$ is just the linear mass density of the string. Thus, we have

$$\left. \begin{array}{lll} n \to \infty, & d \to 0, & \text{such that } (n+1)d = L \\[2mm] m \to 0, & d \to 0, & \text{such that } \dfrac{m}{d} = \rho = \text{constant} \end{array} \right\} \quad \textbf{(13.1)}$$

From Equation 12.154, we have

$$q_j(t) = \sum_r \eta_r(t) \sin\left(j\frac{r\pi}{n+1}\right) \tag{13.2}$$

We can now write

$$j\frac{r\pi}{n+1} = r\pi\frac{jd}{(n+1)d} = r\pi\frac{x}{L} \tag{13.3}$$

where $jd = x$ now specifies the distance along the continuous string. Thus, $q_j(t)$ becomes a continuous function of the variables x and t:

$$q(x, t) = \sum_r \eta_r(t) \sin\left(\frac{r\pi x}{L}\right) \tag{13.4}$$

or

$$q(x, t) = \sum_r \beta_r e^{i\omega_r t} \sin\left(\frac{r\pi x}{L}\right) \tag{13.5}$$

In the case of a loaded string containing n particles, there are n degrees of freedom of motion and therefore n normal modes and n characteristic frequencies. Thus, in Equation 12.154 (or Equation 13.2) the sum is over the range $r = 1$ to $r = n$. But now the number of particles is infinite, so there is an infinite set of normal modes and the sum in Equations 13.4 and 13.5 runs from $r = 1$ to $r = \infty$. There are, then, infinitely many constants (the real and imaginary parts of the β_r) that must be evaluated to completely specify the motion of the continuous string. This is exactly the situation encountered in representing some function as a Fourier series—the infinitely many constants are specified by certain integrals involving the original function (see Equations 3.102). We may view the situation in another way: There are infinitely many arbitrary constants in the solution of the equation of motion, but there are also infinitely many initial conditions available for their evaluation, namely, the continuous functions $q(x, 0)$ and $\dot{q}(x, 0)$. The real and imaginary parts of the β_r can thus be obtained in terms of the initial conditions by a procedure analogous to that used in Section 12.9. Using $\beta_r = \mu_r + i\nu_r$, we have from Equation 13.5,

$$q(x, 0) = \sum_r \mu_r \sin\left(\frac{r\pi x}{L}\right) \tag{13.6a}$$

$$\dot{q}(x, 0) = -\sum_r \omega_r \nu_r \sin\left(\frac{r\pi x}{L}\right) \tag{13.6b}$$

Next, we multiply each of these equations by $\sin(s\pi x/L)$ and integrate from $x = 0$ to $x = L$. We can make use of the trigonometric relation

$$\int_0^L \sin\left(\frac{r\pi x}{L}\right) \sin\left(\frac{s\pi x}{L}\right) dx = \frac{L}{2}\delta_{rs} \tag{13.7}$$

from which we obtain

$$\mu_r = \frac{2}{L} \int_0^L q(x, 0) \sin\left(\frac{r\pi x}{L}\right) dx \tag{13.8a}$$

$$\nu_r = -\frac{2}{\omega_r L} \int_0^L \dot{q}(x, 0) \sin\left(\frac{r\pi x}{L}\right) dx \tag{13.8b}$$

The characteristic frequency ω_r may also be obtained as the limiting value of the result for the loaded string. From Equation 12.152, we have

$$\omega_r = 2\sqrt{\frac{\tau}{md}} \sin\left[\frac{r\pi}{2(n+1)}\right] \tag{13.9}$$

which can be written as

$$\omega_r = \frac{2}{d}\sqrt{\frac{\tau}{\rho}} \sin\left(\frac{r\pi d}{2L}\right) \tag{13.10}$$

When $d \to 0$, we can approximate the sine term by its argument, with the result

$$\omega_r = \frac{r\pi}{L}\sqrt{\frac{\tau}{\rho}} \tag{13.11}$$

E X A M P L E ⬤ 13.1 —

Find the displacement $q(x, t)$ for a "plucked string," where one point of the string is displaced (such that the string assumes a triangular shape) and then released from rest. Consider the case shown in Figure 13-1, in which the center of the string is displaced a distance h.

Solution: The initial conditions are

$$q(x, 0) = \begin{cases} \dfrac{2h}{L}x, & 0 \le x \le L/2 \\ \dfrac{2h}{L}(L - x), & L/2 \le x \le L \end{cases} \tag{13.12}$$

$$\dot{q}(x, 0) = 0$$

Because the string is released from rest, all the ν_r vanish. The μ_r are given by

$$\mu_r = \frac{4h}{L^2} \int_0^{L/2} x \sin\left(\frac{r\pi x}{L}\right) dx + \frac{4h}{L^2} \int_{L/2}^L (L - x) \sin\left(\frac{r\pi x}{L}\right) dx$$

Integrating,

$$\mu_r = \frac{8h}{r^2\pi^2} \sin\frac{r\pi}{2}$$

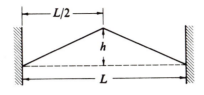

FIGURE 13-1

so that

$$\mu_r = \begin{cases} 0, & r \text{ even} \\ \dfrac{8h}{r^2\pi^2}(-1)^{\frac{1}{2}(r-1)}, & r \text{ odd} \end{cases}$$

Therefore,

$$q(x, t) = \frac{8h}{\pi^2}\left[\sin\left(\frac{\pi x}{L}\right)\cos\omega_1 t - \frac{1}{9}\sin\left(\frac{3\pi x}{L}\right)\cos\omega_3 t + \cdots\right] \quad \textbf{(13.13)}$$

where the ω_r are proportional to r and are given by Equation 13.11.

From Equation 13.13, we see that the fundamental mode (with frequency ω_1) and all the *odd* harmonics (with frequencies ω_3, ω_5, etc.) are excited but that none of the *even* harmonics are involved in the motion. Because the initial displacement was symmetrical, the subsequent motion must also be symmetrical, so none of the even modes (for which the center position of the string is a node) are excited. In general, if the string is plucked at some arbitrary point, none of the harmonics with nodes at that point will be excited.

As we prove in the next section, the energy in each of the excited modes is proportional to the square of the coefficient of the corresponding term in Equation 13.13. Thus, the energy ratios for the fundamental, third harmonic, fifth harmonic, and so on are $1 : \frac{1}{81} : \frac{1}{625} : \cdots$. Therefore the energy in the system (or the intensity of the emitted sound) is dominated by the fundamental. The third harmonic is 19 db* down from the fundamental and the fifth harmonic is down by 28 db.

13.3 ENERGY OF A VIBRATING STRING

Because we have made the assumption that frictional forces are not present, the total energy of a vibrating string must remain constant. We now show this explicitly; moreover, we show that the energy of the string is expressed simply as the sum of contributions from each of the normal modes. According to Equation 13.4,

---

* The *decibel* (db) is a unit of relative sound intensity (or acoustic power). The intensity ratio of a sound with intensity I to a sound with intensity I_0 is given by $10\log(I/I_0)$ db. Thus, for the fundamental (I_0) and third harmonic (I), we have $10 \log(1/81) = -19.1$ db or "19 db down" in intensity. A ratio of 3 db corresponds approximately to a factor of two in relative intensity.

the displacement of the string is given by

$$q(x, t) = \sum_r \eta_r(t) \sin\left(\frac{r\pi x}{L}\right) \qquad (13.14)$$

where the normal coordinates are

$$\eta_r(t) = \beta_r e^{i\omega_r t} \qquad (13.15)$$

As always, the β_r are complex quantities and the physically meaningful normal coordinates are obtained by taking the *real part* of Equation 13.15.

The kinetic energy of the string is obtained by calculating the kinetic energy for an element of the string, $\frac{1}{2}(\rho\,dx)\dot{q}^2$, and then integrating over the length. Thus,

$$T = \frac{1}{2}\rho \int_0^L \left(\frac{\partial q}{\partial t}\right)^2 dx \qquad (13.16)$$

or, using Equation 13.14,

$$T = \frac{1}{2}\rho \int_0^L \left[\sum_r \dot{\eta}_r \sin\left(\frac{r\pi x}{L}\right)\right]^2 dx \qquad (13.17)$$

The square of the series can be expressed as a double sum, this technique ensuring that all cross terms are properly included:

$$T = \frac{1}{2}\rho \sum_{r,s} \dot{\eta}_r \dot{\eta}_s \int_0^L \sin\left(\frac{r\pi x}{L}\right) \sin\left(\frac{s\pi x}{L}\right) dx \qquad (13.18)$$

The integral is now the same as that in Equation 13.7, so

$$T = \frac{\rho L}{4} \sum_{r,s} \dot{\eta}_r \dot{\eta}_s \delta_{rs}$$

$$= \frac{\rho L}{4} \sum_r \dot{\eta}_r^2 \qquad (13.19)$$

In the evaluation of the kinetic energy, we must be careful to take the product of real quantities. We must therefore compute the square of the real part of $\dot{\eta}_r$:

$$(\text{Re } \dot{\eta}_r)^2 = \left(\text{Re } \frac{d}{dt}[(\mu_r + i\nu_r)(\cos \omega_r t + i \sin \omega_r t)]\right)^2$$

$$= (-\omega_r \mu_r \sin \omega_r t - \omega_r \nu_r \cos \omega_r t)^2$$

The kinetic energy of the string is therefore

$$T = \frac{\rho L}{4} \sum_r \omega_r^2 (\mu_r \sin \omega_r t + \nu_r \cos \omega_r t)^2 \qquad (13.20)$$

The potential energy of the string can be calculated easily by writing down the expression for the loaded string and then passing to the limit of a continuous string. (Recall that we consider the potential energy to be only the elastic energy in the string.) For the loaded string,

$$U = \frac{1}{2} \frac{\tau}{d} \sum_j (q_{j-1} - q_j)^2$$

Multiplying and dividing by d,

$$U = \frac{1}{2} \tau \sum_j \left(\frac{q_{j-1} - q_j}{d} \right)^2 d$$

In passing to the limit, $d \to 0$, the term in parentheses becomes just the partial derivative of $q(x, t)$ with respect to x, and the sum (including the factor d) becomes an integral:

$$U = \frac{1}{2} \tau \int_0^L \left(\frac{\partial q}{\partial x} \right)^2 dx \tag{13.21}$$

Using Equation 13.14, we have

$$\frac{\partial q}{\partial x} = \sum_r \frac{r\pi}{L} \eta_r \cos \left(\frac{r\pi x}{L} \right) \tag{13.22}$$

so that

$$U = \frac{1}{2} \tau \int_0^L \left[\sum_r \frac{r\pi}{L} \eta_r \cos \left(\frac{r\pi x}{L} \right) \right]^2 dx \tag{13.23}$$

Again, the squared term can be written as a double sum, and because the trigonometric relation (Equation 13.7) applies for cosines as well as sines, we have

$$U = \frac{\tau}{2} \sum_{r,s} \frac{r\pi}{L} \frac{s\pi}{L} \eta_r \eta_s \int_0^L \cos \left(\frac{r\pi x}{L} \right) \cos \left(\frac{s\pi x}{L} \right) dx$$

$$= \frac{\tau}{2} \sum_{r,s} \frac{r\pi}{L} \frac{s\pi}{L} \eta_r \eta_s \cdot \frac{L}{2} \delta_{rs}$$

$$= \frac{\tau}{2} \sum_r \frac{r^2 \pi^2}{L^2} \cdot \frac{L}{2} \eta_r^2$$

$$= \frac{\rho L}{4} \sum_r \omega_r^2 \eta_r^2 \tag{13.24}$$

where Equation 13.11 has been used in the last line to express the result in terms of ω_r^2. Evaluating the square of the real part of η_r, we have, finally,

$$U = \frac{\rho L}{4} \sum_r \omega_r^2 (\mu_r \cos \omega_r t - \nu_r \sin \omega_r t)^2 \tag{13.25}$$

The total energy is now obtained by adding Equations 13.20 and 13.25, in which the cross terms cancel and the squared terms add to unity:

$$E = T + U$$

$$= \frac{\rho L}{4} \sum_r \omega_r^2 (\mu_r^2 + \nu_r^2) \tag{13.26a}$$

or

$$E = \frac{\rho L}{4} \sum_r \omega_r^2 |\beta_r|^2 \tag{13.26b}$$

The total energy is therefore constant in time and, furthermore, is given by a sum of contributions from each of the normal modes.

The kinetic and potential energies each vary with time, so it is sometimes useful to calculate the *time-averaged* kinetic and potential energies—that is, the averages over one complete period of the fundamental vibration $r = 1$:

$$\langle T \rangle = \frac{\rho L}{4} \sum_r \omega_r^2 \langle (\mu_r \sin \omega_r t + \nu_r \cos \omega_r t)^2 \rangle \tag{13.27}$$

where the slanted brackets denote an average over the time interval $2\pi/\omega_1$. The averages of $\sin^2 \omega_1 t$ or $\cos^2 \omega_1 t$ over this interval are equal to $\frac{1}{2}$. Similarly, the averages of $\sin^2 \omega_r t$ and $\cos^2 \omega_r t$ for $r \geq 2$ are also $\frac{1}{2}$, because the period of the fundamental vibration is always some integer times the period of a higher harmonic vibration. The averages of the cross terms, $\cos \omega_r t \sin \omega_r t$, all vanish. Therefore,

$$\langle T \rangle = \frac{\rho L}{8} \sum_r \omega_r^2 (\mu_r^2 + \nu_r^2)$$

$$= \frac{\rho L}{8} \sum_r \omega_r^2 |\beta_r|^2 \tag{13.28}$$

For the time-averaged potential energy, we have a similar result:

$$\langle U \rangle = \frac{\rho L}{4} \sum_r \omega_r^2 \langle (\mu_r \cos \omega_r t - \nu_r \sin \omega_r t)^2 \rangle$$

$$= \frac{\rho L}{8} \sum_r \omega_r^2 (\mu_r^2 + \nu_r^2)$$

$$= \frac{\rho L}{8} \sum_r \omega_r^2 |\beta_r|^2 \tag{13.29}$$

We therefore have the important result that *the average kinetic energy of a vibrating string is equal to the average potential energy*[*]:

$$\langle T \rangle = \langle U \rangle \tag{13.30}$$

Notice also the simplification that results from the use of normal coordinates: Both $\langle T \rangle$ and $\langle U \rangle$ are simple sums of contributions from each of the normal modes.

13.4 WAVE EQUATION

Our procedure thus far has been to describe the motion of a continuous string as the limiting case of the loaded string for which we have a complete solution; we have not yet written down the fundamental equation of motion for the continuous

---

[*] This result also follows from the virial theorem.

case. We may accomplish this by returning to the loaded string and again using the limit technique—but now on the equation of motion rather than on the solution. Equation 12.131 can be expressed as

$$\frac{m}{d}\ddot{q}_j = \frac{\tau}{d}\left(\frac{q_{j-1}-q_j}{d}\right) - \frac{\tau}{d}\left(\frac{q_j-q_{j+1}}{d}\right)$$ (13.31)

As d approaches zero, we have

$$\frac{q_j-q_{j+1}}{d} \rightarrow \frac{q(x)-q(x+d)}{d} \rightarrow -\frac{\partial q}{\partial x}\bigg|_{x+d/2}$$

which is the derivative at $x + d/2$. For the other term in Equation 13.31, we have

$$\frac{q_{j-1}-q_j}{d} \rightarrow \frac{q(x-d)-q(x)}{d} \rightarrow -\frac{\partial q}{\partial x}\bigg|_{x-d/2}$$

which is the derivative at $x - d/2$. The limiting value of the right-hand side of Equation 13.31 is therefore

$$\lim_{d\to 0} \tau\left(\frac{\frac{\partial q}{\partial x}\bigg|_{x+d/2} - \frac{\partial q}{\partial x}\bigg|_{x-d/2}}{d}\right) = \tau\frac{\partial^2 q}{\partial x^2}\bigg|_x = \tau\frac{\partial^2 q}{\partial x^2}$$

Also in the limit, m/d becomes ρ, so the equation of motion is

$$\rho\ddot{q} = \tau\frac{\partial^2 q}{\partial x^2}$$ (13.32)

or

$$\boxed{\frac{\partial^2 q}{\partial x^2} = \frac{\rho}{\tau}\frac{\partial^2 q}{\partial t^2}}$$ (13.33)

This is the **wave equation** in one dimension. In Section 13.6, we shall discuss the solutions to this equation.

We now want to show that Equation 13.33 can also be easily obtained by considering the forces on a continuous string. Only transverse waves are considered. A portion of the string fixed at both ends, as discussed so far in this chapter, is shown in Figure 13-2.

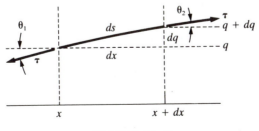

FIGURE 13-2

We assume that the string has a constant mass density ρ (mass/length). We consider a length ds of the string described by $s(x, t)$. The tensions τ on each end of the string are equal in magnitude but not in direction. This imbalance leads to a force and thus an acceleration of the system. We assume that the displacement q (perpendicular to x) is small. The mass dm of the length of string ds is $\rho\, ds$. The horizontal components of the tension are approximately equal and opposite, so we neglect the movement of the string in the x-direction. The force in the q-direction is

$$\Delta F = \rho\, ds\, \frac{\partial^2 q}{\partial t^2} \tag{13.34}$$

where ΔF represents the difference in tension at x and $x + dx$. We use partial derivatives to describe the acceleration, $\partial^2 q / \partial t^2$, because we are not considering the x-dependence of the displacement $q(x, t)$.

The force can be found from the difference in the y-components of the tension.

$$(\Delta F)_y = -\tau \sin \theta_1 + \tau \sin \theta_2$$

$$= -\tau \tan \theta_1 + \tau \tan \theta_2$$

$$= -\tau \left(\frac{\partial q}{\partial x}\right)\bigg|_x + \tau \left(\frac{\partial q}{\partial x}\right)\bigg|_{x + dx}$$

$$= \tau \frac{\partial^2 q}{\partial x^2}\, dx \tag{13.35}$$

where we let $\sin \theta \approx \tan \theta$ because the angles θ are small for small displacements. We now set Equations 13.34 and 13.35 equal, letting $ds \approx dx$:

$$\tau \frac{\partial^2 q}{\partial x^2}\, dx = \rho\, dx\, \frac{\partial^2 q}{\partial t^2}$$

$$\frac{\partial^2 q}{\partial x^2} = \frac{\rho}{\tau} \frac{\partial^2 q}{\partial t^2} \tag{13.36}$$

Equation 13.36 is identical to Equation 13.33 but does not provide the useful information obtained earlier from the normal coordinate method.

13.5 FORCED AND DAMPED MOTION

We can easily determine Lagrange's equations of motion for the vibrating string by using the kinetic energy from Equation 13.19 and the potential energy from Equation 13.24:

$$L = T - U$$

$$= \frac{\rho b}{4} \sum_r \dot{\eta}_r^2 - \frac{\rho b}{4} \sum_r \omega_r^2 \eta_r^2$$

$$= \frac{\rho b}{4} \sum_r (\dot{\eta}_r^2 - \omega_r^2 \eta_r^2) \tag{13.37}$$

where the length of the string has been set equal to b to avoid confusion between the Ls. The ease of the normal coordinate description is apparent. The equations of motion follow from Equation 13.37:

$$\ddot{\eta}_r + \omega_r^2 \eta_r = 0 \tag{13.38}$$

Next, we add a force per unit length $F(x, t)$ acting along the string. We also add a damping force proportional to the velocity. The wave equation (Equation 13.33) now becomes

$$\rho \frac{\partial^2 q}{\partial t^2} + D \frac{\partial q}{\partial t} - \tau \frac{\partial^2 q}{\partial x^2} = F(x, t) \tag{13.39}$$

where each term represents a force per unit length, and D is the damping (resistive) term. Equation 13.39 is solved using normal coordinates. As we did in Section 13.2, we use a solution

$$q(x, t) = \sum_r \eta_r(t) \sin\left(\frac{r\pi x}{b}\right) \tag{13.40}$$

Substitution of Equation 13.40 into Equation 13.39 gives Lagrange's equations of motion—similar to Equation 13.38 but with the damping and forced terms added:

$$\sum_{r=1}^{\infty} \left[\left(\rho\ddot{\eta}_r + D\dot{\eta}_r + \frac{r^2\pi^2\tau}{b^2}\eta_r \right) \sin\left(\frac{r\pi x}{b}\right) \right] = F(x, t) \tag{13.41}$$

The sum over r is again from 1 to ∞ because we are considering a continuous string. The solution of Equation 13.41 parallels that of Section 13.2 (which we do not repeat here in detail) by comparing real and imaginary components. We multiply each side of Equation 13.41 by $\sin(s\pi x/b)$ and integrate over dx from 0 to b (remember that $b = L = $ length of string). Using Equation 13.7, we have

$$\sum_{r=1}^{\infty} \left(\rho\ddot{\eta}_r + D\dot{\eta}_r + \frac{r^2\pi^2\tau}{b^2}\eta_r \right) \frac{b}{2}\delta_{rs} = \int_0^b F(x, t)\sin\left(\frac{s\pi x}{b}\right) dx \tag{13.42}$$

which becomes

$$\ddot{\eta}_s + \frac{D}{\rho}\dot{\eta}_s + \frac{s^2\pi^2\tau}{\rho b^2}\eta_s = \frac{2}{\rho b}\int_0^b F(x,t)\sin\left(\frac{s\pi x}{b}\right)dx \tag{13.43}$$

We now let $f_s(t)$ be the Fourier coefficient of the Fourier expansion of $F(x, t)$, which is on the right side of Equation 13.43:

$$f_s(t) = \int_0^b F(x, t)\sin\left(\frac{s\pi x}{b}\right) dx \tag{13.44}$$

In normal coordinate terms, Equation 13.43 simply becomes

$$\ddot{\eta}_s + \frac{D}{\rho}\dot{\eta}_s + \frac{s^2\pi^2\tau}{\rho b^2}\eta_s = \frac{2}{\rho b}f_s(t) \tag{13.45}$$

It is now apparent that $f_s(t)$ is the component of $F(x, t)$ effective in driving the normal coordinate s.

EXAMPLE **13.2** ---

Reconsider Example 13.1. A sinusoidal driving force of angular frequency ω drives the string at $x = b/2$. Find the displacement.

Solution: The driving force per unit length is

$$F(x, t) = F_0 \cos \omega t, \quad x = b/2 \\ = 0, \quad x \neq b/2 \Bigg\} \tag{13.46}$$

The driving Fourier coefficient becomes

$$f_s(t) = F_0 \cos \omega t \sin \frac{s\pi}{2} \tag{13.47}$$

Notice that $f_s(t) = 0$ for even values of s. Only the odd terms are driven.

If we include a small damping term, Equation 13.45 becomes

$$\ddot{\eta}_s + \frac{D}{\rho}\dot{\eta}_s + \frac{s^2\pi^2\tau}{\rho b}\eta_s = \frac{2}{\rho b}F_0 \cos \omega t \sin \frac{s\pi}{2} \tag{13.48}$$

With the damping term effective, we need not determine a complementary solution, which will be damped out. We need only find a particular (steady-state) solution, as was done in Section 3.6. Equation 13.48 may be compared with Equation 3.53, where

$$\left. \begin{aligned} \frac{D}{\rho} &= 2\beta \\ \frac{s^2\pi^2\tau}{\rho b} &= \omega_0^2 \\ \frac{2F_0 \sin (s\pi/2)}{\rho b} &= A \end{aligned} \right\} \tag{13.49}$$

The solution (see Equation 3.60) for $\eta(t)$ becomes

$$\eta_s(t) = \frac{2F_0 \sin (s\pi/2)\cos(\omega t - \delta)}{\rho b\sqrt{\left(\dfrac{s^2\pi^2\tau}{\rho b} - \omega^2\right)^2 + \dfrac{D}{\rho}\omega^2}} \tag{13.50}$$

where

$$\delta = \tan^{-1}\left[\frac{D\omega}{\rho\left(\dfrac{s^2\pi^2\tau}{\rho b} - \omega^2\right)}\right] \tag{13.51}$$

and the displacement of $q(x, t)$ is

$$q(x, t) = \sum_r \frac{2F_0 \sin \frac{r\pi}{2} \cos (\omega t - \delta) \sin \left(\frac{r\pi x}{b}\right)}{\rho b \sqrt{\left(\frac{r^2 \pi^2 \tau}{\rho b} - \omega^2\right)^2 + \frac{D}{\rho} \omega^2}} \tag{13.52}$$

where we have neglected the part of the solution that is damped out. Equation 13.52 represents many of the features discussed previously. Depending on the driving frequency, only a few of the normal coordinates may dominate because of the resonance effects inherent in the denominator. If the damping term is negligible, the dominant normal coordinate terms are

$$r^2 = \frac{\omega^2 \rho b}{\pi^2 \tau} \tag{13.53}$$

and because of the $\sin (r\pi/2)$ term of Equation 13.52, only odd values of r are effective.

13.6 GENERAL SOLUTIONS OF THE WAVE EQUATION

The one-dimensional wave equation for a vibrating string (see Equation 13.33) is*

$$\frac{\partial^2 \Psi}{\partial x^2} - \frac{\rho}{\tau} \frac{\partial^2 \Psi}{\partial t^2} = 0 \tag{13.54}$$

where ρ is the linear mass density of the string, τ is the tension, and Ψ is called the **wave function**. The dimensions of ρ are $[ML^{-1}]$ and the dimensions of τ are those of a force, namely, $[MLT^{-2}]$. The dimensions of ρ/τ are therefore $[T^2 L^{-2}]$, that is, the dimensions of the reciprocal of a squared velocity. If we write $\sqrt{\tau/\rho} = v$, the wave equation becomes

$$\boxed{\frac{\partial^2 \Psi}{\partial x^2} - \frac{1}{v^2} \frac{\partial^2 \Psi}{\partial t^2} = 0} \tag{13.55}$$

One of our tasks is to give a physical interpretation of the velocity v; it is not sufficient to say that v is the "velocity of propagation" of the wave.

To show that Equation 13.55 does indeed represent a general wave motion, we introduce two new variables,

$$\left. \begin{array}{l} \xi \equiv x + vt \\ \eta \equiv x - vt \end{array} \right\} \tag{13.56}$$

---

* We use the notation $\Psi = \Psi(x, t)$ to denote a *time-dependent* wave function and $\psi = \psi(x)$ to denote a *time-independent* wave function.

Evaluating the derivatives of $\Psi = \Psi(x, t)$, which appear in Equation 13.55, we have

$$\frac{\partial \Psi}{\partial x} = \frac{\partial \Psi}{\partial \xi}\frac{\partial \xi}{\partial x} + \frac{\partial \Psi}{\partial \eta}\frac{\partial \eta}{\partial x} = \frac{\partial \Psi}{\partial \xi} + \frac{\partial \Psi}{\partial \eta}$$

(13.57)

Then,

$$\frac{\partial^2 \Psi}{\partial x^2} = \frac{\partial}{\partial x}\frac{\partial \Psi}{\partial x} = \frac{\partial}{\partial x}\left(\frac{\partial \Psi}{\partial \xi} + \frac{\partial \Psi}{\partial \eta}\right)$$

$$= \frac{\partial}{\partial \xi}\left(\frac{\partial \Psi}{\partial \xi} + \frac{\partial \Psi}{\partial \eta}\right)\frac{\partial \xi}{\partial x} + \frac{\partial}{\partial \eta}\left(\frac{\partial \Psi}{\partial \xi} + \frac{\partial \Psi}{\partial \eta}\right)\frac{\partial \eta}{\partial x}$$

$$= \frac{\partial^2 \Psi}{\partial \xi^2} + 2\frac{\partial^2 \Psi}{\partial \xi \partial \eta} + \frac{\partial^2 \Psi}{\partial \eta^2}$$

(13.58)

Similarly, we find

$$\frac{1}{v}\frac{\partial \Psi}{\partial t} = \frac{\partial \Psi}{\partial \xi} - \frac{\partial \Psi}{\partial \eta}$$

(13.59)

and

$$\frac{1}{v^2}\frac{\partial^2 \Psi}{\partial t^2} = \frac{1}{v}\frac{\partial}{\partial t}\left(\frac{1}{v}\frac{\partial \Psi}{\partial t}\right) = \frac{1}{v}\frac{\partial}{\partial t}\left(\frac{\partial \Psi}{\partial \xi} - \frac{\partial \Psi}{\partial \eta}\right)$$

$$= \frac{\partial^2 \Psi}{\partial \xi^2} - 2\frac{\partial^2 \Psi}{\partial \xi \partial \eta} + \frac{\partial^2 \Psi}{\partial \eta^2}$$

(13.60)

But according to Equation 13.55, the right-hand sides of Equations 13.58 and 13.60 must be equal. This can be true only if

$$\frac{\partial^2 \Psi}{\partial \xi \partial \eta} \equiv 0$$

(13.61)

The most general expression for Ψ that can satisfy this equation is a sum of two terms, one of which depends only on ξ and the other only on η; no more complicated function of ξ and η permits Equation 13.61 to be valid. Thus,

$$\Psi = f(\xi) + g(\eta)$$

(13.62a)

or, substituting for ξ and η,

$$\Psi = f(x + vt) + g(x - vt)$$

(13.62b)

where f and g are *arbitrary* functions of the variables $x + vt$ and $x - vt$, respectively, which are not necessarily of a periodic nature, although they may be.

As time increases, the value of x must also increase in order to maintain a constant value for $x - vt$. The function g therefore retains its original form as time increases if we shift our viewpoint along the x-direction (in a positive sense) with a speed v. Thus, the function g must represent a disturbance that moves to the right

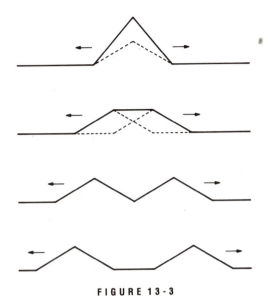

FIGURE 13-3

(i.e., to larger values of x) with a speed v, whereas f represents the propagation of a disturbance to the left. We therefore conclude that Equation 13.55 does indeed describe wave motion and, in general, a **traveling** (or **propagating**) wave.

Let us now attempt to interpret Equation 13.62b in terms of the motion of a stretched string. At time $t = 0$, the displacement of the string is described by

$$q(x, 0) = f(x) + g(x)$$

If we take identical triangular forms for $f(x)$ and $g(x)$, the shape of the string at $t = 0$ is as shown at the top of Figure 13-3. As time increases, the disturbance represented by $f(x + vt)$ propagates to the *left*, whereas the disturbance represented by $g(x - vt)$ propagates to the *right*. This propagation of the individual disturbances to the left and right is illustrated in the lower part of Figure 13-3.

Consider next the left-going disturbance alone. If we terminate the string (at $x = 0$) by attaching it to a rigid support, we find the phenomenon of **reflection**. Because the support is *rigid*, we must have $f(vt) \equiv 0$ for all values of time. This condition cannot be met by the function f alone (unless it trivially vanishes). We can satisfy the condition at $x = 0$ if we consider, in addition to $f(x + vt)$, an imaginary disturbance, $-f(-x + vt)$, which approaches the boundary point from the left, as in Figure 13-4. The disturbance $f(x + vt)$ continues to propagate to the left, even into the imaginary section of the string ($x < 0$), while the disturbance $-f(-x + vt)$ propagates across the boundary and along the real string. The net effect is that the original disturbance is reflected at the support and thereafter propagates to the right.

If the string is terminated by rigid supports at $x = 0$ and also at $x = L$, the disturbance propagates periodically back and forth with a period $2L/v$.

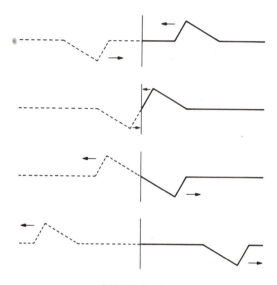

FIGURE 13-4

13.7   SEPARATION OF THE WAVE EQUATION

If we require a general solution of the wave equation that is harmonic (as for the vibrating string or, for that matter, for a large number of problems of physical interest), we can write

$$\Psi(x,t) = \psi(x)e^{i\omega t} \tag{13.63}$$

so that the one-dimensional wave equation (Equation 13.55) becomes

$$\frac{\partial^2 \psi}{\partial x^2} + \frac{\omega^2}{v^2}\psi = 0 \tag{13.64}$$

where ψ is now a function of x only.

The general wave motion of a system is not restricted to a single frequency ω. For a system with n degrees of freedom, there are n possible characteristic frequencies, and for a continuous string there is an infinite set of frequencies.* If we designate the rth frequency by ω_r, the wave function corresponding to this frequency is

$$\Psi_r(x, t) = \psi_r(x)e^{i\omega_r t} \tag{13.65}$$

The complete wave function is a superposition (recall that we are dealing with a *linear* system) of all the particular wave functions (or *modes*). Thus

$$\Psi(x, t) = \sum_r \Psi_r(x, t) = \sum_r \psi_r(x)\, e^{i\omega_r t} \tag{13.66}$$

---

* An infinite set of frequencies would exist for a truly continuous string, but because a real string is composed fundamentally of atoms, there does exist an upper limit for ω (see Section 13.8).

In Equation 13.63, we *assumed* that the wave function was periodic in time. But now we see that this assumption entails no real restriction at all (apart from the usual assumptions regarding the continuity of the functions and the convergence of the series), because the summation in Equation 13.66 actually gives a Fourier representation of the wave function and is therefore the most general expression for the true wave function.*

We now wish to show that Equation 13.65 results naturally from a powerful method that can often be used to obtain solutions to partial differential equations— the method of **separation of variables**. First, we express the solution as

$$\Psi(x, t) \equiv \psi(x) \cdot \chi(t) \tag{13.67}$$

that is, we assume that the variables are *separable* and therefore that the complete wave function can be expressed as the product of two functions, one of which is a spatial function only, and one of which is a temporal function only. It is not guaranteed that we will always find such functions, but many of the partial differential equations encountered in physical problems are separable in at least one coordinate system; some (such as those involving the Laplacian operator) are separable in many coordinate systems. In short, the justification of the method of separation of variables, as is the case with many assumptions in physics, is in its success in producing mathematically acceptable solutions to a problem that eventually are found to properly describe the physical situation, i.e. are "experimentally verifiable."

Substituting $\Psi = \psi\chi$ into Equation 13.55, we have

$$\chi\frac{d^2\psi}{dx^2} - \frac{\psi}{v^2}\frac{d^2\chi}{dt^2} = 0$$

or

$$\frac{v^2}{\psi}\frac{d^2\psi}{dx^2} = \frac{1}{\chi}\frac{d^2\chi}{dt^2} \tag{13.68}$$

But, in view of the definitions of $\psi(x)$ and $\chi(t)$, the left-hand side of Equation 13.68 is a function of x alone, whereas the right-hand side is a function of t alone. This situation is possible only if each part of the equation is equal to the same constant. To be consistent with our previous notation, we choose this constant to be $-\omega^2$. Thus, we have

$$\frac{d^2\psi}{dx^2} + \frac{\omega^2}{v^2}\psi = 0 \tag{13.69a}$$

---

* Euler proved in 1748 that the wave equation for a continuous string is satisfied by an arbitrary function of $x \pm vt$, and Daniel Bernoulli showed in 1753 that the motion of a string is a superposition of its characteristic frequencies. These two results, taken together, indicated that an arbitrary function could be described by a superposition of trigonometric functions. This Euler could not believe, and so he (as well as Lagrange) rejected Bernoulli's superposition principle. The French mathematician Alexis Claude Clairaut (1713–1765) gave a proof in an obscure paper in 1754 that the results of Euler and Bernoulli were actually consistent, but it was not until Fourier gave his famous proof in 1807 that the question was settled.

and

$$\frac{d^2\chi}{dt^2} + \omega^2\chi = 0 \tag{13.69b}$$

These equations are of a familiar form, and we know that the solutions are

$$\psi(x) = Ae^{i(\omega/v)x} + Be^{-i(\omega/v)x} \tag{13.70a}$$

$$\chi(t) = Ce^{i\omega t} + De^{-i\omega t} \tag{13.70b}$$

where the constants A, B, C, D are determined by the boundary conditions. We may write the solution $\Psi(x, t)$ in a shorthand manner as

$$\Psi(x, t) = \psi(x)\chi(t) \sim \exp[\pm i(\omega/v)x]\exp[\pm i\omega t]$$

$$\sim \exp[\pm i(\omega/v)(x \pm vt)] \tag{13.71}$$

This notation means that the wave function Ψ varies as a *linear combination* of the terms

$$\exp[i(\omega/v)(x + vt)]$$

$$\exp[i(\omega/v)(x - vt)]$$

$$\exp[-i(\omega/v)(x + vt)]$$

$$\exp[-i(\omega/v)(x - vt)]$$

The *separation constant* for Equation 13.68 was chosen to be $-\omega^2$. There is nothing in the mathematics of the problem to indicate that there is a unique value of ω; hence, there must exist a set* of equally acceptable frequencies ω_r. To each such frequency, there corresponds a wave function:

$$\Psi_r(x, t) \sim \exp[\pm i(\omega_r/v)(x \pm vt)]$$

The general solution is therefore not only a linear combination of the harmonic terms but also a sum over all possible frequencies:

$$\Psi(x, t) \sim \sum_r a_r \Psi_r$$

$$\sim \sum_r a_r \exp[\pm i(\omega_r/v)(x \pm vt)] \tag{13.72}$$

The general solution of the wave equation therefore leads to a very complicated wave function. There are, in fact, an infinite number of arbitrary constants a_r. This is a general result for partial differential equations; but this infinity of constants must satisfy the physical requirements of the problem (the boundary conditions), and therefore they can be evaluated in the same manner that the coefficients of an infinite Fourier expansion can be evaluated.

---

* At this stage of the development, the set is in fact infinite, because no frequencies have yet been eliminated by boundary conditions.

For much of our discussion, it is sufficient to consider only one of the four possible combinations expressed by Equation 13.71; that is, we select a wave propagating in a particular direction and with a particular phase. Then, we can write, for example,

$$\Psi_r(x, t) \sim \exp[-i(\omega_r/v)(x - vt)]$$

This is the rth Fourier component of the wave function, and the general solution is a summation over all such components. The functional form of each component is, however, the same, and so they can be discussed separately. Thus, we shall usually write, for simplicity,

$$\Psi(x, t) \sim \exp[-i(\omega/v)(x - vt)] \tag{13.73}$$

The general solution must be obtained by a summation over all frequencies that are allowed by the particular physical situation.

It is customary to write the differential equation for $\psi(x)$ as

$$\boxed{\frac{d^2\psi}{dx^2} + k^2\psi = 0} \tag{13.74}$$

which is the time-independent form of the one-dimensional wave equation, also called the **Helmholtz equation**,* and where

$$k^2 \equiv \frac{\omega^2}{v^2} \tag{13.75}$$

The quantity k, called the **propagation constant** or the **wave number** (i.e., proportional to the number of wavelengths per unit length), has dimensions $[L^{-1}]$. The wavelength λ is the distance required for one complete vibration of the wave,

$$\lambda = \frac{v}{\nu} = \frac{2\pi v}{\omega}$$

and thus the relation[†] between k and λ is

$$k = \frac{2\pi}{\lambda}$$

We can therefore write, in general,

$$\Psi_r(x, t) \sim e^{\pm ik_r(x \pm vt)}$$

---

* Hermann von Helmholtz (1821–1894) used this form of the wave equation in his treatment of acoustic waves in 1859.

[†] More properly the wave number should be defined as $k = 1/\lambda$ rather than $2\pi/\lambda$, because $1/\lambda$ is the number of wavelengths per unit distance. However, $k = 2\pi/\lambda$ is more commonly used in theoretical physics, and we follow that usage here.

or, for the simplified wave function,

$$\Psi(x, t) \sim e^{-ik(x-vt)} = e^{i(\omega t - kx)} \tag{13.76}$$

If we superimpose two traveling waves of the type given by Equation 13.76 and if these waves are of equal magnitude (amplitude) but moving in opposite directions, then

$$\Psi = \Psi_+ + \Psi_- = Ae^{-ik(x+vt)} + Ae^{-ik(x-vt)} \tag{13.77}$$

or

$$\Psi = Ae^{-ikx}(e^{i\omega t} + e^{-i\omega t})$$

$$= 2Ae^{-ikx} \cos \omega t$$

the real part of which is

$$\Psi = 2A \cos kx \cos \omega t \tag{13.78}$$

Such a wave no longer has the property that it propagates; the wave form does not move forward with time. There are, in fact, certain positions at which there is no motion. These positions, the **nodes**, result from the complete cancellation of one wave by the other. The nodes of the wave function given by Equation 13.78 occur at $x = (2n + 1)\pi/2k$, where n is an integer. Because there are fixed positions in waves of this type, they are called **standing waves**. Solutions to the problem of the vibrating string are of this form (but with a phase factor attached to the term kx such that the cosine is transformed into a sine function satisfying the boundary conditions).

E X A M P L E   **13.3**   -

Consider a string consisting of two densities, ρ_1 in region 1 where $x < 0$ and ρ_2 in region 2 where $x > 0$. A continuous wave train is incident from the left (i.e., from negative values of x). What are the ratios of the square of the amplitude magnitudes for the reflected and transmitted waves to the incident wave?

Solution: The wave will be both reflected and transmitted at $x = 0$ where the mass density discontinuity occurs. Therefore, in region 1 we have the superposition of the incident and reflected waves, and in region 2 we have only the transmitted wave. If the incident wave is $Ae^{i(\omega t - k_1 x)}$, then we have for the waves $\Psi_1(x, t)$ and $\Psi_2(x, t)$ in regions 1 and 2, respectively (see Equation 13.77)

$$\left. \begin{array}{l} \Psi_1(x, t) = \Psi_{inc} + \Psi_{refl} = Ae^{i(\omega t - k_1 x)} + Be^{i(\omega t + k_1 x)} \\ \Psi_2(x, t) = \Psi_{trans} = Ce^{i(\omega t - k_2 x)} \end{array} \right\} \tag{13.79}$$

In Equation 13.79, we have explicitly taken into account the fact that the waves in both regions have the same frequency. But because the wave velocity on a string is given by

$$v = \sqrt{\frac{T}{\rho}}$$

we have $v_1 \neq v_2$, and therefore $k_1 \neq k_2$. We also have

$$k = \frac{\omega}{v} = \omega \sqrt{\frac{\rho}{\tau}} \tag{13.80}$$

so, in terms of the wave number of the incident wave,

$$k_2 = k_1 \sqrt{\frac{\rho_2}{\rho_1}} \tag{13.81}$$

The amplitude A of the incident wave (see Equation 13.79) is given and is real. We must then obtain the amplitudes B and C of the reflected and transmitted waves to complete the solution of the problem. There are as yet no restrictions on B and C, and they may be complex quantities.

The physical requirements on the problem may be stated in terms of the boundary conditions. These are, simply, that the total wave function $\Psi = \Psi_1 + \Psi_2$ and its derivative must be continuous across the boundary. The continuity of Ψ results from the fact that the string is continuous. The condition on the derivative prevents the occurrence of a "kink" in the string, for if $\partial\Psi/\partial x_{0+} \neq \partial\Psi/\partial x_{0-}$, then $\partial^2\Psi/\partial x^2$ is infinite at $x = 0$; but the wave equation relates $\partial^2\Psi/\partial x^2$ and $\partial^2\Psi/\partial t^2$; and if the former is infinite, this implies an infinite acceleration, which is not allowed by the physical situation. We have, therefore, for all values of the time t,

$$\left. \Psi_1 \right|_{x=0} = \left. \Psi_2 \right|_{x=0} \tag{13.82a}$$

$$\left. \frac{\partial \Psi_1}{\partial x} \right|_{x=0} = \left. \frac{\partial \Psi_2}{\partial x} \right|_{x=0} \tag{13.82b}$$

From Equations 13.79 and 13.82a, we have

$$A + B = C \tag{13.83a}$$

and from Equations 13.79 and 13.82b we obtain

$$-k_1A + k_1B = -k_2C \tag{13.83b}$$

The solution of this pair of equations yields

$$B = \frac{k_1 - k_2}{k_1 + k_2} A \tag{13.84a}$$

and

$$C = \frac{2k_1}{k_1 + k_2} A \tag{13.84b}$$

The wave numbers k_1 and k_2 are both real, so amplitudes B and C are likewise real. Furthermore, k_1, k_2, and A are all positive, so C is always positive. Thus, the transmitted wave is always in phase with the incident wave. Similarly, if $k_1 > k_2$, then the incident and reflected waves are in phase, but they are out of phase for $k_2 > k_1$, that is, for $\rho_2 > \rho_1$.

The **reflection coefficient** R is defined as the ratio of the squared magnitudes of the amplitudes of the reflected and incident waves:

$$R \equiv \frac{|B|^2}{|A|^2} = \left(\frac{k_1 - k_2}{k_1 + k_2}\right)^2 \tag{13.85}$$

Because the energy content of a wave is proportional to the square of the amplitude of the wave function, R represents the ratio of the reflected energy to the incident energy. The quantity $|B|^2$ represents the *intensity* of the reflected wave.

No energy can be stored in the junction of the two strings, so the incident energy must be equal to the sum of the reflected and transmitted energies; that is, $R + T = 1$. Thus,

$$T = 1 - R = \frac{4k_1 k_2}{(k_1 + k_2)^2} \tag{13.86}$$

or

$$T = \frac{k_2}{k_1} \frac{|C|^2}{|A|^2} \tag{13.87}$$

In the study of the reflection and transmission of electromagnetic waves, we find quite similar expressions for R and T.

13.8 PHASE VELOCITY, DISPERSION, AND ATTENUATION

We have seen in Equations 13.71 that the general solution to the wave equation produces, even in the one-dimensional case, a complicated system of exponential factors. For the purposes of further discussion, we restrict our attention to the particular combination

$$\boxed{\Psi(x, t) = A e^{i(\omega t - kx)}} \tag{13.88}$$

This equation describes the propagation to the right (larger x) of a wave possessing a well-defined angular frequency ω. Certain physical situations can be quite adequately approximated by a wave function of this type—for example, the propagation of a monochromatic light wave in space or the propagation of a sinusoidal wave on a long (strictly, infinitely long) string.

If the argument of the exponential in Equation 13.88 remains constant, then the wave function $\Psi(x, t)$ also remains constant. The argument of the exponential is called the **phase** ϕ of the wave,

$$\phi \equiv \omega t - kx \tag{13.89}$$

If we move our viewpoint along the x-axis at a velocity such that the phase at every

point is the same, we always see a stationary wave of the same shape. The velocity
V with which we must move, called the **phase velocity** of the wave, corresponds
to the velocity with which the **wave form** propagates. To ensure ϕ = constant, we
set

$$d\phi = 0 \tag{13.90}$$

or

$$\omega \, dt = k \, dx$$

from which

$$V = \frac{dx}{dt} = \frac{\omega}{k} = v \tag{13.91}$$

so that the phase velocity in this case is just the quantity originally introduced as
the velocity. It is possible to speak of a phase velocity only when the wave function
has the same form throughout its length. This condition is necessary so we can
measure the wavelength by taking the distance between *any* two successive wave
crests (or between *any* two successive corresponding points on the wave). If the
wave form were to change as a function of time or of distance along the wave,
these measurements would not always yield the same results. The wavelength is
not a function of time or space (i.e., that ω is *pure*) only if the wave train is of
infinite length. If the wave train is of finite length, there must be a spectrum of
frequencies present in the wave, each with its own phase velocity. We will often
assign a single frequency and phase velocity to a wave of finite length as a convenient approximation.

Let us return to the example of the loaded string and examine the properties
of the phase velocity in that case. We have previously found (Equation 12.152) that
the frequency for the rth mode of the loaded string when terminated at both ends
is given by

$$\omega_r = 2\sqrt{\frac{\tau}{md}} \sin\left[\frac{r\pi}{2(n+1)}\right] \tag{13.92}$$

where the notation is the same as in Chapter 12. Recall that we take only positive
values for the frequencies. When $r = 1$, there is a node at each end, and none
between; hence, the length of the string is one-half of a wavelength. Similarly, when
$r = 2$, then $L = \lambda$ and, in general, $\lambda_r = 2L/r$. Therefore,

$$\frac{r\pi}{2(n+1)} = \frac{r\pi d}{2d(n+1)} = \frac{r\pi d}{2L} = \frac{\pi d}{\lambda_r} = \frac{k_r d}{2} \tag{13.93}$$

and

$$\omega_r = 2\sqrt{\frac{\tau}{md}} \sin\left(\frac{k_r d}{2}\right) \tag{13.94}$$

Because this expression no longer contains n or L, it applies equally well to a terminated or infinite loaded string.

To study the propagation of a wave in the loaded string, we initiate a disturbance by forcing one of the particles, say, the *zeroth* one, to move according to

$$q_0(t) = Ae^{i\omega t} \tag{13.95}$$

If the string contains many particles,* then any angular frequency less than $2\sqrt{\tau/md}$ is an allowed frequency (actually an eigenfrequency), satisfying Equation 13.95. After the transient effects have subsided and the steady-state conditions are attained, the phase velocity of the wave is given by[†]

$$V = \frac{\omega}{k} = \sqrt{\frac{\tau d}{m}} \frac{|\sin(kd/2)|}{kd/2} = V(k) \tag{13.96}$$

Thus the phase velocity is a function of the wave number; that is, V is frequency-dependent. When $V = V(k)$ for a given medium, that medium is said to be **dispersive**, and the wave exhibits **dispersion**. The best-known example of this phenomenon is the simple optical prism. The index of refraction of the prism depends on the wavelength of the incident light (i.e., the prism is a dispersive medium for optical light); on passing through the prism, the light is separated into a spectrum of wavelengths (i.e., the light wave is dispersed).

For a longitudinal wave propagating down a long, slender rod, most of the energy is associated with the direction of the longitudinal wave propagation. There is, however, a small amount of energy dissipated in a transverse wave moving at right angles. This lateral disturbance causes the phase velocity of the longitudinal wave to be decreased, and the effect depends on wavelength. For large wavelengths, the effect is small; for short wavelengths, especially those approaching the radius of the rod, the velocity dispersion is pronounced.

From Equation 13.96, we see that, as the wavelength becomes very long ($\lambda \rightarrow \infty$ or $k \rightarrow 0$), the phase velocity approaches the constant value

$$V(\lambda \rightarrow \infty) = \sqrt{\frac{\tau d}{m}} \tag{13.97}$$

Otherwise, $V = V(k)$, and the wave is dispersive. We note that the phase velocity

---

* Strictly, we need an infinite number of particles for this type of analysis, but we may approach the ideal conditions as closely as desired by increasing the finite number of particles.

[†] In Equation 13.92 the values of r are required to be $\leq n$ (see Equation 12.144), so we automatically have $\omega_r \geq 0$ because $\sin[r\pi/2(n+1)] \geq 0$ for $0 \leq r \leq n$. We no longer have such a restriction on kd, so $\sin(kd/2)$ can become negative. We continue to consider only positive frequencies by always taking only the magnitude of $\sin(kd/2)$.

This result was obtained by Baden-Powell in 1841, but William Thomson (Lord Kelvin) (1824–1907) realized the full significance only in 1881.

for the continuous string (see Equation 13.55) is

$$V_{\text{cont.}} = v = \sqrt{\frac{\tau}{\rho}} \tag{13.98}$$

and because m/d for the loaded string corresponds to ρ for the continuous string, the phase velocities for the two cases are equal in the long-wavelength limit (but *only* in this limit). This is a reasonable result because as λ becomes large compared with d, the properties of the wave are less sensitive to the spacing between particles, and in the limit, d may vanish without affecting the phase velocity.

In Equation 13.94, the restriction on r is $1 \leq r \leq n$. Then, because $k_r = r\pi/L$, we see that the maximum value of k is

$$k_{\text{max}} = \pi/d \tag{13.99}$$

The corresponding frequency, from Equation 13.96, is $2\sqrt{\tau/md}$. What is the result of forcing the string to vibrate at a frequency *greater* than $2\sqrt{\tau/md}$? For this purpose, we allow k to become complex and investigate the consequences:

$$k \equiv \kappa - i\beta, \qquad \kappa, \beta > 0 \tag{13.100}$$

The expression for ω (Equation 13.94) then becomes

$$\omega = 2\sqrt{\frac{\tau}{md}} \sin\left[\frac{d}{2}(\kappa - i\beta)\right]$$

$$= 2\sqrt{\frac{\tau}{md}} \left(\sin\frac{d\kappa}{2}\cos\frac{i\beta d}{2} - \cos\frac{\kappa d}{2}\sin\frac{i\beta d}{2}\right)$$

$$= 2\sqrt{\frac{\tau}{md}} \left(\sin\frac{\kappa d}{2}\cosh\frac{\beta d}{2} - i\cos\frac{\kappa d}{2}\sinh\frac{\beta d}{2}\right) \tag{13.101}$$

If the frequency is to be a real quantity, the imaginary part of this expression must vanish. Thus, we may have either $\cos(\kappa d/2) = 0$ or $\sinh(\beta d/2) = 0$. But the latter choice requires $\beta = 0$, contrary to the requirement that k be complex. We therefore have

$$\cos\frac{\kappa d}{2} = 0 \tag{13.102}$$

For this case, we must also have

$$\sin\frac{\kappa d}{2} = 1 \tag{13.103}$$

The expression for the angular frequency becomes

$$\omega = 2\sqrt{\frac{\tau}{md}} \cosh\frac{\beta d}{2} \tag{13.104}$$

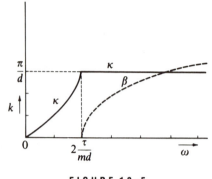

FIGURE 13-5

Thus, we have the result that, for $\omega \leq 2\sqrt{\tau/md}$, the wave number k is real and the relation between ω and k is given by Equation 13.94; whereas, for $\omega > 2\sqrt{\tau/md}$, k is complex with the real part κ fixed by Equation 13.102 at the value $\kappa = \pi/d$ and with the imaginary part β given by Equation 13.104. The situation is shown in Figure 13-5.

What is the physical significance of a complex wave number? Our original wave function was of the form

$$\Psi = Ae^{i(\omega t - kx)}$$

but, if $k = \kappa - i\beta$, then Ψ can be written as

$$\Psi = Ae^{-\beta x}e^{i(\omega t - \kappa x)} \tag{13.105}$$

and the factor $\exp(-\beta x)$ represents a damping, or *attenuation*, of the wave with increasing distance x. We therefore conclude that the wave is propagated without attenuation for $\omega \leq 2\sqrt{\tau/md}$ (this region is called the *passing band* of frequencies), and that attenuation sets in at $\omega_c = 2\sqrt{\tau/md}$ (called the *critical* or *cutoff frequency**) and increases with increasing frequency.

The physical significance of the real and imaginary parts of k is now apparent: β is the attenuation coefficient[†] (and exists only if $\omega > \omega_c$), whereas κ is the wave number in the sense that the phase velocity V' is given by

$$V' = \frac{\omega}{\kappa} = \frac{\omega}{\text{Re } k} \tag{13.106}$$

rather than by $V = \omega/k$. If k is real, these expressions for V and V' are identical.

This example emphasizes the fact that the fundamental definition of the phase velocity is based on the requirement of the constancy of the phase and *not* on the

---

* The occurrence of a cutoff frequency was discovered by Lord Kelvin in 1881.

[†] The reason for writing $k = \kappa - i\beta$ rather than $k = \kappa + i\beta$ in Equation 13.100 is now clear; if $\beta > 0$ for the latter choice, then the amplitude of the wave increases without limit rather than decreasing toward zero.

ratio ω/k. Thus, in general, the phase velocity V and the so-called wave velocity v are distinct quantities. We note also that if ω is real and if the wave number k is complex, then the *wave velocity* v must also be complex so that the product kv yields a real quantity for the frequency through the relation $\omega = kv$. On the other hand, the *phase velocity*, which arises from the requirement that ϕ = constant, is necessarily always a real quantity.

In the preceding discussion, we considered the system to be conservative and argued that this requires ω to be a real quantity.* We found that if ω exceeds the critical frequency ω_c, attenuation results and the wave number becomes complex. If we relax the condition that the system is conservative, the frequency may then be complex and the wave number real. In such a case, the wave is damped in *time* rather than in *space* (see Problem 13-13). Spatial attenuation (ω real, k complex) is of particular significance for traveling waves, whereas temporal attenuation (ω complex, k real) is important for standing waves.

Although attenuation occurs in the loaded string if $\omega > \omega_c$, the system is still conservative and no energy is lost. This seemingly anomalous situation results because the force applied to the particle in the attempt to initiate a traveling wave is (after the steady-state condition of an attenuated wave is set up) exactly 90° out of phase with the velocity of the particle, so that the power transferred, $P = \mathbf{F} \cdot \mathbf{v}$, is zero.

In this treatment of the loaded string, we have tacitly assumed an ideal situation; that is, the system was assumed to be lossless. As a result, we found that there was attenuation for $\omega > \omega_c$ but none for $\omega < \omega_c$. However, every real system is subject to loss, so in fact there is some attenuation even for $\omega < \omega_c$.

13.9 GROUP VELOCITY AND WAVE PACKETS

It was demonstrated in Section 3.10 that the superposition of various solutions of a linear differential equation is still a solution to the equation. Indeed, we formed the general solution to the problem of small oscillations (see Equation 12.43) by summing all the particular solutions. Let us assume, therefore, that we have two almost equal solutions to the wave equation represented by the wave functions Ψ_1 and Ψ_2, each of which has the same amplitude,

$$\left.\begin{array}{l} \Psi_1(x, t) = Ae^{i(\omega t - kx)} \\ \Psi_2(x, t) = Ae^{i(\Omega t - Kx)} \end{array}\right\} \tag{13.107}$$

but whose frequencies and wave numbers differ by only small amounts:

$$\left.\begin{array}{l} \Omega = \omega + \Delta\omega \\ K = k + \Delta k \end{array}\right\} \tag{13.108}$$

---

* See the discussion in Section 12.4 in the paragraph following Equation 12.39.

Forming the solution that consists of the sum Ψ_1 and Ψ_2, we have

$$\Psi(x, t) = \Psi_1 + \Psi_2 = A\left[\exp(i\omega t)\exp(-ikx)\right.$$

$$\left. + \exp\{i(\omega + \Delta\omega)t\}\exp\{-i(k + \Delta k)x\}\right]$$

$$= A\left[\exp\left\{i\left(\omega + \frac{\Delta\omega}{2}\right)t\right\}\exp\left\{-i\left(k + \frac{\Delta k}{2}\right)x\right\}\right]$$

$$\cdot\left[\exp\left\{-i\left(\frac{(\Delta\omega)t - (\Delta k)x}{2}\right)\right\} + \exp\left\{i\left(\frac{(\Delta\omega)t - (\Delta k)x}{2}\right)\right\}\right]$$

The second bracket is just twice the cosine of the argument of the exponential, and the real part of the first bracket is also a cosine. Thus, the real part of the wave function is

$$\Psi(x, t) = 2A\cos\left[\frac{(\Delta\omega)t - (\Delta k)x}{2}\right]\cos\left[\left(\omega + \frac{\Delta\omega}{2}\right)t - \left(k + \frac{\Delta k}{2}\right)x\right] \quad \textbf{(13.109)}$$

This expression is similar to that obtained in the problem of the weakly coupled oscillators (see Section 12.3), in which we found a slowly varying amplitude, corresponding to the term

$$2A\cos\left[\frac{(\Delta\omega)t - (\Delta k)x}{2}\right]$$

which modulates the wave function. The primary oscillation takes place at a frequency $\omega + (\Delta\omega/2)$, which, according to our assumption that $\Delta\omega$ is small, differs negligibly from ω. The varying amplitude gives rise to **beats** (Figure 13-6).

The velocity U (called the **group velocity***) with which the modulations (or groups of waves) propagate is given by the requirement that the phase of the amplitude term be constant. Thus,

$$U = \frac{dx}{dt} = \frac{\Delta\omega}{\Delta k} \quad \textbf{(13.110)}$$

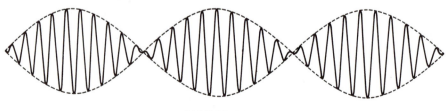

FIGURE 13-6

---

* The concept of group velocity is due to Hamilton, 1839; the distinction between phase and group velocity was made clear by Lord Rayleigh (*Theory of Sound*, 1st edition, 1877; see Ra94).

In a nondispersive medium $\Delta\omega/\Delta k = V$, so the group and phase velocities are identical.* If dispersion is present, however, U and V are distinct.

Thus far, we have considered only the superposition of two waves. If we wish to superpose a system of n waves, we must write

$$\Psi(x, t) = \sum_{r=1}^{n} A_r \exp\left[i(\omega_r t - k_r x)\right] \tag{13.111a}$$

where A_r represents the amplitudes of the individual waves. In the event that n becomes very large (strictly, infinite), the frequencies are continuously distributed, and we may replace the summation by an integration, obtaining[†]

$$\Psi(x, t) = \int_{-\infty}^{+\infty} A(k)e^{i(\omega t - kx)}\, dk \tag{13.111b}$$

where the factor $A(k)$ represents the distribution amplitudes of the component waves with different frequencies, that is, the **spectral distribution** of the waves. The most interesting cases occur when $A(k)$ has a significant value only in the neighborhood of a particular wave number (say, k_0) and becomes vanishingly small for k outside a small range, denoted by $k_0 \pm \Delta k$. In such a case, the wave function can be written as

$$\Psi(x, t) = \int_{k_0 - \Delta k}^{k_0 + \Delta k} A(k)e^{i(\omega t - kx)}\, dk \tag{13.112}$$

A function of this type is called a **wave packet**.[‡] The concept of group velocity can be applied only to those cases that can be represented by a wave packet, that is, to wave functions containing a small range (or *band*) of frequencies.

For the case of the wave packet represented by Equation 13.112, the contributing frequencies are restricted to those lying near $\omega(k_0)$. We can therefore expand $\omega(k)$ about $k = k_0$:

$$\omega(k) = \omega(k_0) + \left(\frac{d\omega}{dk}\right)_{k=k_0} \cdot (k - k_0) + \cdots \tag{13.113a}$$

which we can abbreviate as

$$\omega = \omega_0 + \omega_0'(k - k_0) + \cdots \tag{13.113b}$$

The argument of the exponential in the wave packet integral becomes, approximately,

$$\omega t - kx = (\omega_0 t - k_0 x) + \omega_0'(k - k_0)t - (k - k_0)x$$

---

* This identity is shown explicitly in Equation 13.117.

[†] We have previously made the tacit assumption that $k \geq 0$. However, k is defined by $k^2 = \omega^2/v^2$ (see Equation 13.75), so there is no mathematical reason why we may not also have $k < 0$. We may therefore extend the region of integration to include $-\infty < k < 0$ without mathematical difficulty. This procedure allows the identification of the integral representation of $\Psi(x, t)$ as a Fourier integral.

[‡] The term *wave packet* is due to Erwin Schrödinger.

where we have added and subtracted the term k_0x. Thus,

$$\omega t - kx = (\omega_0 t - k_0 x) + (k - k_0)(\omega_0' t - x) \tag{13.114}$$

and Equation 13.112 becomes

$$\Psi(x, t) = \int_{k_0 - \Delta k}^{k_0 + \Delta k} A(k)\exp[i(k - k_0)(\omega_0' t - x)]\exp[i(\omega_0 t - k_0 x)]\, dk \tag{13.115}$$

The wave packet, expressed in this fashion, may be interpreted as follows. The quantity

$$A(k)\exp[i(k - k_0)(\omega_0' t - x)]$$

constitutes an effective amplitude that, because of the small quantity $(k - k_0)$ in the exponential, varies slowly with time and describes the motion of the wave packet (or envelope of a group of waves) in the same manner that the term

$$2A \cos\left[\frac{(\Delta\omega)t - (\Delta k)x}{2}\right]$$

describes the propagation of the packet formed from two superposed waves. The requirement of constant phase for the amplitude term leads to

$$U = \omega_0' = \left(\frac{d\omega}{dk}\right)_{k=k_0} \tag{13.116}$$

for the group velocity. As stated earlier, only if the medium is dispersive does U differ from the phase velocity V. To show this explicitly, we write Equation 13.116 as

$$\frac{1}{U} = \left(\frac{dk}{d\omega}\right)_0$$

where the subscript zero means "evaluated at $k = k_0$ or, equivalently, at $\omega = \omega_0$." Because $k = \omega/v$,

$$\frac{1}{U} = \left[\frac{d}{d\omega}\left(\frac{\omega}{v}\right)\right]_0 = \frac{v_0 - (\omega\, dv/d\omega)_0}{v_0^2}$$

Thus,

$$U = \frac{v_0}{1 - \frac{\omega_0}{v_0}\left(\frac{dv}{d\omega}\right)_0} \tag{13.117}$$

If the medium is nondispersive, $v = V = $ constant, so $dv/d\omega = 0$ (see Equation 13.91); hence $U = v_0 = V$.

The remaining quantity in Equation 13.115, $\exp[i(\omega_0 t - k_0 x)]$, varies rapidly with time; and if this were the only factor in Ψ, it would describe an infinite wave train oscillating at frequency ω_0 and traveling with phase velocity $V = \omega_0/k_0$.

We should note that an infinite train of waves of a given frequency cannot transmit a signal or carry information from one point to another. Such transmission can be accomplished only by starting and stopping the wave train and thereby impressing a signal on the wave—in other words, by forming a wave packet. As a consequence of this fact, it is the group velocity, not the phase velocity, that corresponds to the velocity at which a signal may be transmitted.*

P R O B L E M S

▷ **13-1.** Discuss the motion of a continuous string when the initial conditions are $\dot{q}(x, 0) = 0$, $q(x, 0) = A \sin(3\pi x/L)$. Resolve the solution into normal modes.

▷ **13-2.** Rework the problem in Example 13.1 in the event that the plucked point is a distance $L/3$ from one end. Comment on the nature of the allowed modes.

▷ **13-3.** Refer to Example 13.1. Show by a numerical calculation that the initial displacement of the string is well represented by the first three terms of the series in Equation 13.13. Sketch the shape of the string at intervals of time of $\frac{1}{8}$ of a period.

13-4. Discuss the motion of a string when the initial conditions are $q(x, 0) = 4x(L - x)/L^2$, $\dot{q}(x, 0) = 0$. Find the characteristic frequencies and calculate the amplitude of the nth mode.

▷ **13-5.** A string with no initial displacement is set into motion by being struck over a length $2s$ about its center. This center section is given an initial velocity v_0. Descibe the subsequent motion.

13-6. A string is set into motion by being struck at a point $L/4$ from one end by a triangular hammer. The initial velocity is greatest at $x = L/4$ and decreases linearly to zero at $x = 0$ and $x = L/2$. The region $L/2 \leq x \leq L$ is initially undisturbed. Determine the subsequent motion of the string. Why are the fourth, eighth, and related harmonics absent? How many decibels down from the fundamental are the second and third harmonics?

13-7. A string is pulled aside a distance h at a point $3L/7$ from one end. At a point $3L/7$ from the other end, the string is pulled aside a distance h in the opposite direction. Discuss the vibrations in terms of normal modes.

---

* The group velocity corresponds to the signal velocity only in nondispersive media (in which case the phase, group, and signal velocities are all equal) and in media of normal dispersion (in which case the phase velocity exceeds the group and signal velocities). In media with anomalous dispersion, the group velocity may exceed the signal velocity (and, in fact, may even become negative or infinite). We need only note here that a medium in which the wave number k is complex exhibits attenuation, and the dispersion is said to be *anomalous*. If k is real, there is no attenuation, and the dispersion is *normal*. What is called anomalous dispersion (due to a historical misconception) is, in fact, normal (i.e., frequent), and so-called normal dispersion is anomalous (i.e., rare). Dispersive effects are quite important in optical and electromagnetic phenomena.

Detailed analyses of the interrelationship among phase, group, and signal velocities were made by Arnold Sommerfeld and by Léon Brillouin in 1914. Translations of these papers are given in the book by Brillouin (Br60).

13-8. Compare, by plotting a graph, the characteristic frequencies ω_r as a function of the mode number r for a loaded string consisting of 3, 5, and 10 particles and for a continuous string with the same values of τ and $m/d = \rho$. Comment on the results.

13-9. In Example 13.2, the complementary solution (transient part) was omitted. If transient effects are included, what are the appropriate conditions for overdamped, critically damped, and underdamped motion? Find the displacement $q(x, t)$ that results when underdamped motion is included in Example 13.2 (assume that the motion is underdamped for all normal modes).

13-10. Consider the string of Example 13.1. Show that if the string is driven at an arbitrary point, none of the normal modes with nodes at the driving point will be excited.

13-11. When a particular driving force is applied to a string, it is observed that the string vibration is purely of the nth harmonic. Find the driving force.

13-12. Determine the complementary solution for Example 13.2.

13-13. Consider the simplified wave function

$$\Psi(x, t) = Ae^{i(\omega t - kx)}$$

Assume that ω and v are complex quantities and that k is real:

$$\omega = \alpha + i\beta$$
$$v = u + iw$$

Show that the wave is damped in time. Use the fact that $k^2 = \omega^2/v^2$ to obtain expressions for α and β in terms of u and w. Find the phase velocity for this case.

13-14. Consider an electrical transmission line that has a uniform inductance per unit length L and a uniform capacitance per unit length C. Show that an alternating current I in such a line obeys the wave equation

$$\frac{\partial^2 I}{\partial x^2} - LC\frac{\partial^2 I}{\partial t^2} = 0$$

so that the wave velocity is $v = 1/\sqrt{LC}$.

13-15. Consider the superposition of two infinitely long wave trains with almost the same frequencies but with different amplitudes. Show that the phenomenon of beats occurs but that the waves never beat to zero amplitude.

13-16. Consider a wave $g(x - vt)$ propagating in the $+x$-direction with velocity v. A rigid wall is placed at $x = x_0$. Describe the motion of the wave for $x < x_0$.

13-17. Treat the problem of wave propagation along a string loaded with particles of two different masses, m' and m'', which alternate in placement; that is,

$$m_j = \begin{cases} m', & \text{for } j \text{ even} \\ m'', & \text{for } j \text{ odd} \end{cases}$$

Show that the $\omega - k$ curve has two branches in this case, and show that there is attenuation for frequencies between the branches as well as for frequencies above the upper branch.

13-18. Sketch the phase velocity $V(k)$ and the group velocity $U(k)$ for the propagation of waves along a loaded string in the range of wave numbers $0 \le k \le \pi/d$. Show that $U(\pi/d) = 0$, whereas $V(\pi/d)$ does not vanish. What is the interpretation of this result in terms of the behavior of the waves?

13-19. Consider an infinitely long continuous string with linear mass density ρ_1 for $x < 0$ and for $x > L$, but density $\rho_2 > \rho_1$ for $0 < x < L$. If a wave train oscillating with an angular frequency ω is incident from the left on the high-density section of the string, find the reflected and transmitted intensities for the various portions of the string. Find a value of L that allows a maximum transmission through the high-density section. Discuss briefly the relationship of this problem to the application of nonreflective coatings to optical lenses.

13-20. Consider an infinitely long continuous string with tension τ. A mass M is attached to the string at $x = 0$. If a wave train with velocity ω/k is incident from the left, show that reflection and transmission occur at $x = 0$ and that the coefficients R and T are given by

$$R = \sin^2\theta, \qquad T = \cos^2\theta$$

where

$$\tan\theta = \frac{M\omega^2}{2k\tau}$$

Consider carefully the boundary condition on the derivatives of the wave functions at $x = 0$. What are the phase changes for the reflected and transmitted waves?

13-21. Consider a wave packet in which the amplitude distribution is given by

$$A(k) = \begin{cases} 1, & |k - k_0| < \Delta k \\ 0, & \text{otherwise} \end{cases}$$

Show that the wave function is

$$\Psi(x, t) = \frac{2 \sin[(\omega_0' t - x) \Delta k]}{\omega_0' t - x} e^{i(\omega_0 t - k_0 x)}$$

Sketch the shape of the wave packet (choose $t = 0$ for simplicity).

13-22. Consider a wave packet with a Gaussian amplitude distribution

$$A(k) = B \exp[-\sigma(k - k_0)^2]$$

where $2/\sqrt{\sigma}$ is equal to the $1/e$ width* of the packet. Using this function for $A(k)$, show that

$$\Psi(x, 0) = B \int_{-\infty}^{+\infty} \exp[-\sigma(k - k_0)^2] \exp(-ikx)\, dk$$

$$= B \sqrt{\frac{\pi}{\sigma}} \exp(-x^2/4\sigma) \exp(-ik_0 x)$$

Sketch the shape of this wave packet. Next, expand $\omega(k)$ in a Taylor series, retain the first two terms, and integrate the wave packet equation to obtain the general result

$$\Psi(x, t) = B \sqrt{\frac{\pi}{\sigma}} \exp[-(\omega_0' t - x)^2/4\sigma] \exp[i(\omega_0 t - k_0 x)]$$

Finally, take one additional term in the Taylor series expression of $\omega(k)$ and show that σ is now replaced by a complex quantity. Find the expression for the $1/e$ width of the packet as a function of time for this case and show that the packet moves with the same group velocity as before but spreads in width as it moves. Illustrate this result with a sketch.

---

* At the points $k = k_0 \pm 1/\sqrt{\sigma}$, the amplitude distribution is $1/e$ of its maximum value $A(k_0)$. Thus $2/\sqrt{\sigma}$ is the width of the curve at the $1/e$ height.

14

SPECIAL THEORY
OF RELATIVITY

14.1 INTRODUCTION

In Section 2.8, it was pointed out that the Newtonian idea of the complete separability of space and time and the concept of the absoluteness of time break down when they are subjected to critical analysis. The final overthrow of the Newtonian system as the ultimate description of dynamics was the result of several crucial experiments, culminating with the work of Michelson and Morley in 1881–1887. The results of these experiments indicated that the speed of light is independent of any relative uniform motion between source and observer. This fact, coupled with the finite speed of light, required a fundamental reorganization of the structure of dynamics. This was provided during the period 1904–1905 by H. Poincaré, H. A. Lorentz, and A. Einstein,* who formulated the **theory of relativity** in order to pro-

---

* Although Albert Einstein (1879–1955) is usually accorded the credit for the formulation of relativity theory (see, however, Wh53, Chapter 2), the basic *formalism* had been discovered by Poincaré and Lorentz by 1904. Einstein was unaware of some of this previous work at the time (1905) of the publication of his first paper on relativity. (Einstein's friends often remarked that "he read little, but thought much.") The important contribution of Einstein to special relativity theory was the replacement of the many *ad hoc* assumptions made by Lorentz and others with but two basic postulates from which all the results could be derived. [The question of precedence in relativity theory is discussed by G. Holton, *Am. J. Phys.* **28**, 627 (1960); see also Am63.] In addition, Einstein later provided the fundamental contribution to the formulation of the *general* theory of relativity in 1916. His first publication on a topic of importance in general relativity—speculations on the influence of gravity on light—was in 1907. It is interesting to note that Einstein's 1921 Nobel Prize was awarded, not for contributions to relativity theory, but for his work on the photoelectric effect.

vide a consistent description of the experimental facts. The basis of relativity theory is contained in two postulates:

I. *The laws of physical phenomena are the same in all inertial reference frames (that is, only the relative motion of inertial frames can be measured; the concept of motion relative to "absolute rest" is meaningless).*

II. *The velocity of light (in free space) is a universal constant, independent of any relative motion of the source and the observer.*

Using these postulates as a foundation, Einstein was able to construct a beautiful, logically precise theory. A wide variety of phenomena that take place at high velocity and cannot be interpreted in the Newtonian scheme are accurately described by relativity theory.

Postulate I, which Einstein called the *principle of relativity*, is the fundamental basis for the theory of relativity. Postulate II, the law of propagation of light, follows from Postulate I if we accept, as Einstein did, that Maxwell's equations are fundamental laws of physics. Maxwell's equations predict the speed of light in vacuum to be c, and Einstein believed this to be the case in all inertial reference frames.

We do not attempt here to give the experimental background for the theory of relativity; such information can be found in essentially every textbook on modern physics and in many others concerned with electrodynamics.* Rather, we simply accept as correct the above two postulates and work out some of their consequences for the area of mechanics.† The discussion here is limited to the case of **special relativity**, in which we consider only inertial reference frames, that is, frames that are in uniform motion with respect to one another. The more general treatment of accelerated reference frames is the subject of the **general theory of relativity.**

14.2 GALILEAN INVARIANCE

In Newtonian mechanics, the concepts of space and time are completely separable; furthermore, time is assumed to be an absolute quantity susceptible of precise definition independent of the reference frame. These assumptions lead to the invariance of the laws of mechanics under coordinate transformations of the following type. Consider two inertial reference frames K and K', which move along their x_1- and x_1'-axes with a uniform relative velocity v (Figure 14-1). The transformation of the coordinates of a point from one system to the other is clearly of the form

$$\left. \begin{array}{l} x_1' = x_1 - vt \\ x_2' = x_2 \\ x_3' = x_3 \end{array} \right\} \tag{14.1a}$$

---

* A particularly good discussion of the experimental necessity for relativity theory can be found in Panofsky and Phillips (Pa62, Chapter 15).

† Relativistic effects in electrodynamics are discussed in Marion and Heald (Ma80, Chapter 13).

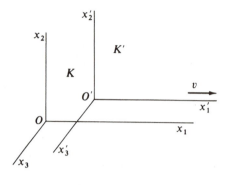

FIGURE 14-1

Also, we have

$$t' = t \tag{14.1b}$$

Equations 14.1 defines a **Galilean transformation**. Furthermore, the element of length in the two systems is the same and is given by

$$ds^2 = \sum_j dx_j^2$$
$$= \sum_j dx_j'^2 = ds'^2 \tag{14.2}$$

The fact that Newton's laws are invariant with respect to Galilean transformations is termed **the principle of Newtonian relativity** or **Galilean invariance**. Newton's equations of motion in the two systems are

$$F_j = m\ddot{x}_j$$
$$= m\ddot{x}_j' = F_j' \tag{14.3}$$

The form of the law of motion is then *invariant* to a Galilean transformation. The individual terms are not invariant, however, but they transform according to the same scheme and are said to be *covariant*.

We can easily show that the Galilean transformation is inconsistent with Postulate II. Consider a light pulse emanating from a flashbulb positioned in frame K'. The velocity transformation is found from Equation 14.1a, where we consider the light pulse only along x_1:

$$\dot{x}_1' = \dot{x}_1 - v \tag{14.4}$$

In system K', the velocity is measured as $\dot{x}_1' = c$; Equation 14.4 therefore indicates the speed of the light pulse to be $\dot{x}_1 = c + v$, clearly in violation of Postulate II.

14.3 LORENTZ TRANSFORMATION

The principle of Galilean invariance predicts that the velocity of light is different in two inertial reference frames that are in relative motion. This result is in contradiction to the second postulate of relativity. Therefore, a new transformation law that renders physical laws *relativistically* covariant must be found. Such a transformation law is the **Lorentz transformation**. The original use of the Lorentz transformation preceded the development of Einsteinian relativity theory,* but it also follows from the basic postulates of relativity; we derive it on this basis in the following discussion.

If a light pulse from a flashbulb is emitted from the common origin of the systems K and K' (see Figure 14-1) when they are coincident, then according to Postulate II, the wavefronts observed in the two systems must be described† by

$$
\left.
\begin{aligned}
\sum_{j=1}^{3} x_j^2 - c^2 t^2 = 0 \\
\sum_{j=1}^{3} x_j'^2 - c^2 t'^2 = 0
\end{aligned}
\right\}
\tag{14.5}
$$

We can already see that Equations 14.5, which are consistent with the two postulates of the theory of relativity, cannot be reconciled with the Galilean transformations of Equations 14.1. The Galilean transformation allows a spherical light wavefront in one system but requires the center of the spherical wavefront in the second system to move at velocity v with respect to the first system. The interpretation of Equations 14.5, according to Postulate II, is that each observer believes that his spherical wavefront has its center fixed at his own coordinate origin as the wavefront expands.

We are faced with a quandary. We must abandon either the two relativity postulates or the Galilean transformation. Much experimental evidence, including the Michelson-Morley experiment and the aberration of starlight, requires the two postulates. However, the belief in the Galilean transformation is entrenched in our minds by our everyday experience. The Galilean transformation had produced satisfactory results, including those of the preceding chapters of this book, for centuries. Einstein's great contribution was to realize that the Galilean transformation was *approximately* correct, but that we needed to reexamine our concepts of space and time.

Notice that we do not assume $t = t'$ in Equations 14.5. Each system, K and K', has its own clocks, and we assume that a clock may be located at any point in

---

* The transformation was originally postulated by Hendrik Anton Lorentz (1853–1928) in 1904 to explain certain electromagnetic phenomena, but the formulas had been set up as early as 1900 by J. J. Larmor. The complete generality of the transformation was not realized until Einstein *derived* the result. W. Voigt was actually the first to use the equations in a discussion of oscillatory phenomena in 1887.

† See Appendix G.

space. These clocks are all identical, run the same way, and are synchronized. Because the flashbulb goes off when the origins are coincident and the systems move only in the x_1-direction with respect to each other, by direct observation we have

$$\left. \begin{array}{l} x_2' = x_2 \\ x_3' = x_3 \end{array} \right\} \tag{14.6}$$

At time $t = t' = 0$, when the flashbulb goes off, the motion of the origin $0'$ of K' is measured in K to be

$$x_1 - vt = 0 \tag{14.7}$$

and in system K', the motion of O' is

$$x_1' = 0 \tag{14.8}$$

At time $t = t' = 0$ we have $x_1' = x_1 - vt$, but we know that Equation 14.1a is incorrect. Let us assume the next simplest transformation, namely,

$$x_1' = \gamma(x_1 - vt) \tag{14.9}$$

where γ is some constant that may depend on v and some constants, but not on the coordinates x_1, x_1', t, or t'. Equation 14.9 is a linear equation and assures us that each event in K corresponds to one and only one event in K'. This additional assumption in our derivation will be vindicated if we can produce a transformation that is consistent with *all* the experimental results. Notice that γ must normally be very close to 1 to be consistent with the classical results discussed in earlier chapters.

We can use the preceding arguments to describe the motion of the origin O of system K in both K and K' to also determine

$$x_1 = \gamma'(x_1' + vt') \tag{14.10}$$

where we only have to change the relative velocities of the two systems.

Postulate I demands that the laws of physics be the same in both reference systems such that $\gamma = \gamma'$. By substituting x_1' from Equation 14.9 into Equation 14.10, we can solve the remaining equation for t':

$$t' = \gamma t + \frac{x_1}{\gamma v}(1 - \gamma^2) \tag{14.11}$$

Postulate II demands that the speed of light be measured to be the same in both systems. Therefore, in both systems we have similar equations for the position of the flashbulb light pulse:

$$\left. \begin{array}{l} x_1 = ct \\ x_1' = ct' \end{array} \right\} \tag{14.12}$$

Algebraic manipulation of Equations 14.9–14.12 gives (see Problem 14-1)

$$\gamma = \frac{1}{\sqrt{1 - v^2/c^2}} \tag{14.13}$$

The complete transformation equations can now be written as

$$x_1' = \frac{x_1 - vt}{\sqrt{1 - v^2/c^2}}$$

$$x_2' = x_2$$

$$x_3' = x_3$$

$$t' = \frac{t - \frac{vx_1}{c^2}}{\sqrt{1 - v^2/c^2}}$$

(14.14)

These equations are known as the Lorentz (or Lorentz-Einstein) transformation in honor of the Dutch physicist H. A. Lorentz, who first showed that the equations are necessary so that the laws of electromagnetism have the same form in all inertial reference frames. Einstein showed that these equations are required for all the laws of physics.

The inverse transformation can easily be obtained by replacing v by $-v$ and exchanging primed and unprimed quantities in Equations 14.14.

$$x_1 = \frac{x_1' + vt'}{\sqrt{1 - v^2/c^2}}$$

$$x_2 = x_2'$$

$$x_3 = x_3'$$

$$t = \frac{t' + \frac{vx_1'}{c^2}}{\sqrt{1 - v^2/c^2}}$$

(14.15)

As required, these equations reduce to the Galilean equations (Equations 14.1) when $v \to 0$ (or when $c \to \infty$).

In electrodynamics, the fields propagate with the speed of light, so Galilean transformations are never allowed. Indeed, the fact that the electrodynamic field equations (**Maxwell's equations**) are not covariant to Galilean transformations was a main factor in the realization of the need for a new theory. It seems rather extraordinary that Maxwell's equations, which are a complete set of equations for the electromagnetic field and are *covariant to Lorentz transformations*, were deduced from experiment long before the advent of relativity theory.

The velocities measured in each of the systems are denoted by u.

$$u_i = \frac{dx_i}{dt}$$

$$u_i' = \frac{dx_i'}{dt'}$$

(14.16)

Using Equations 14.14, we determine

$$u_1' = \frac{dx_1'}{dt'} = \frac{dx_1 - v\,dt}{dt - \frac{v}{c^2}\,dx_1}$$

$$u_1' = \frac{u_1 - v}{1 - \frac{u_1 v}{c^2}} \qquad\qquad (14.17a)$$

Similarly, we determine

$$u_2' = \frac{u_2}{\gamma\left(1 - \frac{u_1 v}{c^2}\right)} \qquad\qquad (14.17b)$$

$$u_3' = \frac{u_3}{\gamma\left(1 - \frac{u_1 v}{c^2}\right)} \qquad\qquad (14.17c)$$

Now we can determine whether Postulate II is satisfied directly. An observer in system K measures the speed of the light pulse from the flashbulb to be $u_1 = c$ in the x_1-direction. From Equation 14.17a, an observer in K' measures

$$u_1' = \frac{c - v}{1 - \frac{v}{c}} = c\left(\frac{c - v}{c - v}\right) = c$$

as required by Postulate II, independent of the relative system speed v.

E X A M P L E   14.1 --

Determine the relativistic length contraction* using the Lorentz transformation.

Solution: Consider a rod of length l lying along the x_1-axis of an inertial frame K. An observer in system K' moving with uniform speed v along the x_1-axis (as in Figure 14-1) measures the length of the rod in the observer's own coordinate system by determining *at a given instant of time t'* the difference in the coordinates of the ends of the rod, $x_1'(2) - x_1'(1)$. According to the transformation equations (Equations 14.14),

$$x_1'(2) - x_1'(1) = \frac{[x_1(2) - x_1(1)] - v[t(2) - t(1)]}{\sqrt{1 - v^2/c^2}} \qquad (14.18)$$

---

* The contraction of length in the direction of motion was proposed by G. F. FitzGerald (1851–1901) in 1892 as a possible explanation of the Michelson-Morley ether-drift experiment. This hypothesis was adopted almost immediately by Lorentz, who proceeded to apply it in his theory of electrodynamics.

where $x_1(2) - x_1(1) = l$. Note that times $t(2)$ and $t(1)$ are the times in the K system at which the observations are made; they do not correspond to the instants in K' at which the observer measures the rod. In fact, because $t'(2) = t'(1)$, Equations 14.14 give

$$t(2) - t(1) = [x_1(2) - x_1(1)]\frac{v}{c^2}$$

The length l' as measured in the K' system is therefore

$$l' = x_1'(2) - x_1'(1)$$

Equation 14.18 now becomes

length contraction $\boxed{l' = l\sqrt{1 - v^2/c^2}}$ **(14.19)**

Thus, to an observer in motion relative to an object, the dimensions of objects are contracted by a factor $\sqrt{1 - \beta^2}$ in the direction of motion, in which $\beta \equiv v/c$.

An interesting consequence of the FitzGerald-Lorentz contraction of length was reported in 1959 by James Terrell.* Consider a cube of side l moving with uniform velocity **v** with respect to an observer some distance away. Figure 14-2a

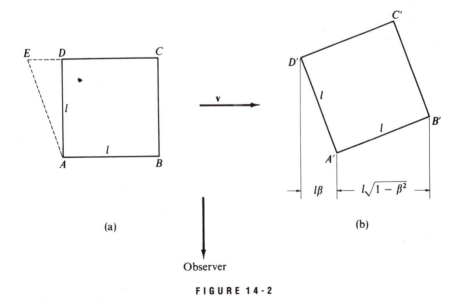

(a)

(b)

Observer

FIGURE 14-2

---

* J. Terrell, *Phys. Rev.* **116**, 1041 (1959).

shows the projection of the cube on the plane containing the velocity vector **v** and the observer. The cube moves with its side AB perpendicular to the observer's line of sight. We wish to determine what the observer "sees"; that is, at a given instant of time in the observer's rest frame, we wish to determine the relative orientation of the corners A, B, C, and D. The traditional view (which went unquestioned for more than 50 years!) was that the only effect is a foreshortening of the sides AB and CD such that the observer sees a distorted cube of height l but of length $l\sqrt{1 - \beta^2}$. Terrell pointed out that this interpretation overlooks certain facts: For light from corners A and D to reach the observer at the same instant, the light from D, which must travel a distance l farther than that from A, must have been emitted when corner D was at position E. The length DE is equal to $(l/c)v = l\beta$. Therefore, the observer sees not only face AB, which is perpendicular to the line of sight, but also face AD, which is *parallel* to the line of sight. Also, the length of the side AB is foreshortened in the normal way to $l\sqrt{1 - \beta^2}$. The net result (Figure 14-2b) corresponds exactly to the view the observer would have if the cube were rotated through an angle $\sin^{-1}\beta$. Therefore, the cube is not distorted; it undergoes an *apparent* rotation. Similarly, the customary statement* that a moving sphere appears as an ellipsoid is incorrect; it appears still as a sphere.[†] Computers can be used to show extremely interesting results of the type we have been discussing (Figure 14-3).

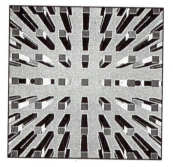 

FIGURE 14-3 The array of rectangular bars is seen from above at rest in the figure on the left. In the right figure the bars are moving to the right with $v = 0.9c$. The bars appear to contract and rotate. *Quoted from P.-K. Hsiung and R.H.P. Dunn, see Science News 137, 232 (1990).*

EXAMPLE 14.2 -

Use the Lorentz transformation to determine the time dilation effect.

Solution: Consider a clock fixed at a certain position (x_1) in the K system that produces signal indications with the interval

$$\Delta t = t(2) - t(1)$$

---

* See, for example, Joos and Freeman (Jo50, p. 242).

[†] An interesting discussion of apparent rotations at high velocity is given by V. F. Weisskopf, *Phys. Today* **13**, no. 9, 24 (1960), reprinted in Am63.

According to the Lorentz transformation (Equations 14.14), an observer in the moving system K' measures a time interval $\Delta t'$ (on the same clock) of

$$\Delta t' = t'(2) - t'(1)$$

$$= \frac{\left[t(2) - \dfrac{vx_1(2)}{c^2}\right] - \left[t(1) - \dfrac{vx_1(1)}{c^2}\right]}{\sqrt{1 - v^2/c^2}}$$

Because $x_1(2) = x_1(1)$ and because the clock is fixed in the K system, we have

$$\Delta t' = \frac{t(2) - t(1)}{\sqrt{1 - v^2/c^2}}$$

$$\boxed{\Delta t' = \frac{\Delta t}{\sqrt{1 - v^2/c^2}}} \qquad \textbf{(14.20)}$$

Thus, to an observer in motion relative to the clock, the time intervals appear to be lengthened. This is the origin of the phrase "moving clocks run more slowly." Because the measured time interval on the moving clock is lengthened, the clock actually ticks slower. Notice that the clock is fixed in the K system, $x_1(1) = x_1(2)$, but not in the K' system, $x_1'(1) \neq x_1'(2)$.

The argument in the previous example can be reversed and the clock fixed in the K' system. The same result occurs; moving clocks run slower. The effect is called **time dilation**. It is important to note that the physical system is unimportant. The same effect occurs for a tuning fork, an hourglass, a quartz crystal, and a heartbeat. The problem is one of simultaneity. Events simultaneous in one system may not be simultaneous in another one moving with respect to the first. The same clock may be viewed from n different reference frames and found to be running at n different rates, simultaneously. Space and time are intricately interwoven. We shall return to this point later.

The time measured on a clock fixed in a system present at two events is called the **proper time** and given the symbol τ. For example, $\Delta t = \Delta \tau$ when a clock fixed in system K is present for both events, $x(1)$ and $x(2)$. Equation 14.20 becomes

$$\Delta t' = \gamma \Delta \tau \qquad \textbf{(14.21)}$$

Notice that the proper time is always the minimum measurable time difference between two events. Moving observers always measure a longer time period.

14.4 EXPERIMENTAL VERIFICATION OF THE SPECIAL THEORY

The special theory of relativity explains the difficulties existing before 1900 with optics and electromagnetism. For example, the problems with stellar aberration and the Michelson-Morley experiment are solved by assuming no ether but requiring the Lorentz transformation.

But what about the new startling predictions of the special theory—length contraction and time dilation? These topics are addressed every day in the accelerator laboratories of nuclear and particle physics, where particles are accelerated to speeds close to that of light, and relativity must be considered. Other experiments can be performed with natural phenomena. We examine two of these.

Muon Decay

When cosmic rays enter the Earth's outer atmosphere, they interact with particles and create cosmic showers. Many of the particles in these showers are π-mesons, which decay to other particles called muons. Muons are also unstable and decay according to the radioactive decay law, $N = N_0 \exp(-0.693 \, t/t_{1/2})$, where N_0 and N are the number of muons at time $t = 0$ and t, respectively, and $t_{1/2}$ is the half-life. However, enough muons reach the Earth's surface that we can detect them easily.

Let us assume that we mount a detector on top of a 2,000-m mountain and count the number of muons traveling at a speed near $v = 0.98c$. Over a given period of time, we count 10^3 muons. The half-life of muons is known to be 1.52 $\times 10^{-6}$ s in their own rest frame (system K'). We move our detector to sea level and measure the number of muons (having $v = 0.98c$) detected during an equal period of time. What do we expect?

Determined classically, muons traveling at a speed of $0.98c$ cover the 2,000 m in 6.8×10^{-6} s, and 45 muons should survive the flight from 2,000 m to sea level according to the radioactive decay law. But experimental measurement indicates that 542 muons survive, a factor of 12 more.

This phenomenon must be treated relativistically. The decaying muons are moving at a high speed relative to the experimenters fixed on the Earth. We therefore observe the muons' clock to be running slower. In the muons' rest frame, the time period of the muons' flight is not $\Delta t = 6.8 \times 10^{-6}$ s but rather $\Delta t/\gamma$. For $v = 0.98c$, $\gamma = 5$, so we measure the flight time on a clock at rest in the muons' system to be 1.36×10^{-6} s. The radioactive decay law predicts that 538 muons survive, much closer to our measurement and within the experimental uncertainties. An experiment similar to this has verified the time dilation prediction.*

E X A M P L E **14.3**

Examine the muon decay just discussed from the perspective of an observer moving with the muon.

---

* The experiment was reported by B. Rossi and D. B. Hall in the *Phys. Rev.*, **59**, 223 (1941). A film entitled "Time Dilation—An Experiment with μ-Mesons" by D. H. Frisch and J. H. Smith is available from the Education Development Center; Newton, Mass. See also D. H. Frisch and J. H. Smith, *Am. J. Phys.*, **31**, 342 (1963).

Solution: The half-life of the muon according to its own clock is 1.52×10^{-6}s. But an observer moving with the muon would not measure the distance from the top of the mountain to sea level to be 2,000 m. According to that observer, the distance would be only 400 m. At a speed of $0.98c$, it takes the muon only 1.36×10^{-6} s to travel the 400 m. An observer in the muon system would predict 538 muons to survive, in agreement with an observer on the Earth.

Muon decay is an excellent example of a natural phenomenon that can be described in two systems moving with respect to each other. One observer sees time dilated and the other observer sees length contracted. Each, however, predicts a result in agreement with experiment.

Atomic Clock Time Measurements

An even more direct confirmation of special relativity was reported by two American physicists, J. C. Hafele and Richard E. Keating, in 1972.* They used four extremely accurate cesium atomic clocks. Two clocks were flown on regularly scheduled commercial jet airplanes around the world, one eastward and one westward; the other two reference clocks stayed fixed on Earth at the U.S. Naval Observatory. A well-defined, hyperfine transition in the ground state of the ^{133}Cs atom has a frequency of 9,192,631,770 Hz and can be used as an accurate measurement of a time period.

The time measured on the two moving clocks was compared with that of the two reference clocks. The eastward trip lasted 65.4 hours with 41.2 flight hours. The westward trip, a week later, took 80.3 hours with 48.6 flight hours. The predictions are complicated by the rapid rotation of the Earth and by a gravitational effect from the general theory of relativity.

We can gain some insight to the expected effect by neglecting the corrections and calculating the time difference as if the Earth were not rotating. The circumference of the Earth is about 4×10^7 m, and a typical jet airplane speed is almost 300 m/s. A clock fixed on the ground measures a flight time T_0 of

$$T_0 = \frac{4 \times 10^7 \text{m}}{300 \text{ m/s}} = 1.33 \times 10^5 \text{s} \ (\approx 37 \text{ hr}) \tag{14.22}$$

Because the moving clock runs more slowly, the observer on the Earth would say that the moving clock measures only $T = T_0\sqrt{1 - \beta^2}$. The time difference is

$$\Delta T = T_0 - T = T_0(1 - \sqrt{1 - \beta^2})$$

$$\approx \frac{1}{2}\beta^2 T_0 \tag{14.23}$$

---

* See J. C. Hafele and Richard E. Keating, *Science*, **177**, 166–170 (1972).

where only the first and second terms of the power series expansion for $\sqrt{1 - \beta^2}$ are kept because β^2 is so small.

$$\Delta T = \frac{1}{2} \left(\frac{300 \text{ m/s}}{3 \times 10^8 \text{ m/s}} \right)^2 (1.33 \times 10^5 \text{ s})$$

$$= 6.65 \times 10^{-8} \text{ s} = 66.5 \text{ ns} \qquad (14.24)$$

This time difference is greater than the uncertainty of the measurement. Notice that in this case, the clock left on Earth actually measures more time in seconds than the moving clock. This seems at variance with our earlier comments (see Equation 14.21 and discussion). But the time period referred to in Equation 14.21 is the time between two ticks, in this case, a transition in ^{133}Cs, which we measure in seconds. It is easy to remember that moving clocks run more slowly, so that in seconds the measured time difference involves fewer ticks and, according to the definition of a second, fewer seconds.

The actual predictions and observations for the time difference are

Travel	Predicted	Observed
Eastward	−40 ± 23 ns	−59 ± 10 ns
Westward	275 ± 21 ns	273 ± 7 ns

Again, the special theory of relativity is verified within the experimental uncertainties. A negative sign indicates that the time on the moving clock is less than the Earth reference clock. The moving clocks lost time (ran slower) during the eastward trip and gained time (ran faster) during the westward trip. This difference is caused by the rotation of the Earth, indicating that the flying clocks actually ticked faster or slower than the reference clocks on Earth. The overall positive time difference is a result of the gravitational potential effect (which we do not discuss here).

We have only briefly described two of the many experiments that have verified the special theory of relativity. There are no known experimental measurements that are inconsistent with the special theory of relativity. Einstein's work in this regard has so far withstood the test of time.

14.5   RELATIVISTIC DOPPLER EFFECT

The Doppler effect in sound is represented by an increased pitch of sound as a source approaches a receiver and a decrease of pitch as the source recedes. The change in frequency of the sound depends on whether the source or receiver is moving. This effect seems to violate Postulate I of the theory of relativity until we realize that there is a special frame for sound waves because there is a medium (e.g., air or water) in which the waves travel. In the case of light, however, there is no such medium. Only relative motion of source and receiver is meaningful in

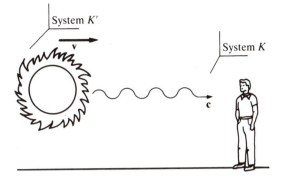

(a) Source and receiver approaching

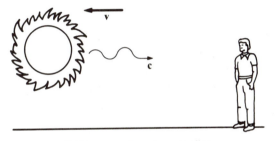

(b) Source and receiver receding

FIGURE 14-4

this context, and we should therefore expect some differences in the relativistic Doppler effect for light from the normal Doppler effect of sound.

Consider a source of light (e.g., a star) and a receiver approaching one another with relative speed v (Figure 14-4a). First, consider the receiver fixed in system K and the light source in system K' moving toward the receiver with speed v. During time Δt as measured by the receiver, the source emits n waves. During that time Δt, the total distance between the front and rear of the waves is

$$\text{length of wave train} = c\Delta t - v\Delta t \qquad (14.25)$$

The wavelength is then

$$\lambda = \frac{c\Delta t - v\Delta t}{n} \qquad (14.26)$$

and the frequency is

$$\nu = \frac{c}{\lambda} = \frac{cn}{c\Delta t - v\Delta t} \qquad (14.27)$$

According to the source, it emits n waves of frequency ν_0 during the proper time $\Delta t'$:

$$n = \nu_0 \Delta t' \qquad (14.28)$$

This proper time $\Delta t'$ measured on a clock in the source system is related to the time Δt measured on a clock fixed in system K of the receiver by

$$\Delta t' = \frac{\Delta t}{\gamma} \tag{14.29}$$

The clock moving with the source measures the proper time, because it is present at both the beginning and end of the waves.

Substituting Equation 14.29 into Equation 14.28, which in turn is substituted for n in Equation 14.27, gives

$$\nu = \frac{1}{(1 - v/c)} \frac{\nu_0}{\gamma}$$

$$= \frac{\sqrt{1 - v^2/c^2}}{1 - v/c} \nu_0 \tag{14.30}$$

which can be written as

$$\nu = \frac{\sqrt{1 + \beta}}{\sqrt{1 - \beta}} \nu_0 \qquad \text{source and receiver approaching} \tag{14.31}$$

It is left for the reader (Problem 14-14) to show that Equation 14.31 is also valid when the source is fixed and the receiver approaches it with speed v.

Next, we consider the case in which the source and receiver recede from each other with velocity v (Figure 14-4b). The derivation is similar to the one just presented—with one small exception. In Equation 14.25, the distance between the beginning and end of the waves becomes

$$\text{length of wave train} = c\,\Delta t + v\,\Delta t \tag{14.32}$$

This change in sign is propagated through Equations 14.30 and 14.31, giving

$$\nu = \frac{\sqrt{1 - v^2/c^2}}{1 + v/c} \nu_0$$

$$\nu = \frac{\sqrt{1 - \beta}}{\sqrt{1 + \beta}} \nu_0 \qquad \text{source and receiver receding} \tag{14.33}$$

Equations 14.31 and 14.33 can be combined into one equation,

$$\nu = \frac{\sqrt{1 + \beta}}{\sqrt{1 - \beta}} \nu_0 \qquad \text{relativistic Doppler effect} \tag{14.34}$$

if we agree to use a $+$ sign for β ($+ v/c$) when the source and receiver are approaching each other and a $-$ sign for β when they are receding.

The relativistic Doppler effect is important in astronomy. Equation 14.34 indicates that, if the source is receding at high speed from an observer, then a lower frequency (or longer wavelength) is observed for certain spectral lines or characteristic frequencies. This is the origin of the term *red shift*; the wavelengths of visible light are shifted toward longer wavelengths (red) if the source is receding from us. Astronomical observations indicate that the universe is expanding. The farther away a star is, the faster it appears to be moving away (or the greater its red shift). These data are consistent with the "big bang" origin of the universe, which is estimated to have occurred some 15 billion years ago.

EXAMPLE **14.4** ---

During a spaceflight to a distant star, an astronaut and her twin brother on the Earth send radio signals to each other at annual intervals. What is the frequency of the radio signals each twin receives from the other during the flight to the star if the astronaut is moving at $v = 0.8c$? What is the frequency during the return flight at the same speed?

Solution: We use Equation 14.34 to determine the frequency of radio signals that each receives from the other. The frequency $v_0 = 1$ signal/year. On the leg of the trip away from the earth, $\beta = -0.8$ and Equation 14.34 gives

$$v = \frac{\sqrt{1 - 0.8}}{\sqrt{1 + 0.8}} \, v_0$$

$$= \frac{v_0}{3}$$

The radio signals are received one every 3 years.

On the return trip, however, $\beta = +0.8$ and Equation 14.34 gives $v = 3v_0$, so the radio signals are received every 4 months. In this way, the twin on the Earth can monitor the progress of his astronaut twin.

--- ●

14.6 TWIN PARADOX

Consider twins who choose different career paths. Mary becomes an astronaut, and Frank decides to be a stockbroker. At age 30, Mary leaves on a mission to a planet in a nearby star's system. Mary will have to travel at a high speed to reach the planet and return. According to Frank, Mary's biological clock will tick more slowly during her trip, so she will age more slowly. He expects Mary to look and appear younger than he does when she returns. According to Mary, however, Frank will appear to be moving rapidly with respect to her system, and she thinks Frank will be younger when she returns. This is the paradox. Which twin, if either, is younger when Mary (the moving twin) returns to Earth where Frank (the fixed twin) has remained? Because the two expectations are so contradictory, doesn't Nature have a way to prove they will be the same age?

This paradox has existed almost since Einstein first published his special theory of relativity. Variations of the argument have been presented many times. The correct answer is that Mary, the astronaut, will return younger than her twin brother, Frank, who remains busy on Wall Street. The correct analysis is as follows. According to Frank, Mary's spaceship blasts off and quickly reaches a coasting speed of $v = 0.8c$, travels a distance of 8 ly (ly = a light year, the distance light travels in 1 year) to the planet, and quickly decelerates for a short visit to the planet. The acceleration and deceleration times are negligible compared with the total travel time of 10 years to the planet. The return trip also takes 10 years, so on Mary's return to Earth, Frank will be $30 + 10 + 10 = 50$ years old. Frank calculates that Mary's clock is ticking slower and that each leg of the trip takes only $10\sqrt{1 - 0.8^2} = 6$ years. Mary therefore is only $30 + 6 + 6 = 42$ years old when she returns. Frank's clock is (almost) in an inertial system.

When Mary performs the time measurements on her clock, they may be invalid according to the special theory because her system is not in an inertial frame of reference moving at a constant speed with respect to the Earth. She accelerates and decelerates at both the Earth and the planet, and to make valid time measurements to compare with Frank's clock, she must account for this acceleration and deceleration. The instantaneous rate of Mary's clock is still given by Equation 14.20, because the instantaneous rate is determined by the instantaneous speed v.* Thus, there is no paradox if we obey the two postulates of the special theory. It is also clear which twin is in the inertial frame of reference. Mary will actually feel the forces of acceleration and deceleration. Frank feels no such forces. When Mary returns home, her twin brother has invested her 20 years of salary, making her a rich woman at the young age of 42. She was paid a 20-year salary for a job that took her only 12 years!

E X A M P L E 14.5 -

Mary and Frank send radio signals to each other at 1-year intervals after she leaves the Earth. Analyze the times of receipt of the radio messages.

Solution: In Example 14.4, we calculated that such radio signals are received every 3 years on the trip out and every $\frac{1}{3}$ year on the trip back. First, we examine the signals Mary receives from Frank. During the 6-year trip to the planet, Mary receives only two radio messages, but on the 6-year return trip, she receives eighteen signals, so she correctly concludes that her twin brother Frank has aged 20 years and is now 50 years old.

In Frank's system, Mary's trip to the planet takes 10 years. By the time Mary reaches the planet, Frank receives 10/3 signals (i.e., three signals plus one-third of the time to the next one). However, Frank continues to receive a signal every 3 years for the 8 years it takes the last signal Mary sends when she reaches the planet

---

* See the clock hypothesis of W. Rindler (Ri82, p. 31).

to travel to Frank. Thus, Frank receives signals every 3 years for 8 more years (total of 18 years) for a total of six radio signals from the period of travel to the planet. Frank has no way of knowing that Mary has stopped and turned around until the radio message, which takes 8 years, is received. Of the remaining 2 years of Mary's journey according to Frank ($20 - 18 = 2$), Frank receives signals every $\frac{1}{3}$ year, or six more signals. Frank correctly determines that Mary has aged $6 + 6 = 12$ years during her journey because he receives a total of twelve signals.

Thus, both twins agree about their own ages and about each other's. Mary is 42 and Frank is 50 years old.

14.7   RELATIVISTIC MOMENTUM

Newton's Second Law, $\mathbf{F} = d\mathbf{p}/dt$, is covariant under a Galilean transformation. Therefore, we do not expect it to keep its form under a Lorentz transformation. We can foresee difficulties with Newton's laws and the conservation laws unless we make some necessary changes. According to Newton's Second Law, for example, an acceleration at high speeds might cause a particle's velocity to exceed c, an impossible condition according to the special theory of relativity.

We begin by examining the conservation of linear momentum in a force-free (no external forces) collision. There are no accelerations. Observer A at rest in system K holds a ball of mass m, as does observer B in system K' moving to the right with relative speed v with respect to system K, as in Figure 14-1. The two observers throw their (identical) balls along their respective x_2-axes, which results in a perfectly elastic collision. The collision, according to observers in the two systems, is shown in Figure 14-5. Each observer measures the speed of his or her ball to be u_0.

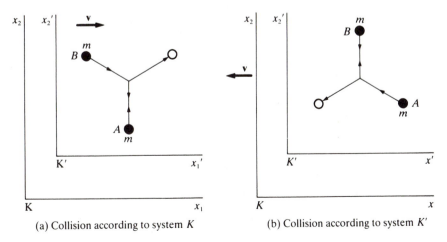

(a) Collision according to system K (b) Collision according to system K'

FIGURE 14-5

We first examine the conservation of momentum according to system K. The velocity of the ball thrown by observer A has components

$$\left.\begin{array}{l} u_{A1} = 0 \\ u_{A2} = u_0 \end{array}\right\} \qquad (14.35)$$

The momentum of ball A is in the x_2-direction:

$$p_{A2} = mu_0 \qquad (14.36)$$

The collision is perfectly elastic, so the ball returns down with speed u_0. The change in momentum observed in system K is

$$\Delta p_{A2} = 2mu_0 \qquad (14.37)$$

Does Equation 14.37 also represent the change in momentum of the ball thrown by observer B in the moving system K'? We use the inverse velocity transformation of Equations 14.17 (i.e., we interchange primes and unprimes and let $v \rightarrow -v$) to determine

$$\left.\begin{array}{l} u_{B1} = v \\ u_{B2} = -u_0\sqrt{1 - v^2/c^2} \end{array}\right\} \qquad (14.38)$$

where $u'_{B1} = 0$ and $u'_{B2} = -u_0$. The momentum of ball B and its change in momentum during the collision become

$$p_{B2} = -mu_0\sqrt{1 - v^2/c^2} \qquad (14.39)$$

$$\Delta p_{B2} = -2mu_0\sqrt{1 - v^2/c^2} \qquad (14.40)$$

Equations 14.37 and 14.40 do not add to zero: *Linear momentum is not conserved according to the special theory if we use the conventions for momentum of classical physics.* Rather than abandoning the law of conservation of momentum, we look for a solution that allows us to retain both it and Newton's Second Law.

As we did for the Lorentz transformation, we assume the simplest possible change. We assume that the classical form of momentum $m\mathbf{u}$ is multiplied by a constant that may depend on speed $k(u)$:

$$\mathbf{p} = k(u)m\mathbf{u} \qquad (14.41)$$

In Example 14.6, we show that the value

$$k(u) = \frac{1}{\sqrt{1 - u^2/c^2}} \qquad (14.42)$$

allows us to retain the conservation of linear momentum. Notice that the *form* of Equation 14.42 is the same as that found for the Lorentz transformation. In fact, the constant $k(u)$ is given the same label: γ. However, this γ contains the speed of the particle u, whereas the Lorentz transformation contains the relative speed v between the two inertial reference frames. This distinction must be kept in mind; it often causes confusion.

We can make a plausible calculation for the relativistic momentum if we use the proper time τ (see Equation 14.21) rather than the normal time t. In this case,

$$\mathbf{p} = m\frac{d\mathbf{x}}{d\tau} = m\frac{d\mathbf{x}}{dt}\frac{dt}{d\tau} \tag{14.43}$$

$$= m\frac{d\mathbf{x}}{dt}\frac{1}{\sqrt{1 - u^2/c^2}} \tag{14.44}$$

$$\boxed{\mathbf{p} = \frac{m\mathbf{u}}{\sqrt{1 - u^2/c^2}} = \gamma m\mathbf{u}} \quad \text{relativistic momentum} \tag{14.45}$$

where we retain $\mathbf{u} = d\mathbf{x}/dt$ as used classically. Although all observers do not agree as to $d\mathbf{x}/dt$, they do agree as to $d\mathbf{x}/d\tau$, where the proper time $d\tau$ is measured by the moving object itself. The relation $dt/d\tau$ is obtained from Equation 14.21, where the speed u has been used in γ to represent the speed of a reference frame fixed in the object that is moving with respect to a fixed frame.

Equation 14.45 is our new definition of momentum, called **relativistic momentum.** Notice that it reduces to the classical result for small values of u/c. It was fashionable in past years to call the mass in Equation 14.45 the **rest mass** m_0 and to call the term

$$m = \frac{m_0}{\sqrt{1 - u^2/c^2}} \quad \left(\begin{array}{c} \textbf{old-fashioned} \\ \textbf{notation} \end{array}\right) \tag{14.46}$$

the **relativistic mass.** The term *rest mass* resulted from Equation 14.46 when $u = 0$, and the classical form of momentum was thus retained: $\mathbf{p} = m\mathbf{u}$. Scientists spoke of the mass increasing at high speeds. We prefer to keep the concept of mass as an invariant, intrinsic property of an object. The use of the two terms *relativistic* and *rest mass* is now considered old-fashioned. *We therefore always refer to the mass m, which is the same as the rest mass.* The use of relativistic mass often leads to mistakes when using classical expressions.

E X A M P L E   14.6 --

Show that linear momentum is conserved in the x_2-direction for the collision shown in Figure 14-5 if relativistic momentum is used.

Solution: We can modify the classical expressions for momentum already obtained for the two balls. The momentum for ball A becomes (from Equation 14.36)

$$p_{A2} = \frac{mu_0}{\sqrt{1 - u_0^2/c^2}} \tag{14.47}$$

and

$$\Delta p_{A2} = \frac{2mu_0}{\sqrt{1 - u_0^2/c^2}} \tag{14.48}$$

Before modifying Equation 14.39 for the momentum of ball B, we must first find the speed of ball B as measured in system K. We use Equations 14.38 to determine

$$u_B = \sqrt{u_{B1}^2 + u_{B2}^2}$$

$$= \sqrt{v^2 + u_0^2(1 - v^2/c^2)} \tag{14.49}$$

The momentum p_{B2} is found by modifying Equation 14.39:

$$p_{B2} = - mu_0\gamma \sqrt{1 - v^2/c^2}$$

where

$$\gamma = \frac{1}{\sqrt{1 - u_B^2/c^2}}$$

$$p_{B2} = \frac{- mu_0\sqrt{1 - v^2/c^2}}{\sqrt{1 - u_B^2/c^2}} \tag{14.50}$$

Using u_B from Equation 14.49 gives

$$p_{B2} = \frac{- mu_0\sqrt{1 - v^2/c^2}}{\sqrt{(1 - u_0^2/c^2)(1 - v^2/c^2)}}$$

$$= \frac{- mu_0}{\sqrt{1 - u_0^2/c^2}} \tag{14.51}$$

$$\Delta p_{B2} = \frac{- 2mu_0}{\sqrt{1 - u_0^2/c^2}} \tag{14.52}$$

Equations 14.48 and 14.52 add to zero, as required for the conservation of linear momentum.

14.8   ENERGY

With a new definition of linear momentum (Equation 14.45) in hand, we turn our attention to energy and force. We keep our former definition (Equation 2.86) of kinetic energy as being the work done on a particle. The work done is defined in Equation 2.84 to be

$$W_{12} = \int_1^2 \mathbf{F} \cdot d\mathbf{r} = T_2 - T_1 \tag{14.53}$$

Equation 2.2 for Newton's Second Law is modified to account for the new definition of linear momentum:

$$\mathbf{F} = \frac{d\mathbf{p}}{dt} = \frac{d}{dt}(\gamma m\mathbf{u}) \tag{14.54}$$

If we start from rest, $T_1 = 0$, and the velocity **u** is initially along the direction of the force.

$$W = T = \int \frac{d}{dt} (\gamma m\mathbf{u}) \cdot \mathbf{u} \, dt \tag{14.55}$$

$$= m \int_0^u u \, d(\gamma u) \tag{14.56}$$

Equation 14.56 is integrated by parts to obtain

$$T = \gamma m u^2 - m \int_0^u \frac{u \, du}{\sqrt{1 - u^2/c^2}}$$

$$= \gamma m u^2 + mc^2 \sqrt{1 - u^2/c^2} \, \Big|_0^u$$

$$= \gamma m u^2 + mc^2 \sqrt{1 - u^2/c^2} - mc^2 \tag{14.57}$$

With algebraic manipulation Equation 14.57 becomes

$$\boxed{T = \gamma mc^2 - mc^2} \qquad \text{relativistic kinetic energy} \tag{14.58}$$

Equation 14.58 seems to resemble in no way our former result for kinetic energy, $T = \frac{1}{2}mu^2$. However, Equation 14.58 must reduce to $\frac{1}{2}mu^2$ for small values of velocity.

EXAMPLE 14.7 -

Show that Equation 14.58 reduces to the classical result for small speeds, $u \ll c$.

Solution: The first term of Equation 14.58 can be expanded in a power series:

$$T = mc^2 (1 - u^2/c^2)^{-1/2} - mc^2$$

$$= mc^2 \left(1 + \frac{1}{2} \frac{u^2}{c^2} + \cdots \right) - mc^2 \tag{14.59}$$

where all terms of power $(u/c)^4$ or greater are neglected because $u \ll c$.

$$T = mc^2 + \frac{1}{2} mu^2 - mc^2$$

$$= \frac{1}{2} mu^2 \tag{14.60}$$

which is the classical result.

- ●

It is important to note that neither $\frac{1}{2}mu^2$ nor $\frac{1}{2}\gamma mu^2$ gives the correct relativistic value for the kinetic energy.

The term mc^2 in Equation 14.58 is called the **rest energy** and is denoted by E_0.

$$\boxed{E_0 \equiv mc^2} \quad \text{rest energy} \tag{14.61}$$

Equation 14.58 is rewritten

$$\gamma mc^2 = T + mc^2$$

Thus,

$$E = T + E_0 \tag{14.62}$$

where

$$\boxed{E \equiv \gamma mc^2 = T + E_0} \quad \text{total energy} \tag{14.63}$$

The total energy, $E = \gamma mc^2$, is defined as the sum of kinetic energy and the rest energy. Equations 14.58–14.63 are the origin of Einstein's famous relativistic result of the equivalence of mass and energy (energy $= mc^2$). These equations are consistent with this interpretation. Note that when a body is not in motion ($u = 0 = T$), Equation 14.63 indicates that the total energy is equal to the rest energy.

If mass is simply another form of energy, then we must combine the classical conservation laws of mass and energy into one conservation law of mass-energy represented by Equation 14.63. This law is easily demonstrated in the atomic nucleus, where the mass of constituent particles is converted to the energy that binds the individual particles together.

EXAMPLE 14.8 --

Use the atomic masses of the particles involved to calculate the binding energy of a deuteron.

Solution: A deuteron is composed of a neutron and a proton. We use atomic masses, because the electron masses cancel.

$$\text{mass of neutron} = 1.008665 \text{ u}$$

$$\text{mass of proton } (^1\text{H}) = \underline{1.007825 \text{ u}}$$

$$\text{sum} = 2.016490 \text{ u}$$

$$\text{mass of deuteron } (^2\text{H}) = 2.014102 \text{ u}$$

$$\text{difference} = 0.002388 \text{ u}$$

This difference in mass-energy is equal to the binding energy holding the neutron and proton together as a deuteron. The mass units are atomic mass units (u), which can be converted to kilograms if necessary. However, the conversion of mass to energy is facilitated by the well-known relation between mass and energy:

$$1 \ uc^2 = 931.5 \ \text{MeV} \tag{14.64}$$

The binding energy of the deuteron is therefore

$$0.002388 \ uc^2 \times 931.5 \frac{\text{MeV}}{uc^2} = 2.22 \ \text{MeV}$$

Nuclear experiments of the form $\gamma + {}^2\text{H} \rightarrow n + p$ indicate that gamma rays of energy just greater than 2.22 MeV are required to break the deuteron apart into a neutron and a proton. Conversely, when a neutron and proton join at rest to form a deuteron, 2.22 MeV of energy is released for kinetic energy of the deuteron and gamma ray.

Because physicists believe that momentum is a more fundamental concept than kinetic energy (for example, there is no general law of conservation of kinetic energy), we would like a relation for mass-energy that includes momentum rather than kinetic energy. We begin with Equation 14.45 for momentum:

$$p = \gamma m u$$
$$p^2 c^2 = \gamma^2 m^2 u^2 c^2$$
$$= \gamma^2 m^2 c^4 \left(\frac{u^2}{c^2} \right) \tag{14.65}$$

It is easy to show that

$$\frac{u^2}{c^2} = 1 - \frac{1}{\gamma^2} \tag{14.66}$$

so Equation 14.65 becomes

$$p^2 c^2 = \gamma^2 m^2 c^4 \left(1 - \frac{1}{\gamma^2} \right)$$
$$= \gamma^2 m^2 c^4 - m^2 c^4$$
$$= E^2 - E_0^2$$

$$\boxed{E^2 = p^2 c^2 + E_0^2} \tag{14.67}$$

Equation 14.67 is a very useful kinematic relationship. It relates the total energy of a particle to its momentum and rest energy.

Notice that a photon has no mass, so that Equation 14.67 gives

$$E = pc \qquad \text{photon} \tag{14.68}$$

There is no such thing as a photon at rest.

14.9 SPACETIME AND FOUR-VECTORS

In Section 14.3 (Equation 14.5), we noticed that the quantities

$$\left.\begin{aligned} \sum_{j=1}^{3} x_j^2 - c^2 t^2 = 0 \\[2mm] \sum_{j=1}^{3} x_j'^2 - c^2 t'^2 = 0 \end{aligned}\right\}$$

are invariant because the speed of light is the same in all inertial systems in relative motion. Consider two events separated by space and time. In system K,

$$\Delta x_i = x_i(\text{event 2}) - x_i(\text{event 1})$$
$$\Delta t = t(\text{event 2}) - t(\text{event 1})$$

The interval Δs^2 is invariant in all inertial systems in relative motion (see Problem 14-34):

$$\Delta s^2 = \sum_{j=1}^{3} (\Delta x_j)^2 - c^2 \Delta t^2 \tag{14.69}$$

$$\Delta s^2 = \Delta s'^2 = \sum_{j=1}^{3} (\Delta x_j')^2 - c^2 \Delta t'^2 \tag{14.70}$$

Equation 14.69 can be written as a differential equation:

$$ds^2 = dx_1^2 + dx_2^2 + dx_3^2 - c^2\, dt^2 \tag{14.71}$$

Consider the system K', where the particle is instantaneously at rest. Because $dx_1' = dx_2' = dx_3' = 0$ in this case, $dt' = d\tau$, the proper time interval discussed above (Equation 14.21). Equation 14.70 becomes

$$-c^2\, d\tau^2 = dx_1^2 + dx_2^2 + dx_3^2 - c^2\, dt^2 \tag{14.72}$$

Using the Lorentz transformation, Equation 14.72 gives a similar result to Equation 14.21:

$$d\tau = \frac{dt}{\gamma} \tag{14.73}$$

The proper time τ is, along with the length quantity Δs^2, another Lorentz invariant quantity.

A useful concept in special relativity is that of the **light cone**. The invariant length Δs^2 suggests adding ct as a fourth dimension to the three space dimensions x_1, x_2, and x_3. In Figure 14-6, we plot ct versus one of the Euclidean space coordinates. The origin of (x, ct) is the present $(0, 0)$. The solid lines represent the paths taken in the past and in the future by light. A particle traveling the path from A to B is said to be moving along its **worldline**. For times $t < 0$, the particle has been in the lower cone, the past. Similarly, for $t > 0$ the particle will move in the upper cone, the future. It is not possible for us to know that a particle is outside the range $x = ct$; this region, called "elsewhere," requires $v > c$.

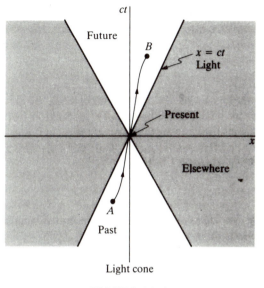

FIGURE 14-6

There are two possibilities concerning the value of Δs^2. If $\Delta s^2 > 0$, the two events have a **spacelike interval**. One can always find an inertial frame traveling with $v < c$ such that the two events occur at different space coordinates but at the same time. When $\Delta s^2 < 0$, the two events are said to have a **timelike interval**. One can always find a suitable inertial frame in which the events occur at the same point in space but at different times. In the case $\Delta s^2 = 0$, the two events are separated by a light ray.

Only events separated by a timelike interval can be causally connected. The present event in the light cone can be causally related only to events in the past region of the light cone. Events with a spacelike interval cannot be causally connected. Space and time, although distinct, are nonetheless intricately related.

The previous discussion of space and time suggests using ct as a fourth dimensional parameter. We continue this line of thought by defining $x_4 \equiv ict$ and $x_4' \equiv ict'$. The use of the imaginary number $i(\sqrt{-1})$ does not indicate that this component is imaginary. The imaginary number simply allows us to represent the relations in concise, mathematical form. The rest of this section could just as well be carried out without the use of i (e.g., $x_4 = ct$), but the mathematics would be more cumbersome. The useful results are in terms of real, physical quantities.

Using $x_4 = ict$ and $x_4' = ict'$, we can write Equations 14.5 as*

---

* In accordance with standard convention, we use Greek indices (usually μ or ν) to indicate summations that run from 1 to 4; in relativity theory, Latin indices are usually reserved for summations that run from 1 to 3.

$$\left.\begin{array}{c} \sum\limits_{\mu=1}^{4} x_{\mu}^{2} = 0 \\[2mm] \sum\limits_{\mu=1}^{4} x_{\mu}'^{2} = 0 \end{array}\right\} \tag{14.74}$$

From these equations, it is clear that the two sums must be proportional, and because the motion is symmetrical between the systems, the proportionality constant is unity.* Thus,

$$\sum_{\mu} x_{\mu}^{2} = \sum_{\mu} x_{\mu}'^{2} \tag{14.75}$$

This relation is analogous to the three-dimensional, distance-preserving, orthogonal rotations we have studied previously (see Section 1.4) and indicates that the Lorentz transformation corresponds to a rotation in a *four-dimensional* space (called **world space** or **Minkowski space**[†]). The Lorentz transformations are then orthogonal transformations in Minkowski space:

$$x_{\mu}' = \sum_{\nu} \lambda_{\mu\nu} x_{\nu} \tag{14.76}$$

where the $\lambda_{\mu\nu}$ are the elements of the Lorentz transformation matrix. From Equations 14.14, the transformation λ is

$$\lambda = \begin{pmatrix} \gamma & 0 & 0 & i\beta\gamma \\ 0 & 1 & 0 & 0 \\ 0 & 0 & 1 & 0 \\ -i\beta\gamma & 0 & 0 & \gamma \end{pmatrix} \tag{14.77}$$

A quantity is called a **four-vector** if it consists of four components, each of which transforms according to the relation[‡]

$$A_{\mu}' = \sum_{\nu} \lambda_{\mu\nu} A_{\nu} \tag{14.78}$$

where the $\lambda_{\mu\nu}$ define a Lorentz transformation. Such a four-vector[§] is

$$\mathbb{X} = (x_1, x_2, x_3, ict) \tag{14.79a}$$

or

$$\boxed{\mathbb{X} = (\mathbf{x}, ict)} \tag{14.79b}$$

---

* A "proof" is given in Appendix G.

[†] Herman Minkowski (1864–1909) made important contributions to the mathematical theory of relativity and introduced *ict* as a fourth component.

[‡] We do not distinguish here between *covariant* and *contravariant* vector components; see, for example, Bergmann (Be46, Chapter 5).

[§] Four-vectors are denoted exclusively by openface capital letters.

where the notation of the last line means that the first three (space) components of $\mathbf{X}$ define the ordinary three-dimensional position vector $\mathbf{x}$ and that the fourth component is ict. Similarly, the differential of $\mathbf{X}$ is a four-vector:

$$d\mathbf{X} = (d\mathbf{x}, ic\, dt) \tag{14.80}$$

In Minkowski space, the four-dimensional element of length is invariant. Its magnitude is unaffected by a Lorentz transformation, and such a quantity is called a **four-scalar** or **world scalar**. Equation 14.71 can be written as

$$ds = \sqrt{\sum_\mu dx_\mu^2} \tag{14.81}$$

and Equation 14.72 as

$$d\tau = \frac{i}{c}\sqrt{\sum_\mu dx_\mu^2} = \frac{i}{c}\, ds \tag{14.82}$$

The proper time $d\tau$ is invariant because it is simply i/c times the element of length ds. The ratio of the four-vector $d\mathbf{X}$ to the invariant $d\tau$ is therefore also a four-vector, called the four-vector velocity $\mathbb{V}$:

$$\boxed{\mathbb{V} = \frac{d\mathbf{X}}{d\tau} = \left(\frac{d\mathbf{x}}{d\tau}, ic\frac{dt}{d\tau}\right)} \tag{14.83}$$

The components of the ordinary velocity $\mathbf{u}$ are

$$u_j = \frac{dx_j}{dt}$$

so, using Equations 14.71 and 14.82, $d\tau$ can be expressed as

$$d\tau = dt\sqrt{1 - \frac{1}{c^2}\sum_j \frac{dx_j^2}{dt^2}}$$

or

$$d\tau = dt\sqrt{1 - \beta^2} \tag{14.84}$$

as we found in Equation 14.73. The four-vector velocity can therefore be written as

$$\boxed{\mathbb{V} = \frac{1}{\sqrt{1 - \beta^2}}(\mathbf{u}, ic)} \tag{14.85}$$

where $\mathbf{u}$ represents the three space components of ordinary velocity, u_1, u_2, u_3. (Remember that the particle's velocity is now denoted by $\mathbf{u}$ to distinguish it from the moving frame velocity **v**.) The four-vector momentum is now simply the mass times four-vector velocity,* because mass is invariant:

$$\mathbb{P} = m\mathbb{V} \tag{14.86}$$

$$\boxed{\mathbb{P} = \left(\frac{m\mathbf{u}}{\sqrt{1 - \beta^2}}, ip_4\right)} \tag{14.87}$$

where

$$p_4 \equiv \frac{mc}{\sqrt{1 - \beta^2}} \tag{14.88}$$

The first three components of the four-vector momentum $\mathbb{P}$ are the components of the relativistic momentum (Equation 14.45):

$$P_j = p_j = \gamma m u_j, \qquad j = 1, 2, 3 \tag{14.89}$$

Using Equation 14.63, the fourth component of the momentum is related to the total energy E:

$$p_4 = \gamma mc = \frac{E}{c} \tag{14.90}$$

The four-vector momentum can therefore be written as

$$\mathbb{P} = \left(\mathbf{p}, i\frac{E}{c}\right) \tag{14.91}$$

where **p** stands for the three space components of momentum. Thus, in relativity theory, momentum and energy are linked in a manner similar to that which joins the concepts of space and time. If we apply the Lorentz transformation matrix (Equation 14.77) to the momentum $\mathbb{P}$, we find

$$\boxed{\begin{aligned} p_1' &= \frac{p_1 - (v/c^2)E}{\sqrt{1 - \beta^2}} \\ p_2' &= p_2 \\ p_3' &= p_3 \\ E' &= \frac{E - vp_1}{\sqrt{1 - \beta^2}} \end{aligned}} \tag{14.92}$$

---

* A four-vector multiplied by a four-scalar is also a four-vector.

EXAMPLE 14.9 ---

Using the methods of this section, derive Equation 14.67.

Solution: If we place the origin of the moving system K' fixed on the particle, we have $u = v$. The square of the four-vector velocity (Equation 14.85) is invariant:

$$\mathbb{V}^2 = \sum_\mu V_\mu^2 = \frac{v^2 - c^2}{1 - \beta^2} = - c^2 \qquad (14.93)$$

Hence, the square of the four-vector momentum is also invariant:

$$\mathbb{P}^2 = \sum_\mu P_\mu^2 = m^2 \mathbb{V}^2 = - m^2 c^2 \qquad (14.94)$$

From Equation 14.91, we also have, using $\mathbf{p} \cdot \mathbf{p} = p^2 = p_1^2 + p_2^2 + p_3^2$,

$$\mathbb{P}^2 = p^2 - \frac{E^2}{c^2} \qquad (14.95)$$

Combining the last two equations gives Equation 14.67.

$$E^2 = p^2 c^2 + m^2 c^4 = p^2 c^2 + E_0^2$$

-- ●

If we define an angle ϕ such that $\beta = \sin \phi$, the relativistic relations between velocity, momentum, and energy can be obtained by trigonometric relations involving the so-called "relativistic triangle" (Figure 14-7).

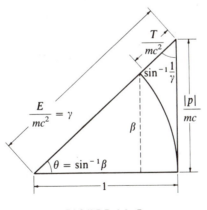

FIGURE 14-7

EXAMPLE 14.10 ---

Derive the velocity addition rule.

Solution: Suppose that there are three inertial reference frames, K, K', and K'', which are in collinear motion along their respective x_1-axes. Let the velocity of K' relative to K be v_1 and let the velocity of K'' relative to K' be v_2. The speed of K'' relative to K cannot be $v_1 + v_2$, because it must be possible to propagate a signal between any two inertial frames, and if both v_1 and v_2 are greater than $c/2$ (but less than c), then $v_1 + v_2 > c$. Therefore, the rule for the addition of velocities in relativity must be different from that in Galilean theory. The relativistic velocity addition rule can be obtained by considering the Lorentz transformation matrix connecting K and K''. The individual transformation matrices are

$$\lambda_{K' \to K} = \begin{pmatrix} \gamma_1 & 0 & 0 & i\beta_1\gamma_1 \\ 0 & 1 & 0 & 0 \\ 0 & 0 & 1 & 0 \\ -i\beta_1\gamma_1 & 0 & 0 & \gamma_1 \end{pmatrix}$$

$$\lambda_{K'' \to K'} = \begin{pmatrix} \gamma_2 & 0 & 0 & i\beta_2\gamma_2 \\ 0 & 1 & 0 & 0 \\ 0 & 0 & 1 & 0 \\ -i\beta_2\gamma_2 & 0 & 0 & \gamma_2 \end{pmatrix}$$

The transformation from K'' to K is just the product of these two transformations:

$$\lambda_{K'' \to K} = \lambda_{K'' \to K'}\,\lambda_{K' \to K} = \begin{pmatrix} \gamma_1\gamma_2(1 + \beta_1\beta_2) & 0 & 0 & i\gamma_1\gamma_2(\beta_1 + \beta_2) \\ 0 & 1 & 0 & 0 \\ 0 & 0 & 1 & 0 \\ -i\gamma_1\gamma_2(\beta_1 + \beta_2) & 0 & 0 & \gamma_1\gamma_2(1 + \beta_1\beta_2) \end{pmatrix}$$

So that the elements of this matrix correspond to those of the normal Lorentz matrix (Equation 14.77), we must identify β and γ for the $K'' \to K$ transformation as

$$\left.\begin{aligned} \gamma &= \gamma_1\gamma_2(1 + \beta_1\beta_2) \\ \beta\gamma &= \gamma_1\gamma_2(\beta_1 + \beta_2) \end{aligned}\right\} \tag{14.96}$$

from which we obtain

$$\beta = \frac{\beta_1 + \beta_2}{1 + \beta_1\beta_2} \tag{14.97}$$

If we multiply this last expression by c, we have the usual form of the velocity (speed) addition rule:

$$\boxed{v = \frac{v_1 + v_2}{1 + (v_1 v_2/c^2)}} \tag{14.98}$$

It follows that if $v_1 < c$ and $v_2 < c$, then $v < c$ also.

Even though *signal* velocities can never exceed c, there are other types of velocity that can be greater than c. For example, the *phase velocity* of a light wave in a medium for which the index of refraction is less than unity is greater than c, but the phase velocity does not correspond to the signal velocity in such a medium; the signal velocity is indeed less than c. Or consider an electron gun that emits a beam of electrons. If the gun is rotated, then the electron beam describes a certain path on a screen placed at some appropriate distance. If the angular velocity of the gun and the distance to the screen are sufficiently large, then the velocity of the spot traveling across the screen can be *any* velocity, arbitrarily large. Thus, the *writing speed* of an oscilloscope can exceed c, but again the writing speed does not correspond to the signal velocity; that is, information cannot be transmitted from one point on the screen to another by means of the electron beam. In such a device, a signal can be transmitted only from the gun to the screen, and this transmission takes place at the velocity of the electrons in the beam (i.e., $< c$).

E X A M P L E   14.11 ---

Derive the relativistic Doppler effect if the angle between the light source and direction of relative motion of the observer is θ (Figure 14-8).

Solution: This example can easily be solved using the momentum-energy four-vector by treating the light as a photon with total energy $E = h\nu$. The light source is at rest in system K and emits a single frequency ν_0.

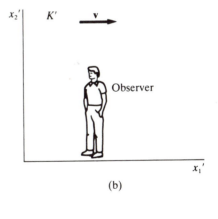

(b)

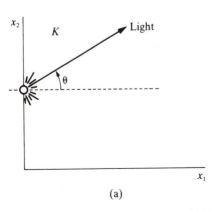

(a)

FIGURE 14-8

$$E = h\nu_0 \tag{14.99}$$

$$p = \frac{E}{c} = \frac{h\nu_0}{c} \tag{14.100}$$

The observer moving to the right in system K' measures the energy E' for a photon of frequency ν'. From Equation 14.92, we have

$$E' = \gamma(h\nu_0 - \nu p_1) \tag{14.101}$$

$$h\nu' = \gamma\left(h\nu_0 - \frac{\nu h\nu_0}{c}\cos\theta\right) \tag{14.102}$$

where $p_1 = p\cos\theta$. Equation 14.102 reduces to

$$\nu' = \gamma\nu_0(1 - \beta\cos\theta) \tag{14.103}$$

which is equivalent to Equation 14.34, depending on the value of θ. For an early time, the observer is far to the left of the source, and as the observer approaches the source ($\theta = \pi$),

$$\nu' = \nu_0\frac{\sqrt{1+\beta}}{\sqrt{1-\beta}} \qquad \text{observer approaching source} \tag{14.104}$$

as in Equation 14.31. At a much later time, the observer is receding ($\theta = 0$) and

$$\nu' = \nu_0\frac{\sqrt{1-\beta}}{\sqrt{1+\beta}} \qquad \text{observer receding from source} \tag{14.105}$$

as in Equation 14.33. When the observer just passes the source ($\theta = \pi/2$),

$$\nu' = \frac{\nu_0}{\sqrt{1-\beta^2}} \qquad \text{observer passing source} \tag{14.106}$$

We can also treat the case where the observer is at rest and the source is moving (see Problem 14-18). We still obtain Equations 14.104–14.106 because, according to the principle of relativity, it is not possible to distinguish between the motion of the observer and the motion of the source.

14.10 LAGRANGIAN FUNCTION IN SPECIAL RELATIVITY

Lagrangian and Hamiltonian dynamics (discussed in Chapter 7) must be adjusted in light of the new concepts presented here. We can extend the Lagrangian formalism into the realm of special relativity in the following way. For a single (nonrelativistic) particle moving in a velocity-independent potential, the rectangular momentum components (see Equation 7.150) may be written as

$$p_i = \frac{\partial L}{\partial u_i} \tag{14.107}$$

According to Equation 14.87, the relativistic expression for the ordinary (i.e., space) momentum component is

$$p_i = \frac{mu_i}{\sqrt{1 - \beta^2}} \tag{14.108}$$

We now require that the *relativistic* Lagrangian, when differentiated with respect to u_i as in Equation 14.107, yield the momentum components given by Equation 14.108:

$$\frac{\partial L}{\partial u_i} = \frac{mu_i}{\sqrt{1 - \beta^2}} \tag{14.109}$$

This requirement involves only the *velocity* of the particle, so we expect that the velocity-*independent* part of the relativistic Lagrangian is unchanged from the non-relativistic case. The velocity-*dependent* part, however, may no longer be equal to the kinetic energy. We therefore write

$$L = T^* - U \tag{14.110}$$

where $U = U(x_i)$ and $T^* = T^*(u_i)$. The function T^* must satisfy the relation

$$\frac{\partial T^*}{\partial u_i} = \frac{mu_i}{\sqrt{1 - \beta^2}} \tag{14.111}$$

It can be easily verified that a suitable expression for T^* (apart from a possible constant of integration that can be suppressed) is

$$T^* = -mc^2\sqrt{1 - \beta^2} \tag{14.112}$$

Hence, the relativistic Lagrangian can be written as

$$\boxed{L = -mc^2\sqrt{1 - \beta^2} - U} \tag{14.113}$$

and the equations of motion are obtained in the standard way from Lagrange's equations.

Notice that the Lagrangian is *not* given by $T - U$, because the relativistic expression for the kinetic energy (Equation 14.58) is

$$T = \frac{mc^2}{\sqrt{1 - \beta^2}} - mc^2 \tag{14.114}$$

The Hamiltonian (see Equation 7.153) can be calculated from

$$H = \sum_i u_i p_i - L$$

$$= \sum_i \frac{p_i^2 c^2}{\gamma mc^2} + \frac{mc^2}{\gamma} + U$$

where we have used Equations 14.108 and 14.113 and changed $\sqrt{1 - \beta^2}$ to γ^{-1}.

Thus,

$$H = \frac{p^2 c^2}{\gamma m c^2} + \frac{m c^2}{\gamma} + U = \frac{1}{\gamma m c^2}(p^2 c^2 + m^2 c^4) + U$$

$$= \frac{E^2}{\gamma m c^2} + U$$

$$= E + U = T + U + E_0 \tag{14.115}$$

The relativistic Hamiltonian is equal to the total energy defined in Section 14.8 *plus* the potential energy. It differs from the total energy used previously in Chapter 7 by now including the rest energy.

14.11 RELATIVISTIC KINEMATICS

In the event that the velocities in a collision process are not negligible with respect to the velocity of light, it becomes necessary to use *relativistic* kinematics. In the discussion in Chapter 9, we took advantage of the properties of the center-of-mass coordinate system in deriving many of the kinematic relations. Because mass and energy are interrelated in relativity theory, it no longer is meaningful to speak of a "center-of-mass" system; in relativistic kinematics, one uses a "center-of-momentum" coordinate system instead. Such a system possesses the same essential property as the previously used center-of-mass system—the total linear momentum in the system is zero. Therefore, if a particle of mass m_1 collides with a particle of mass m_2, then in the center-of-momentum system we have

$$p_1' = p_2' \tag{14.116}$$

Using Equation 14.87, the space components of the momentum four-vector can be written as

$$m_1 u_1' \gamma_1' = m_2 u_2' \gamma_2' \tag{14.117}$$

where, as before, $\gamma \equiv 1/\sqrt{1 - \beta^2}$ and $\beta \equiv u/c$.

In a collision problem, it is convenient to associate the laboratory coordinate system with the inertial system K and the center-of-momentum system with K'. A simple Lorentz transformation then connects the two systems. To derive the relativistic kinematic expressions, the procedure is to obtain the center-of-momentum relations and then perform a Lorentz transformation back to the laboratory system. We choose the coordinate axes so that m_1 moves along the x-axis in K with speed u_1. Because m_2 is initially at rest in K, $u_2 = 0$. In K', m_2 moves with speed u_2' and so K' moves with respect to K also with speed u_2' and in the same direction as the initial motion of m_1.

Using the fact that $\beta \gamma = \sqrt{\gamma^2 - 1}$, we have

$$p_1' = m_1 u_1' \gamma_1' = m_1 c \beta_1' \gamma_1'$$

$$= m_1 c \sqrt{\gamma_1'^2 - 1} = m_2 c \sqrt{\gamma_2'^2 - 1}$$

$$= p_2' \tag{14.118}$$

which expresses the equality of the momenta in the center-of-momentum system.

According to Equation 14.92, the transformation of the momentum p_1 (from K to K') is

$$p_1' = \left(p_1 - \frac{u_2'}{c^2}E_1\right)\gamma_2' \tag{14.119}$$

We also have

$$\left.\begin{array}{l} p_1 = m_1 u_1 \gamma_1 \\ E_1 = m_1 c^2 \gamma_1 \end{array}\right\} \tag{14.120}$$

so Equation 14.118 can be used to obtain

$$m_1 c\sqrt{\gamma_1'^2 - 1} = (m_1 c\beta_1\gamma_1 - \beta_2' m_1 c\gamma_1)\gamma_2'$$

$$= m_1 c\left(\gamma_2'\sqrt{\gamma_1^2 - 1} - \gamma_1\sqrt{\gamma_2'^2 - 1}\right)$$

$$= m_2 c\sqrt{\gamma_2'^2 - 1} \tag{14.121}$$

These equations can be solved for γ_1' and γ_2' in terms of γ_1:

$$\gamma_1' = \frac{\gamma_1 + \dfrac{m_1}{m_2}}{\sqrt{1 + 2\gamma_1\left(\dfrac{m_1}{m_2}\right) + \left(\dfrac{m_1}{m_2}\right)^2}} \tag{14.122a}$$

$$\gamma_2' = \frac{\gamma_1 + \dfrac{m_2}{m_1}}{\sqrt{1 + 2\gamma_1\left(\dfrac{m_2}{m_1}\right) + \left(\dfrac{m_2}{m_1}\right)^2}} \tag{14.122b}$$

Next, we write the equations of the transformation of the momentum components from K' back to K after the scattering. We now have both x- and y-components:

$$p_{1,x} = \left(p_{1,x}' + \frac{u_2'}{c^2}E_1'\right)\gamma_2'$$

$$= (m_1 c\beta_1'\gamma_1' \cos\theta + m_1 c\beta_2'\gamma_1')\gamma_2'$$

$$= m_1 c\gamma_1'\gamma_2'(\beta_1' \cos\theta + \beta_2') \tag{14.123a}$$

(Note that, because the transformation is from K' to K, a plus sign occurs before the second term, in contrast to Equation 14.119.) Also,

$$p_{1,y} = m_1 c\beta_1'\gamma_1' \sin\theta \tag{14.123b}$$

The tangent of the laboratory scattering angle ψ is given by $p_{1,y}/p_{1,x}$; therefore, dividing Equation 14.123b by Equation 14.123a, we obtain

$$\tan\psi = \frac{1}{\gamma_2'}\frac{\sin\theta}{\cos\theta + (\beta_2'/\beta_1')}$$

Using Equation 14.117 to express β_2'/β_1', the result is

$$\tan \psi = \frac{1}{\gamma_2'} \frac{\sin \theta}{\cos \theta + (m_1 \gamma_1'/m_2 \gamma_2')} \tag{14.124}$$

For the recoil particle, we have

$$p_{2,x} = \left(p_{2,x}' + \frac{u_2'}{c^2} E_2' \right) \gamma_2'$$

$$= (-m_2 c \beta_2' \gamma_2' \cos \theta + m_2 c \beta_2' \gamma_2') \gamma_2'$$

$$= m_2 c \beta_2' \gamma_2'^2 (1 - \cos \theta) \tag{14.125a}$$

where a minus sign occurs in the first term because $p_{2,x}'$ is directed opposite to $p_{1,x}'$. Also,

$$p_{2,y} = -m_2 c \beta_2' \gamma_2' \sin \theta \tag{14.125b}$$

As before, the tangent of the laboratory recoil angle ζ is given by $p_{2,y}/p_{2,x}$:

$$\tan \zeta = -\frac{1}{\gamma_2'} \frac{\sin \theta}{1 - \cos \theta} \tag{14.126}$$

The overall minus sign indicates that if m_1 is scattered toward positive values of y, then m_2 recoils in the negative y-direction.

A case of special interest is that in which $m_1 = m_2$. From Equations 14.122, we find

$$\gamma_1' = \gamma_2' = \sqrt{\frac{1 + \gamma_1}{2}}, \qquad m_1 = m_2 \tag{14.127}$$

The tangents of the scattering angles become

$$\tan \psi = \sqrt{\frac{2}{1 + \gamma_1}} \cdot \frac{\sin \theta}{1 + \cos \theta} \tag{14.128}$$

$$\tan \zeta = -\sqrt{\frac{2}{1 + \gamma_1}} \cdot \frac{\sin \theta}{1 - \cos \theta} \tag{14.129}$$

The product is therefore

$$\tan \psi \tan \zeta = -\frac{2}{1 + \gamma_1}, \qquad m_1 = m_2 \tag{14.130}$$

(The minus sign is of no essential importance; it only indicates that ψ and ζ are measured in opposite directions.)

We previously found that in the nonrelativistic limit there was always a right angle between the final velocity vectors in the scattering of particles of equal mass. Indeed, in the limit $\gamma_1 \rightarrow 1$, Equations 14.128 and 14.129 become equal to Equations 9.69 and 9.73, respectively, and so $\psi + \zeta = \pi/2$. Equation 14.130, however, shows that in the relativistic case $\psi + \zeta < \pi/2$; thus, the included angle in the scattering is always smaller than in the nonrelativistic limit. For equal scattering and recoil angles ($\psi = \zeta$), Equation 14.130 becomes

$$\tan \psi = \left(\frac{2}{1 + \gamma_1}\right)^{1/2}, \qquad m_1 = m_2$$

and the included angle between the directions of the scattered and recoil particles is

$$\phi = \psi + \zeta = 2\psi$$

$$= 2 \tan^{-1} \left(\frac{2}{1 + \gamma_1}\right)^{1/2}, \qquad m_1 = m_2 \qquad \textbf{(14.131)}$$

Figure 14-9 shows ϕ as a function of γ_1 up to $\gamma_1 = 20$. At $\gamma_1 = 10$, the included angle is approximately 46°. This value of γ_1 corresponds to an initial velocity that is 99.5% of the velocity of light. According to Equation 14.58, the kinetic energy is given by $T_1 = m_1 c^2 (\gamma_1 - 1)$; therefore, a proton with $\gamma_1 = 10$ would have a kinetic energy of approximately 8.4 GeV, whereas an electron with the same velocity would have $T_1 \cong 4.6$ MeV.*

By using the transformation properties of the fourth component of the momentum four-vector (i.e., the total energy), it is possible to obtain the relativistic analogs of all the energy equations we have previously derived in the nonrelativistic limit.

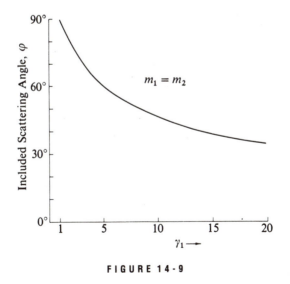

FIGURE 14-9

---

* These units of energy are defined in Problem 14-39: 1 GeV $= 10^3$ MeV $= 10^9$ eV $= 1.602 \times 10^{-3}$ erg $= 1.602 \times 10^{-10}$ J.

P R O B L E M S

14-1. Prove Equation 14.13 by using Equations 14.9–14.12.

14-2. Show that the transformation equations connecting the K' and K systems (Equations 14.14) can be expressed as

$$x_1' = x_1 \cosh \alpha - ct \sinh \alpha$$
$$x_2' = x_2, \qquad x_3' = x_3$$
$$t' = t \cosh \alpha - \frac{x_1}{c} \sinh \alpha$$

where $\tanh \alpha = v/c$. Show that the Lorentz transformation corresponds to a rotation through an angle $i\alpha$ in four-dimensional space.

14-3. Show that the equation

$$\nabla^2 \Psi - \frac{1}{c^2} \frac{\partial^2 \Psi}{\partial t^2} = 0$$

is invariant under a Lorentz transformation but not under a Galilean transformation. (This is the wave equation that describes the propagation of light waves in free space.)

14-4. Show that the expression for the FitzGerald-Lorentz contraction (Equation 14.19) can also be obtained if the observer in the K' system measures the time necessary for the rod to pass a fixed point in that system and then multiplies the result by v.

14-5. What is the apparent shape of a cube moving with a uniform velocity directly *toward* or *away from* an observer?

14-6. Consider two events that take place at different points in the K system at the same instant t. If these two points are separated by a distance Δx, show that in the K' system the events are not simultaneous but are separated by a time interval $\Delta t' = -v\gamma \Delta x/c^2$.

14-7. Two clocks located at the origins of the K and K' systems (which have a relative speed v) are synchronized when the origins coincide. After a time t, an observer at the origin of the K system observes the K' clock by means of a telescope. What does the K' clock read?

14-8. In his 1905 paper (see the translation in Lo23), Einstein states: "We conclude that a balance-clock at the equator must go more slowly, by a very small amount, than a precisely similar clock situated at one of the poles under otherwise identical conditions." Neglect the fact that the Equator clock does not undergo uniform motion and show that after a century the clocks will differ by approximately 0.0038 s.

14-9. Consider a relativistic rocket whose velocity with respect to a certain inertial frame is v and whose exhaust gases are emitted with a constant velocity V with respect to the rocket. Show that the equation of motion is

$$m \frac{dv}{dt} + V \frac{dm}{dt} (1 - \beta^2)^{3/2} = 0$$

where $m = m(t)$ is the mass of the rocket in its rest frame and $\beta = v/c$.

14-10. Show by algebraic methods that Equations 14.15 follow from Equations 14.14.

14-11. A stick of length l is fixed at an angle θ from its x_1-axis in its own rest system K.

What is the length and orientation of the stick as measured by an observer moving along x_1 with speed v?

14-12. A racer attempting to break the land speed record rockets by two markers spaced 100 m apart on the ground in a time of 0.4 μs as measured by an observer on the ground. How far apart do the two markers appear to the racer? What elapsed time does the racer measure? What speeds do the racer and ground observer measure?

14-13. A muon is moving with speed $v = 0.999c$ vertically down through the atmosphere. If its half-life in its own rest frame is 1.5 μs, what is its half-life as measured by an observer on the Earth?

14-14. Show that Equation 14.31 is valid when a receiver approaches a fixed light source with speed v.

14-15. A star is known to be moving away from the Earth at a speed of 4×10^4 m/s. This speed is determined by measuring the shift of the H_α line ($\lambda = 656.3$ nm). By how much and in what direction is the shift of the wavelength of the H_α line?

14-16. A photon is emitted at an angle θ' by a star (system K') and then received at an angle θ on the Earth (system K). The angles are measured from a line between the star and the Earth. The star is receding at speed v with respect to the Earth. Find the relation between θ and θ'; this effect is called the *aberration of light*.

14-17. A spectral line of wavelength λ on the Earth is found to increase by 50% on a far distant galaxy. What is the speed of the galaxy relative to the Earth?

14-18. Solve Example 14.11 for the case of the observer at rest and the source moving. Show that the results are the same as those given in Example 14.11.

14-19. Equation 14.34 indicates that a red (blue) shift occurs when a source and observer are receding (approaching) with respect to one another in purely radial motion (i.e., $\beta = \beta_r$). Show that, if there is also a relative tangential speed β_t, Equation 14.34 becomes

$$\frac{\lambda_0}{\lambda} = \frac{\nu}{\nu_0} = \frac{\sqrt{1 - \beta_r^2 - \beta_t^2}}{1 - \beta_r}$$

and that the condition for always having a red shift (i.e., no blue shift), $\lambda > \lambda_0$ or $\nu < \nu_0$, is*

$$\beta_t^2 > 2\beta_r(1 - \beta_r)$$

14-20. An astronaut travels to the nearest star system, 4 light years away, and returns at speed $0.3c$. How much has the astronaut aged relative to those people remaining on the Earth?

14-21. The expression for the ordinary force is

$$\mathbf{F} = \frac{d}{dt}\left(\frac{m\mathbf{u}}{\sqrt{1 - \beta^2}}\right)$$

Take $\mathbf{u}$ to be in the x_1-direction and compute the components of the force. Show that

$$F_1 = m_l \dot{u}_1, \qquad F_2 = m_t \dot{u}_2, \qquad F_3 = m_t \dot{u}_3$$

---

* See J. J. Dykla, *Am. J. Phys.* **47**, 381 (1979).

where m_l and m_t are, respectively, the *longitudinal mass* and the *transverse mass*:

$$m_l = \frac{m}{(1 - \beta^2)^{3/2}}, \qquad m_t = \frac{m}{\sqrt{1 - \beta^2}}$$

14-22. The average rate at which solar radiant energy reaches the Earth is approximately 1.4×10^3 W/m². Assume that all this energy results from the conversion of mass to energy. Calculate the rate at which the solar mass is being lost. If this rate is maintained, calculate the remaining lifetime of the sun. (Pertinent numerical data can be found in Table 8-1.)

14-23. Show that the momentum and the kinetic energy of a particle are related by $p^2c^2 = 2Tmc^2 + T^2$.

14-24. What is the minimum proton energy needed in an accelerator to produce antiprotons $\bar{p}$ by the reaction

$$p + p \rightarrow p + p + (p + \bar{p})$$

The mass of a proton and antiproton is m_p.

14-25. A particle of mass m, kinetic energy T, and charge q is moving perpendicular to a magnetic field B as in a cyclotron. Find the relation for the radius r of the particle's path in terms of m, T, q, and B.

14-26. Show that an isolated photon cannot be converted into an electron-positron pair, $\gamma \rightarrow e^- + e^+$. (The conservation laws allow this to happen only near another object.)

14-27. Electrons and positrons collide from opposite directions head-on with equal energies in a storage ring to produce protons by the reaction

$$e^- + e^+ \rightarrow p + \bar{p}$$

The rest energy of a proton and antiproton is 938 MeV. What is the minimum kinetic energy for each particle to produce this reaction?

14-28. Calculate the range of speeds for a particle of mass m in which the classical relation for kinetic energy, $\frac{1}{2}mv^2$, is within 1% of the correct relativistic value. Find the values for an electron and a proton.

14-29. The 2-mile long Stanford Linear Accelerator accelerates electrons to 50 GeV $(50 \times 10^9$ eV). What is the speed of the electrons at the end?

14-30. A free neutron is unstable and decays into a proton and an electron. How much energy other than the rest energies of the proton and electron is available if a neutron at rest decays? (This is an example of nuclear beta decay. Another particle, called a neutrino— actually an antineutrino $\bar{\nu}$ is also produced.)

14-31. A neutral pion π^0 moving at speed $v = 0.98c$ decays in flight into two photons. If the two photons emerge on each side of the pion's direction with equal angles θ, find the angle θ and energies of the photons. The rest energy of π^0 is 135 MeV.

14-32. In nuclear and particle physics, momentum is usually quoted in MeV/c to facilitate calculations. Calculate the kinetic energy of an electron and proton if each has a momentum of 1000 MeV/c.

14-33. A neutron ($m_n = 939.6$ MeV/c^2) at rest decays into a proton ($m_p = 938.3$ MeV/c^2), and electron ($m_e = 0.5$ MeV/c^2), and an antineutrino ($m_{\bar{\nu}} \approx 0$). The three particles emerge at symmetrical angles in a plane, 120° apart. Find the momentum and kinetic energy of each particle.

14-34. Show that Δs^2 is invariant in all inertial systems moving at relative velocities to each other.

14-35. A spacecraft passes Saturn with a speed of $0.9c$ relative to Saturn. A second spacecraft is observed to pass the first one (going in the same direction) at relative speed of $0.2c$. What is the speed of the second spacecraft relative to Saturn?

14-36. We define the four-vector force $\mathbb{F}$ (called the Minkowski force) by differentiating the four-vector momentum with respect to proper time.

$$\mathbb{F} = \frac{d\mathbb{P}}{d\tau}$$

Show that the four-vector force transformation is

$$F_1' = \gamma(F_1 + i\beta F_4)$$
$$F_2' = F_2$$
$$F_3' = F_3$$
$$F_4' = \gamma(F_4 - i\beta F_1)$$

14-37. Consider a one-dimensional, relativistic harmonic oscillator for which the Lagrangian is

$$L = mc^2\left(1 - \sqrt{1 - \beta^2}\right) - \tfrac{1}{2}kx^2$$

Obtain the Lagrange equation of motion and show that it can be integrated to yield

$$E = mc^2 + \tfrac{1}{2}ka^2$$

where a is the maximum excursion from equilibrium of the oscillating particle. Show that the period

$$\tau = 4\int_{x=0}^{x=a} dt$$

can be expressed as

$$\tau = \frac{2a}{\kappa c}\int_0^{\pi/2} \frac{1 + 2\kappa^2\cos^2\phi}{\sqrt{1 + \kappa^2\cos^2\phi}} d\phi$$

Expand the integrand in powers of $\kappa \equiv (a/2)\sqrt{k/mc^2}$ and show that, to first order in κ,

$$\tau \cong \tau_0\left(1 + \frac{3}{16}\frac{ka^2}{mc^2}\right)$$

where τ_0 is the nonrelativistic period for small oscillations, $2\pi\sqrt{m/k}$.

14-38. Show that the relativistic form of Newton's Second Law becomes

$$F = m\frac{du}{dt}\left(1 - \frac{u^2}{c^2}\right)^{-3/2}$$

14-39. A common unit of energy used in atomic and nuclear physics is the electron volt (eV), the energy acquired by an electron in falling through a potential difference of one volt: 1 MeV $= 10^6$ eV $= 1.602 \times 10^{-13}$ J. In these units, the mass of an electron is $m_e c^2 = 0.511$

MeV and that of a proton is $m_pc^2 = 938$ MeV. Calculate the kinetic energy and the quantities β and γ for an electron and for a proton each having a momentum of 100 MeV/c. Show that the electron is "relativistic" whereas the proton is "nonrelativistic."

14-40. Consider an inertial frame K that contains a number of particles with masses m_α, ordinary momentum components $p_{\alpha,j}$, and total energies E_α. The center-of-mass system of such a group of particles is defined to be that system in which the net ordinary momentum is zero. Show that the velocity components of the center-of-mass system with respect to K are given by

$$\frac{v_j}{c} = \frac{\sum_\alpha p_{\alpha,j} c}{\sum_\alpha E_\alpha}$$

14-41. Show that the relativistic expression for the kinetic energy of a particle scattered through an angle ψ by a target particle of equal mass is

$$\frac{T_1}{T_0} = \frac{2\cos^2\psi}{(\gamma_1 + 1) - (\gamma_1 - 1)\cos^2\psi}$$

The expression evidently reduces to Equation 9.89a in the nonrelativistic limit $\gamma_1 \to 1$. Sketch $T_1(\psi)$ for neutron-proton scattering for incident neutron energies of 100 MeV, 1 GeV, and 10 GeV.

14-42. The energy of a light quantum (or photon) is expressed by $E = h\nu$, where h is Planck's constant and ν is the frequency of the photon. The momentum of the photon is $h\nu/c$. Show that, if the photon scatters from a free electron (of mass m_e), the scattered photon has an energy

$$E' = E\left[1 + \frac{E}{m_e c^2}(1 - \cos\theta)\right]^{-1}$$

where θ is the angle through which the photon scatters. Show also that the electron acquires a kinetic energy

$$T = \frac{E^2}{m_e c^2}\left[\frac{1 - \cos\theta}{1 + \dfrac{E}{m_e c^2}(1 - \cos\theta)}\right]$$

"Better is the end of a thing than the beginning thereof."—Ecclesiastes

A

TAYLOR'S
THEOREM

A theorem of considerable importance in mathematical physics is **Taylor's theorem**,* which relates to the expansion of an arbitrary function in a power series. In many instances, it is necessary to use this theorem to simplify a problem to a tractable form.

Consider a function $f(x)$ with continuous derivatives of all orders within a certain interval of the independent variable x. If this interval includes $x_0 \leq x \leq x_0 + h$, we may write

$$I \equiv \int_{x_0}^{x_0+h} f'(x)dx = f(x_0 + h) - f(x_0) \tag{A.1}$$

where $f'(x)$ is the derivative of $f(x)$ with respect to x. If we make the change of variable

$$x = x_0 + h - t \tag{A.2}$$

we have

$$I = \int_0^h f'(x_0 + h - t) \, dt \tag{A.3}$$

---

* First published in 1715 by the English mathematician Brook Taylor (1685–1731).

Integrating by parts

$$I = tf'(x_0 + h - t) \Big|_0^h + \int_0^h tf''(x_0 + h - t)\, dt$$

$$= hf'(x_0) + \int_0^h tf''(x_0 + h - t)\, dt \tag{A.4}$$

Integrating the second term by parts, we find

$$I = hf'(x_0) + \frac{h^2}{2!} f''(x_0) + \int_0^h \frac{t^2}{2!} f'''(x_0 + h - t)\, dt \tag{A.5}$$

Continuing this process, we generate an infinite series for I. From the definition of I, we then have

$$\boxed{f(x_0 + h) = f(x_0) + hf'(x_0) + \frac{h^2}{2!} f''(x_0) + \cdots} \tag{A.6}$$

This is the Taylor series expansion* of the function $f(x_0 + h)$. A more common form of the series results if we set $x_0 = 0$ and $h = x$ [i.e., the function $f(x)$ is expanded about the origin]:

$$\boxed{\begin{aligned} f(x) = f(0) + xf'(0) + \frac{x^2}{2!} f''(0) + \frac{x^3}{3!} f'''(0) + \cdots \\ + \frac{x^n}{n!} f^{(n)}(0) + \cdots \end{aligned}} \tag{A.7}$$

where

$$f^{(n)}(0) \equiv \frac{d^n}{dx^n} f(x) \Big|_{x=0} \tag{A.8}$$

Equation A.7 is usually called the **Maclaurin's series**[†] for the function $f(x)$.

The series expansions given in Equations A.6 and A.7 possess two important properties. Under very general conditions, they may be differentiated or integrated term by term, and the resulting series converge to the derivative or integral of the original function.

---

* The *remainder* term of a series that is terminated after a finite number of terms is discussed, for example, by Kaplan (Ka84).

[†] Discovered by James Stirling in 1717 and published by Colin Maclaurin in 1742.

EXAMPLE **A.1** --

Find the Taylor series expansion of e^x.

Solution: Because the derivative of exp (x) of any order is just exp (x), the exponential series is

$$\boxed{e^x = 1 + x + \frac{x^2}{2!} + \frac{x^3}{3!} + \cdots}$$ (A.9)

This result is of considerable importance and will be used often.

-- ●

EXAMPLE **A.2** --

Find the Taylor series expansion of $\sin x$.

Solution: To expand $f(x) = \sin x$, we need

$$f(x) = \sin x, \qquad f(0) = 0$$
$$f'(x) = \cos x, \qquad f'(0) = 1$$
$$f''(x) = -\sin x, \qquad f''(0) = 0$$
$$f'''(x) = -\cos x, \qquad f'''(0) = -1$$

Therefore,

$$\boxed{\sin x = x - \frac{x^3}{3!} + \frac{x^5}{5!} - \cdots}$$ (A.10)

Similarly,

$$\boxed{\cos x = 1 - \frac{x^2}{2!} + \frac{x^4}{4!} - \cdots}$$ (A.11)

-- ●

EXAMPLE **A.3** --

Use the Taylor series expansion of $(1 + t)^{-1}$ to integrate

$$\int_0^x \frac{dt}{1+t}$$

Solution: A series expansion can often be profitably used in the evaluation of a definite integral. (This is particularly true for those cases in which the indefinite integral cannot be found in closed form.)

$$\int_0^x \frac{dt}{1+t} = \int_0^x (1 - t^2 - t^3 + \cdots)\, dt, \qquad |t| < 1$$

Integrating term by term, we find

$$\int_0^x \frac{dt}{1+t} = x - \frac{x^2}{2} + \frac{x^3}{3} - \cdots \tag{A.12}$$

Because

$$\frac{d}{dx} \ln(1 + x) = \frac{1}{1 + x} \tag{A.13}$$

we also have the result

$$\ln(1 + x) = x - \frac{x^2}{2} + \frac{x^3}{3} - \cdots \tag{A.14}$$

E X A M P L E **A.4** ----

Taylor's series can be used to restructure a function as well as to approximate it. For some applications, such a restructuring may be more useful to work with. We may, for example, want to expand the polynomial $f(x) = 4 + 6x + 3x^2 + 2x^3 + x^4$ about $x = 2$ rather than $x = 0$. First, we compute the various derivatives and evaluate them at $x = 2$:

$$f(2) = 60$$
$$f'(2) = (6 + 6x + 6x^2 + 4x^3)|_{x=2}$$
$$= 74$$
$$f''(2) = (6 + 12x + 12x^2)|_{x=2}$$
$$= 78$$
$$f'''(2) = (12 + 24x)|_{x=2}$$
$$= 60$$
$$f^{iv}(2) = 24$$
$$f^{v}(2) = 0$$

Using Equation A.3 with $h = (x - 2)$

$$f(x) = 60 + 74(x - 2) + 39(x - 2)^2 + 10(x - 2)^3 + (x - 2)^4 \tag{A.15}$$

E X A M P L E **A.5** ----

There are a great many important integrals arising in physics that cannot be integrated in closed form, that is, in terms of elementary functions (polynomials, exponentials, logarithms, trigonometric functions, and their inverses). Integrals with integrands

$$e^{-x^2}, \quad \frac{e^{-x}}{x}, \quad x \tan x, \quad \sin x^2, \quad 1/\ln x, \quad (\sin x)/x, \quad \text{or } 1/\sqrt{1-x^3}$$

are a few such examples. Nevertheless, the values of the integrals or good approximations of their values are needed. A Taylor series expansion of all or part of the integrand followed by a term-by-term integration of the resulting series produces an answer as precise as is wished. As an example, solve the following integral:

$$\int_1^x \frac{e^t}{t}\,dt \tag{A.16}$$

Solution: Using Equation A.9,

$$\int_1^x \frac{e^t}{t}\,dt = \int_1^x \frac{\left(1 + t + \frac{t^2}{2!} + \frac{t^3}{3!} + \cdots\right)\,dt}{t} \tag{A.17}$$

$$= \int_1^x \frac{dt}{t} + \int_1^x dt + \int_1^x \frac{t}{2!}\,dt + \int_1^x \frac{t^2}{3!}\,dt + \cdots$$

$$= \ln x - (x-1) + \frac{1}{4}(x^2 - 1) + \frac{1}{18}(x^3 - 1) + \cdots \tag{A.18}$$

--●

P R O B L E M S

A-1.   Show by division and by direct expansion in a Taylor series that

$$\frac{1}{1-x} = 1 + x + x^2 + x^3 + \cdots + x^n + \cdots$$

For what range of x is the series valid?

A-2.   Expand $\cos x$ about the point $x = \pi/4$.

A-3.   Use a series expansion to show that

$$\int_0^1 \frac{e^x - e^{-x}}{x}\,dx = 2.1145 \ldots .$$

A-4.   Use a Taylor series to expand $\sin^{-1} x$. Verify the result by expanding the integral in the relation

$$\sin^{-1} x = \int_0^x \frac{dt}{\sqrt{1-t^2}}$$

A-5. Evaluate to three decimal places:

$$\int_0^1 \exp(-x^2/2)\, dx$$

Compare the result with that determined from tables of the probability integral.

A-6. Show that if $f(x) = (1 + x)^n$ (with $|x| < 1$) is expanded in a Taylor series, the result is the same as a binomial expansion.

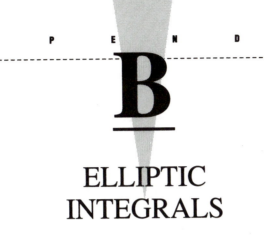

ELLIPTIC
INTEGRALS

There is a large and important class of integrals called **elliptic integrals** that cannot be evaluated in closed form in terms of elementary functions. Elliptic integrals occur in many physical situations; for example, see the exact solution to the plane pendulum in Section 4.4. Any integral of the form

$$\int (a \sin \theta + b \cos \theta + c)^{\pm 1/2} \, d\theta, \quad \text{or} \quad \int R(x, \sqrt{y}) \, dx \qquad \textbf{(B.1)}$$

where R is a rational function, $y = ax^4 + bx^3 + cx^2 + dx + e$ with distinct linear factors and a, b, c, d, and e constants with not both a, b zero is an elliptic integral. It is customary, however, to transform all elliptic integrals into one or more of three standard forms. These standard forms have been much studied and tabulated. Several handbooks are available with tables of values for them.*

ELLIPTIC INTEGRALS OF THE FIRST KIND

$$F(k\phi) = \int_0^\phi \frac{d\theta}{\sqrt{1 - k^2 \sin^2 \theta}}, \quad k^2 < 1 \qquad \textbf{(B.2a)}$$

---

* One of the best of these is Abramowitz and Stegun (Ab65). See also extensive numerical tables in Adams and Hippisley (Ad22) and short tables in Dwight (Dw61).

or if $z = \sin \theta$

$$\bar{F}(k,x) = \int_0^x \frac{dz}{\sqrt{(1 - z^2)(1 - k^2 z^2)}}, \qquad k^2 < 1 \qquad \text{(B.2b)}$$

ELLIPTIC INTEGRALS OF THE SECOND KIND

$$E(k, \phi) = \int_0^\phi \sqrt{1 - k^2 \sin^2 \theta} \, d\theta, \qquad k^2 < 1 \qquad \text{(B.3a)}$$

or if $z = \sin \theta$

$$\bar{E}(k,x) = \int_0^x \sqrt{\frac{1 - k^2 z^2}{1 - z^2}} \, dz, \qquad k^2 < 1 \qquad \text{(B.3b)}$$

ELLIPTIC INTEGRALS OF THE THIRD KIND

$$\Pi(n, k, \phi) = \int_0^\phi \frac{d\theta}{(1 + n \sin^2 \theta) \sqrt{1 - k^2 \sin^2 \theta}} \qquad \text{(B.4a)}$$

or if $z = \sin \theta$

$$\bar{\Pi}(n, k, x) = \int_0^x \frac{dz}{(1 + nz^2) \sqrt{(1 - z^2)(1 - k^2 z^2)}} \qquad \text{(B.4b)}$$

These standard forms obey the following identities, which are often helpful:

$$\left.\begin{array}{l} F(k, \phi) = F(k, \pi) - F(k, \pi - \phi) \\ E(k, \phi) = E(k, \pi) - E(k, \pi - \phi) \end{array}\right\} \qquad \text{(B.5)}$$

and

$$\left.\begin{array}{l} F(k, m\pi + \phi) = mF(k, \pi) + F(k, \phi) \\ E(k, m\pi + \phi) = mE(k, \pi) + E(k, \phi) \end{array}\right\} \qquad \text{(B.6)}$$

where m is an integer.

If tables are not handy or if ϕ or x is needed as a variable, the standard integrals may be approximated by expanding the integrand in an infinite series and integrating term by term. For example, consider

$$E(k, \phi) = \int_0^\phi \sqrt{1 - k^2 \sin^2 \theta} \, d\theta$$

Using the binomial theorem on the integrand

$$(1 - k^2 \sin^2 \theta)^{1/2} = 1 - \frac{1}{2} k^2 \sin^2 \theta - \frac{1}{8} k^4 \sin^4 \theta - \cdots$$

so

$$E(k, \phi) = \int_0^\phi \left[1 - \frac{1}{2} k^2 \sin^2\theta - \frac{1}{8} k^4 \sin^4\theta - \cdots \right.$$

$$\left. - \frac{1 \cdot 3 \cdot 5 \cdots (2n-3)}{2 \cdot 4 \cdot 6 \cdots (2n)} k^{2n} \sin^{2n}\theta - \cdots \right] d\theta$$

$$= \phi - \frac{k^2}{2} \int_0^\phi \sin^2\theta \, d\theta - \cdots - \frac{1 \cdot 3 \cdot 5 \cdots (2n-3)}{2 \cdot 4 \cdot 6 \cdots (2n)} k^{2n}$$

$$\times \int_0^\phi \sin^{2n}\theta \, d\theta - \cdots \tag{B.7}$$

Similarly, the binomial theorem can be used to expand $(1 - k^2\sin^2\theta)^{-1/2}$ to yield

$$F(k, \phi) = \phi + \frac{1}{2} k^2 \int_0^\phi \sin^2\theta \, d\theta + \frac{3}{8} k^4 \int_0^\phi \sin^4\theta \, d\theta + \cdots$$

$$+ \frac{1 \cdot 3 \cdot 5 \cdots (2n-1)}{2 \cdot 4 \cdot 6 \cdots (2n)} k^{2n} \int_0^\phi \sin^{2n}\theta \, d\theta + \cdots \tag{B.8}$$

E X A M P L E   **B.1**   -

Put the integral $\int_{\phi_1}^{\phi_2} \sqrt{1 - k^2 \sin^2\theta} \, d\theta$ **into standard form.**

Solution: Recall from calculus that for any integral $\int_a^b f(x) \, dx$ it is possible to write

$$\int_a^b f(x) \, dx = \int_a^c f(x) \, dx + \int_c^b f(x) \, dx$$

so

$$\int_{\phi_1}^{\phi_2} \sqrt{1 - k \sin^2\theta} \, d\theta = \int_{\phi_1}^0 \sqrt{1 - k^2 \sin^2\theta} \, d\theta + \int_0^{\phi_2} \sqrt{1 - k^2 \sin^2\theta} \, d\theta$$

But there is another property of integrals:

$$\int_a^b f(x) \, dx = - \int_b^a f(x) \, dx$$

so

$$\int_{\phi_1}^{\phi_2} \sqrt{1 - k^2 \sin^2\theta} \, d\theta = \int_0^{\phi_2} \sqrt{1 - k^2 \sin^2\theta} \, d\theta - \int_0^{\phi_1} \sqrt{1 - k^2 \sin^2\theta} \, d\theta$$

or

$$\int_{\phi_1}^{\phi_2} \sqrt{1 - k^2 \sin^2\theta} \, d\theta = E(k, \phi_2) - E(k, \phi_1) \tag{B.9}$$

The terms on the right can be looked up in a handbook.

- ●

EXAMPLE **B.2**

Transform the elliptic integral

$$\int_0^\phi \frac{d\theta}{\sqrt{1 - n^2 \sin^2\theta}}, \qquad \text{where } n^2 > 1$$

into a standard form.

Solution: To reduce this integral to standard form, the radical must be transformed to $\sqrt{1 - k^2 \sin^2\theta}$, with $k^2 < 1$. To do this, consider the transformation $n \sin\theta = \sin\beta$. Differentiating, we have

$$n \cos\theta \, d\theta = \cos\beta \, d\beta$$

so

$$d\theta = \frac{\cos\beta \, d\beta}{n \cos\theta}$$

Using the identity $\sin^2\theta + \cos^2\theta = 1$ leads to

$$\cos\theta = \sqrt{1 - \sin^2\theta} = \sqrt{1 - \left(\frac{\sin\beta}{n}\right)^2}$$

Also, $\cos\beta = \sqrt{1 - \sin^2\beta}$, and $\sqrt{1 - n^2 \sin^2\theta} = \sqrt{1 - \sin^2\beta}$. Hence the integral becomes

$$\int_0^\phi \frac{d\theta}{\sqrt{1 - n^2 \sin^2\theta}} = \int_0^{\sin^{-1}(n \sin\phi)} \frac{\sqrt{1 - \sin^2\beta} \, d\beta}{n\sqrt{1 - \left(\frac{\sin\beta}{n}\right)^2} \sqrt{1 - \sin^2\beta}}$$

$$= \frac{1}{n} \int_0^{\sin^{-1}(n \sin\phi)} \frac{d\beta}{\sqrt{1 - \left(\frac{1}{n^2}\right)\sin^2\beta}}$$

so

$$\int_0^\phi \frac{d\theta}{\sqrt{1 - n^2 \sin^2\theta}} = \frac{1}{n}\int_0^{\sin^{-1}(n \sin\phi)} \frac{d\beta}{\sqrt{1 - \left(\frac{1}{n^2}\right)\sin^2\beta}} \qquad \textbf{(B.10)}$$

where $1/n^2 < 1$. The integral on the right is now in standard form.

EXAMPLE **B.3**

Transform the elliptic integral

$$\int_0^\phi \frac{d\theta}{\sqrt{\cos 2\theta}}$$

into a standard form.

Solution: Let $\mu = \sin\theta$; then $d\mu = \cos\theta\,d\theta$. Because $\cos^2\theta + \sin^2\theta = 1$, $\cos\theta = \sqrt{1 - \sin^2\theta} = \sqrt{1 - \mu^2}$, so $d\theta = d\mu/\sqrt{1 - \mu^2}$. By another trigonometric identity, $\cos 2\theta = 1 - 2\sin^2\theta = 1 - 2\mu^2$. Thus $\sqrt{\cos 2\theta} = \sqrt{1 - 2\mu^2}$, and

$$\int_0^\phi \frac{d\theta}{\sqrt{\cos 2\theta}} = \int_0^{\sin\phi} \frac{d\mu}{\sqrt{1 - \mu^2}\sqrt{1 - 2\mu^2}}$$

Let $z = \sqrt{2}\,\mu$; then $dz = \sqrt{2}\,d\mu$, so

$$\int_0^\phi \frac{d\theta}{\sqrt{\cos 2\theta}} = \frac{1}{\sqrt{2}}\int_0^{\sqrt{2}\sin\phi} \frac{dz}{\sqrt{(1 - z^2)(1 - \frac{1}{2}z^2)}} \qquad \text{(B.11)}$$

The integral on the right is in standard form.

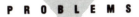

P R O B L E M S

B-1. Evaluate the following integrals using a set of tables.
(a) $F(0.27, \pi/3)$ (b) $E(0.27, \pi/3)$
(c) $F(0.27, 7\pi/4)$ (d) $E(0.27, 7\pi/4)$

B-2. Reduce to standard form:

(a) $\displaystyle\int_0^{\pi/6} \frac{d\theta}{\sqrt{1 - 4\sin^2\theta}}$ (b) $\displaystyle\int_{-1/4}^{3/4} \sqrt{\frac{25 - 4z^2}{1 - z^2}}\,dz$

B-3. Find the binomial expansion of $(1 - k^2\sin^2\theta)^{-1/2}$ and then derive Equation B.8.

C

ORDINARY DIFFERENTIAL EQUATIONS OF SECOND ORDER*

C.1   LINEAR HOMOGENEOUS EQUATIONS

By far, the most important type of ordinary differential equation encountered in problems in mathematical physics is the second-order linear equation with constant coefficients. Equations of this type have the form

$$\frac{d^2y}{dx^2} + a\frac{dy}{dx} + by = f(x) \tag{C.1a}$$

or, denoting derivatives by primes,

$$y'' + ay' + by = f(x) \tag{C.1b}$$

A particularly important class of such equations are those for which $f(x) = 0$. These equations (called **homogeneous equations**) are important not only in themselves but also as *reduced* equations in the solution of the more general type of equation (Equation C.1).

We consider the linear homogeneous second-order equation with constant coefficients first[†]:

$$y'' + ay' + by = 0 \tag{C.2}$$

---

* A standard treatise on differential equations is that of Ince (In27). A listing of many types of equations and their solutions is given by Murphy (Mu60). A modern viewpoint is contained in the book by Hochstadt (Ho64).

[†] The first published solution of an equation of this type was by Euler in 1743, but the solution appears to have been known to Daniel and Johann Bernoulli in 1739.

These equations have the following important properties:

a. If $y_1(x)$ is a solution of Equation C.2, then $c_1y_1(x)$ is also a solution.

b. If $y_1(x)$ and $y_2(x)$ are solutions, then $y_1(x) + y_2(x)$ is also a solution (principle of *superposition*).

c. If $y_1(x)$ and $y_2(x)$ are *linearly* independent solutions, then the *general* solution to the equation is given by $c_1y_1(x) + c_2y_2(x)$. (The general solution always contains two arbitrary constants.)

The functions $y_1(x)$ and $y_2(x)$ are **linearly independent** if and only if the equation

$$\lambda y_1(x) + \mu y_2(x) \equiv 0 \tag{C.3}$$

is satisfied only by $\lambda = \mu = 0$. If Equation C.3 can be satisfied with λ and μ different from zero, then $y_1(x)$ and $y_2(x)$ are said to be **linearly dependent**.

The general condition (i.e., the necessary and sufficient condition) that a set of functions $y_1, y_2, y_3, \cdots$ be linearly dependent is that the **Wronskian determinant** of these functions vanish identically:

$$W = \begin{vmatrix} y_1 & y_2 & y_3 & \cdots & y_n \\ y_1' & y_2' & y_3' & \cdots & y_n' \\ y_1'' & y_2'' & y_3'' & \cdots & y_n'' \\ \vdots & & & & \\ y_1^{(n-1)} & y_2^{(n-1)} & y_3^{(n-1)} & \cdots & y_n^{(n-1)} \end{vmatrix} = 0 \tag{C.4}$$

where $y^{(n)}$ is the nth derivative of y with respect to x.

The properties (a) and (b) above can be verified by direction substitution, but (c) is only asserted here to yield the general solution. These properties apply *only* to the homogeneous equation (Equation C.2) and *not* to the general equation (Equation C.1).

Equations of the type C.2 are reducible through the substitution

$$y = e^{rx} \tag{C.5}$$

Now

$$y' = re^{rx}, \qquad y'' = r^2e^{rx} \tag{C.6}$$

Using these expressions for y' and y'' in Equation C.2, we find an algebraic equation called the **auxiliary equation**:

$$r^2 + ar + b = 0 \tag{C.7}$$

The solution of this quadratic in r is

$$r = -\frac{a}{2} \pm \frac{1}{2}\sqrt{a^2 - 4b} \tag{C.8}$$

We first assume that the two roots, denoted by r_1 and r_2, are not identical and write the solution as

$$y = e^{r_1x} + e^{r_2x} \tag{C.9}$$

Because the Wronskian determinant of $\exp(r_1 x)$ and $\exp(r_2 x)$ does not vanish, these functions are linearly independent. Thus, the general solution is

$$y = c_1 e^{r_1 x} + c_2 e^{r_2 x}, \qquad r_1 \neq r_2 \tag{C.10}$$

If it happens that $r_1 = r_2 = r$, then it can be verified by direct substitution that $x \exp(rx)$ is also a solution, and because $\exp(rx)$ and $x \exp(rx)$ are linearly independent, the general solution for identical roots is given by

$$y = c_1 e^{rx} + c_2 x e^{rx}, \qquad r_1 = r_2 \equiv r \tag{C.11}$$

EXAMPLE C.1

Solve the equation

$$y'' - 2y' - 3y = 0 \tag{C.12}$$

Solution: The auxiliary equation is

$$r^2 - 2r - 3 = (r - 3)(r + 1) = 0 \tag{C.13}$$

The roots are

$$r_1 = 3, \qquad r_2 = -1 \tag{C.14}$$

The general solution is therefore

$$y = c_1 e^{3x} + c_2 e^{-x} \tag{C.15}$$

EXAMPLE C.2

Solve the equation

$$y'' + 4y' + 4y = 0 \tag{C.16}$$

Solution: The auxiliary equation is

$$r^2 + 4r + 4 = (r + 2)^2 = 0 \tag{C.17}$$

The roots are equal, and $r = -2$. The general solution is therefore

$$y = c_1 e^{-2x} + c_2 x e^{-2x} \tag{C.18}$$

If the roots r_1 and r_2 of the auxiliary equation are imaginary, the solutions given by $c_1 \exp(r_1 x)$ and $c_2 \exp(r_2 x)$ are still correct.

To give the solutions entirely in terms of real quantities, we use the Euler relations to express the exponentials. Then,

$$\left.\begin{array}{l} e^{r_1 x} = e^{\alpha x} e^{i\beta x} = e^{\alpha x}(\cos \beta x + i \sin \beta x) \\ e^{r_2 x} = e^{\alpha x} e^{-i\beta x} = e^{\alpha x}(\cos \beta x - i \sin \beta x) \end{array}\right\} \qquad \textbf{(C.19)}$$

and the general solution is

$$\begin{aligned} y &= c_1 e^{r_1 x} + c_2 e^{r_2 x} \\ &= e^{\alpha x}[(c_1 + c_2)\cos \beta x + i(c_1 - c_2)\sin \beta x] \end{aligned} \qquad \textbf{(C.20)}$$

Now c_1 and c_2 are arbitrary, but these constants may be complex. However, not all four elements can be independent (because there would be *four* arbitrary constants rather than *two*). The number of independent elements can be reduced to the required *two* by making c_1 and c_2 complex conjugates. Then the combinations $A \equiv c_1 + c_2$ and $B \equiv i(c_1 - c_2)$ become a pair of arbitrary, real constants. Using these quantities in the solution, we have

$$y = e^{\alpha x}(A \cos \beta x + B \sin \beta x) \qquad \textbf{(C.21)}$$

Equation C.21 may be put into a form that is sometimes more convenient by multiplying and dividing by $\mu = \sqrt{A^2 + B^2}$:

$$y = \mu e^{\alpha x}[(A/\mu)\cos \beta x + (B/\mu)\sin \beta x] \qquad \textbf{(C.22)}$$

Next, we define an angle δ (see Figure C-1) such that

$$\sin \delta = A/\mu, \qquad \cos \delta = B/\mu, \qquad \tan \delta = A/B \qquad \textbf{(C.23)}$$

Then, the solution becomes

$$\begin{aligned} y &= \mu e^{\alpha x}(\sin \delta \cos \beta x + \cos \delta \sin \beta x) \\ &= \mu e^{\alpha x} \sin(\beta x + \delta) \end{aligned}$$

Depending on the exact definition of the phase δ, we may write the solution alternatively as

$$\boxed{\begin{aligned} y &= \mu e^{\alpha x} \sin(\beta x + \delta) \\ y &= \mu e^{\alpha x} \cos(\beta x + \delta) \end{aligned}}$$

$\textbf{(C.24a)}$

$\textbf{(C.24b)}$

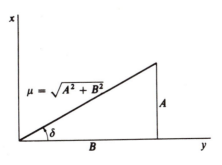

FIGURE C-1

EXAMPLE C.3 ---

Solve the equation

$$y'' + 2y' + 4y = 0 \tag{C.25}$$

Solution: The auxiliary equation is

$$r^2 + 2r + 4 = 0 \tag{C.26}$$

with

$$r = \frac{-2 \pm \sqrt{4 - 16}}{2} = -1 \pm i\sqrt{3} \tag{C.27}$$

Hence,

$$\alpha = -1, \qquad \beta = \sqrt{3} \tag{C.28}$$

and the general solution is

$$y = e^{-x}(c_1 \cos \sqrt{3}x + c_2 \sin \sqrt{3}x) \tag{C.29}$$

or

$$y = \mu e^{-x} \sin [(\sqrt{3}x + \delta)] \tag{C.30}$$

Summarizing, then, there are three possible types of general solutions to homogeneous second-order linear differential equations, as indicated in Table C-1.

TABLE C-1

Roots of the auxiliary equations	General solution
Real, unequal ($r_1 \neq r_2$)	$c_1 e^{r_1 x} + c_2 e^{r_2 x}$
Real, equal ($r_1 = r_2 \equiv r$)	$c_1 e^{rx} + c_2 x e^{rx}$
Imaginary ($\alpha \pm i\beta$)	$e^{\alpha x}(c_1 \cos \beta x + c_2 \sin \beta x)$
	or
	$\mu e^{\alpha x} \sin (\beta x + \delta)$

C.2 LINEAR INHOMOGENEOUS EQUATIONS

To solve the general (i.e., inhomogeneous) second-order linear differential equation, consider the following. Let $y = u$ be the *general solution* of

$$y'' + ay' + by = 0 \tag{C.31}$$

and let $y = v$ be *any* solution of

$$y'' + ay' + by = f(x) \tag{C.32}$$

Then, $y = u + v$ is a solution of Equation C.32, because

$$y'' + ay' + by = (u'' + au' + bu) + (v'' + av' + bv)$$
$$= 0 + f(x)$$

Because u contains the two arbitrary constants c_1 and c_2, the combinations $u + v$ satisfies all the requirements of the general solution to Equation C.32. The function u is the **complementary function** and v is the **particular integral** of the equation. Because a general method of finding u has been given above, it only remains to find, by inspection or by trial, some function v that satifies

$$v'' + av' + bv = f(x) \tag{C.33}$$

EXAMPLE  C.4 — — — — — — — — — — — — — — — — — —

Solve the equation

$$y'' + 5y' + 6y = x^2 + 2x \tag{C.34}$$

Solution: The auxiliary equation is

$$r^2 + 5r + 6 = (r + 3)(r + 2) = 0 \tag{C.35}$$

$$r_1 = -3, \qquad r_2 = -2 \tag{C.36}$$

so the complementary function is

$$u = c_1 e^{-3x} + c_2 e^{-2x} \tag{C.37}$$

Because the right-hand side of the original equation is a second-degree polynomial, we guess a particular integral of the form

$$v = Ax^2 + Bx + C \tag{C.38}$$

Then,

$$v' = 2Ax + B \tag{C.39}$$

$$v'' = 2A \tag{C.40}$$

Substituting into the differential equation, we have

$$2A + 5(2Ax + B) + 6(Ax^2 + Bx + C) = x^2 + 2x \tag{C.41}$$

or

$$(6A)x^2 + (10A + 6B)x + (2A + 5B + 6C) = x^2 + 2x \tag{C.42}$$

Equation coefficients of like powers of x:

$$\left. \begin{aligned} 6A &= 1 \\ 10A + 6B &= 2 \\ 2A + 5B + 6C &= 0 \end{aligned} \right\} \tag{C.43}$$

Solving,

$$A = \frac{1}{6}, \qquad B = \frac{1}{18}, \qquad C = -\frac{11}{108} \tag{C.44}$$

Hence,

$$v = \frac{1}{6}x^2 + \frac{1}{18}x - \frac{11}{108}$$

$$= \frac{18x^2 + 6x - 11}{108} \tag{C.45}$$

The general solution is therefore

$$y = u + v = c_1 e^{-3x} + c_2 e^{-2x} + \frac{18x^2 + 6x - 11}{108} \tag{C.46}$$

The type of solution illustrated in this example is called the **method of undetermined coefficients**.

●

E X A M P L E  **C.5**

Solve the equation

$$y'' + 4y = 3x \cos x \tag{C.47}$$

Solution: The auxiliary equation is

$$r^2 + 4 = (r + 2i)(r - 2i) = 0 \tag{C.48}$$

with roots

$$\left. \begin{array}{l} r_1 = \alpha + i\beta = 0 + 2i \\ r_2 = \alpha - i\beta = 0 - 2i \end{array} \right\} \tag{C.49}$$

so

$$\alpha = 0, \qquad \beta = 2 \tag{C.50}$$

and the complementary function is

$$u = e^{\alpha x}(c_1 \cos \beta x + c_2 \sin \beta x)$$

$$= c_1 \cos 2x + c_2 \sin 2x \tag{C.51}$$

To find a particular integral, we note that from $x \cos x$ and its derivatives it is possible to generate only terms involving the following functions:

$$x \cos x, \qquad x \sin x, \qquad \cos x, \qquad \sin x$$

Therefore, because these functions are linearly independent, the trial particular integral is

$$v = Ax \cos x + Bx \sin x + C \cos x + D \sin x \qquad \text{(C.52)}$$

$$v' = A(\cos x - x \sin x) + B(\sin x + x \cos x)$$

$$-C \sin x + D \cos x \qquad \text{(C.53)}$$

$$v'' = -A(2 \sin x + x \cos x) + B(2 \cos x - x \sin x)$$

$$- C \cos x - D \sin x \qquad \text{(C.54)}$$

Substituting into the original differential equation,

$$(3D - 2A)\sin x + (2B + 3C)\cos x + 3(A - 1)x \cos x + (3B)x \sin x = 0 \quad \text{(C.55)}$$

The coefficient of each term must vanish (because of the linear independence of the terms):

$$3D = 2A, \qquad 2B = -3C, \qquad A = 1, \qquad 3B = 0 \qquad \text{(C.56)}$$

from which

$$A = 1, \qquad B = 0, \qquad C = 0, \qquad D = \frac{2}{3} . \qquad \text{(C.57)}$$

The general solution is therefore

$$y = c_1 \sin 2x + c_2 \cos 2x + x \cos x + \frac{2}{3} \sin x \qquad \text{(C.58)}$$

--- ●

If the right-hand side, $f(x)$, of the general equation (Equation C.1 or C.32) is such that $f(x)$ and its first two derivatives (only second-order equations are being considered) contain only linearly independent functions, then a linear combination of these functions constitutes the trial particular integral. In the event that the trial function contains a term that already appears in the complementary function, use the term multiplied by x; if this combination also appears in the complementary function, use the term multiplied by x^2. No higher powers are needed because only second-order equations are being considered and only $\exp(rx)$ or $x \exp(rx)$ occur as solutions to the reduced equation; $(x^2) \exp(rx)$ never occurs.

P R O B L E M S

C-1. Solve the following homogeneous second-order equations:
(a) $y'' + 2y' - 3y = 0$ \qquad (b) $y'' + y = 0$
(c) $y'' - 2y' + 2y = 0$ \qquad (d) $y'' - 2y' + 5y = 0$

C-2. Solve the following inhomogeneous equations by the method of undetermined coefficients:

(a) $y'' + 2y' - 8y = 16x$ 　　　　　　 (b) $y'' - 2y' + y = 2e^{2x}$

(c) $y'' + y = \sin x$ 　　　　　　　　 (d) $y'' - 2y' + y = 3xe^{x}$

(e) $y'' - 4y' + 5y = e^{2x} + 4 \sin x$

C-3. Use a Taylor series expansion to obtain the solution of

$$y'' + y^2 = x^2$$

that obeys the conditions $y(0) = 1$ and $y'(0) = 0$. (Differentiate the equation successively to obtain the derivatives that occur in the Taylor series.)

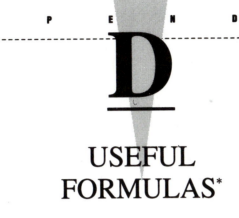

USEFUL FORMULAS*

D.1 BINOMIAL EXPANSION

$$(1 + x)^n = 1 + nx + \frac{n(n - 1)}{2!}x^2 + \frac{n(n - 1)(n - 2)}{3!}x^3$$

$$+ \cdots + \binom{n}{r}x^r + \cdots, \qquad |x| < 1 \qquad \text{(D.1)}$$

$$(1 - x)^n = 1 - nx + \frac{n(n - 1)}{2!}x^2 - \frac{n(n - 1)(n - 2)}{3!}x^3$$

$$+ \cdots + (-1)^r\binom{n}{r}x^r + \cdots, \qquad -|x| < 1 \qquad \text{(D.2)}$$

where the **binomial coefficient** is

$$\binom{n}{r} \equiv \frac{n!}{(n - r)!r!} \qquad \text{(D.3)}$$

Some particularly useful cases of the above are

$$(1 \pm x)^{1/2} = 1 \pm \frac{1}{2}x - \frac{1}{8}x^2 \pm \frac{1}{16}x^3 - \cdots \qquad \text{(D.4)}$$

---

* An extensive list may be found, for example, in Dwight (Dw61).

$$(1 \pm x)^{1/3} = 1 \pm \frac{1}{3}x - \frac{1}{9}x^2 \pm \frac{5}{81}x^3 - \cdots \qquad \textbf{(D.5)}$$

$$(1 \pm x)^{-1/2} = 1 \mp \frac{1}{2}x + \frac{3}{8}x^2 \mp \frac{5}{16}x^3 + \cdots \qquad \textbf{(D.6)}$$

$$(1 \pm x)^{-1/3} = 1 \mp \frac{1}{3}x + \frac{2}{9}x^2 \mp \frac{14}{81}x^3 + \cdots \qquad \textbf{(D.7)}$$

$$(1 \pm x)^{-1} = 1 \mp x + x^2 \mp x^3 + \cdots \qquad \textbf{(D.8)}$$

$$(1 \pm x)^{-2} = 1 \mp 2x + 3x^2 \mp 4x^3 + \cdots \qquad \textbf{(D.9)}$$

$$(1 \pm x)^{-3} = 1 \mp 3x + 6x^2 \mp 10x^3 + \cdots \qquad \textbf{(D.10)}$$

For convergence of *all* the above series, we must have $|x| < 1$.

D.2 TRIGONOMETRIC RELATIONS

$$\sin(A \pm B) = \sin A \cos B \pm \cos A \sin B \qquad \textbf{(D.11)}$$

$$\cos(A \pm B) = \cos A \cos B \mp \sin A \sin B \qquad \textbf{(D.12)}$$

$$\sin 2A = 2 \sin A \cos A = \frac{2 \tan A}{1 + \tan^2 A} \qquad \textbf{(D.13)}$$

$$\cos 2A = 2 \cos^2 A - 1 \qquad \textbf{(D.14)}$$

$$\sin^2 \frac{A}{2} = \frac{1}{2}(1 - \cos A) \qquad \textbf{(D.15)}$$

$$\cos^2 \frac{A}{2} = \frac{1}{2}(1 + \cos A) \qquad \textbf{(D.16)}$$

$$\sin^2 A = \frac{1}{2}(1 - \cos 2A) \qquad \textbf{(D.17)}$$

$$\sin^3 A = \frac{1}{4}(3 \sin A - \sin 3A) \qquad \textbf{(D.18)}$$

$$\sin^4 A = \frac{1}{8}(3 - 4 \cos 2A + \cos 4A) \qquad \textbf{(D.19)}$$

$$\cos^2 A = \frac{1}{2}(1 + \cos 2A) \qquad \textbf{(D.20)}$$

$$\cos^3 A = \frac{1}{4}(3 \cos A + \cos 3A) \qquad \textbf{(D.21)}$$

$$\cos^4 A = \frac{1}{8}(3 + 4 \cos 2A + \cos 4A) \qquad \textbf{(D.22)}$$

$$\tan(A + B) = \frac{\tan A + \tan B}{1 - \tan A \tan B} \qquad \textbf{(D.23)}$$

$$\tan^2 \frac{A}{2} = \frac{1 - \cos A}{1 + \cos A} \tag{D.24}$$

$$\sin x = \frac{e^{ix} - e^{-ix}}{2i} \tag{D.25}$$

$$\cos x = \frac{e^{ix} + e^{-ix}}{2} \tag{D.26}$$

$$e^{ix} = \cos x + i \sin x \tag{D.27}$$

D.3 TRIGONOMETRIC SERIES

$$\sin x = x - \frac{x^3}{3!} + \frac{x^5}{5!} - \frac{x^7}{7!} + \cdots \tag{D.28}$$

$$\cos x = 1 - \frac{x^2}{2!} + \frac{x^4}{4!} - \frac{x^6}{6!} + \cdots \tag{D.29}$$

$$\tan x = x + \frac{x^3}{3} + \frac{2}{15}x^5 + \cdots, \qquad |x| < \pi/2 \tag{D.30}$$

$$\sin^{-1} x = x + \frac{x^3}{6} + \frac{3}{40}x^5 + \cdots, \qquad \begin{cases} |x| < 1 \\ |\sin^{-1} x| < \pi/2 \end{cases} \tag{D.31}$$

$$\cos^{-1} x = \frac{\pi}{2} - x - \frac{x^3}{6} - \frac{3}{40}x^5 - \cdots, \qquad \begin{cases} |x| < 1 \\ 0 < \cos^{-1} x < \pi \end{cases} \tag{D.32}$$

$$\tan^{-1} x = x - \frac{x^3}{3} + \frac{x^5}{5} - \frac{x^7}{7} + \cdots, \qquad |x| < 1 \tag{D.33}$$

D.4 EXPONENTIAL AND LOGARITHMIC SERIES

$$e^x = 1 + x + \frac{x^2}{2!} + \frac{x^3}{3!} + \cdots = \sum_{n=0}^{\infty} \frac{x^n}{n!} \tag{D.34}$$

$$\ln(1 + x) = x - \frac{x^2}{2} + \frac{x^3}{3} - \frac{x^4}{4} + \cdots, \qquad |x| < 1, \qquad x = 1 \tag{D.35}$$

$$\ln[\sqrt{(x^2/a^2) + 1} + (x/a)] = \sinh^{-1} x/a \tag{D.36}$$

$$= -\ln[\sqrt{(x^2/a^2) + 1} - (x/a)] \tag{D.37}$$

D.5 COMPLEX QUANTITIES

Cartesian form: $z = x + iy$, complex conjugate $z^* = x - iy$, $i = \sqrt{-1}$ $\qquad$ **(D.38)**

Polar form:

$$z = |z|e^{i\theta} \tag{D.39}$$

$$z^* = |z|e^{-i\theta} \tag{D.40}$$

$$zz^* = |z|^2 = x^2 + y^2 \tag{D.41}$$

Real part of z:

$$\text{Re } z = \frac{1}{2}(z + z^*) = x \tag{D.42}$$

Imaginary part of z:

$$\text{Im } z = -\frac{1}{2}(z - z^*) = y \tag{D.43}$$

Euler's formula:

$$e^{i\theta} = \cos\theta + i\sin\theta \tag{D.44}$$

D.6   HYPERBOLIC FUNCTIONS

$$\sinh x = \frac{e^x - e^{-x}}{2} \tag{D.45}$$

$$\cosh x = \frac{e^x + e^{-x}}{2} \tag{D.46}$$

$$\tanh x = \frac{e^{2x} - 1}{e^{2x} + 1} \tag{D.47}$$

$$\sin ix = i \sinh x \tag{D.48}$$

$$\cos ix = \cosh x \tag{D.49}$$

$$\sinh ix = i \sin x \tag{D.50}$$

$$\cosh ix = \cos x \tag{D.51}$$

$$\sinh^{-1} x = \tanh^{-1}\left(\frac{x}{\sqrt{x^2 + 1}}\right) \tag{D.52}$$

$$= \ln(x + \sqrt{x^2 + 1}) \tag{D.53}$$

$$= \cosh^{-1}(\sqrt{x^2 + 1}), \quad \begin{cases} > 0, & x > 0 \\ < 0, & x < 0 \end{cases} \tag{D.54}$$

$$\cosh^{-1} x = \pm\tanh^{-1}\left(\frac{\sqrt{x^2 - 1}}{x}\right), \quad x > 1 \tag{D.55}$$

$$= \pm\ln(x + \sqrt{x^2 - 1}), \quad x > 1 \tag{D.56}$$

$$= \pm\sinh^{-1}(\sqrt{x^2 - 1}), \quad x > 1 \tag{D.57}$$

$$\frac{d}{dy}\sinh y = \cosh y \tag{D.58}$$

$$\frac{d}{dy}\cosh y = \sinh y \tag{D.59}$$

$$\sinh(x_1 + x_2) = \sinh x_1 \cosh x_2 + \cosh x_1 \sinh x_2 \qquad \textbf{(D.60)}$$

$$\cosh(x_1 + x_2) = \cosh x_1 \cosh x_2 + \sinh x_1 \sinh x_2 \qquad \textbf{(D.61)}$$

$$\cosh^2 x - \sinh^2 x = 1 \qquad \textbf{(D.62)}$$

P R O B L E M S

D-1. Is it possible to ascribe a meaning to the inequality $z_1 < z_2$? Explain. Does the inequality $|z_1| < |z_2|$ have a different meaning?

D-2. Solve following equations:
(a) $z^2 + 2z + 2 = 0$ (b) $2z^2 + z + 2 = 0$

D-3. Express the following in polar form:
(a) $z_1 = i$ (b) $z_2 = -1$
(c) $z_3 = 1 + i\sqrt{3}$ (d) $z_4 = 1 + 2i$
(e) Find the product $z_1 z_2$ (f) Find the product $z_1 z_3$
(g) Find the product $z_3 z_4$

D-4. Express $(z^2 - 1)^{-1/2}$ in polar form.

D-5. If the function $w = \sin^{-1} z$ is defined as the inverse of $z = \sin w$, then use the Euler relation for $\sin w$ to find an equation for $\exp(iw)$. Solve this equation and obtain the result

$$w = \sin^{-1} z = -i \ln(iz + \sqrt{1 - z^2})$$

D-6. Show that

$$y = Ae^{ix} + Be^{-ix}$$

can be written as

$$y = C \cos(x - \delta)$$

where A and B are *complex* but where C and δ are *real*.

D-7. Show that
(a) $\sinh(x_1 + x_2) = \sinh x_1 \cosh x_2 + \cosh x_1 \sinh x_2$
(b) $\cosh(x_1 + x_2) = \cosh x_1 \cosh x_2 + \sinh x_1 \sinh x_2$

E

USEFUL
INTEGRALS*

E.1   ALGEBRAIC FUNCTIONS

$$\int \frac{dx}{a^2 + x^2} = \frac{1}{a} \tan^{-1}\left(\frac{x}{a}\right), \qquad \left|\tan^{-1}\left(\frac{x}{a}\right)\right| < \frac{\pi}{2} \qquad \text{(E.1)}$$

$$\int \frac{x\,dx}{a^2 + x^2} = \frac{1}{2} \ln(a^2 + x^2) \qquad \text{(E.2)}$$

$$\int \frac{dx}{x(a^2 + x^2)} = \frac{1}{2a^2} \ln\left(\frac{x^2}{a^2 + x^2}\right) \qquad \text{(E.3)}$$

$$\int \frac{dx}{a^2x^2 - b^2} = \frac{1}{2ab} \ln\left(\frac{ax - b}{ax + b}\right) \qquad \text{(E.4a)}$$

$$= -\frac{1}{ab} \coth^{-1}\left(\frac{ax}{b}\right) \qquad \text{(E.4b)}$$

$$= -\frac{1}{ab} \tanh^{-1}\left(\frac{ax}{b}\right), \qquad a^2x^2 < b^2 \qquad \text{(E.4c)}$$

---

* This list is confined to those (nontrivial) integrals that arise in the text and in the problems. Extremely useful compilations are, for example, Pierce and Foster (Pi57) and Dwight (Dw61).

$$\int \frac{dx}{\sqrt{a + bx}} = \frac{2}{b}\sqrt{a + bx} \tag{E.5}$$

$$\int \frac{dx}{\sqrt{x^2 + a^2}} = \ln(x + \sqrt{x^2 + a^2}) \tag{E.6}$$

$$\int \frac{x^2\, dx}{\sqrt{a^2 - x^2}} = -\frac{x}{2}\sqrt{a^2 - x^2} + \frac{a^2}{2}\sin^{-1}\frac{x}{a} \tag{E.7}$$

$$\int \frac{dx}{\sqrt{ax^2 + bx + c}} = \frac{1}{\sqrt{a}}\ln(2\sqrt{a}\sqrt{ax^2 + bx + c} + 2ax + b), \quad a > 0 \tag{E.8a}$$

$$= \frac{1}{\sqrt{a}}\sinh^{-1}\left(\frac{2ax + b}{\sqrt{4ac - b^2}}\right), \qquad \begin{cases} a > 0 \\ 4ac > b^2 \end{cases} \tag{E.8b}$$

$$= -\frac{1}{\sqrt{-a}}\sin^{-1}\left(\frac{2ax + b}{\sqrt{b^2 - 4ac}}\right), \qquad \begin{cases} a < 0 \\ b^2 > 4ac \\ |2ax + b| < \sqrt{b^2 - 4ac} \end{cases} \tag{E.8c}$$

$$\int \frac{x\, dx}{\sqrt{ax^2 + bx + c}} = \frac{1}{a}\sqrt{ax^2 + bx + c} - \frac{b}{2a}\int \frac{dx}{\sqrt{ax^2 + bx + c}} \tag{E.9}$$

$$\int \frac{dx}{x\sqrt{ax^2 + bx + c}} = -\frac{1}{\sqrt{c}}\sinh^{-1}\left(\frac{bx + 2c}{|x|\sqrt{4ac - b^2}}\right), \qquad \begin{cases} c > 0 \\ 4ac > b^2 \end{cases} \tag{E.10a}$$

$$= \frac{1}{\sqrt{-c}}\sin^{-1}\left(\frac{bx + 2c}{|x|\sqrt{b^2 - 4ac}}\right), \qquad \begin{cases} c < 0 \\ b^2 > 4ac \end{cases} \tag{E.10b}$$

$$= -\frac{1}{\sqrt{c}}\ln\left(\frac{2\sqrt{c}}{x}\sqrt{ax^2 + bx + c} + \frac{2c}{x} + b\right), \qquad c > 0 \tag{E.10c}$$

$$\int \sqrt{ax^2 + bx + c}\, dx = \frac{2ax + b}{4a}\sqrt{ax^2 + bx + c}$$

$$+ \frac{4ac - b^2}{8a}\int \frac{dx}{\sqrt{ax^2 + bx + c}} \tag{E.11}$$

E.2 TRIGONOMETRIC FUNCTIONS

$$\int \sin^2 x\, dx = \frac{x}{2} - \frac{1}{4}\sin 2x \tag{E.12}$$

$$\int \cos^2 x\, dx = \frac{x}{2} + \frac{1}{4}\sin 2x \tag{E.13}$$

$$\int \frac{dx}{a + b \sin x} = \frac{2}{\sqrt{a^2 - b^2}} \tan^{-1}\left[\frac{a \tan (x/2) + b}{\sqrt{a^2 - b^2}}\right], \qquad a^2 > b^2 \qquad \text{(E.14)}$$

$$\int \frac{dx}{a + b \cos x} = \frac{2}{\sqrt{a^2 - b^2}} \tan^{-1}\left[\frac{(a - b) \tan (x/2)}{\sqrt{a^2 - b^2}}\right], \qquad a^2 > b^2 \qquad \text{(E.15)}$$

$$\int \frac{dx}{(a + b \cos x)^2} = \frac{b \sin x}{(b^2 - a^2)(a + b \cos x)} - \frac{a}{b^2 - a^2}\int \frac{dx}{a + b \cos x} \qquad \text{(E.16)}$$

$$\int \tan x \, dx = -\ln |\cos x| \qquad \text{(E.17a)}$$

$$\int \tanh x \, dx = \ln \cosh x \qquad \text{(E.17b)}$$

$$\int e^{ax} \sin x \, dx = \frac{e^{ax}}{a^2 + 1} (a \sin x - \cos x) \qquad \text{(E.18a)}$$

$$\int e^{ax} \sin^2 x \, dx = \frac{e^{ax}}{a^2 + 4} \left(a \sin^2 x - 2 \sin x \cos x + \frac{2}{a}\right) \qquad \text{(E.18b)}$$

$$\int_{-\infty}^{\infty} e^{-ax^2} \, dx = \sqrt{\pi/a} \qquad \text{(E.18c)}$$

E.3   GAMMA FUNCTIONS

$$\Gamma(n) = \int_0^{\infty} x^{n-1} e^{-x} \, dx \qquad \text{(E.19a)}$$

$$= \int_0^1 [\ln(1/x)]^{n-1} \, dx \qquad \text{(E.19b)}$$

$$\Gamma(n) = (n - 1)!, \qquad \text{for } n = \text{positive integer} \qquad \text{(E.19c)}$$

$$n\Gamma(n) = \Gamma(n + 1) \qquad \text{(E.20)}$$

$$\Gamma\left(\frac{1}{2}\right) = \sqrt{\pi} \qquad \text{(E.21)}$$

$$\Gamma(1) = 1 \qquad \text{(E.22)}$$

$$\Gamma\left(1\frac{1}{4}\right) = 0.906 \qquad \text{(E.23)}$$

$$\Gamma\left(1\frac{3}{4}\right) = 0.919 \qquad \text{(E.24)}$$

$$\Gamma(2) = 1 \qquad \text{(E.25)}$$

$$\int_0^1 \frac{dx}{\sqrt{1-x^n}} = \frac{\sqrt{\pi}}{n} \frac{\Gamma\left(\dfrac{1}{n}\right)}{\Gamma\left(\dfrac{1}{n}+\dfrac{1}{2}\right)} \tag{E.26}$$

$$\int_0^1 x^m (1-x^2)^n \, dx = \frac{\Gamma(n+1)\,\Gamma\left(\dfrac{m+1}{2}\right)}{2\Gamma\left(n+\dfrac{m+3}{2}\right)} \tag{E.27a}$$

$$\int_0^{\pi/2} \cos^n x \, dx = \frac{\sqrt{\pi}}{2} \frac{\Gamma\left(\dfrac{n+1}{2}\right)}{\Gamma\left(\dfrac{n}{2}+1\right)}, \qquad n > -1 \tag{E.27b}$$

F

DIFFERENTIAL RELATIONS IN DIFFERENT COORDINATE SYSTEMS

F.1   RECTANGULAR COORDINATES

$$\mathbf{grad}\ U = \nabla U = \sum_i \mathbf{e}_i \frac{\partial U}{\partial x_i} \qquad \text{(F.1)}$$

$$\text{div}\ \mathbf{A} = \nabla \cdot \mathbf{A} = \sum_i \frac{\partial A_i}{\partial x_i} \qquad \text{(F.2)}$$

$$\mathbf{curl}\ \mathbf{A} = \nabla \times \mathbf{A} = \sum_{i,j,k} \varepsilon_{ijk} \frac{\partial A_k}{\partial x_j} \mathbf{e}_i \qquad \text{(F.3)}$$

$$\nabla^2 U = \nabla \cdot \nabla U = \sum_i \frac{\partial^2 U}{\partial x_i^2} \qquad \text{(F.4)}$$

F.2   CYLINDRICAL COORDINATES

Refer to Figures F-1 and F-2.

$$x_1 = r \cos \phi, \qquad x_2 = r \sin \phi, \qquad x_3 = z \qquad \text{(F.5)}$$

$$r = \sqrt{x_1^2 + x_2^2}, \qquad \phi = \tan^{-1} \frac{x_2}{x_1}, \qquad z = x_3 \qquad \text{(F.6)}$$

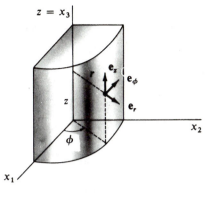

FIGURE F-1

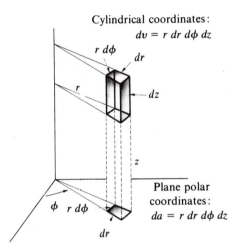

FIGURE F-2

$$ds^2 = dr^2 + r^2 \, d\phi^2 + dz^2 \tag{F.7}$$

$$dv = r \, dr \, d\phi \, dz \tag{F.8}$$

$$\mathbf{grad} \; \psi = \nabla \psi = \mathbf{e}_r \frac{\partial \psi}{\partial r} + \mathbf{e}_\phi \frac{1}{r} \frac{\partial \psi}{\partial \phi} + \mathbf{e}_z \frac{\partial \psi}{\partial z} \tag{F.9}$$

$$\text{div } \mathbf{A} = \frac{1}{r} \frac{\partial}{\partial r} (rA_r) + \frac{1}{r} \frac{\partial A_\phi}{\partial \phi} + \frac{\partial A_z}{\partial z} \tag{F.10}$$

$$\mathbf{curl} \; \mathbf{A} = \mathbf{e}_r \left(\frac{1}{r} \frac{\partial A_z}{\partial \phi} - \frac{\partial A_\phi}{\partial z} \right) + \mathbf{e}_\phi \left(\frac{\partial A_r}{\partial z} - \frac{\partial A_z}{\partial r} \right)$$

$$+ \mathbf{e}_z \left(\frac{1}{r} \frac{\partial}{\partial r} (rA_\phi) - \frac{1}{r} \frac{\partial A_r}{\partial \phi} \right) \tag{F.11}$$

$$\nabla^2\psi = \frac{1}{r}\frac{\partial}{\partial r}\left(r\frac{\partial\psi}{\partial r}\right) + \frac{1}{r^2}\frac{\partial^2\psi}{\partial\phi^2} + \frac{\partial^2\psi}{\partial z^2}$$ **(F.12)**

F.3   SPHERICAL COORDINATES

Refer to Figures F-3 and F-4.

$$x_1 = r\sin\theta\cos\phi, \qquad x_2 = r\sin\theta\sin\phi, \qquad x_3 = r\cos\theta \qquad \textbf{(F.13)}$$

$$r = \sqrt{x_1^2 + x_2^2 + x_3^2}, \qquad \theta = \cos^{-1}\frac{x_3}{r}, \qquad \phi = \tan^{-1}\frac{x_2}{x_1} \qquad \textbf{(F.14)}$$

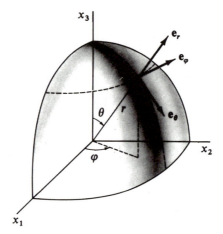

FIGURE F-3

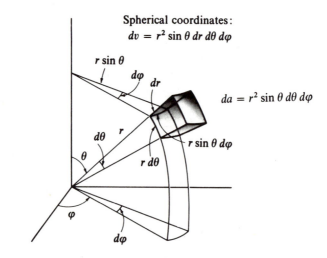

Spherical coordinates:
$$dv = r^2\sin\theta\, dr\, d\theta\, d\varphi$$

$$da = r^2\sin\theta\, d\theta\, d\varphi$$

FIGURE F-4

$$ds^2 = dr^2 + r^2\,d\theta^2 + r^2\sin^2\theta\,d\phi^2 \tag{F.15}$$

$$dv = r^2\sin\theta\,dr\,d\theta\,d\phi \tag{F.16}$$

$$\textbf{grad } \psi = \boldsymbol{\nabla}\psi = \mathbf{e}_r\frac{\partial\psi}{\partial r} + \mathbf{e}_\theta\frac{1}{r}\frac{\partial\psi}{\partial\theta} + \mathbf{e}_\phi\frac{1}{r\sin\theta}\frac{\partial\psi}{\partial\phi} \tag{F.17}$$

$$\text{div } \mathbf{A} = \frac{1}{r^2}\frac{\partial}{\partial r}(r^2 A_r) + \frac{1}{r\sin\theta}\frac{\partial}{\partial\theta}(A_\theta\sin\theta) + \frac{1}{r\sin\theta} + \frac{\partial A_\phi}{\partial\phi} \tag{F.18}$$

$$\begin{aligned}
\textbf{curl } \mathbf{A} = {}& \mathbf{e}_r\frac{1}{r\sin\theta}\left[\frac{\partial}{\partial\theta}(A_\phi\sin\theta) - \frac{\partial A_\theta}{\partial\phi}\right] \\
&+ \mathbf{e}_\theta\frac{1}{r\sin\theta}\left[\frac{\partial A_r}{\partial\phi} - \sin\theta\frac{\partial}{\partial r}(rA_\phi)\right] \\
&+ \mathbf{e}_\phi\frac{1}{r}\left[\frac{\partial}{\partial r}(rA_\theta) - \frac{\partial A_r}{\partial\theta}\right]
\end{aligned} \tag{F.19}$$

$$\nabla^2\psi = \frac{1}{r^2}\frac{\partial}{\partial r}\left(r^2\frac{\partial\psi}{\partial r}\right) + \frac{1}{r^2\sin\theta}\frac{\partial}{\partial\theta}\left(\sin\theta\frac{\partial\psi}{\partial\theta}\right) + \frac{1}{r^2\sin^2\theta}\frac{\partial^2\psi}{\partial\phi^2} \tag{F.20}$$

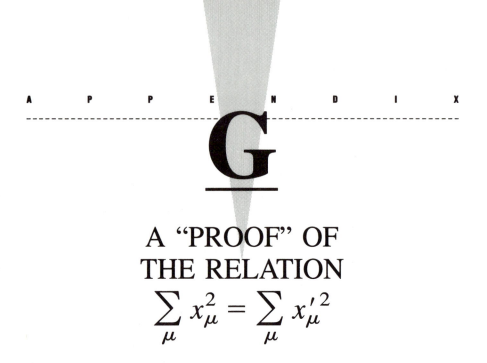

A "PROOF" OF THE RELATION $\sum_{\mu} x_{\mu}^2 = \sum_{\mu} x_{\mu}'^2$

Consider the two inertial systems K and K' that are moving relative to one another with a speed v. At the instant when the two origins coincide ($t = 0$, $t' = 0$), let a light pulse be emitted from the common origin. The equations that describe the propagation of the wave fronts are required, by the second Einstein postulate, to be of the same form in the two systems:

$$\sum_j x_j^2 - c^2 t^2 = \sum_{\mu} x_{\mu}^2 \equiv s^2 = 0, \qquad \text{in } K \tag{G.1a}$$

$$\sum_j x_j'^2 - c^2 t'^2 = \sum_{\mu} x_{\mu}'^2 \equiv s'^2 = 0, \qquad \text{in } K' \tag{G.1b}$$

These equations state that the vanishing of the four-dimensional interval between two events in one inertial reference frame implies the vanishing of the interval between the same two events in any other inertial reference frame. But we need more than this; we must show, in fact, that $s^2 = s'^2$ in general.

If we require that the motion of a particle observed to be *linear* in the system K also be linear in the system K', then the equations of transformation that connect the x_{μ} and the x_{μ}' must themselves be linear. In such a case, the quadratic forms s^2 and s'^2 can be connected by, at most, a proportionality factor:

$$s'^2 = \kappa s^2 \tag{G.2a}$$

The factor κ could conceivably depend on the coordinates, the time, and the relative speed of the two systems. As pointed out in Section 2.3, the space and time associated with an inertial reference frame are *homogeneous*, so the relation between

s^2 and s'^2 cannot be different at different points in space nor at different instants of time. Therefore, the factor κ cannot depend on either the coordinates or the time. A dependence on v is still allowed, however, but the *isotropy* of space forbids a dependence on the *direction* of v. We have therefore reduced the possible dependence of s'^2 on s^2 to a factor that involves at most the magnitude of the speed v; that is, we have

$$s'^2 = \kappa(v)s^2 \tag{G.2b}$$

If we make the transformation from K' back to K, we have the result

$$s^2 = \kappa(-v)s'^2$$

where $-v$ occurs because the velocity of K relative to K' is the negative of the velocity of K' relative to K. But we have already argued that the factor κ can depend only on the *magnitude* of v. We therefore have the two equations

$$\left. \begin{array}{l} s'^2 = \kappa(v)s^2 \\ s^2 = \kappa(v)s'^2 \end{array} \right\} \tag{G.3}$$

Combining these equations, we conclude that $\kappa^2 = 1$, or $\kappa(v) = \pm 1$. The value of $\kappa(v)$ must not be a discontinuous function of v; that is, if we change v at some rate, κ cannot suddenly jump from $+1$ to -1. In the limit of zero velocity, the systems K and K' become identical, so that $\kappa(v = 0) = +1$. Hence,

$$\kappa = +1 \tag{G.4}$$

for all values of the velocity, and we have, finally,

$$s^2 = s'^2 \tag{G.5}$$

This important result states that the four-dimensional interval between two events is the same in all inertial reference frames.

H

NUMERICAL SOLUTION
FOR EXAMPLE 2.7

In this appendix, we show the MathCad solution that produced Figures 2-8 and 2-9 for Example 2.7. This program was written for MathCad for Windows, version 4.0.

g := 9.8   acceleration of gravity

$th := 60 \cdot \left(\dfrac{\pi}{180}\right)$ initial angle

vo := 600

u := vo · cos(th)   initial velocity

v := v · sin(th)   initial horizontal velocity

i := 1 .. 6   initial vertical velocity

   $k_i :=$

0.0000001
0.01
0.02
0.04
0.08
0.005

table of drag coefficients

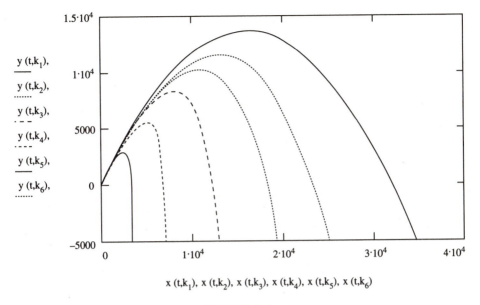

$$\frac{y\,(t,k_1),}{y\,(t,k_2),}$$
....

$$y\,(t,k_3),$$
- - -

$$y\,(t,k_4),$$
- - - -

$$\frac{y\,(t,k_5),}{}$$

$$y\,(t,k_6),$$
....

x (t,k_1), x (t,k_2), x (t,k_3), x (t,k_4), x (t,k_5), x (t,k_6)

FIGURE 2-8

t := 0, 1 .. 130   range of time values

$$x(t,K) := \left(\left(\frac{u}{K}\right)\right) \cdot (1 - \exp(-K \cdot t))$$ calculate horizontal position

$$y(t,K) := -g \cdot \frac{t}{K} + \frac{K \cdot v + g}{(K)^2} \cdot (1 - \exp(-K \cdot t))$$ calculate vertical position

[Now plot y(t,k_j) versus x(t,k_j) to produce Figure 2-8.]

Now set up an equation to solve Equation 2.45 for *T* for any value of the retarding force constant *k*.

$$f(k,T) := \text{root}\left[T - \frac{k \cdot v + g}{g \cdot k} \cdot (1 - \exp(-k \cdot T)), T\right]$$ **(2.45)**

j := 1, 2 .. 81   Set up range of values to calculate; 80 values.

$K_j := -0.001 + 0.001 \cdot j + 0.00000001$ This will allow us to calculate over a range of k values from 0 to 0.08.

$Tr_0 := 100$ The time value for k = 0 is 106 s. This is a guess to get the calculation started.

$Tr_j := f(K_j, Tr_{j-1})$ We now determine the solution for the time T for all the values of k. Solve Equation 2.45.

$Tr_1 = 106.074$

This is the value of T for k1. We do not bother to calculate all the others here.

Now we want to calculate the range R for all the values of T (as a function of k) that we have just found. To do this, we need to solve Equation 2.43 for each of the values of k and t = T that we have just found.

$x := 100$

This is the guess for the first value of x. The actual value of the guess does not matter.

$$f(k,T) := root\left[x - \frac{u}{k} \cdot (1 - exp(-k \cdot T)), x \right]$$

This is the Equation 2.43 that we need to solve to find the range R.

$R_j := f(K_j, Tr_j)$

Now calculate the range R for all the values.

$R_1 = 3.182 \cdot 10^4$

We just list the first value and plot the remainder. This is the range for no air resistance, that is k = 0.

Now let's calculate and plot the range determined from the approximate calculation. Calculate Equation 2.55.

$$Rp_j := R_1 \cdot \left(1 - \frac{4 \cdot K_j \cdot v}{3 \cdot g} \right)$$

[Now plot R_j and Rp_j versus K_j to produce Figure 2-9.]

Plot approximate and numerical solutions. Figure 2-9.

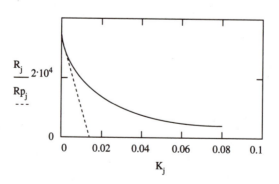

FIGURE 2-9

The following texts are particularly recommended as general sources of collateral reading material.

A. General Theoretical Physics

Blass (B162), *Theoretical Physics.*
Lindsay and Margenau (Li36), *Foundations of Physics.*
Wangsness (Wa63), *Introduction to Theoretical Physics.*

B. Elementary Mechanics

Baierlein (Ba83), *Newtonian Dynamics.*
Barger and Olsson (Ba73), *Classical Mechanics.*
Davis (Da86), *Classical Mechanics.*
Fowles (Fo86), *Analytical Mechanics.*
French (Fr71), *Newtonian Mechanics.*
Rossberg (Ro83), *Analytical Mechanics.*

C. Intermediate Mechanics

Becker (Be54), *Introduction to Theoretical Mechanics.*
Desloge (De82), *Classical Mechanics, Vol.1.*
Lindsay (Li61), *Physical Mechanics.*
Slater and Frank (Sl47), *Mechanics.*
Symon (Sy71), *Mechanics.*

D. Advanced Mechanics

Desloge (De82), *Classical Mechanics, Vol. 2.*
Goldstein (Go80), *Classical Mechanics.*
Landau and Lifshitz (La76), *Mechanics.*
McCuskey (Mc59), *An Introduction to Advanced Dynamics.*

E. Mathematical Methods

Abramowitz and Stegun (Ab65), *Handbook of Mathematical Functions.*
Arfken (Ar85), *Mathematical Methods for Physicists.*
Byron and Fuller (By69), *Mathematics of Classical and Quantum Physics.*

Churchill (Ch78), *Fourier Series and Boundary Value Problems.*
Davis (Da61), *Introduction to Vector Analysis.*
Dennery and Krzywicki (De67), *Mathematics for Physicists.*
Dwight (Dw61), *Tables of Integrals and Other Mathematical Data.*
Kaplan (Ka84), *Advanced Calculus.*
Mathews and Walker (Ma70), *Mathematical Methods of Physics.*
Pipes and Harvill (Pi70), *Applied Mathematics for Engineers and Physicists.*

F. Special Relativity

Einstein (Ei61), *Relativity.*
French (Fr68), *Special Relativity.*
Resnick (Re72), *Basic Concepts in Relativity and Early Quantum Theory.*
Rindler (Ri82), *Introduction to Special Relativity.*
Taylor and Wheeler (Ta66), *Spacetime Physics.*

G. Chaos

Baker and Gollub (Ba90), *Chaotic Dynamics.*
Bessoir and Wolf (Be91), *Chaos Simulations.*
Hilborn (Hi94), *Chaos and Nonlinear Dynamics.*
Moon (Mo92), *Chaotic and Fractal Dynamics.*
Rasband (Ra90), *Chaotic Dynamics of Nonlinear Systems.*
Rollins (Ro90), *Chaotic Dynamics Workbench.*
Sprott and Rowlands (Sp92), *Chaos Demonstrations.*

H. Numerical Methods

DeJong (De91), *Introduction to Computational Physics.*
Johnson and Reiss (Jo82), *Numerical Analysis.*
Press, Teukolsky, Vetterling, and Flannery (Pr92), *Numerical Recipes.*

Ab65   M. Abramowitz and I. Stegun, *Handbook of Mathematical Functions*. Dover, New York, 1965.

Ad22   E. P. Adams and R. L. Hippisley, *Smithsonian Mathematical Formulae and Tables of Elliptical Functions*. Smithsonian Institution, Washington, D. C., 1922.

Am63   American Association of Physics Teachers, *Special Relativity Theory, Selected Reprints*. American Institute of Physics, New York, 1963.

An49   A. A. Andronow and C. E. Chaikin, *Theory of Oscillations*, transl. of the 1937 Russian ed. Princeton University Press, Princeton, New Jersey, 1949.

Ar85   G. Arfkin, *Mathematical Methods for Physicists*, 3rd ed. Academic Press, Orlando, Florida, 1985.

Ba73   V. Barger and M. Olsson, *Classical Mechanics*. McGraw-Hill, New York, 1973.

Ba83   R. Baierlein, *Newtonian Dynamics*. McGraw-Hill, New York, 1983.

Ba90   G. L. Baker and J. P. Gollub, *Chaotic Dynamics*. Cambridge, New York, 1990.

Be46   P. G. Bergman, *Introduction to the Theory of Relativity*, Prentice-Hall, Englewood Cliffs, New Jersey, 1946 (reprinted by Dover, New York, 1976).

Be54   R. A. Becker, *Introduction to Theoretical Mechanics*. McGraw-Hill, New York, 1954.

Be91   T. Bessoir and A. Wolf, *Chaos Simulations*. American Institute of Physics, College Park, 1991. Available from Physics Academic Software, Box 8202, North Carolina State University, Raleigh, North Carolina 27695-8202.

Bl62   G. A. Blass, *Theoretical Physics*. Appleton-Century-Crofts, New York, 1962.

Br60   L. Brillouin, *Wave Propagation and Group Velocity*. Academic Press, New York, 1960.

Br68   T. C. Bradbury, *Theoretical Mechanics*. Wiley, New York, 1968 (reprinted by Krieger, Melbourne, Florida, 1981).

By69   F. Byron and R. Fuller, *Mathematics of Classical and Quantum Physics*. Addison-Wesley, Reading, Massachusetts, 1969.

Ch78   R. V. Churchill, *Fourier Series and Boundary Value Problems*, 3rd ed. McGraw-Hill, New York, 1978.

Co53   R. Courant and D. Hilbert, *Methods of Mathematical Physics*, Vol. 1. Wiley (Interscience), New York, 1953.

Co60   H. C. Corben and P. Stehle, *Classical Mechanics*, 2nd ed. Wiley, New York, 1960.

Da61   H. F. Davis, *Introduction to Vector Analysis*. Allyn & Bacon, Boston, 1961.

Da63   H. F. Davis, *Fourier Series and Orthogonal Functions*. Allyn & Bacon, Boston, 1963.

Da86   A. D. Davis, *Classical Mechanics*. Academic Press, Orlando, Florida, 1986.

De62   J. W. Dettman, *Mathematical Methods in Physics and Engineering*. McGraw-Hill, New York, 1962.

De67   P. Dennery and A. Krzywicki, *Mathematics for Physicists*. Harper & Row, New York, 1967.

De82   E. A. Desloge, *Classical Mechanics*, Vols. 1 and 2. Wiley, New York, 1982.

De91   M. L. DeJong, *Introduction to Computational Physics*. Addison-Wesley, Reading, Massachusetts, 1991.

Dw61   H. B. Dwight, *Tables of Integrals and Other Mathematical Data*, 4th ed. Macmillan, New York, 1961.

Ed30   Sir A. S. Eddington, *The Nature of the Physical World*. Macmillan, New York, 1930.

Ei61   A. Einstein, *Relativity*, 15th ed. Crown, New York, 1961.

Fe59   N. Feather, *The Physics of Mass, Length, and Time*. Edinburgh University Press, Edinburgh, 1959.

Fe65   R. P. Feynman and A. R. Hibbs, *Quantum Mechanics and Path Integrals*. McGraw-Hill, New York, 1965.

Fo86   G. R. Fowles, *Analytical Mechanics*, 4th ed. Saunders, Philadelphia, 1986.

Fr68   A. P. French, *Special Relativity*. W. W. Norton, New York, 1968.

Fr71   A. P. French, *Newtonian Mechanics*. W. W. Norton, New York, 1971.

Go80   H. Goldstein, *Classical Mechanics*, 2nd ed. Addison-Wesley, Reading, Massachusetts, 1980.

Ha62   J. Haag, *Oscillatory Motions*. Wadsworth, Belmont, California, 1962.

He95   J. B. Marion and M. Heald, *Classical Electromagnetic Radiation*, 3rd ed. Academic Press, New York, 1995.

Hi94   R. C. Hilborn, *Chaos and Nonlinear Dynamics*. Oxford, New York, 1994.

Ho64   H. Hochstadt, *Differential Equations—A Modern Approach*. Holt, New York, 1964.

In27   E. L. Ince, *Ordinary Differential Equations*. Longmans, Green, New York, 1927 (reprinted by Dover, New York, 1944).

Jo50   G. Joos and I. M. Freeman, *Theoretical Physics*, 2nd ed. Hafner, New York, 1950.

Jo82   Lee W. Johnson and R. Dean Reiss, *Numerical Analysis*. Addison-Wesley, Reading, Massachusetts, 1982.

Ka76   M. Kaplan, *Modern Spacecraft Dynamics and Control*. Wiley, New York, 1976.

Ka84   W. Kaplan, *Advanced Calculus*, 3rd ed. Addison-Wesley, Reading, Massachusetts, 1984.

La49   C. Lanczos, *The Variational Principles of Mechanics*. University of Toronto Press, Toronto, 1949.

La76   L. D. Landau and E. M. Lifshitz, *Mechanics*, 3rd ed. Pergammon, New York, 1976.

Li36   R. B. Lindsay and H. Margenau, *Foundations of Physics*. Wiley, New York, 1936 (reprinted by Dover, New York, 1957; reprinted by Ox Bow, Woodbridge, Connecticut, 1981).

Li51   R. B. Lindsay, *Concepts and Methods of Theoretical Physics*. Van Nostrand, Princeton, New Jersey, 1951.

Li61   R. B. Lindsay, *Physical Mechanics*, 3rd ed. Van Nostrand, Princeton, New Jersey, 1961.

Lo23   H. A. Lorentz, A. Einstein, H. Minowski, and H. Weyl, *The Principle of Relativity*, original papers. Translated in 1923; reprinted by Dover, New York, 1952.

Ma59   J. B. Marion, T. I. Arnette, and H. C. Owens, "Tables for the Transformation Between the Laboratory and Center-of-mass Coordinate Systems and for the Calculation of the Energies of Reaction Products." Oak Ridge National Lab. Rept. ORNL-2574, 1959.

Ma60   E. Mach, *The Science of Mechanics*, 6th Am. ed. Open Court, LaSalle, Illinois, 1960 (original German edition published 1883).

Ma65   J. B. Marion, *Principles of Vector Analysis*. Academic Press, New York, 1965.

Ma70   J. Matthews and R. Walker, *Mathematical Methods of Physics*, 2nd ed. Benjamin, New York, 1970.

Ma77   H. Margenau, *The Nature of Physical Reality*, 2nd ed. McGraw-Hill, New York, 1977.

Mc59   S. W. McCusky, *An Introduction to Advanced Dynamics*. Addison-Wesley, Reading, Massachusetts, 1959.

Mi47   N. Minorsky, *Introduction to Non-linear Mechanics*. Edwards, Ann Arbor, Michigan, 1947.

Mo53   P. M. Morse and H. Feshbach, *Methods of Theoretical Physics*, 2 vols. McGraw-Hill, New York, 1953.

Mo53a   P. Morrison, "A Survey of Nuclear Reactions," in *Experimental Nuclear Physics* (E. Segré, ed.), Vol. II. Wiley, New York, 1953.

Mo58   F. R. Moulton, *An Introduction to Celestial Mechanics*, 2nd ed. Macmillan, New York, 1958.

Mo92   Frances Moon, *Chaotic and Fractal Dynamics*. Wiley, New York, 1992.

Mu60   G. M. Murphy, *Ordinary Different Equations and Their Solutions*. Van Nostrand, Princeton, New Jersey, 1960.

Ob83   J. E. Oberg, *Mission to Mars*. New American, New York, 1983.

Pa62   W. K. H. Panofsky and M. Phillips, *Classical Electricity and Magnetism*, 2nd ed. Addison-Wesley, Reading, Massachusetts, 1962.

Pi57   B. O. Pierce and R. M. Foster, *A Short Table of Integrals*, 4th ed. Ginn, Boston, 1957.

Pi70   L. Pipes and L. Harvill, *Applied Mathematics for Engineers and Physicists*. McGraw-Hill, New York, 1970.

Pr92   W. H. Press, S. A. Teukolsky, W. T. Vetterling. B. P. Flannery, *Numerical Recipes*, 2nd ed. Cambridge, New York, 1992.

Ra90   S. N. Rasband, *Chaotic Dynamics of Nonlinear Systems*. Wiley, New York, 1990.

Ra94   J. W. S. Rayleigh, *The Theory of Sound*, 2nd ed., 2 vols. Macmillan, London, 1894 (reprinted by Dover, New York, 1945).

Re72   R. Resnick, *Basic Concepts in Relativity and Early Quantum Theory*. Wiley, New York, 1972.

Rh82   Rheinmetall GmbH, *Handbook on Weaponry*. Düsseldorf, 1982.

Ri82   W. Rindler, *Introduction to Special Relativity*. Clarendon, Oxford, 1982.

Ro83   K. Rossberg, *Analytical Mechanics*. Wiley, New York, 1983.

Ro90   R. W. Rollins, *Chaotic Dynamics Workbench*. American Institute of Physics, College Park, 1990. Available from Physics Academic Software, Box 8202, North Carolina State University, Raleigh, North Carolina 27695-8202.

Se58   F. W. Sears, *Mechanics, Wave Motion, and Heat*. Addison-Wesley, Reading, Massachusetts, 1958.

Sl47   J. C. Slater and N. H. Frank, *Mechanics*. McGraw-Hill, New York, 1947.

So50   A. Sommerfeld, *Mechanics*. Academic Press, New York, 1950.

Sp92   J. C. Sprott and G. Rowlands, *Chaos Demonstrations*. American Institute of Physics, College Park, 1992. Available from Physics Academic Software, Box 8202, North Carolina State University, Raleigh, North Carolina 27695-8202.

Sy71   K. R. Symon, *Mechanics*, 3rd ed. Addison-Wesley, Reading, Massachusetts, 1971.

Ta66   E. F. Taylor and J. A. Wheeler, *Spacetime Physics*. Freeman, San Francisco, 1966.

Tr68   C. Truesdell, *Essays in the History of Mechanics*. Springer-Verlag, New York, 1968.

Tu04   H. H. Turner, *Astronomical Discovery*, Arnold, London, 1904 (reprinted by the University of California Press, Berkeley, California, 1963).

Wa63   R. K. Wangsness, *Introduction to Theoretical Physics*. Wiley, New York, 1963.

We61   J. Weber, *General Relativity and Gravitational Waves*. Wiley (Interscience), New York, 1961.

Wh37   E. T. Whittaker, *A Treatise on the Analytical Dynamics of Particles and Rigid Bodies*, 4th ed. Cambridge University Press, London and New York, 1937 (reprinted by Dover, New York, 1944).

Wh53   E. T. Whittaker, *A History of the Theories of Aether and Electricity*; *Vol. II*: *The Modern Theories*. Nelson, London, 1953 (reprinted by Harper and Bros., New York, 1960).

Chapter 1

10. (a) $\mathbf{v} = 2b\omega \cos \omega t\, \mathbf{i} - b\omega \sin \omega t\, \mathbf{j}$ (b) $90°$

$\mathbf{a} = -\omega^2 \mathbf{r}$

$|\mathbf{v}| = b\omega[3 \cos^2 \omega t + 1]^{\frac{1}{2}}$

12. $h = \dfrac{|\mathbf{a} \cdot \mathbf{b} \times \mathbf{c}|}{|\mathbf{a} \times \mathbf{b} + \mathbf{b} \times \mathbf{c} + \mathbf{c} \times \mathbf{a}|}$

$A = \dfrac{1}{2}|(\mathbf{b} - \mathbf{a}) \times (\mathbf{c} - \mathbf{b})| = \dfrac{1}{2}|(\mathbf{a} - \mathbf{c}) \times (\mathbf{b} - \mathbf{a})| = \dfrac{1}{2}|(\mathbf{c} - \mathbf{b}) \times (\mathbf{a} - \mathbf{c})|$

14. (a) -104 (b) $\begin{pmatrix} 9 & 7 \\ 13 & 9 \\ 5 & 2 \end{pmatrix}$ (c) $\begin{pmatrix} -5 & -5 \\ 3 & -5 \\ 25 & 14 \end{pmatrix}$ (d) $\begin{pmatrix} 0 & -3 & -4 \\ 3 & 0 & 6 \\ 4 & -6 & 0 \end{pmatrix}$

26. $\mathbf{a} \cdot \mathbf{e}_r = -\dfrac{3v^2}{4\,k}$; $|\mathbf{a}| = \dfrac{3}{4}\dfrac{v^2}{k} \cdot \sqrt{\dfrac{2}{1 + \cos \theta}}$; $\dot{\theta} = \dfrac{v}{\sqrt{2kr}}$

34. $\displaystyle\int (\mathbf{A} \times \ddot{\mathbf{A}})\,dt = (\mathbf{A} \times \dot{\mathbf{A}}) + \mathbf{C}$, where $\mathbf{C}$ is a constant vector.

36. $\pi c^2 d$

38. $-\pi$

Chapter 2

2. $F_\theta = mR(\ddot{\theta} - \dot{\phi}^2 \sin \theta \cos \theta)$

$F_\phi = mR(2\dot{\theta}\dot{\phi} \cos \theta + \ddot{\phi} \sin \theta)$

4. 13.2 m $\cdot$ s^{-1}

6. (a) 210 m behind   (b) can be no more than 0.68 s late

14. (a) $d = \dfrac{2v_0^2 \cos \alpha \sin (\alpha - \beta)}{g \cos^2 \beta}$ (b) $\dfrac{\pi}{4} + \dfrac{\beta}{2}$ (c) $d_{max} = \dfrac{v_0^2}{g(1 + \sin \beta)}$

16. $\dfrac{2v_0}{g \sin \alpha}$

18. (a) 35.2 m$\cdot$s^{-1} (b) $40.7°$; 1.1 m

20. 17.4°

22. (c) $\dot{x}(t) = C_1 \cos \omega_c t + C_2 \sin \omega_c t + \dfrac{E_y}{B}$

$\dot{y}(t) = -C_1 \sin \omega_c t + C_2 \cos \omega_c t$

24. $\mu_k = 0.18$; $v_B = 15.6$ m/s

26. 2.3 m; 1.1 m

28. $h_{\text{marble}} = h \left(\dfrac{3-a}{1+a} \right)^2$; $h_{\text{superball}} = h \left(\dfrac{1-3a}{1+a} \right)^2$ where $a = m/M$

30. 71 m

32. $\sin \theta_0 = \dfrac{1 \pm \mu_k \sqrt{3 + 4\mu_k^2}}{2(1 + \mu_k^2)}$

34. (a) $y = -\dfrac{m}{\alpha} \left[v + \dfrac{mg}{\alpha} \ln \left(1 - \dfrac{\alpha v}{mg} \right) \right]$ (b) $y = -\dfrac{m}{2\beta} \ln \left(1 - \dfrac{\beta v^2}{mg} \right)$

36. $R = \dfrac{v_0^2}{g} \cos \theta \left(\sin \theta + \sqrt{\sin^2 \theta + \dfrac{2gh}{v_0^2}} \right)$

38. (a) $F(x) = -mna^2 x^{-(2n+1)}$
   (b) $x(t) = [(n+1)at]^{\frac{1}{n+1}}$
   (c) $F(t) = -mna^2[(n+1)at]^{-2n-1/n-1}$

40. (a) $a_t = \dfrac{2A\alpha^2 \sin \alpha t}{\sqrt{5 - 4 \cos \alpha t}}$; $a_n = \dfrac{A\alpha^2 |2 \cos \alpha t - 1|}{\sqrt{5 - 4 \cos \alpha t}}$

   (b) $\dfrac{n\pi}{\alpha}$ where $n = $ integer

42. Stable if $R > b/2$; unstable if $R \le b/2$

48. e^{-1}

52. $\dfrac{v_B^2}{2g}$

54. 25 s

56. 273 s

58. (a) 3700 km   (b) 890 km   (c) 950 km   (d) 8900 km

Chapter 3

2. (a) 6.9×10^{-2} s^{-1} (b) $\dfrac{10}{2\pi}(1 - 2.40 \times 10^{-5})$s^{-1} (c) 1.0445

4. $<T> = <U> = \dfrac{mA^2\omega_0^2}{4}$ $\bar{U} = \dfrac{1}{2}\bar{T} = \dfrac{mA^2\omega_0^2}{6}$

6. 2.74 rad·s^{-1}

12. $\ddot{\theta} = -\dfrac{g}{l} \sin \theta$

14. $x(t) = (\cosh \beta t - \sinh \beta t)[(A_1 + A_2)\cosh \omega_2 t + (A_1 - A_2)\sinh \omega_2 t]$

$\dot{x}(t) = (\cosh \beta t - \sinh \beta t)[(A_1\omega_2 - A_1\beta)(\cosh \omega_2 t + \sinh \omega_2 t)$

$\qquad - (A_2\beta + A_2\omega_2)(\cosh \omega_2 t - \sinh \omega_2 t)]$

28. $I(t) \cong \varepsilon_0 C\omega^2(RC - t), \ \omega \to 0$

30. $\dfrac{R_1[R_2(R_2 + R_1) + \omega^2 L_2^2] + i[R_1\omega L_2 + (\omega L_1 - 1/\omega C)((R_1 + R_2)^2 + \omega^2 L_2^2)]}{(R_1 + R_2)^2 + \omega^2 L_2^2}$

32. $F(t) = \dfrac{4}{\pi}\sin t + \dfrac{4}{3\pi}\sin 3t + \dfrac{4}{5\pi}\sin 5t + \cdots$

34. $F(t) = \dfrac{2}{\pi} - \dfrac{4}{3\pi}\cos 2\omega t - \dfrac{4}{15\pi}\cos 4\omega t - \cdots$

36. (a) $x(t) = \dfrac{H(0)}{\omega_0^2}\left(1 - e^{-\beta t}\cosh \omega_2 t - \dfrac{\beta e^{-\beta t}}{\omega_2}\sinh \omega_2 t\right)$

(b) $x(t) = \dfrac{b}{\omega_2}e^{-\beta t}\sinh \omega_2 t; \ t > 0$

38. $x(t) = \begin{cases} 0 & t < 0 \\ 4[1 - \cos(0.5t)] \text{ m} & 0 < t < 4\pi \\ 0 & t > 4\pi \end{cases}$

40. $x(t) = e^{-\beta(t - t_0)}\left[x_0 \cos \omega_1(t - t_0) + \left(\dfrac{\dot{x}_0}{\omega_1} + \dfrac{\beta x_0}{\omega_1} + \dfrac{b}{\omega_1}\right)\sin \omega_1(t - t_0)\right]; \qquad t > t_0$

42. $x(t) = \dfrac{F_0}{m}\dfrac{\omega}{[(\beta - \gamma)^2 + (\omega + \omega_1)^2][(\beta - \gamma)^2 + (\omega - \omega_1)^2]}$

$\qquad \times \left[e^{-\gamma t}\left[2(\gamma - \beta)\cos \omega t + ([\beta - \gamma]^2 + \omega_1^2 - \omega^2)\dfrac{\sin \omega t}{\omega}\right]\right.$

$\qquad \left. + e^{-\beta t}\left[2(\beta - \gamma)\cos \omega_1 t + ([\beta - \gamma]^2 + \omega^2 - \omega_1^2)\dfrac{\sin \omega_1 t}{\omega_1}\right]\right]$

Chapter 4

6. $\dot{\theta} = \sqrt{\dfrac{2}{ml^2}}[E - mgl(1 - \cos \theta)]^{1/2}$

8. $\tau = 4\sqrt{\dfrac{2mA}{F_0}}$

10. Only 0.6 and 0.7 are chaotic.

14. $n = 30$

16. (a) geometric sequence

22. Transitions at $B_1 = 9.8\text{-}9.9$, $B_2 = 11.6\text{-}11.7$, and $B_3 = 13.3\text{-}13.4$. Behavior: (i) one period per three drive cycles when $B < B_1$, (ii) chaotic when $B_1 < B < B_2$, (iii) mixed chaotic/one period per drive cycle (depending on initial conditions) when $B_2 < B < B_3$, and (iv) one period per drive cycle when $B > B_3$.

Chapter 5

2. $\rho = \dfrac{C}{2\pi G r}$ where $C = \dfrac{\partial \phi}{\partial r} = \text{const.}$

6. $\mathbf{g} = -\dfrac{GM}{r^2}\mathbf{e}_r$

8. $g_z = -2\pi G\rho\left(\sqrt{a^2 + (z_0 - l)^2} - \sqrt{a^2 + z_0^2} + l\right)$

10. $\phi(R) \cong -\dfrac{GM}{R}\left[1 - \dfrac{1}{2}\dfrac{a^2}{R^2}\left(1 - \dfrac{3}{2}\sin^2\theta\right)\right]$

16. $F_z = 2\pi\rho_s GM$

Chapter 6

8. (a) $a_1 = b_1 = c_1 = \dfrac{2}{\sqrt{3}}R$ (b) $a_1 = a\dfrac{2}{\sqrt{3}}; b_1 = b\dfrac{2}{\sqrt{3}}; c_1 = c\dfrac{2}{\sqrt{3}}$

10. $R = \frac{1}{2}H$

Chapter 7

4. $m\ddot{r} - mr\dot{\theta}^2 + Ar^{\alpha-1} = 0;$ $\dfrac{d}{dt}(mr^2\dot{\theta}) = 0;$ yes; yes

6. $2m\ddot{S} + m\ddot{\xi}\cos\alpha - mg\sin\alpha = 0$
  $(m + M)\ddot{\xi} + m\ddot{S}\cos\alpha = 0$

10. (a) $y(t) = -\dfrac{g}{4}t^2$ (b) $y(t) = \dfrac{Ml}{m}(1 - \cosh\gamma t)$

12. $r(t) = r_0\cosh\alpha t + \dfrac{g}{2\alpha^2}(\sin\alpha t - \sinh\alpha t)$

14. (a) $\ddot{\theta} + \dfrac{a + g}{b}\sin\theta = 0$ (b) $2\pi\sqrt{\dfrac{b}{a + g}}$

16. $\ddot{\theta} + \dfrac{g}{b}\sin\theta - \dfrac{a}{b}\omega^2\sin\omega t\cos\theta = 0$

18. $\omega = \sqrt{\dfrac{g\sin\theta_0}{l - R\theta_0}};$ $\theta_0 = \dfrac{\pi}{2}$

22. $L = \dfrac{1}{2}m\dot{x}^2 - \dfrac{k}{x}e^{-t/\tau};$ $H = \dfrac{p_x^2}{2m} + \dfrac{k}{x}e^{-t/\tau}$

24. $L = \dfrac{1}{2}m(\alpha^2 + l^2\dot{\theta}^2) + mgl\cos\theta$

  $H = \dfrac{p_\theta^2}{2ml^2} - \dfrac{1}{2}m\alpha^2 - mgl\cos\theta$

26. (a) $H = \dfrac{p_\theta^2}{2ml^2} - mgl \cos \theta;$ $\dot\theta = \dfrac{p_\theta}{ml^2};$ $\dot p_\theta = -mgl \sin \theta$

   (b) $H = \dfrac{p_x^2}{2(m_1 + m_2 + I/a^2)} - m_1 gx - m_2 g(l - x)$

   $p_x = \left(m_1 + m_2 + \dfrac{I}{a^2}\right)\dot x$

   $\dot p_x = g(m_1 - m_2)$

28. $p_r = m\dot r; \ \dot p_r = \dfrac{p_\theta^2}{mr^3} - \dfrac{k}{r^2}; p_\theta = mr^2\dot\theta; \dot p_\theta = 0$

32. $H = \dfrac{1}{2m}\left(p_r^2 + \dfrac{p_\theta^2}{r^2} + \dfrac{p_\phi^2}{r^2 \sin^2 \theta}\right) - \dfrac{k}{r};$ $\dot p_r = -\dfrac{k}{r^2} + \dfrac{p_\theta^2}{mr^3} + \dfrac{p_\phi^2}{mr^3 \sin^2 \theta};$

   $\dot p_\theta = \dfrac{p_\phi^2 \cot \theta}{mr^2 \sin^2 \theta}; \dot p_\phi = 0$

34. (a) $\ddot x = aR(\ddot\theta \sin \theta + \dot\theta^2 \cos \theta); \ddot\theta = \dfrac{\ddot x \sin \theta + g \cos \theta}{R};$ where $a \equiv \dfrac{m}{M+m}$

   (b) $\lambda = -\dfrac{mMg(3 \sin \theta - a \sin^3 \theta - 2 \sin \theta_0)}{(M+m)(1 - a \sin^2 \theta)^2}$

Chapter 8

4. $<U> = -\dfrac{k}{a}; <T> = \dfrac{k}{2a}$

10. Parabola; yes

12. 76 days

14. $F(r) = -\dfrac{l^2}{\mu}\left(\dfrac{6k}{r^4} + \dfrac{1}{r^3}\right)$

22. No

24. (a) 1590 km   (b) 1900 km

28. 2380 m/s

30. $\Delta v = 3.23$ km/s; parabola

32. Stable if $r < a$

38. $\Delta v = 5275$ m/s (opposite to direction of motion); 146 days

40. Carrying the waste out of the solar system requires less energy than crashing it into the sun.

42. 2.57×10^{11} J

Chapter 9

2. On the axis; $\dfrac{3}{4} h$ from vertex

4. $\bar x = \dfrac{2a}{\theta} \sin \dfrac{\theta}{2}; \bar y = 0$

6. $\mathbf{r}_{cm} = \dfrac{F_0}{4m} t^2 \mathbf{i}; \mathbf{v}_{cm} = \dfrac{F_0}{2m} t \mathbf{i}; \mathbf{a}_{cm} = \dfrac{F_0}{2m}\mathbf{i}$

8. $\bar{x} = 0; \bar{y} = \dfrac{a}{3\sqrt{2}}$

10. $\dfrac{v_0}{g} \sin \theta \sqrt{\dfrac{2E}{m_1 + m_2}} \left(\sqrt{\dfrac{m_1}{m_2}} + \sqrt{\dfrac{m_2}{m_1}} \right)$

12. (a) yes   (b) 11 m/s

14. No

20. $\sqrt{ga}$

22. $\omega = \dfrac{\omega_0}{1 - \dfrac{a}{b}\theta}; T = mb\omega_0\omega$

24. $\mathbf{N} = \dfrac{kr}{v_0}(\mathbf{r}_1 - \mathbf{r}_2) \times (\dot{\mathbf{r}}_1 - \dot{\mathbf{r}}_2)$

26. $\dfrac{4m_1m_2}{(m_1 + m_2)^2}$

28. (a) $(-0.09\mathbf{i} + 1.27\mathbf{j})\text{N} \cdot \text{s}$ (b) $(-9\mathbf{i} + 127\mathbf{j})\text{N}$

30. $v_1 = v_2 = \dfrac{u_1}{3}$

32. $v_1 = v_2 = \dfrac{u_1}{\sqrt{2}};$ $\theta = 45°$

34. $\dfrac{m_1}{m_2} = 3 \pm 2\sqrt{2};$ $\dfrac{u_2}{u_1} = -(1 \pm \sqrt{2})$ with $\begin{cases} + : \alpha < 0 \\ - : \alpha > 0 \end{cases}$

38. $v_1 = \dfrac{u_1(m_1 \sin^2 \alpha - \varepsilon m_2)}{m_1 \sin^2 \alpha + m_2};$ along $\mathbf{u}_1$

   $v_2 = \dfrac{(\varepsilon + 1)m_1u_1 \sin \alpha}{m_1 \sin^2 \alpha + m_2};$ straight up

40. 4.3 m/s, 36° from normal

42. $\mu ag \left(1 + \dfrac{u_1^2}{ag} \right)$

44. $\sigma(\theta) = \dfrac{a^2}{4}; \sigma_t = \pi a^2$

46. $\sigma_{\text{LAB}}(\psi) \cong \dfrac{\left(\dfrac{m_1^2 k}{2m_2^2 T_0} \right)^2}{\left[1 - \sqrt{1 - \left(\dfrac{m_1}{m_2}\psi \right)^2} \right]^2 \sqrt{1 - \left(\dfrac{m_1}{m_2}\psi \right)^2}}$

Chapter 10

2. The location is given by $\tan \theta = \dfrac{ar_0}{v^2}$, where θ is the angle between the radius

and the horizontal; $|\mathbf{a}_r| = a + \sqrt{a^2 + \dfrac{v^4}{r_0^2}}$

4. $v_0 = 0.5\omega R$, in y direction; a circle

6. paraboloid $\left(z = \dfrac{\omega^2}{2g} r^2 + \text{const.} \right)$

12. 0.0018 rad $= 6$ min

16. (a) 77 km (b) 8.9 km (c) 10 km (d) 160 km (all to the west)

Chapter 11

2. $I_1 = I_2 = \dfrac{3}{20} M(R^2 + 4h^2); I_3 = \dfrac{3}{10} MR^2;$

$I_1' = I_2' = \dfrac{3}{20} M\left(R^2 + \dfrac{1}{4} h^2 \right); I_3' = I_3$

4. $I = \dfrac{1}{3} ml^2; a = \dfrac{\ell}{\sqrt{3}}$

14. $I_1 = I_2 = \dfrac{83}{320} Mb^2; I_3 = \dfrac{2}{5} Mb^2$

20. $\sqrt{\dfrac{3g}{b}}$

24. (a) $\sqrt{\sqrt{3}\dfrac{g}{a}}$ (b) $\sqrt{\dfrac{12}{5\sqrt{3}}\dfrac{g}{a}}$

Chapter 12

8. $\omega_1 = \sqrt{2 + \sqrt{2}}\sqrt{\dfrac{g}{l}}; \omega_2 = \sqrt{2 - \sqrt{2}}\sqrt{\dfrac{g}{l}}$

10. $m\ddot{x}_1 + b\dot{x}_1 + (\kappa + \kappa_{12})x_1 - \kappa_{12}x_2 = F_0 \cos \omega t$
$m\ddot{x}_2 + b\dot{x}_2 + (\kappa + \kappa_{12})x_2 - \kappa_{12}x_1 = 0$

16. $\theta_0 = -\dfrac{1}{2}\phi_0$, Mode 1; $\theta_0 = \phi_0$, Mode 2

18. $\omega_1 = 0; \omega_2 = \sqrt{\dfrac{g}{Mb}(M + m)}$

20. $\mathbf{a}_1 = \left(\dfrac{2}{\sqrt{14}}, \dfrac{1}{\sqrt{14}}, -\dfrac{3}{\sqrt{14}} \right);$ $\mathbf{a}_2 = \left(\dfrac{4}{\sqrt{42}}, -\dfrac{5}{\sqrt{42}}, \dfrac{1}{\sqrt{42}} \right)$

22. $\omega_1 = 2\sqrt{\dfrac{\kappa}{M}}; \omega_2 = 2\sqrt{\dfrac{3\kappa}{M + m}}; \omega_3 = 2\sqrt{\dfrac{3\kappa}{M}}$

Chapter 13

4. $\omega_n = \dfrac{n\pi}{L}\sqrt{\dfrac{\tau}{\rho}}$

The amplitude of the nth mode is given by $\mu_n = \begin{cases} 0, & n \text{ even} \\ \dfrac{32}{n^3 \pi^3}, & n \text{ odd} \end{cases}$

6. The second harmonic is down 4.4 db; the third, 13.3 db.

12. $\eta_s(t) = e^{-Dt/2\rho}\left[A_1 \exp\left(\sqrt{\dfrac{D^2}{4\rho^2} - \dfrac{s^2\pi^2\tau}{\rho b}}\, t\right) + A_2 \exp\left(-\sqrt{\dfrac{D^2}{4\rho^2} - \dfrac{s^2\pi^2\tau}{\rho b}}\, t\right)\right]$

20. $\phi_{B_1} - \phi_{A_1} = \tan^{-1}(\cot\theta)$ $\qquad \phi_{A_2} - \phi_{A_1} = -\theta$

Chapter 14

12. 55.3 m; 0.22 μs; 2.5×10^8 m/s; 2.5×10^8 m/s

16. $\cos\theta = \dfrac{\cos\theta' - \beta}{1 - \beta\cos\theta'}$

20. The astronaut ages 25.4 years; those on Earth age 26.7 years.

22. 4.4×10^9 kg/s; 1.4×10^{13} years

24. $7\ m_pc^2$, including the rest mass of the proton (kinetic energy is $6\ m_pc^2$).

28. $v \le 0.115c$

30. 0.8 MeV

32. $T_{\text{electron}} = 999.5$ MeV

$T_{\text{proton}} = 433$ MeV